Quantitative Bioimaging

Quantitative Bioimaging

An Introduction to Biology, Instrumentation, Experiments, and Data Analysis for Scientists and Engineers

Raimund J. Ober
Centre for Cancer Immunology
Faculty of Medicine
University of Southampton
Southampton, United Kingdom

E. Sally Ward
Centre for Cancer Immunology
Faculty of Medicine
University of Southampton
Southampton, United Kingdom

Jerry Chao
Astero Technologies LLC
College Station, Texas, United States

CRC Press
Taylor & Francis Group
Boca Raton London New York

CRC Press is an imprint of the
Taylor & Francis Group, an **informa** business

First edition published 2020
by CRC Press
6000 Broken Sound Parkway NW, Suite 300, Boca Raton, FL 33487-2742

and by CRC Press
2 Park Square, Milton Park, Abingdon, Oxon, OX14 4RN

CRC Press is an imprint of Taylor & Francis Group, LLC

ISBN: 9781138598980 (hbk)
ISBN: 9780367615451 (pbk)
ISBN: 9780429469893 (ebk)

Typeset in Computer Modern font
by KnowledgeWorks Global Ltd.

Visit the eResources: www.routledge.com/9781138598980

To our families.

Contents

Preface

The investigation of cellular processes is undergoing revolutionary developments that are propelled by modern technologies that allow the study of questions in ways that not long ago were unimaginable. One of these technologies is imaging and, in particular, fluorescence microscopy, which has the specificity to investigate the roles and interactions of different molecular species in a cellular environment. This specificity has been made possible by new fluorescent labeling techniques, foremost amongst them those based on fluorescent proteins, which permit the specific labeling of receptors and other molecules in cells. Importantly, imaging in general has become a significantly more powerful technology with the advent and continued development of, for example, modern light sources and high-sensitivity image detectors. Light sources such as lasers have enabled or facilitated various methods of sample illumination that can improve the image quality. At the same time, modern detectors, combined with computer control, have allowed the recording of vast amounts of image data at high spatial and temporal resolution. Additionally, advances in data acquisition have been made more valuable by the development of sophisticated data analysis algorithms that enable the extraction of biological quantities of interest from the acquired images. In the case of fluorescence microscopy, the use of appropriate illumination techniques in conjunction with suitable detectors and data analysis algorithms have enabled the visualization and accurate quantitation of highly dynamic cellular processes at the level of individual, specifically labeled molecules of interest. Indeed, imaging at the single molecule level has been one of the more recent high points in the field of quantitative bioimaging.

The fascination of quantitative bioimaging is that it is a paradise for multi- and interdisciplinary research as important aspects of different disciplines are necessary for mastering the technique. A detailed understanding of the biological processes that are investigated is required to develop a well-thought-out experimental plan. Knowledge of chemistry is needed as approaches from chemistry form the basis for the labeling of molecules of interest with fluorescent tags. A good understanding of optics is crucial as it is the discipline that informs us about the principles of image formation. Familiarity with statistical data analysis is important as the theory and mathematics involved underlie the ever more sophisticated image analysis algorithms that need to be employed to analyze the acquired image data.

Classically, textbooks on quantitative bioimaging and microscopy focus on one of the underlying disciplines, for example the optical principles or image analysis. There are also textbooks that concentrate on the practical aspects of using specific imaging techniques. Here we are pursuing a different approach as we present, in a fair amount of depth, the various constituent scientific components of our composite discipline. The discussion of two important biological problems forms a thread throughout a significant part of the book and allows us to introduce experimental considerations using concrete biological questions.

Quantitative bioimaging, and microscopy in particular, is a vast field, and it is not possible to cover all important aspects in significant detail. As we believe strongly that it is important to understand the fundamental principles behind cellular microscopy experiments and their analysis, we have opted to present some key aspects within that context in detail and depth. Invariably there are personal preferences in our choices of emphasis and other authors might have chosen other topics to highlight. We believe that students who study the material as presented will have the tools available to consider by themselves any of the other exciting areas of quantitative bioimaging with relative ease.

The book is divided into four parts. The main material is introduced in Parts II, III, and IV. Part I serves as a brief overview and provides short introductions to the various areas covered by the book. The two biological problems that motivate and are used to illustrate many of the concepts are briefly presented. This is followed by an introduction of the basics of microscopy, an elementary discussion of resolution, and a motivation of the stochastic modeling that underpins the data analysis approach discussed in Part IV.

Part II is devoted to an extensive introduction of molecular biology that is required for an understanding of some of the biological questions that are examined using bioimaging techniques. A special emphasis is placed on antibody technology as it plays an important role in one of the key biological problems that we consider here, and in the labeling of cellular samples. This is followed by a discussion of the basic principles of fluorescence and cell biology. A detailed discussion of the two motivating biological problems and the preparation of cellular samples concludes this part.

Optics and microscopy are the topics of Part III, which begins with a presentation of different designs of microscopes. This is followed by a detailed description of various types of microscopy experiments and an extensive discussion of detectors. Geometrical optics and diffraction are then introduced in detail. In contrast to many other texts we begin our treatment by discussing topics that are relevant for the experimentalist, including basic light paths, microscopy experiments, and detectors. These topics provide motivation for a more rigorous treatment of geometrical optics and diffraction. We hope that this approach will make the material more accessible to those readers who have a less quantitative background.

The fourth and final part of the book is devoted to the analysis of microscopy data with a particular focus on parameter estimation approaches. Parameter estimation plays a central role in many areas, including the deconvolution of images and the extraction of biologically relevant quantities of interest from single molecule microscopy experiments.

While there is a natural order in the material presented, and while we cover the whole spectrum of disciplines that form the basis to cellular microscopy, an instructor has significant flexibility in both the general approach to, and the specifics of, selecting material from the book to provide the content for a one- or two-semester course. A course could be approached from a more applied cell biological point of view by covering in depth Part II and the first chapters of Part III. For a more technical approach, the more advanced parts of Part II could be omitted, and the course could concentrate on Part III and all or part of Part IV.

Parts II and III are written such that readers with different backgrounds should find them accessible. A student with a strong background in biology will most likely want to skip the introductory sections of Part II that were written to give, for example, an engineer the necessary background in biology to be able to read the more advanced sections of that part. Similarly, a reader with a strong background in optics will want to skip some of the more elementary sections of Part III that were written to make the more advanced material accessible to the non-expert.

Each part of the book concludes with a section of notes and a section of exercises. The notes provide references, historical contexts, and suggestions for further reading.

The book includes an appendix that provides the background and derivations for some of the mathematical results presented or utilized in the main text. This appendix is available online at http://www.routledge.com/9781138598980.

Information and updates relating to this book may be found online at https://www.wardoberlab.com.

Acknowledgments

It is a great pleasure to acknowledge the many individuals who have helped to bring this project about. First and foremost, we would like to thank the members of our laboratory who have provided assistance in areas such as the acquisition and processing of data, the preparation of figures, the sourcing of critical information, and the critiquing of various parts of the book as the project developed over time. In particular, we would like to acknowledge Dongyoung Kim, Sreevidhya Ramakrishnan, Sungyong You, Wooseok Kim, Ramraj Velmurugan, Anish Abraham, Taiyoon Lee, Eungjoo Lee, Siva Devanaboyina, and Dilip Challa.

The material of this book has been used as course material for different classes taught at the University of Texas at Dallas and Texas A&M University. The feedback from the students in those classes has been very important for improving the presentation of the material.

A sincere thank-you is also due to Ed Cohen, Charles DiMarzio, and Sripad Ram for the critical reading of different parts of the draft and their many suggestions for improvement. Obviously all remaining problems are due to our failings.

We would also like to acknowledge the National Institutes of Health for funding, through multiple grants, the research that has led to many of the results and insights presented in this book.

Part I

Introduction

Overview

The presentation of the material of this textbook is divided into three main parts, each of which is devoted to one of the pillars that make up modern imaging of biological samples such as cells. Concepts from biology and chemistry are presented in Part II, material on optics and image formation forms the subject of Part III, and methods for the statistical analysis of the acquired image data are presented in Part IV. This first part, in contrast, consists of a brief introduction and overview that serve to provide motivation and the basic background for each of the other parts of the book.

We start Part I with a brief discussion of some of the most basic concepts by contrasting the beginnings of cellular microscopy in the 17th century with currently available approaches. We then move to an introduction of two biological questions that will serve as motivation for many of the later developments, and that give rise to a discussion of basic microscopy experiments. An overview of the components of a microscope and image formation follows. We will then have the necessary background to review the classical notion of Rayleigh's resolution criterion. This is followed by an introduction of the elementary concepts underlying the stochastic modeling of microscopy image data. A reader who wishes to review some of the basic notions of probability theory that are helpful for this text can consult a tutorial with exercises in the Online Appendix (Section A.1).

The material in this introductory part will be significantly expanded on in the subsequent parts of the book. A reader who wishes to get a full exposure to all the essential aspects of cellular microscopy will want to read all three subsequent parts. However, this introductory part does provide a sufficient overview of the material such that the later parts can, at least to some extent, be independently read.

1

Then and Now

Imaging of biological phenomena, such as cellular processes, is a fascinating topic that requires tools from different scientific areas. A quick glance at the history of microscopy can help to illustrate particular challenges and how more recent technological developments have significantly changed the approaches that are available to us.

Exploring cells and small organisms by microscopy has fascinated researchers for centuries. The use of simple optical devices, such as lenses, goes back to the ancient Greeks and probably further. However, the first significant studies of biological systems with microscopes are attributed to the Englishman Robert Hooke (1635–1703) and the Dutchman Antonie van Leeuwenhoek (1632–1723). Robert Hooke (Fig. 1.1(a)) was a scientist who had a major influence in several areas of science. Hooke's law of elasticity is just one of many examples. Figure 1.1(b) shows a photograph and a depiction of a microscope that he used, and Fig. 1.1(c) shows the sketch of an image that he obtained when he used a microscope to investigate the bark of cork. In fact, Hooke is credited with introducing the notion of a "cell" to describe the arrangements of the structures that he saw in the bark. Antonie van Leeuwenhoek (Fig. 1.2(a)) constructed simple, yet powerful, one-lens microscopes (Fig. 1.2(b)) and conducted extensive studies of biological systems. The first observation of bacteria is ascribed to him, and he is often referred to as the father of microbiology. Much of the power of his work is due to his innovative design of microscope lenses. Figure 1.2(c) shows a drawing of a cross section of an optic nerve based on his observations.

The light microscope has, over the centuries, become an indispensable tool for most researchers in biology. Developments in optics over time have led to major improvements in the microscope's performance, even though many of the underlying optical principles

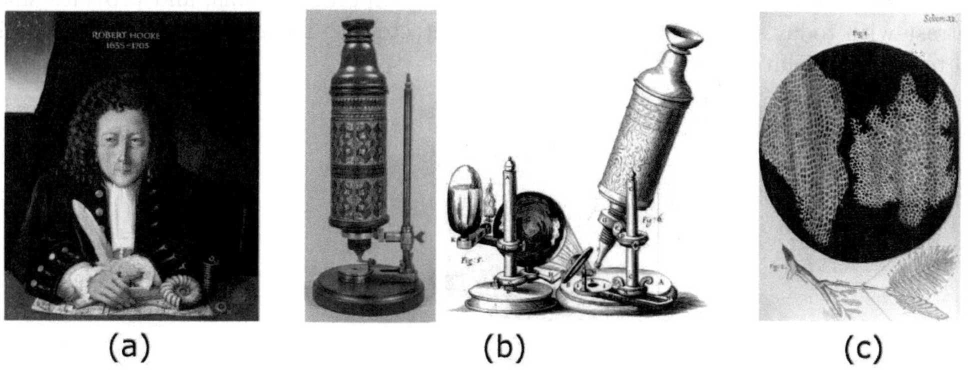

FIGURE 1.1
Robert Hooke. (a) Modern portrait of Robert Hooke based on historical reports. A portrait of his time is said to have been lost when Isaac Newton moved Hooke's office after Hooke's death. (b) Microscope designed by Robert Hooke (left) and a depiction of it from his book *Micrographia* (right). (c) Page from *Micrographia* showing the "cellular" structure of bark. [See Figure Credits.]

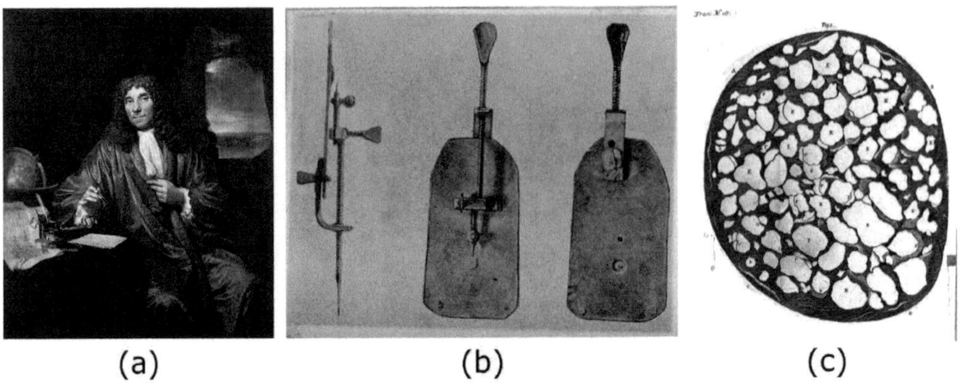

FIGURE 1.2
Antonie van Leeuwenhoek. (a) Portrait of van Leeuwenhoek by his contemporary Jan Verkolje. (b) Side, back, and front views of a microscope by van Leeuwenhoek. (c) Drawing of a transverse section of an optic nerve based on van Leeuwenhoek's observations. [See Figure Credits.]

have remained largely unchanged. Other aspects of microscopy, however, have undergone developments that have fundamentally changed what is attainable with the technique. An important example of this is the preparation of the sample using molecular approaches, which only came about in the 20th century. Not only can molecular biological methods be used to reveal the outlines of particular subcellular structures, they can, very importantly, be used to identify the location of specific molecules within a cell. Another example is the advent of lasers as a light source, which has led to the development of microscopy modalities that, for example, utilize specific illumination schemes to produce images with much higher information content. A third example is the development of modern image detectors which can record images with high fidelity and high speed for subsequent detailed quantitative analysis. Contemporary reports suggest that van Leeuwenhoek was not a skilled artist and needed the help of assistants to draw sketches of the observations he made. This highlights how difficult the recording of the results of imaging experiments was, and provides sharp contrast with today's use of software-operated detectors to record images at high spatial and temporal resolution.

2

Introduction to Two Problems in Cellular Biology

Cells form the building blocks of all complex organisms. Although cells have been studied for centuries, it is surprising how much remains unknown. Over recent decades the revolution in genetics and molecular biology has had a major impact on cell biology, since it is now possible to investigate the molecular makeup of a cell and, even more importantly, the dynamic processes that occur within a cell. Cells are highly structured and are characterized by a fascinating but still poorly understood array of dynamic processes. We will now discuss, in general terms, two important problems in medicine and biology that require significant investigation using light microscopy. These examples will serve to introduce different types of microscopy experiments that we will refer to throughout the following chapters. The two problems are antibody trafficking and the behavior of the transporter of iron, transferrin. The biological concepts presented here will be discussed in more detail in Part II.

2.1 Antibody trafficking

The immune system provides the body's defense mechanism against foreign invaders, such as bacteria, viruses, and toxins. Such bacteria and viruses are generically known as *pathogens*. Protein molecules called antibodies constitute a major component of the immune system. They are Y-shaped molecules (Fig. 2.1(a)) that are produced by the immune system through complex processes, so that they can recognize and bind to pathogens in the body. Once antibodies are bound to these foreign objects, immune cells are recruited that engulf or produce toxic molecules to destroy them (Note 2.1).

Recent progress in the antibody engineering field has led to the exploitation of such processes for drug design. Here antibodies are "designed" in the laboratory and produced with the aim that they bind to specified targets, such as cancer cells. The cancer cells that are bound by the engineered antibodies (Fig. 2.1(b)) are then also attacked by the body's own immune cells and destroyed. This and similar approaches have led to promising new antibody-based drugs for the treatment of different diseases, such as various types of cancers that include breast and colon tumors. In the field of immunotherapy, antibodies can also be used to block immunoregulatory molecules and stimulate the immune response against tumors. An additional area of active interest is to generate antibodies for treating infectious diseases caused by bacteria or viruses. Furthermore, antibodies can be used to remove or change the activity of cells that cause autoimmune diseases. Autoimmune diseases occur when a person's immune system goes awry and attacks one or more components of the body (e.g., myasthenia gravis and systemic lupus erythematosus).

Antibodies also play a major role in diagnostics, biosensors, and laboratory research, since they can be used to specifically detect and label other molecules such as proteins. For example, in cellular microscopy, antibodies are of great significance since they can be used to label various cellular components for imaging. This labeling strategy will be discussed

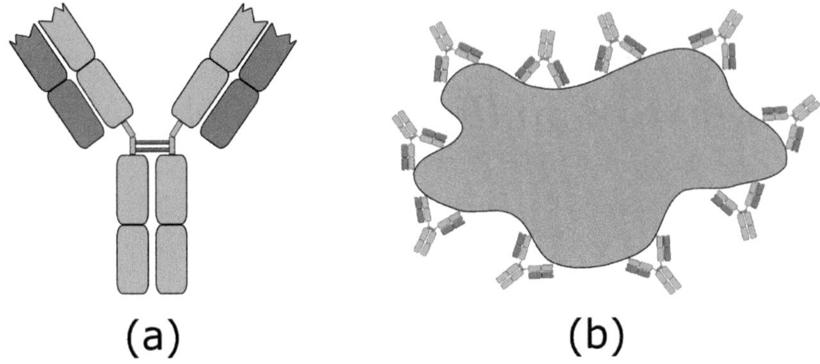

FIGURE 2.1
Antibody binding marks pathogens for destruction by the immune system. (a) An antibody is a Y-shaped molecule. The structural elements that enable the antibody to bind to targets are located at the tips of its arms. (b) A pathogen, such as a bacterium or a cancer cell, is depicted with antibodies binding to its surface.

extensively in Chapter 9. Antibodies can also be used for the detection and imaging of tumors in the body using a variety of whole body imaging approaches.

There are different types of antibodies, with those of the immunoglobulin G (IgG) class being the most abundant. We will learn more about the different antibody classes in Chapter 6. It is the IgG subclass of antibodies that is often used as the platform for drug design.

A question that is important for the functioning of the immune system, in addition to the efficacy of an antibody-based therapeutic, is how long IgG molecules persist in the bloodstream and tissues. For example, a cancer therapeutic is frequently injected into the bloodstream of a cancer patient. If the drug (or antibody) is cleared very rapidly from the bloodstream and tissues, it is necessary to have more frequent, repeated injections than when the therapeutic remains at a high level for a prolonged period of time. It is therefore attractive to use antibodies of the IgG class that have longer persistence in the body compared with other antibody classes.

Receptors are proteins that are usually located on the membranes of cells or cellular compartments. They carry out a diverse array of functions that include the sensing of changes in the external environment (e.g., an increase in growth factor levels) and the transport of molecules within and across cells. A receptor known as FcRn was shown in the mid-1990s to be responsible for regulating the persistence of IgG molecules in the body. The strength of interaction between injected IgG molecules and FcRn within cells governs how fast the injected antibodies are cleared from the body. At the extreme end, in genetically modified mice without FcRn ("knockout" mice), injected IgG molecules are cleared out very quickly from the body (Fig. 2.2) (Note 2.2).

These findings immediately prompted the question as to how FcRn is responsible for the maintenance of IgG levels in the body. Being able to understand this mechanism allows for more specific designs of drugs that exploit the properties of this receptor. In fact, we will see that such insights have led to the development of prototype drugs that have the promise to be used for the treatment of autoimmune diseases. Trying to understand the mechanism of action of FcRn brings us immediately to cell biological questions, with the first one being where the receptor is located in (human) cells. We know there are different types of compartments within cells that carry out distinct functions. Is FcRn located in specific compartments within the cell? If so, the location would suggest particular functions. Such

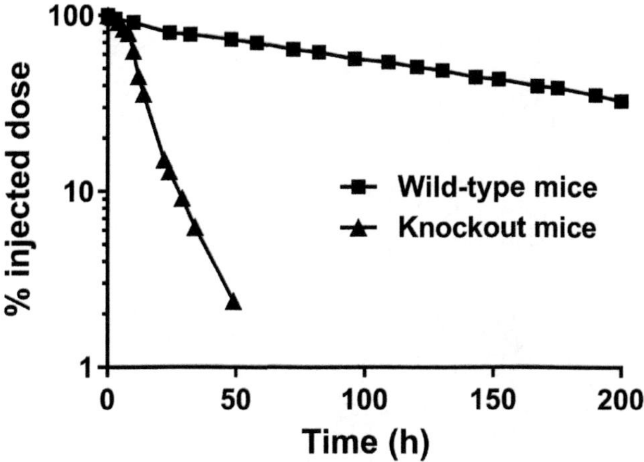

FIGURE 2.2
Levels of injected antibody (IgG) in the body as a function of time. Compared with the antibody in wild-type mice, the antibody in knockout mice that do not express FcRn is cleared very rapidly from the body. [See Figure Credits.]

questions can be answered with prototypical experiments in cellular microscopy, such as localization and association experiments.

2.2 Localization experiments

The most basic, although not always straightforward, microscopy experiment is to answer the question where in the cell the protein of interest is located. In our case we would like to know the cellular location of FcRn. In each of the leftmost images of Fig. 2.3, a part of a cell is shown in which FcRn is visualized by suitable labeling of FcRn and detection with a fluorescence microscopy experiment. In each case, the membrane that encloses the cell is not labeled, and therefore the contours of the cell are not easily visible. We see from both images that FcRn is not typically located on the membrane that surrounds the cell but is instead inside the cell, frequently associated with ring-like structures. This particular experiment was carried out using human endothelial cells. Endothelial cells are cells that line the vasculature and are therefore important for the transport of antibody molecules across endothelial layers and for the maintenance of antibody levels in the bloodstream. Note that each of these images provides a two-dimensional (2D) cross section of a cell. More elaborate experiments, however, can be carried out to obtain a three-dimensional (3D) view of the distribution of molecules of interest within a cell (Section 11.2).

2.3 Association experiments

Through a localization experiment, we can obtain information about the distribution of the molecules (usually, but not always, proteins) of interest in the cells or tissues of the

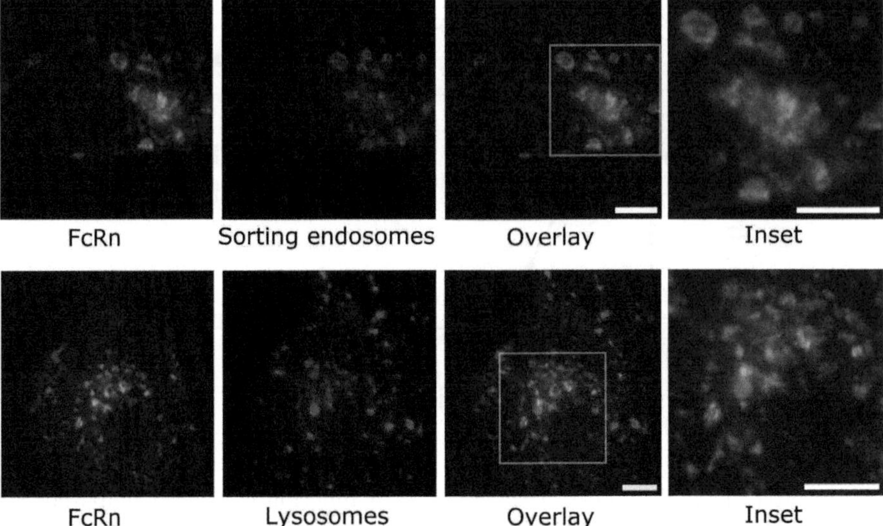

FIGURE 2.3
Fluorescence microscopy images showing the distribution of FcRn in human microvascular endothelial (HMEC-1) cells. In both rows, the images show a significant portion of a single HMEC-1 cell. In the top row, FcRn and sorting endosomes in the same field of view are shown separately in green and red, respectively. In the overlay of the green and red images, and even more clearly in the inset from the overlay image, FcRn can be seen from the good agreement between the green and red signals to be localized in the sorting endosomes. In the bottom row, in the same field of view, FcRn is shown in green and lysosomes in red. In this case, it can be seen from the poor overlap between the green and red signals in the overlay image that FcRn is not detected in lysosomes. Scale bars: 5 μm. [See Figure Credits.]

body. This type of analysis, however, is often not satisfactory since it will not allow us to determine the relationship of the molecules of interest to the various known subcellular compartments in the cell. As we will discuss in much more detail in Chapter 9, cells are highly structured with many different compartments (bound by membranes) and structures that serve different purposes. The sketch in Fig. 2.4 shows various compartments of a typical cell. In particular, compartments referred to as *sorting endosomes* are responsible for sorting substances that are transported into the cell into different pathways. For example, the sorting of proteins into compartments called *lysosomes* results in the degradation of the proteins, since lysosomes are the degradative compartments of cells. The sorting of proteins into the correct pathway in cells is essential for the cell to carry out its functions. Indeed, defects in these activities can lead to a multitude of different diseases that include Alzheimer's and atherosclerosis.

If we can determine on which of the intracellular structures our receptor of interest is located, we can learn something about the possible functions of the receptor in the cell. A type of experiment, the *association experiment*, is a standard experiment in cell biological applications. Its purpose is to determine where the particular protein of interest is located within a cell or tissue section. This is achieved by labeling known cellular structures and determining where the molecule of interest is located with respect to these known structures. We know from the localization experiment shown in Fig. 2.3 that FcRn is often localized in circular structures in a cell. An association experiment is then carried out to identify these

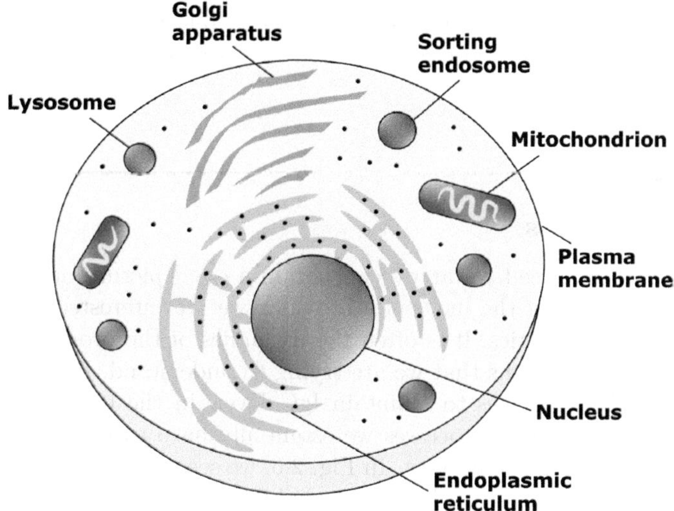

FIGURE 2.4
Sketch of a typical cell depicting different compartments.

round structures. The bottom row of Fig. 2.3 shows an image of lysosomes that are labeled with a fluorescent marker that is specific for lysosomes. In the overlay with the image of fluorescently labeled FcRn, we see little overlap between the signals from the two fluorescent markers, suggesting that the round FcRn-labeled structures are not lysosomes. However, in the top row of Fig. 2.3, the overlay of fluorescently labeled FcRn with an image taken with a fluorescent marker for sorting endosomes shows very good agreement between the two different fluorescent signals. This demonstrates that the round structures identified in the localization experiment are in fact sorting endosomes. This suggests that FcRn may in some ways be associated with sorting processes in the cells.

Figure 2.5 shows another association experiment. Here FcRn is shown in relation to tubulin and actin structures that form some of the basic transportation tracks in cells along which compartments can move.

Association experiments are also referred to as *colocalization experiments*, since the

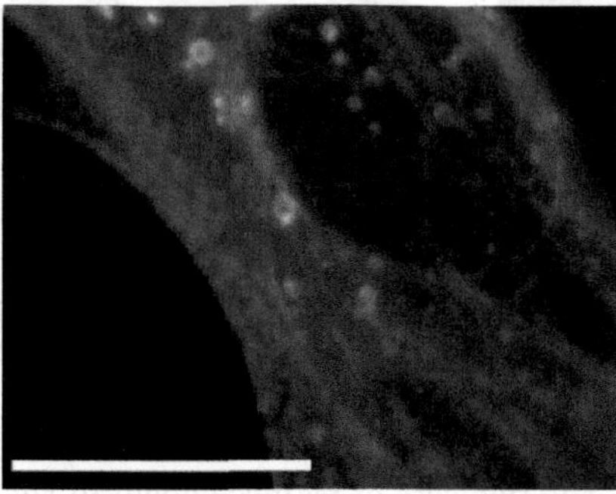

FIGURE 2.5
Overlay of fluorescence microscopy images acquired in three color channels, showing a portion of an HMEC-1 cell. FcRn is shown in green, tubulin in red, and actin in blue. Tubulin and actin form the tracks on which much of cellular transport occurs. Scale bar: 10 μm.

determination of the molecule of interest's location in relation to a cellular structure is based on whether two different fluorescent markers are located in the same areas of the cell.

2.4 Dynamic studies

With the approaches discussed so far we can obtain a good picture of the spatial organization in the cell or tissue of the molecules in which we are interested. However, we have only mentioned static properties. It is often the dynamics of the processes in the cell that reveal the biological mechanisms that we are trying to understand. In our example, we are interested in how FcRn manages to maintain IgG levels in the bloodstream or even the whole body. To reveal dynamic processes we essentially have to produce "movies" of the cell over time periods that are of interest. In Fig. 2.6, we see a sorting endosome at different time points. Tubular structures can be observed that contain both IgG and the receptor, FcRn. This confirms that FcRn is involved in the sorting process of IgGs in the cell. In fact, Fig. 2.6 shows the behavior of mutated IgGs that no longer bind to FcRn. These IgGs are located in the interior space or "vacuole" of sorting endosomes, and the tubules that contain FcRn leave the sorting endosome without co-transporting these IgGs. This difference in sorting behavior directly relates to the differences in persistence in the body of these two antibodies; the mutated antibody is not transported by FcRn and persists for a very short time period compared with its parent, unmutated antibody that binds to FcRn.

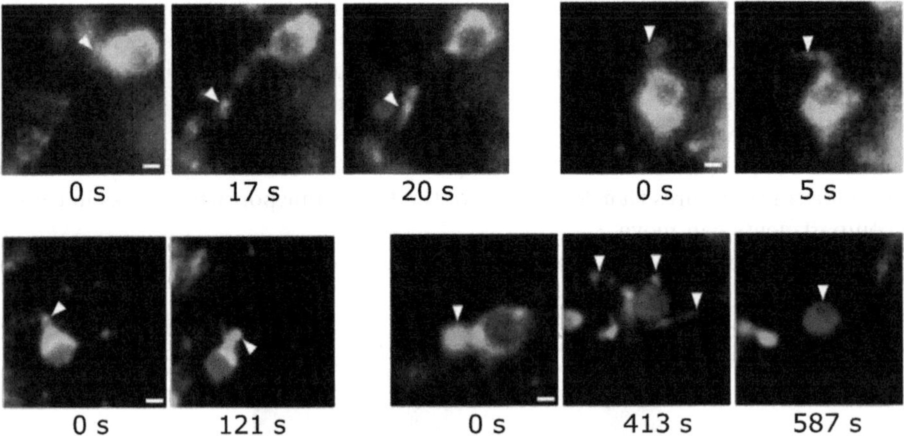

FIGURE 2.6
Live cell fluorescence microscopy images showing the sorting of IgGs by FcRn in endosomes of HMEC-1 cells. Frames from movies of four sorting endosomes are shown with the time indicated below each frame. In each of the two sets of frames shown in the top row, FcRn (green) and IgG (red) are seen to move away (see arrows) together in tubules from the sorting endosome. In each of the two sets of frames shown in the bottom row, the IgG (red) is of a mutated form that does not bind to FcRn (green), and therefore does not leave the sorting endosome in tubules containing FcRn (see arrows). Instead, the mutated IgG remains in the vacuole of the endosome. Note that at time point 587 s in the set of frames on the right, the sorting endosome no longer has a detectable level of FcRn, and only the mutated IgG is detected in the structure (see arrow). Scale bars: 1 μm. [See Figure Credits.]

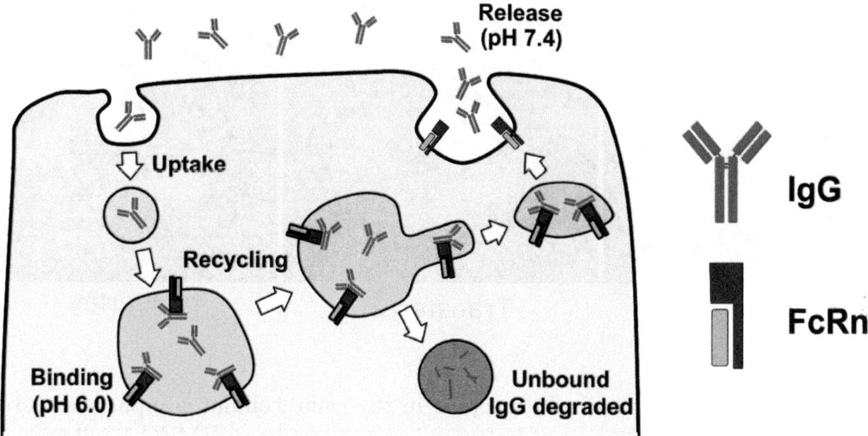

FIGURE 2.7
Recycling and degradation of IgG in cells. Following uptake by the cell, IgG molecules that are bound to FcRn in acidic sorting endosomes are recycled back to the external surface of the cell, where the neutral pH facilitates their dissociation from FcRn. IgG molecules that are not bound to FcRn are eventually degraded by the cell in compartments called lysosomes. [See Figure Credits.]

Imaging experiments such as the ones described here suggest the cell biological model depicted in Fig. 2.7 for the functioning of the receptor, FcRn, which can be summarized as follows. IgG molecules are taken up into the cell and enter sorting endosomes. If the IgG molecules do not bind to FcRn, as is the case for the mutated IgG molecules shown in Fig. 2.6, they are transported into the lysosomal compartments of the cell, where they are degraded. This leads to a decay in antibody levels in the body. However, if the IgG molecules bind to FcRn in sorting endosomes, they are transported back to the outside of the cell, where they are released as FcRn does not bind IgG molecules at the neutral pH that is present outside a cell. This is a mechanism that salvages IgG molecules from degradation in the cells' lysosomal compartments and provides an explanation for how the absence of FcRn results in a significantly shorter persistence of IgG molecules in the body as seen in Fig. 2.2.

2.5 Iron transport, transferrin, and the transferrin receptor

The second biological system that will be used in this text for illustrative purposes is that of the iron transport system. Iron is essential for many biological processes. For example, most commonly it is known as a critical component of a red blood cell protein called hemoglobin, which transports oxygen throughout the body. Iron deficiency leads to reduced hemoglobin levels and low red blood cell numbers, resulting in anemia. Although many different causes of anemia have been identified, iron deficiency is the most common. Iron also plays an important role in the activity of a number of so-called *enzymes*, which are the proteins responsible for interconverting substances in the body. Defects in the activity of these enzymes can lead to severe metabolic problems. In addition, iron is associated with a subset of proteins that are involved in energy generation in cells. The wide array of processes

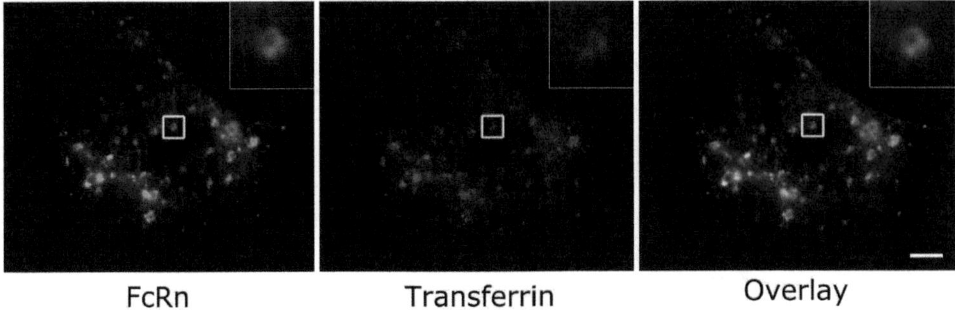

FIGURE 2.8
Localization of FcRn and transferrin receptor in the same cellular compartments revealed by fluorescence microscopy. Images of FcRn and transferrin in an HMEC-1 cell are shown in green and red, respectively. Since transferrin is internalized into the cell after binding to its receptor, the transferrin signal is also indicative of the transferrin receptor. The correspondence between the green and red colors in the overlay of the two images demonstrates that both FcRn and the transferrin receptor localize to the same ring-like structures (sorting endosomes). The inset zooms in on such a ring-like structure, in which signals corresponding to both FcRn and transferrin receptor can be seen. Scale bar: 5 μm.

in which iron plays a role means that it is essential to regulate the levels of this metal in cells.

It is well established that a protein called *transferrin* plays a major role in maintaining the correct iron balance in the body. This protein has an associated transferrin receptor that is, like FcRn, membrane-bound. Given the important function of transferrin and its receptor in maintaining the correct iron balance in the body, delineating how this transport system works has fascinated biologists for many years. For example, where is the transferrin receptor located in cells? This question can be addressed by carrying out localization and association experiments. Such studies show that the transferrin receptor is present in ring-like sorting endosomes that overlap with those occupied by FcRn (Fig. 2.8). By contrast with FcRn, however, the transferrin receptor is present on the plasma membrane of the cell at much higher levels. Here, microscopy experiments can again be used to show that the transferrin receptor binds to transferrin at the cell surface.

As shown in Section 2.4, fluorescence microscopy allows us to determine the location of proteins as a function of time. We can, therefore, use fluorescence microscopy to analyze what happens to transferrin and its receptor following their binding at the cell surface. Such studies have suggested the trafficking model illustrated in Fig. 2.9, which can be described as follows. The transferrin receptor carries iron-bound transferrin into sorting endosomes. The local environment in the sorting endosomes triggers the release of iron from transferrin to produce an unloaded form of transferrin. The iron can then enter the interior of the cell, whilst the iron-free transferrin is sorted with the transferrin receptor to return to the membrane of the cell, where it is released back into the extracellular environment. As with FcRn, the sorting of transferrin and its receptor occurs in tubules that leave the endosomes and are observable by imaging the cells over time (Fig. 2.10). Thus, transferrin and its receptor have a critical function in scavenging iron and delivering it to cells.

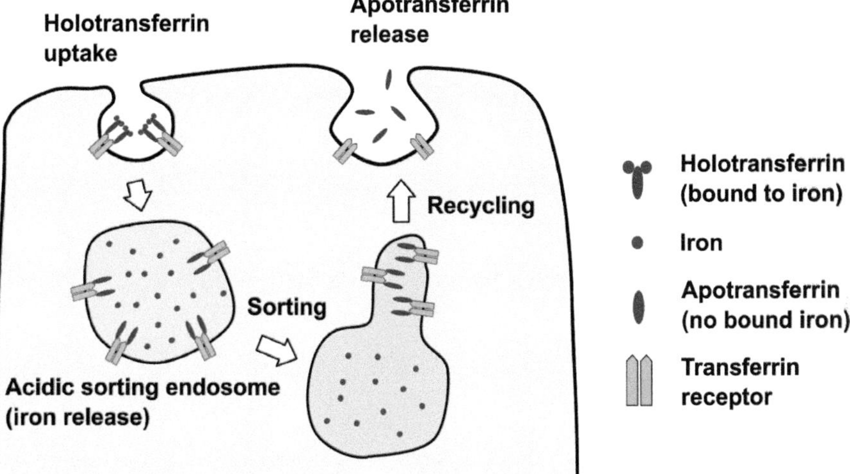

FIGURE 2.9
Trafficking of transferrin, transferrin receptor, and iron in cells. Following uptake of holotransferrin (iron-bound transferrin) via binding to transferrin receptor, iron is released from the receptor-transferrin complex in the sorting endosome and enters the cell, whilst the receptor-transferrin complex recycles back to the external surface of the cell and the transferrin is released as apotransferrin (transferrin without bound iron) from the complex. [See Figure Credits.]

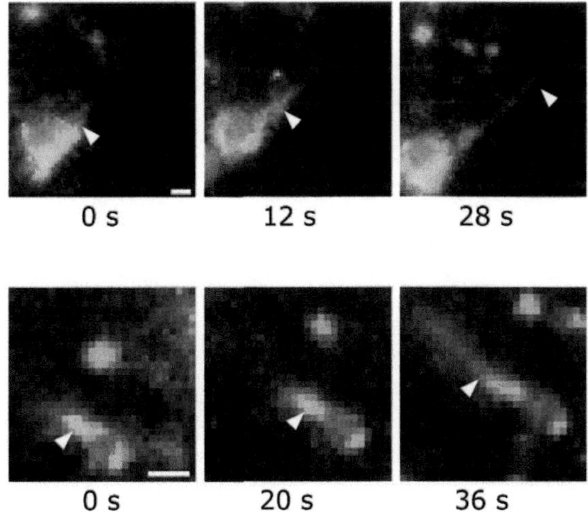

FIGURE 2.10
Live cell fluorescence microscopy images showing transferrin (red) and FcRn (green) in tubular structures (see arrows) that extend from a sorting endosome. The top and bottom rows show frames from movies of two different HMEC-1 cells, with the time indicated below each frame. In each case, the interior of the sorting endosome is occupied by dextran (blue), a polymer of glucose molecules which is known to accumulate in endosomes when cells are incubated with it, and which does not traffick with transferrin and FcRn. Scale bars: 1 μm. [See Figure Credits.]

3

Basics of Microscopy Techniques

In the prior chapter, we presented problems in cell biology that are representative of research topics which can be addressed using modern microscopy approaches. In the current chapter, we will provide an overview of some of the most important microscopy tools that are available to investigate these questions.

3.1 Optical microscopy for cell biology

A microscope is an optical tool that allows us to investigate a sample at a significantly higher magnification than what is possible with the human eye. Given sufficient resolving power, we can expect to reveal many details of the examined structures that are not visible with the naked eye. What one can see with a microscope depends on the specific modality that is used for the imaging. There are two main modalities that are of importance in cellular microscopy. The first is transmitted light microscopy. Here light is transmitted through the object of interest, and we detect those parts of the object that lead to a slight modification of the light path. The second modality is fluorescence microscopy, which relies on imaging parts of a cell that are labeled with a fluorescent tag. The importance of fluorescence microscopy lies in the fact that the microscopist has great flexibility to select what is to be labeled and thereby what it is that can be imaged. This makes this technique especially important for modern molecular cell biology, where a major focus is the determination of the locations, dynamics, and interactions of molecules within their cellular context. Therefore, fluorescence microscopy is often the experimental method of choice for cell biologists. In fact, the images shown in the prior chapter (Figs. 2.3, 2.5, 2.6, 2.8, and 2.10) are all acquired using fluorescence microscopy.

3.2 Transmitted light microscopy

Transmitted light is not only used in microscopy to reveal information about objects that otherwise would not be visible. Children are intrigued to discover, for example, that they can see structures in their hands and fingers if they shine a strong light through their hands (Fig. 3.1).

Cells are translucent and therefore a similar principle can be applied as in the above example to visualize structures within a cell. In the transmitted light imaging modality the cell sample is placed on the microscope stage and is illuminated by a strong light source. Opposite to the light source the detector, often a camera or the eyes of the microscopist, register the resulting image of the cell. As with the example of the child's hand, variations

FIGURE 3.1
A child's hand when a strong light is shone through it.

in the makeup of the cellular structures can induce differences in the light path of the transmitted light that then reveal part of the cellular structure.

Figure 3.2 shows a sketch of the basic components of a transmitted light microscope. The illuminating light source is frequently a halogen lamp, i.e., a lamp similar to the lamps in most cars' headlights. The light is focused on the sample. On the other side of the sample the objective collects the light that passes through the sample. Further optics then focus the image onto the detector, be it a camera or the microscopist's eyes.

The transmitted light images of human microvascular endothelial cells are shown in Fig. 3.3. In each image we can see the outline of the cell and several internal compartments.

We should note that there are several different implementations of a transmitted light microscope. They all vary in the specifics of the light that is transmitted and the aspects of the transmitted light, such as changes in its phase as introduced by the sample, that are recorded by the detector. For descriptions of two such techniques which are commonly employed, see Section 10.8.

3.3 Fluorescence microscopy

Transmitted light microscopy has been a tool in biology going back to the very beginnings of microscopy. A major advantage of transmitted light microscopy is that sample preparation

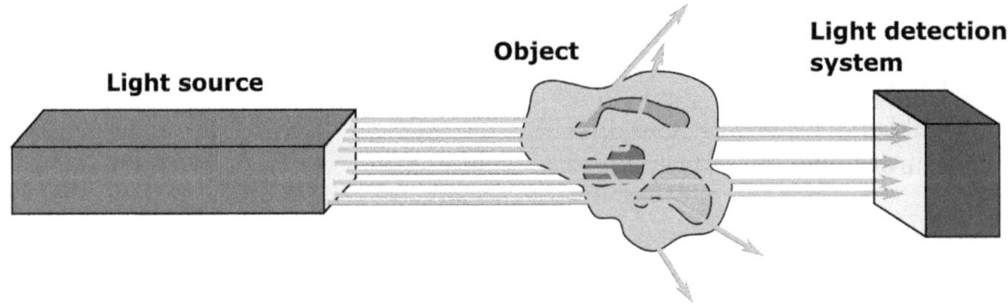

FIGURE 3.2
Sketch of the light path of a transmitted light microscope.

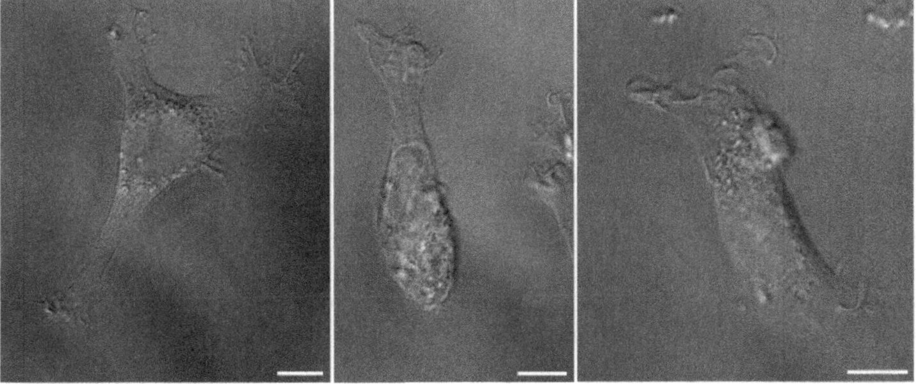

FIGURE 3.3
Images of human microvascular endothelial (HMEC-1) cells acquired with a transmitted light microscope. Scale bars: 10 μm.

is relatively straightforward. This advantage, however, turns into a disadvantage in that the sample cannot be modified to reveal different aspects of the cellular structure or dynamics. This is of particular relevance in modern molecular cell biology where questions are asked such as where the proteins and molecules of interest are located in the cells and how they interact with each other.

Such information can, however, be obtained by labeling the molecule of interest with a specific label that can then be detected. This principle has been applied in many imaging contexts. In cellular microscopy, the use of fluorescent labels has proved to be particularly successful and fluorescence microscopy has become the technique of choice in many cell biology investigations.

Figure 3.4 provides a comparison of transmitted light microscopy and fluorescence microscopy by showing that while the former can be used to obtain a general view of the outline of a cell and its internal structures, the latter can be used to reveal the distribution of a specific protein of interest inside a cell.

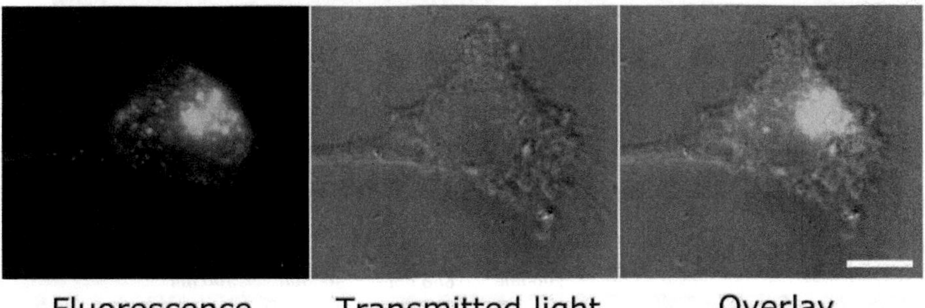

FIGURE 3.4
A fluorescence image and a transmitted light image of a human microvascular endothelial (HMEC-1) cell. Whereas the transmitted light image shows the outline of the cell and its internal compartments, the fluorescence image shows the distribution of the protein receptor FcRn within the cell. An overlay of the two images is also shown. Scale bar: 10 μm.

3.3.1 Fluorescence

The Electromagnetic Spectrum

Before discussing fluorescence and fluorescent labels it is important to review the electromagnetic spectrum since one of the most important aspects of fluorescence microscopy is the large variety of wavelengths of the different fluorescent probes. Fluorescence microscopy is primarily carried out in the visible spectrum. The range of visible light, as defined through the wavelength range in which the human eye can detect light, is usually given as 380 nm to 780 nm (Note 3.1). Figure 3.5 shows an illustration of the wavelength range that is important for fluorescence microscopy, which roughly ranges from around 350 nm at the near-ultraviolet end of the spectrum to about 750 nm at the near-infrared end. The violet region ranges up to around 450 nm, followed successively by the blue region that ranges up to around 495 nm, the green region that ranges up to around 570 nm, the yellow region that ranges up to around 590 nm, the orange region that ranges up to around 620 nm, and finally the red region that ranges up to 780 nm.

Fluorescence and Stokes shift

Fluorescence is the central photophysical effect that is at the heart of many of the modern microscopy techniques. Some molecules have fluorescence properties, meaning that when light of a certain wavelength range is used to excite the molecule, within just a few pico or nanoseconds light of a higher wavelength range is emitted (Fig. 3.6) by the molecule. The difference between the maxima of the exciting and emitting wavelength ranges is known as the *Stokes shift*, which we will revisit in Section 8.3.

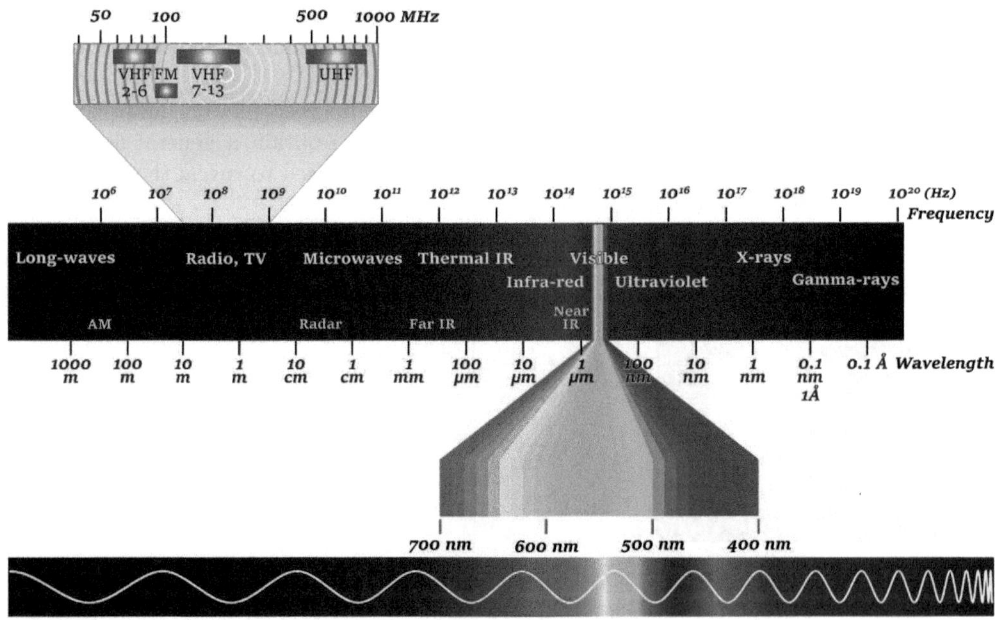

FIGURE 3.5
Electromagnetic spectrum.

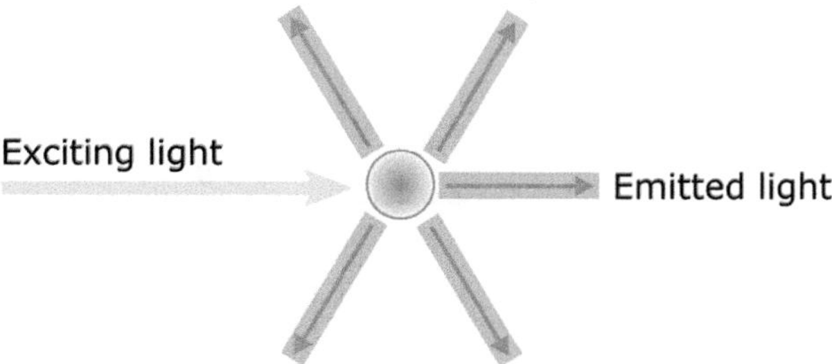

FIGURE 3.6
Excitation light (orange-yellow) impacts a fluorescent molecule, resulting in fluorescence emission at a higher wavelength (orange) from the molecule.

Fluorescence labeling

The principle of fluorescence labeling is simple. However, different fluorescent labels impose different practical limitations and require the use of different labeling techniques, and the choice of appropriate labels represents a critical part of experimental design, as we will see, for example, in Sections 8.7 and 10.3. If the microscopist wants to localize a particular protein in a cell, a *fluorophore*, i.e., the fluorescent agent, is "attached" to the protein. The cell is then illuminated by light of the wavelength that excites the fluorophore. Following excitation, the fluorophore will emit light in the fluorophore's emission band of wavelengths. At the same time, the cell is observed by eye or using a camera, with the microscope being set up such that only the emitted light reaches the observer. This image that is "seen" by eye or by the camera shows those parts of the cell at which the fluorophore is present, thereby allowing us to visualize the location of the protein of interest in the cell. Figures 2.3, 2.5, 2.6, 2.8, and 2.10 show examples of images acquired in this way. Specific cellular compartments can be detected using the same general principle. A common approach is to attach a fluorescent label to an antibody that can specifically recognize a molecular marker of the compartment of interest. In Fig. 2.3, for example, the sorting endosomes and lysosomes are detected using fluorescently labeled antibodies that bind specifically to their molecular markers.

3.3.2 Layout of an epifluorescence widefield microscope

The emission light path of an *epifluorescence* widefield microscope is essentially identical to that of a transmitted light microscope. As in a transmitted light microscope, light from the sample, in this case the fluorescence emitted by the fluorophore, is collected by the objective and then projected onto the detector. It is the delivery of the illuminating or exciting light that is very different in the epifluorescence configuration. In contrast to the transmitted light configuration, here the illuminating light is delivered to the sample through the objective and therefore from the side of the sample from which the emitted light is also collected (Note 3.2).

An important component in the light path of epifluorescence microscopes (Fig. 3.7) is the filter cube that is located immediately behind the objective in the light path. The optical components in this filter cube, i.e., the excitation and emission filters and the dichroic beam splitter, are designed to only allow light into the sample that falls in the range of wavelengths that can excite the fluorophores in the sample, and to only allow light that is emitted by the fluorophores to be detected by the camera. The crucial role of the filter cube assembly is therefore to filter out unwanted light. We will discuss the various design considerations

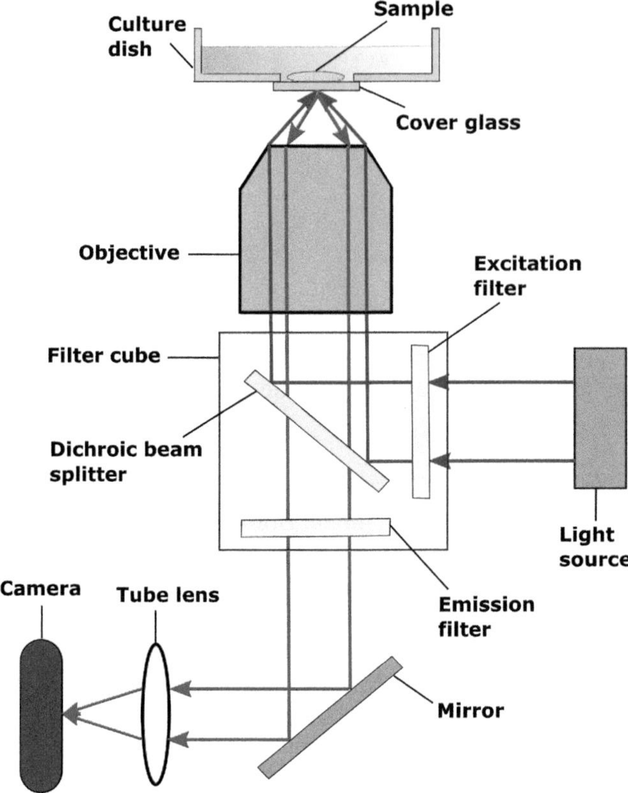

FIGURE 3.7
Sketch of the light path of an epifluorescence microscope.

for the filter cube in detail in Section 10.7. A reader who is interested in finding out more concerning filter cubes at this stage can move ahead to that section. For now, however, this is as much as we need to know in order to be able to proceed.

A variety of light sources are used in epifluorescence microscopy, depending on the specific application. Light sources can be based on lamps such as xenon or mercury arc lamps, lasers or photodiodes, or light-emitting diodes (LEDs).

The term "widefield" is used to suggest that a relatively large or wide region of the cell is illuminated, in contrast to a *point illumination* microscope configuration in which a very small excitation volume is scanned through the sample, such that the image is built up point by point. We will discuss confocal microscopy, a point illumination technique, in Section 10.4.

3.4 Inverted versus upright microscope

A newcomer to microscopy is often puzzled by the difference between an *upright* and an *inverted* microscope (Fig. 3.8). In fact, in most cases, there is no fundamental difference besides a different spatial organization of the optical components. In an upright microscope the objective is situated above the sample, whereas in an inverted microscope it is situated

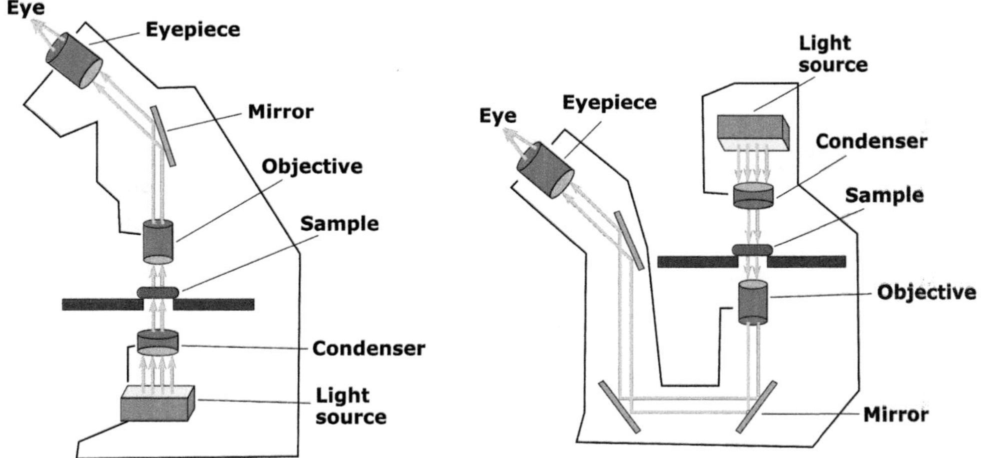

FIGURE 3.8
Sketch of an upright (left) and an inverted (right) transmitted light microscope with their light paths.

below the sample. Upright microscopes are sometimes preferred because they often have a smaller footprint. However, inverted microscopes are essentially required when imaging samples that are suspended in any form of liquid and that need to be accessible from below.

3.5 Components of commercial microscopes

So far we have discussed the basic layout of a transmitted light and an epifluorescence widefield microscope. Commercially available microscopes typically have many additional features to allow for the proper handling of the sample and the imaging. Such microscopes are often placed on a vibration isolation table in order to minimize the impact of environmental vibrations on the imaging. It is surprising what impact slamming a door can have on the quality of an image if the microscope is not isolated from outside vibrations. A modern microscopy setup is also often equipped with a computer that controls the camera and other hardware components during an experiment.

Figure 3.9(a) shows an an example of a commercial inverted microscope, and Fig. 3.9(b) provides a sketch of such a microscope that shows its major components. A typical inverted microscope is equipped with both a halogen lamp for transmitted light illumination and an arc lamp for fluorescence excitation. The focus level of the object in a microscope can usually be changed with a manual or automated focus knob. The lowest increments by which the focus can be changed varies based on the technique that is used, but can be on the order of nanometers for a piezo device. A piezo device is a device that creates movement using material that changes shape upon application of an electric field.

The sample is held in position in a sample holder on the sample stage. This sample stage is movable either manually or in an automated fashion to position the sample at the correct location with respect to the stationary field of view of the microscope. The sample holder is designed in such a way that an objective can be positioned within a few millimeters of the sample. The objectives are screwed into the *nosepiece*, which allows different objectives

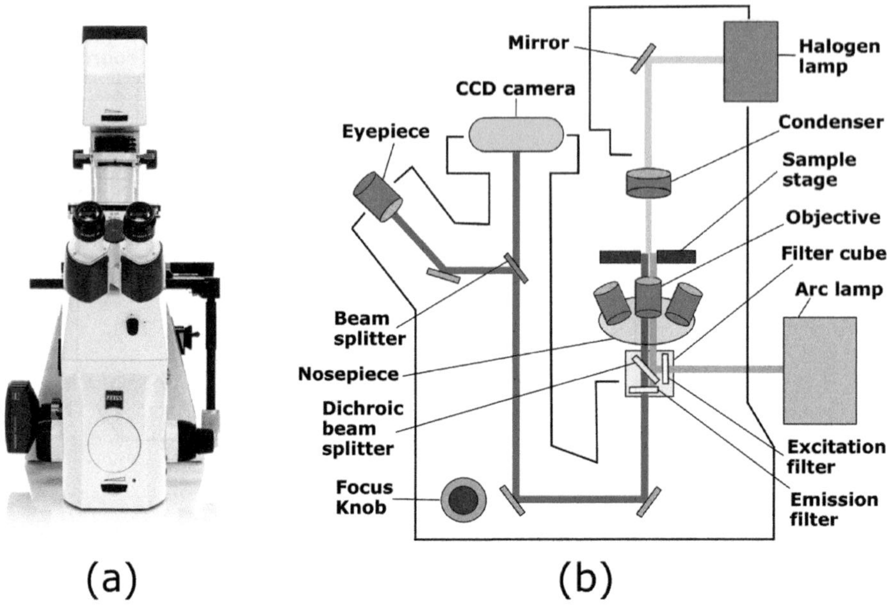

FIGURE 3.9
A typical commercial inverted microscope. (a) The Zeiss Axio Observer.A1, an inverted microscope. (b) Sketch of an inverted microscope showing its major components. When used for fluorescence microscopy, a filter cube containing the excitation and emission filters and the dichroic beam splitter ensures that an appropriate range of wavelengths of light from an arc lamp is used to illuminate the sample, and that the desired range of wavelengths from the fluorescence collected by the objective is passed through and guided to the eyepiece and/or camera. When used for transmitted light microscopy, light from a halogen lamp is focused onto the sample by a condenser lens system, and the light transmitted by the sample is collected by the objective and directed to the eyepiece and/or camera. [See Figure Credits.]

to be easily rotated into the light path. Next in the light path and close to the nosepiece, a filter cube can typically be inserted for use in fluorescence microscopy.

High-grade microscopes usually have stereo eyepieces and are also equipped with a highly sensitive camera. For more advanced applications even multiple cameras can be used.

3.5.1 Light sources

An optical microscope (Fig. 3.7) has a light source that needs to be adapted to its imaging needs. The light source can be a lamp, a laser, or an LED. The general characteristics of a suitable light source are that it is very bright, that it must excite at the correct wavelength, and that it is stable over time in terms of light power and light spectrum. A lamp-based source is typically a halogen lamp, a mercury arc lamp, or a xenon arc lamp. Mercury arc lamps are gas-discharge lamps comprising a glass enclosure that contains liquid mercury and an inert gas. The enclosure has a pair of closely positioned electrodes between which an electric arc is formed upon introduction of a current. This arc generates heat to vaporize the mercury, which in turn produces light upon ionization. As more mercury is vaporized, a high-pressure internal atmosphere is created which has the effect of making the lamp's light output even brighter. Since the arc is small, it provides a point source of illumination that

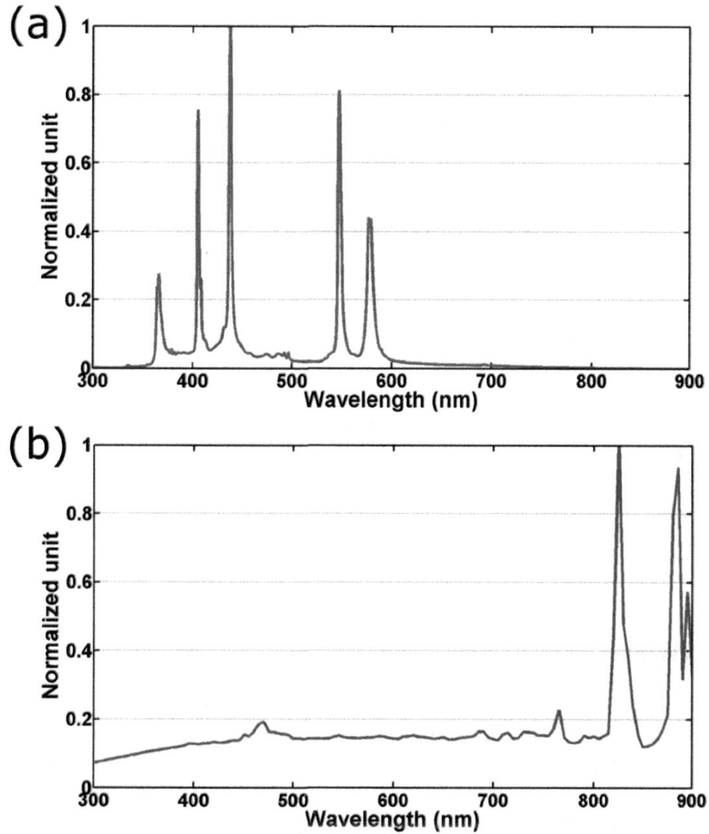

FIGURE 3.10
The spectrum of (a) a mercury arc lamp and (b) a xenon arc lamp. [See Figure Credits.]

is ideal for microscopy. As shown in Fig. 3.10(a), the light spectrum of a mercury arc lamp is characterized by very sharp peaks at wavelengths below 600 nm. Xenon arc lamps are gas-discharge lamps that also comprise a glass enclosure with a pair of electrodes, but they contain pure xenon gas. They operate on the same underlying principle of light production via the transmission of electricity through ionized gas. Compared with mercury arc lamps, xenon lamps have a much more uniform spectrum in the visible spectrum (Fig. 3.10(b)) but contain some very strong peaks above the visible spectrum.

Lasers are light sources that are often used in settings where their main characteristics are of importance, e.g., their typically strong output of coherent light at a single wavelength. These properties make lasers particularly useful in confocal microscopy, where a well-focused excitation beam has to be generated for point excitation. Disadvantages of lasers are that they are frequently expensive, and that because they produce light of a single wavelength, one laser line can only excite a limited set of fluorophores that have a suitable excitation spectrum (Section 8.6).

LEDs are light sources made from semiconductor material comprising a p-n junction. When voltage is applied, electrons from the n-side of the semiconductor and electron holes from the p-side of the semiconductor flow into the junction. Light is emitted when an electron combines with a hole and transitions to a lower energy level. The wavelength range of the emitted light is dependent on the semiconductor material from which the LED is made, and has a bandwidth typically in the tens of nanometers. Among the advantages of

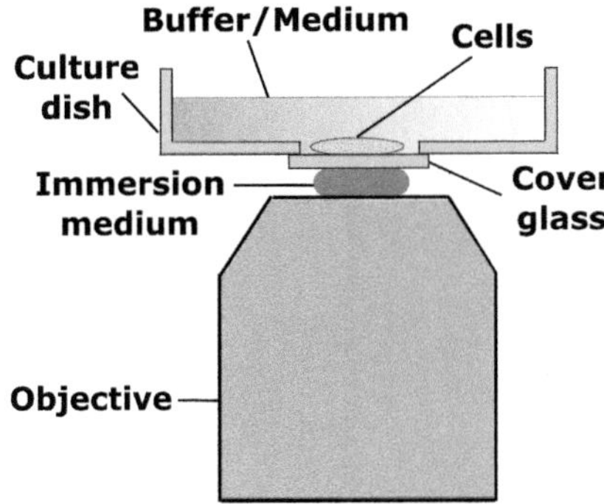

FIGURE 3.11
Sketch of an immersion objective with immersion medium (typically oil) between the front optical element of the objective and the cover glass.

an LED light source are its stable and finely adjustable light intensity, long lifetime, and relatively low cost.

3.5.2 Objectives

The central optical element of a microscope is the objective, which typically comprises a complex system of lenses. Matching its complex design is also often its high cost, which can in some cases compete with that of a small new car. The objective not only determines the magnification of the microscope, but also has a major impact on the resolving power of the microscope, i.e., its capability to reveal small details.

Most objectives are designed to be used with specific choices of *cover glass* and *immersion medium* (Fig. 3.11). A cover glass is a small plate of glass whose optical properties, such as its thickness and refractive index, are well matched to the properties of the objective. An immersion medium is the designated medium that fills the space between the objective's front element and the cover glass. These objectives fall into two categories — the so-called *air* or *dry* objectives, which are designed to simply have air as the immersion medium, and objectives that are designed to be used with a liquid immersion medium. A liquid immersion medium is typically oil, water, or glycerol. As we will see in Section 10.6.1, due to the relatively high refractive index of oil, oil objectives can have significantly higher resolving power than air and water objectives.

There are so-called *dipping* objectives which are designed to work without a cover glass. Dipping objectives are mainly used with upright microscopes to "dip" into the medium of the sample that is being observed.

Figure 3.12 shows a typical objective. Manufacturers have engravings on the sleeve of an objective that specify important characteristics, such as the magnification that it will produce of the sample that is imaged, and whether it is an air objective or is to be used with a particular liquid immersion medium. Importantly, the engravings will also specify the objective's *numerical aperture*, which we will denote by the notation n_a. (Note that in other books and sources, the numerical aperture is also often abbreviated as NA.) Commercial objectives have numerical apertures that vary typically from 1.4 at the high end (with some highly specialized objectives having even higher values) to close to 0 at the low end of the quality range. A high numerical aperture implies high resolving power of the objective. In Section 4.2, we will discuss some implications of the numerical aperture on the performance

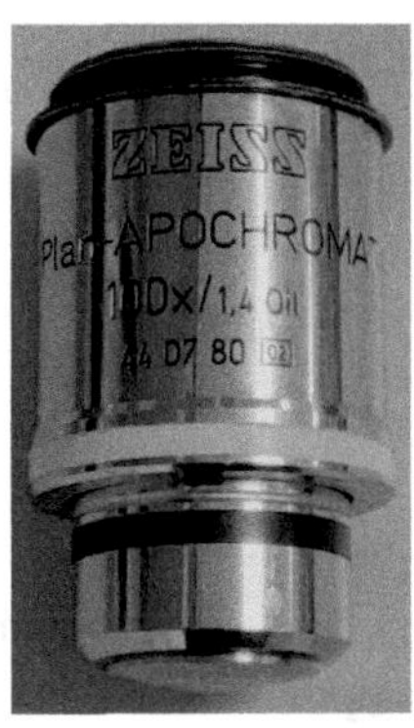

FIGURE 3.12
Objective designed for oil immersion, with a numerical aperture of 1.4 and a magnification of 100×. This particular objective is manufactured by Carl Zeiss. Other companies (e.g., Leica, Nikon, Olympus) offer objectives with similar characteristics.

of a microscopy system. The physical meaning of the numerical aperture in relation to the light-gathering capability of the objective will be discussed in Section 10.6.1.

3.6 Fixed and live cell experiments

Cellular microscopy experiments can usually be classified depending on whether or not they study *fixed cells* or *live cells*. Fixed cell experiments investigate cellular samples that are not live and have been prepared with sample preservation methods that stabilize/"fix" the sample so that the cellular components are immobilized. A common fixative used for this purpose is paraformaldehyde solution, which causes the formation of a stable meshwork of a cell's contents by the cross-linking of the molecules in the cell. In contrast, in live cell experiments, cultured cells that are functional and live are imaged to investigate dynamic cellular events. Such experiments are typically more complex than fixed cell experiments due to the fragile nature of the cellular sample and the more demanding imaging that is necessary to capture the dynamics of interest. Incubation chambers may be necessary to keep the cells alive, and heating devices can be required so as to image at the physiological temperature of 37°C.

3.7 Sample preparation

To be studied by a microscope the cellular sample needs to be mounted on a cover glass. For live cell samples, the cover glass is often inserted beneath a hole in the bottom of a culture dish and attached to the dish bottom using an adhesive, forming a small well in which the cellular sample sits (Fig. 3.13(a)). The live cells are usually cultured on the cover glass in the culture dish prior to imaging, and are subsequently imaged while incubated in a medium appropriate for the maintenance of the cells.

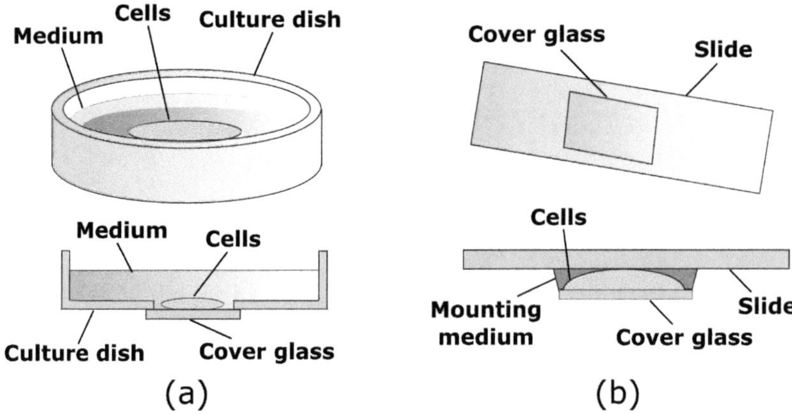

FIGURE 3.13
Sketches of (a) a live cell sample in a culture dish with attached cover glass and (b) a fixed cell sample with cover glass and microscope slide. Both types of samples can be imaged with an upright or an inverted microscope. Use of an upright microscope with a culture dish, however, requires the use of a dipping objective. Note also that using an upright microscope with a microscope slide requires inverting the sample shown in (b) so that the cover glass faces upwards.

For fixed cell applications the cells are covered by a *microscope slide*, which is a somewhat thicker and larger piece of glass. The fixed cellular sample therefore ends up being sandwiched between the cover glass and the microscope slide (Fig. 3.13(b)). A fixed cell sample is typically embedded in a mounting medium, which can be as simple as a standard buffered saline solution. Some mounting media are specially formulated to help preserve the sample for long-term storage. Extended preservation, however, often also requires sealing the edges of the cover glass with an adhesive to protect the cells from oxidative damage and loss of moisture. For samples that contain fluorescent labels, a mounting medium containing an anti-fading reagent can be used to reduce the loss of fluorescence by the fluorescent labels during both imaging and long-term storage of the prepared sample. Note that fixed cell samples can also be prepared and imaged in a culture dish, though this is typically done only when long-term preservation of the sample is not required.

3.8 A note regarding safety

As with many other laboratory research techniques, there are safety considerations when carrying out cellular microscopy experiments. We do not go into these considerations here in any detail, as most institutions have their own safety policies. We would, however, want to mention two areas that are of particular importance. One relates to eye protection as potentially powerful light sources such as lasers can be used, some of which operate at wavelengths that are not visible by eye, but can be so powerful that they produce fire hazards if focused on flammable substances. The other area of concern relates to the samples that are being studied. Some live cells, for example, could harbor highly infectious agents or could be infectious themselves. Certain substances used in the preparation of a sample could also be toxic. We will not discuss these topics further, but instead refer the microscopist to his or her local safety policies and the literature for further information.

4

Introduction to Image Formation and Analysis

At first glance, one might expect that a microscope does nothing else but simply produce a magnified image of the sample. The image formation process is, however, much more complex. An excellent illustration is to consider the image of a very small point object under a microscope. Not only does this provide an interesting example of how a microscope images objects, it is also central for much of the theory regarding image formation in microscopes. Besides image formation, we will introduce in this chapter the concept of resolution in microscopy and the modeling and analysis of the image data acquired by modern detectors.

4.1 Image formation and point spread functions

Image formation in a microscope is governed to a large extent by the phenomenon of *diffraction*. We will discuss this in great detail in Chapter 14. Diffraction in the context of light microscopy comes about due to the changes the electromagnetic light waves undergo as they pass through the aperture of the objective. This leads to complex interference patterns, i.e., interactions of parts of the wave with itself that bring about an image of the object that is not completely faithful.

We will first see this by considering the image of a *point source* in a microscope. Such an image is referred to as a *point spread function*. A point source is a mathematical abstraction by which we mean an infinitesimally small object that is also typically assumed to be quite bright. It will turn out that such a notion is very useful to analyze image formation in an optical system and in a microscope in particular. This is because, in many cases, a general object can be thought to be made up of point sources.

Of course, an infinitesimally small object does not exist in nature. However, very small objects, such as polymer-based spherical particles or *microspheres*, commonly referred to as *beads*, are often excellent approximations and are therefore used to experimentally investigate image formation for point sources. Beads are used for many purposes in microscopy besides the investigation of the point spread function. They are often used as fiducial markers, for example, for the detection of drift in a sample during the acquisition of a time sequence of images, and for the spatial alignment of different images acquired of the same field of view. Beads are commercially available in many different sizes, typically ranging from tens of nanometers to tens of microns in diameter.

In Figs. 4.1(a) and 4.1(b), we see the image of a fluorescent bead (i.e., one that is loaded with many molecules of a fluorescent dye) under a microscope, as captured by a camera and as displayed in a mesh representation, respectively. Despite the simplicity of the object, its image has a very complex structure, which is characterized through a strong central component and additional concentric rings.

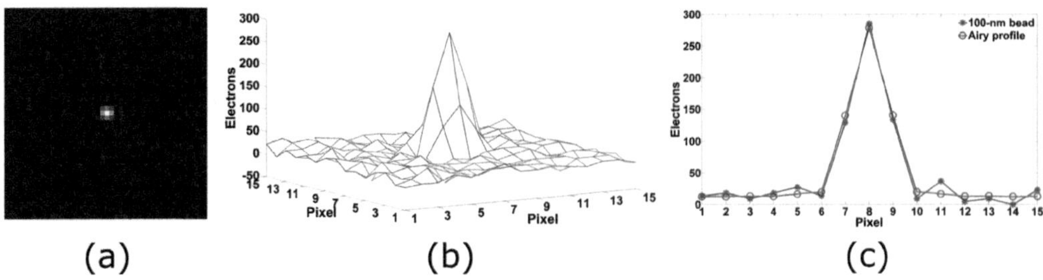

FIGURE 4.1
Image of a 100-nm Fluoresbrite yellow green bead manufactured by Polysciences, Inc. (a) Image acquired using a Zeiss microscope. (b) Center 15×15-pixel region from the image in (a), displayed in units of electrons and in a mesh representation. (c) Center 15 pixels from the middle column of the image in (a), displayed in units of electrons and overlaid with the corresponding 15 pixels from an Airy profile that has been fitted to the bead image.

In fact, it is possible to give an analytical expression for the in-focus, or 2D, point spread function as

$$psf(x,y) = \frac{J_1^2\left(\frac{\alpha}{M}\sqrt{x^2+y^2}\right)}{\pi\left(x^2+y^2\right)}, \quad (x,y) \in \mathbb{R}^2, \tag{4.1}$$

which we refer to as the *Airy profile*. Here J_1 is the first order Bessel function of the first kind, M is the system magnification, and $\alpha = \frac{2\pi n_a}{\lambda}$, with n_a the numerical aperture of the objective and λ the wavelength of the detected light. In Fig. 4.1(c), a slice is shown of the overlay of the bead image with an Airy profile that has been fitted to it. The qualifiers "in-focus" and "2D" denote that we are considering a point source that is "in focus", rather than "out of focus". Similarly, here we are only concerned with a 2D image rather than the more general 3D image. The more general scenario will be studied in detail in Section 14.5.5. Figure 4.2(a) shows the image of a point source that is computed using this analytical form of the point spread function, with values of the function represented as intensity. Figure 4.2(b) gives a plot of the slice across the center of the image in Fig. 4.2(a), and shows the actual numerical values of the function. In both cases, one can see the Airy profile's strong central peak and the diffraction rings around the peak. In Fig. 4.2(c), the center slice is plotted over a range of values that better shows the diffraction rings of the profile.

The expression for the Airy profile shows its dependence on the numerical aperture n_a of the objective, the wavelength λ, and the magnification M. To illustrate the effect of changing the value of n_a or λ, Fig. 4.3 shows that a higher numerical aperture leads to a significant narrowing of the main lobe of the point spread function. Similarly, it shows that a shorter wavelength produces a profile that is narrower than the profile produced by a longer wavelength.

We should note that, for our current purposes, the scaling factor of the Airy profile is not really relevant. For reasons that will become clear later (Section 15.1.2), we have used here a scaling factor such that the 2D integral over the point spread function equals 1, i.e., $\int_{-\infty}^{\infty}\int_{-\infty}^{\infty} psf(x,y)dxdy = 1$. In other texts, different scaling factors are sometimes used, such as one where the point at the origin equals 1. Of course, one scaling factor can easily be replaced by another one.

In the expression for the Airy profile we see that the first order Bessel function of the first kind plays an important role. Bessel functions will accompany us throughout this book, and we will make use of a few of their basic properties in the more advanced parts of the book on diffraction and data analysis.

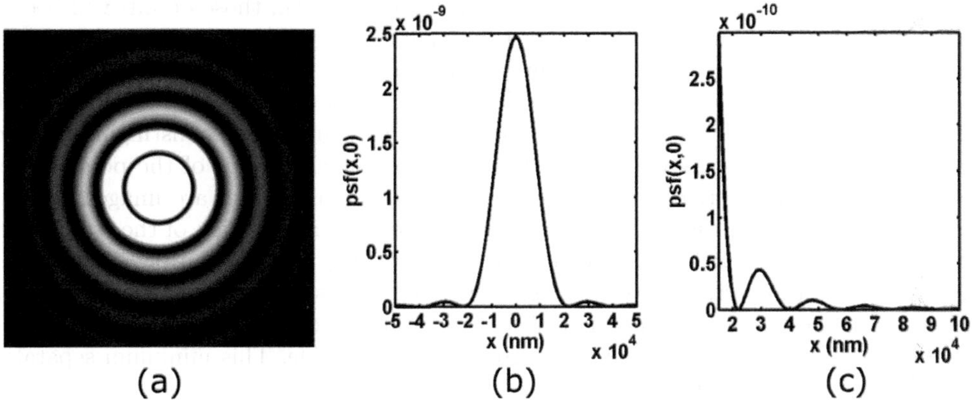

FIGURE 4.2
Airy profile psf for a point source imaged at wavelength $\lambda = 500$ nm with an objective with magnification $M = 100$ and numerical aperture $n_a = 1.4$. (a) Intensity representation, with both x and y ranging from -100 μm to 100 μm. A nonlinear scaling is used to improve the visualization of the rings of the Airy pattern. (b) Center slice: $x = -50$ μm to 50 μm. (c) Center slice: $x = 15$ μm to 100 μm.

4.2 Resolution: an elementary introduction

Resolution is a central issue in imaging and microscopy in particular. In very general terms, the notion of resolution is used to denote the capability of a microscope to discern fine details in the sample that is imaged. We know that certain details in a sample cannot be

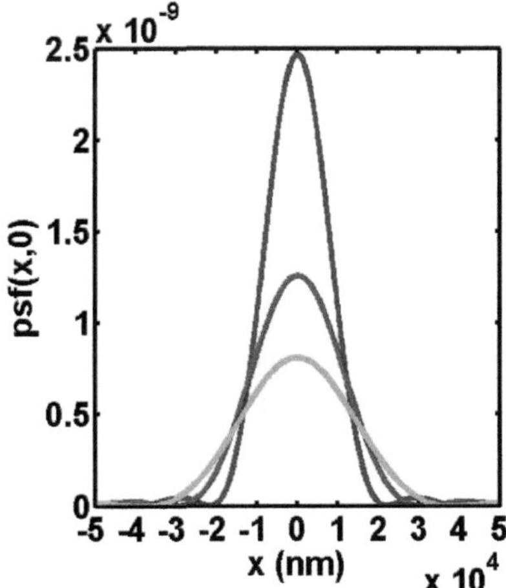

FIGURE 4.3
Comparison of Airy profiles for three different specifications. The blue profile is computed with a numerical aperture of $n_a = 1.4$ and a wavelength of $\lambda = 500$ nm, the red profile is computed with $n_a = 1.4$ and $\lambda = 700$ nm, and the green profile is computed with $n_a = 0.8$ and $\lambda = 500$ nm. In all three cases, the magnification is $M = 100$.

seen by eye, but even using a very basic microscope for children, those details will become clear. So the question immediately arises, how far can we go? Are there any details that cannot be revealed, even using the best microscope design?

The idea behind *Rayleigh's criterion* seeks to address this question, by reducing it to the simplest, special case. This is to consider when two point sources can be distinguished when imaged by a microscope. As we have seen above, the in-focus images of the point sources are given by their respective Airy profiles. When both point sources are imaged together their image is the superposition of the two Airy profiles. So the question of the resolvability of the point sources is, in fact, a question of when two Airy profiles can be distinguished. The approach in the derivation of Rayleigh's criterion is to say, somewhat arbitrarily, that these two Airy profiles can be distinguished if they are no closer than where the peak of the second profile is located at the first zero of the first profile. This minimum separation of two Airy profiles is illustrated in Fig. 4.4.

Now, this criterion can be quantified, since the first zero of the Airy profile can be fairly precisely determined. As can be seen in Fig. 4.5, the Bessel function J_1 has zeros at 0, and at approximately $\pm 1.220\pi$, $\pm 2.233\pi$, $\pm 3.238\pi$, $\cdots$. Therefore, the Airy profile $psf(x, y)$, which is nonzero at $(0, 0)$, will have zeros for all points $(x, y) \in \mathbb{R}^2$ such that $r := \sqrt{x^2 + y^2} \approx \frac{0.61M\lambda}{n_a}, \approx \frac{1.11M\lambda}{n_a}, \approx \frac{1.62M\lambda}{n_a}, \cdots$. The first zero of the point spread function of the first point source thus coincides with the maximum of the point spread function of the second point source when the peaks of the two point spread functions are approximately $\frac{0.61M\lambda}{n_a}$ apart. Note that this expression depends on the magnification M, as this distance is given in image space, i.e., between the images of the point sources. The distance between the point sources themselves in object space is obtained by dividing by M, which yields $\frac{0.61\lambda}{n_a}$ as the distance below which the point sources are taken to be unresolvable. For example, in Fig. 4.4 where we have $\lambda = 500$ nm, $n_a = 1.4$, and $M = 100$, the minimal separation distance is ~21785 nm in image space, and ~218 nm in object space. It is the expression in object space that is referred to as Rayleigh's criterion. The fact that this criterion depends on the wavelength λ of the detected light and the numerical aperture

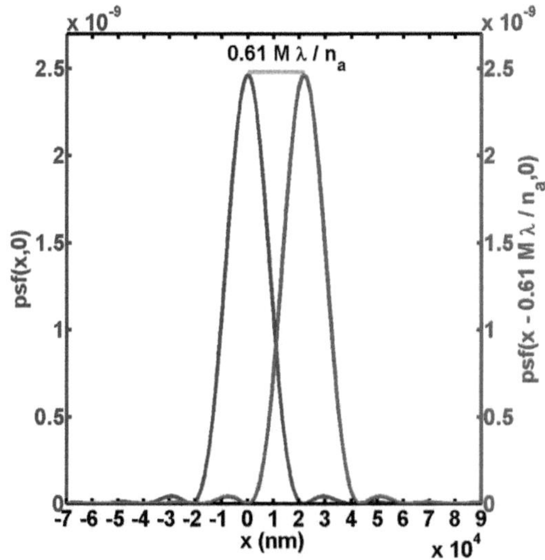

FIGURE 4.4
Illustration of Rayleigh's criterion using a numerical aperture of $n_a = 1.4$, a wavelength of $\lambda = 500$ nm, and a magnification of $M = 100$.

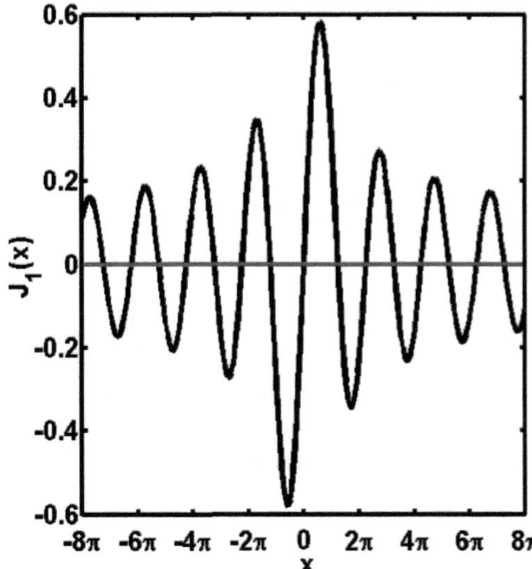

FIGURE 4.5
First order Bessel function of the first kind, J_1. The zeros of J_1 occur at values of x where J_1 intersects the red line $y = 0$.

of the objective, but not on the magnification M of the microscope, importantly indicates that resolution and magnification are very different concepts.

As we have seen in Fig. 4.3, the Airy profile becomes narrower with decreasing wavelength λ and with increasing numerical aperture n_a. It is therefore not surprising that Rayleigh's criterion shows the same dependence on these two parameters. Suppose the two point sources emit red light of, for instance, wavelength $\lambda = 650$ nm, and we have an air objective with numerical aperture $n_a = 0.9$. Then the criterion predicts a resolution limit of ~441 nm. If, on the other hand, the emitted light is blue with wavelength $\lambda = 450$ nm, and we have a high-powered oil objective with numerical aperture $n_a = 1.4$, then the resolution limit decreases to ~196 nm. In Exercise 4.4 we will consider more examples, including ones with low-powered objectives that illustrate why microscopists are prepared to pay significant sums of money to purchase objectives with very high numerical apertures.

Rayleigh's criterion has proved to be an important tool for experimental design. Microscopists have used it extensively to evaluate which features can be imaged and which microscopy configuration, especially which objective, should be used to reveal the features of interest. It is, however, very important to point out that the criterion is a heuristic one. In fact, distances well below Rayleigh's criterion have been measured (Section 20.3). In Lord Rayleigh's times microscopists observed the sample by eye, whereas today we have highly sensitive image detectors available that record the acquired images and allow them to be carefully analyzed using very advanced data analysis approaches. Not surprisingly, it is therefore possible to obtain significantly improved results over those suggested by Rayleigh's criterion. In Section 20.2, we will in fact introduce a "modern" approach to the resolution problem that shows that there is in fact no actual resolution barrier. Rather, how well two objects can be resolved depends crucially on the amount of data that has been collected.

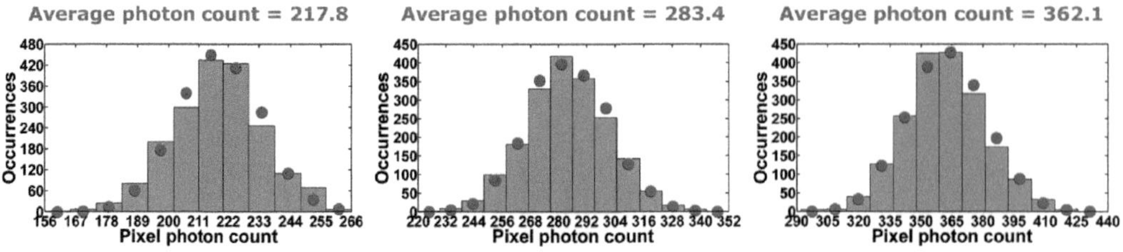

FIGURE 4.6
Histograms of photon counts in a pixel over 1900 repeat exposures of a CCD chip. Each histogram shows the photon counts corresponding to a different level of light that impinges on the pixel, as represented by the average photon count. Each histogram is overlaid with the Poisson probability distribution (red dots) with mean given by the average photon count, summed over the intervals corresponding to the bins of the histogram.

4.3 Modeling and analyzing the data

The power of modern microscopy does not only lie in the fascinating experiments that can be carried out due to modern molecular biology techniques and sample preparation approaches. Equally important are the availability of highly sensitive image detectors and, as a result, the possibility to use advanced data analysis approaches to analyze the acquired image data.

The invention of the charge-coupled device (CCD) detector put into practice the photoelectric effect that was proposed by Albert Einstein, by which light is converted into electric signals (Note 4.3). In an image detector a large number of light-sensitive *pixels* detect the photons that impinge on them and transform the photons into electrons. For each of the pixels the acquired electrons are then translated into corresponding digital counts that make up the digital image. The digital image can then be stored on a computer, displayed on a digital monitor, and most importantly for us, analyzed using digital signal and image processing approaches.

It is also important to point out that scientific research has benefited tremendously from the success of digital photography in the consumer market. The commercial success made it possible that, despite the enormous costs of the production facilities, CCD chips of very high quality can now be produced at a very low cost. In fact, the very same imaging CCD chips can be found in both consumer digital cameras and scientific-grade cameras.

What is then the type of signal that is acquired in each of the pixels? To analyze this we consider the statistics that we obtain by detecting a photon flux of light. Figure 4.6 gives examples of the numbers of photons that are collected in one particular pixel through 1900 repeat exposures of a CCD chip exposed to a constant light source. Three sets of results are shown that correspond to three different levels of light intensity from the source. In each case, the photon counts are displayed in a histogram, which clearly demonstrates that the photon count in the given pixel is not the same in every repeat exposure, and that it can in fact vary very considerably from one exposure to another, despite the fact that the light source does not change in intensity over the 1900 repeat exposures. This shows that light emission is a random or stochastic process which needs to be described using the language of probability theory. In each plot of Fig. 4.6, the histogram is overlaid with a plot of the probability distribution of a Poisson random variable whose mean equals the average of the photon counts that make up the histogram. A good match between the histogram and the

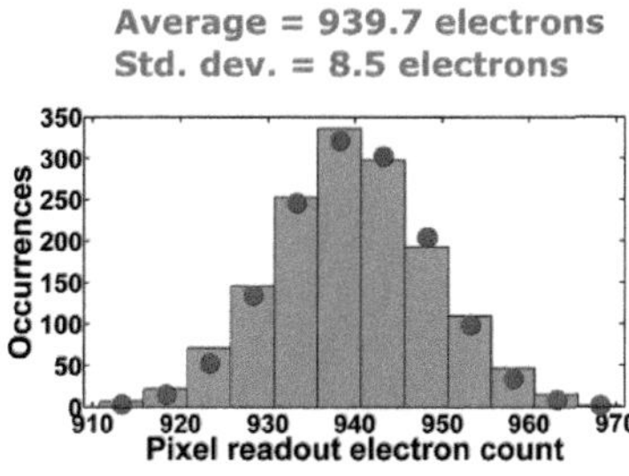

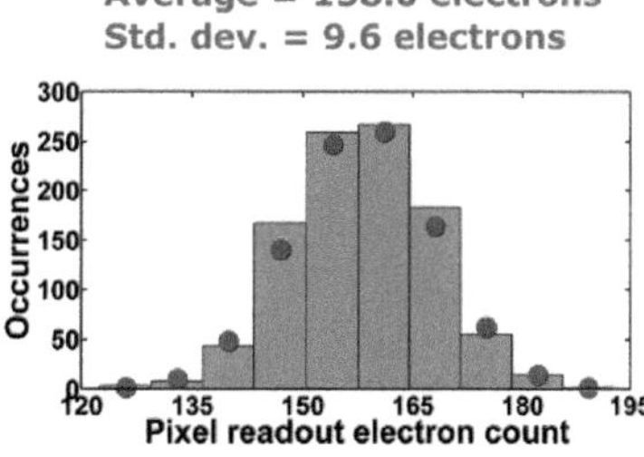

FIGURE 4.7
Histograms of readout noise electron counts acquired in a CCD pixel over consecutive images obtained while light is blocked from hitting the detector. Each histogram shows the electrons counts obtained for a pixel in a different CCD detector, and is overlaid with the Gaussian probability distribution (red dots) with mean and standard deviation given by the average and standard deviation of the electron counts, integrated over the intervals corresponding to the bins of the histogram. The histogram on the left and right comprise 1500 and 1000 measurements, respectively, of the electron count.

probability distribution, which has been summed over the intervals matching the bins of the histogram, is seen in each case, demonstrating that light emission can be modeled using Poisson statistics.

Poisson random variables have very important properties. For example, the mean of the Poisson random variable equals its variance. This has the implication that the variability of the photon count of a light source increases with its intensity. Therefore, however intense the light source, we cannot ignore the stochastic nature of the photon emission process.

Another interesting property of Poisson random variables is that the sum of two random variables is again a Poisson random variable. This means that we can "add" several light sources and still have Poisson statistics for the resulting signal.

A further important random variable that we will need to consider is the Gaussian random variable. The reason is that the readout process in an image detector is not noise-free. In a pixel, the acquired signal is corrupted by readout noise that is typically modeled as a Gaussian random variable. Figure 4.7 shows two examples of the numbers of readout noise electrons that are acquired in a particular CCD pixel in consecutive images obtained with the light blocked from hitting the detector. Each example corresponds to readout noise measured for a different CCD camera, and in each case, a histogram of the electron counts measured in the pixel is overlaid with a Gaussian probability distribution whose mean and standard deviation are given by the average and standard deviation of the electron counts. In each case, the good agreement of the histogram and the Gaussian distribution, which has been integrated over the intervals corresponding to the bins of the histogram, lends support to the modeling of readout noise as a Gaussian random variable. Importantly, the readout noise does not depend on the signal level. This implies that with low-level light sources the impact of the readout noise is very pronounced, whereas with higher-level light sources the readout noise loses its relevance. This is, of course, a well-known phenomenon from digital photography. Images taken in a low-light environment without a flash, such as in a dark room, appear very grainy, whereas images taken in sunlight show no such grainy appearance. In Section 12.7.1, we will show how we can determine the readout noise level from measured data.

A review of the basics of random variables in general, and also of the basics of Poisson and Gaussian random variables, is given in Section A.1 of the Online Appendix.

Notes

Chapter 2

1. The immunological processes involved in the body's defense against foreign objects are described in more detail in [71].
2. For more details of the biology of FcRn, the reader is referred to [23, 84].

Chapter 3

1. To learn more about light detection by the human eye, see [43].
2. The terminology "epifluorescence", which refers to the method of illuminating the sample and capturing the emitted fluorescence from the same side of the sample, is applicable to both the upright and the inverted microscope configuration. The terminology, however, arose from consideration of an upright microscope, wherein the light source is situated above (or "epi", from the Greek preposition for "above") the sample such that the illuminating light passes through the objective to reach the sample.

Chapter 4

1. In 2009, physicists Willard S. Boyle and George E. Smith received the Nobel Prize in Physics for their invention in 1969 of the CCD detector at Bell Labs. However, there is some controversy about who actually did invent the CCD detector, as the patent for the first CCD detector was awarded to another physicist at Bell Labs, Michael F. Tompsett.
2. Lord Rayleigh (born John William Strutt), the British physicist who defined Rayleigh's criterion, won the 1904 Nobel Prize in Physics for his discovery of the noble gas argon.
3. See [37] for the publication in which Albert Einstein uses light as discrete packets of energy (now termed "photons") as a basis for explaining the photoelectric effect.

Exercises

Chapter 4

1. Use a computational software program to plot the Airy profile for different wavelengths from 350 nm to 750 nm and different numerical apertures from 0.1 to 1.4. Observe how the width of the point spread function and the positions of its zeros change depending on the values of these parameters.
2. For a wavelength of 520 nm and numerical apertures of 0.4, 0.8, 0.9, 1, 1.2, and 1.4, determine the location of the first zero of the Airy profile.
3. For a numerical aperture of 1.4 and wavelengths of 350 nm, 405 nm, 488 nm, 520 nm, 650 nm, and 720 nm, determine the location of the first zero of the Airy profile.
4. Calculate Rayleigh's resolution limit for
   (a) different wavelengths in the visible spectrum, and
   (b) different numerical apertures ranging from 0.1 to 1.4.
5. Find an alternative expression for the Airy profile so that its value at $(0,0)$ is 1. More specifically, find the constant C such that for the function

$$f(x,y) = C\frac{J_1^2\left(\frac{\alpha}{M}\sqrt{x^2+y^2}\right)}{x^2+y^2}, \quad (x,y) \in \mathbb{R},$$

we have that $f(0,0) = 1$.

Part II

Biology and Chemistry

Overview

Having presented a very brief overview of the principles and examples of cellular microscopy, we now come to the parts of the book in which we introduce the various concepts in more detail. This second part is devoted to a discussion of the relevant materials from biology, chemistry, and physical chemistry. We begin by providing an introduction to genes and proteins. These sections will form a revision for anyone with a basic background in modern biology. For a reader with a classical engineering training, these sections are expected to provide the background that is necessary for the remainder of this part of the book. We then follow with a survey of antibody structure and biology. This is not only interesting in its own right, but relevant for our discussion of microscopy as antibodies represent one of the primary tools for labeling cellular structures. We then discuss the use of molecular biological tools to modify and clone genes for the generation of genetically modified proteins. This is of particular importance to fluorescence microscopy experiments in which a specific protein of interest is made observable by a genetic modification that joins it to a fluorescent protein. We next proceed to a presentation of the fundamentals of fluorescence and fluorescence labeling before closing this second part with an introduction of the basics of cell biology, which includes a detailed discussion of the preparation of cell samples for investigation with fluorescence microscopy.

5

From genes to proteins

This chapter introduces in a condensed form some of the basics of how genetic information determines the structure of proteins. Proteins are molecules of major interest in biology as they are necessary for the functioning of cells. The topics covered here include DNA and genes, the synthesis of proteins from genes, and the four levels of protein structure. Techniques for the determination of the 3D structure of proteins are also introduced.

5.1 Bonds

Bonding forces hold macromolecules together. Atoms associate with each other in chemical compounds through a variety of bonds that have different strengths. The bonds that occur most frequently between atoms in biological macromolecules are covalent bonds, ionic bonds, hydrogen bonds, van der Waals forces, and hydrophobic forces.

In covalent bonds, negatively charged electrons are shared between the positively charged nuclei (made up of protons and neutrons) of atoms (Fig. 5.1(a)). By contrast, ionic bonds involve the transfer of electrons from one atom to another. Consequently, the two atoms that form an ionic bond have positive and negative charges and attract each other (Fig. 5.1(b)).

Hydrogen bonds occur when a hydrogen atom attached to an electronegative atom (e.g., oxygen, nitrogen) gains a small positive charge. The two atoms therefore have different charges and a dipole is created. If another electronegative atom is nearby that also has a dipole, a hydrogen bond is formed (Fig. 5.1(c)). Hydrogen bonds are about 5–10%

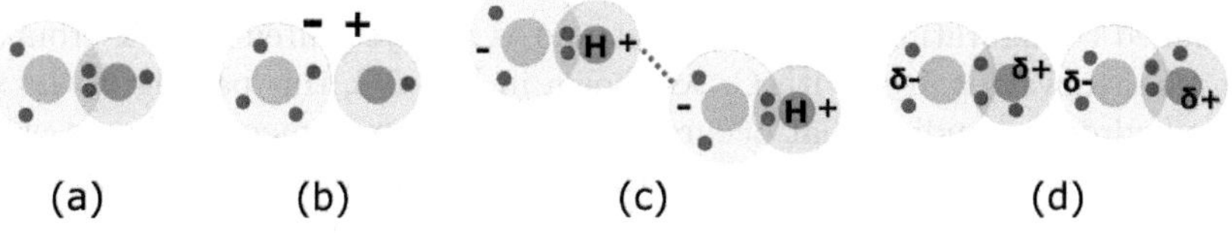

FIGURE 5.1
Illustration of different types of chemical bonds. Two types of atoms, one with electrons shown in red and one with electrons shown in blue, are used for the illustration. (a) In a covalent bond, electrons are shared between two atoms. (b) In an ionic bond, electrons are transferred from one atom to the other, and the resulting positively and negatively charged ions become electrostatically attracted to each other. (c) A hydrogen bond occurs when a hydrogen atom bound to an electronegative atom acquires a small positive charge and becomes electrostatically attracted to a nearby electronegative atom. (d) Van der Waals forces involve attractions between the transient or permanent dipoles in two atoms that are in proximity to each other. The labels $\delta-$ and $\delta+$ denote slight negative and positive charges, respectively.

the strength of covalent bonds, but constitute a major bonding type in biological macromolecules such as DNA, RNA, and proteins. In addition, the high stability of water as a liquid is due to the hydrogen bonds formed between the constituent water (H_2O) molecules.

Van der Waals forces are named after the Dutch physicist Johannes Diderik van der Waals and are similar conceptually to hydrogen bonds insofar as they involve attractive forces between dipoles in atoms that are in proximity to each other (Fig. 5.1(d)). These dipoles can be induced (transient) or permanent. These forces are highly dependent on the distance between the participating dipoles and are weaker than hydrogen bonds.

The hydrophobic ("water phobia") effect is due to the tendency of nonpolar (i.e., uncharged) atoms or molecules to exclude highly polar water molecules. An obvious example of this is the inability of water and oil to mix homogeneously; even emulsions comprise small droplets of oil in a water suspension. Thus, in biological systems hydrophobic forces drive nonpolar molecules or atoms to aggregate together to exclude water molecules and make important contributions to many macromolecular assemblies that include proteins and the membranes that surround cells. The impact of this on protein folding is discussed further in Section 5.4.

5.2 DNA and genes

Genes are the basic units of heredity in living organisms. Genes are made of deoxyribonucleic acid (DNA) and are located inside an organelle called the nucleus in eukaryotic cells. DNA comprises four different building blocks with bases called adenine, thymine, guanine, and cytosine (abbreviated A, T, G, and C). The order of these bases in a DNA strand will determine gene sequences, which in turn encode proteins. Since proteins are fundamentally important in determining essentially every cellular function, including the development of an organism, genes and the proteins they encode are the master regulators of life. During cell division, the genes of cells need to be duplicated to pass on the hereditary units to the daughter cells. How this is achieved puzzled biologists for many years, and a major step towards understanding how this process is carried out was the solution of the 3D structure of DNA (Note 5.1). This milestone revealed that DNA is arranged in two strands that are "complementary" and wrap around each other to form a double helix (Fig. 5.2). In this double-stranded configuration, A from one strand is always paired with T from the other strand, and G is always paired with C. As shown in Fig. 5.3, this pairing of the strands is enforced by hydrogen bonding that occurs between the complementary bases, which stack between the sugar-phosphate backbones of the strands. The figure also shows that the two ends of a DNA strand are designated as $3'$ and $5'$. This terminology is based on the standard numbering of the carbon atoms in the sugar component of the backbone (Note 5.2). The two strands of the double helix run in opposite directions.

The double-stranded nature of DNA provides a mechanism for duplication: prior to cell division, the strands are separated and used as templates to generate new paired strands (Fig. 5.4). The duplication machinery for generating copies of DNA is highly complex. First, the DNA strands need to be separated. This separation creates torsional stress since the two strands are wrapped around each other with a helical twist (an analogy would be the separation of two intertwined ropes that are constrained at their ends). The stress is relieved by the nicking of one strand by an enzyme, i.e., a molecule that catalyzes a chemical process, followed by untwisting and resealing. As the strands are separated, they are copied by enzymes called polymerases. These polymerases have to be highly accurate, because insertion of errors during copying could result in alterations in the encoded proteins

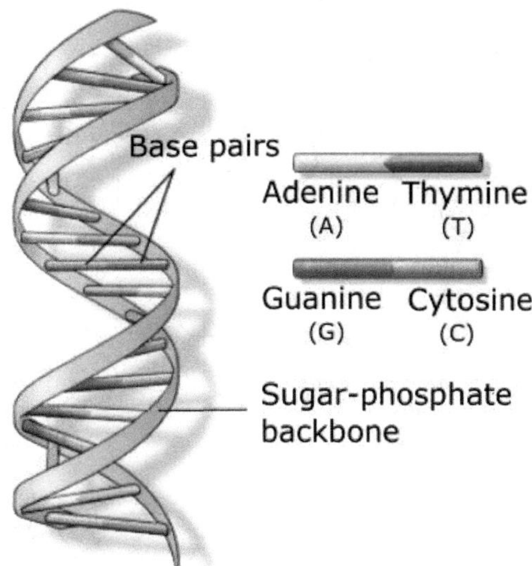

FIGURE 5.2
Schematic representation of double-stranded helical DNA showing the sugar-phosphate backbone and the base pairing between adenine/thymine and guanine/cytosine. [See Figure Credits.]

with potentially disastrous consequences for the organism. On the other hand, a very low error rate does occur, which in turn results in mutations that are selected for or against by evolutionary pressure. The error rate of the enzymes that copy DNA during duplication is greatly diminished by a process called proofreading or editing. This process, which like most cellular processes consumes energy, allows the enzyme to check that the correct base is inserted prior to moving along the DNA template strand to add the next base.

How are genes converted to proteins, which are the controllers of nearly all cellular activities? First, each protein is encoded by a unique gene sequence, which is represented by a stretch of DNA. In higher eukaryotes and some yeast, genes are split up into coding sequences ("exons") and noncoding sequences ("introns") (Section 5.3). Proteins are made up of amino acids, of which there are 20 different kinds. These different amino acids have distinct chemical properties, and their order within a protein gives the protein its particular activities (discussed further below). Importantly, the protein is not directly templated off the DNA. Instead an intermediate molecule called ribonucleic acid (RNA) is made, which comprises essentially the same building blocks as DNA but with two differences. First, the so-called sugar component of a DNA building block is the sugar deoxyribose, whereas that of an RNA building block is the sugar ribose. Second, the base thymine is replaced by the base uracil, which is chemically quite similar. The names of the RNA building block bases are abbreviated to A, U, G, and C. Another important difference between RNA and DNA is that RNA does not, in general, exist in double-stranded form (except in some viruses). In all complex cell types, it is made as a single strand, and is made in multiple copies for each DNA segment/gene that is being transcribed (the process by which enzymes generate RNA copies of DNA using DNA as template is called *transcription*; Fig. 5.5). Also, the generation of RNA from a DNA template is not as accurate as the duplication of DNA. The primary reason for this is that multiple copies of RNA are made and therefore a low mutation rate is acceptable for a cell to survive. Also, RNA does not represent the heritable material of a cell.

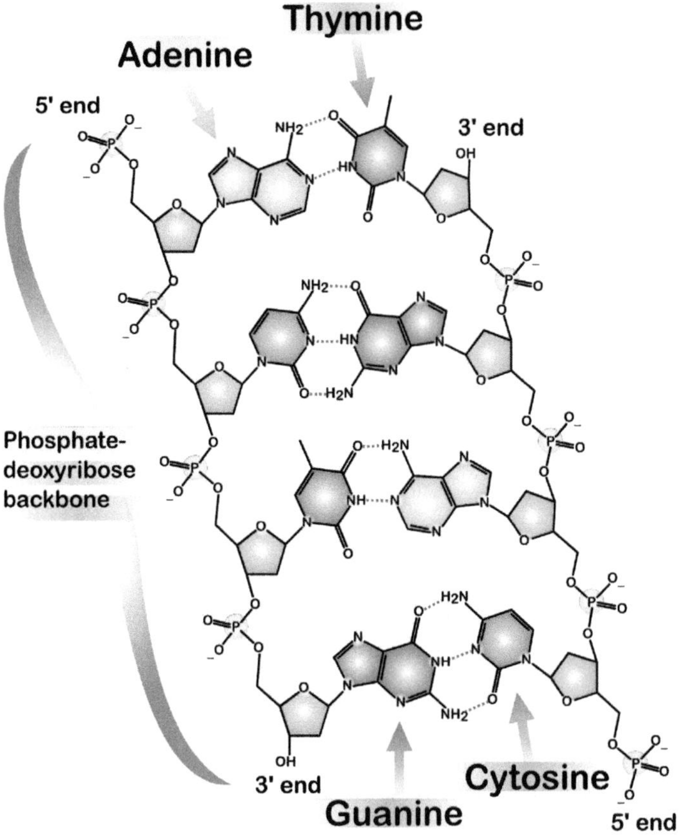

FIGURE 5.3
Detailed representation of DNA showing the chemical structure of the sugar-phosphate backbone, the four bases adenine, thymine, guanine, and cytosine, and the hydrogen bonds (dotted lines) between the paired bases. The 3′ and 5′ ends of each strand of the DNA are also labeled. [See Figure Credits.]

5.3 How are proteins made?

To generate proteins in cells, the following steps occur. One, the region of DNA that encodes the protein is copied into an RNA molecule in the cell (Fig. 5.5). The appropriate regions of DNA to copy are flagged by start and stop sites. The start sites are called promoter regions; the stop sites are called termination sites. These regions are specific DNA sequences that are recognized by RNA polymerase, the enzyme that copies DNA into RNA (Fig. 5.5). In multicellular eukaryotes and some unicellular eukaryotes, but not in prokaryotic cells such as bacteria, the DNA region to copy comprises both introns and exons. Introns, or intervening sequences, do not encode the protein, whereas exons encode the expressed protein. The introns are removed from the RNA in the nucleus by a process called splicing, which needs to be precise so that the spliced exons encode the correct protein. Two, this RNA is then exported from the nucleus into the cytoplasm of a cell (for a sketch of a cell, see Fig. 2.4). Three, the RNA is then used as a template to make proteins in a process called protein synthesis. Protein synthesis is itself a complex series of events that involves many

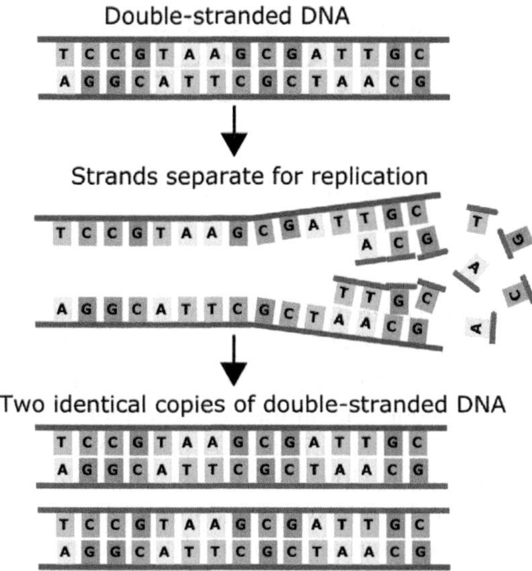

FIGURE 5.4
High-level schematic depicting the replication of double-stranded DNA. The synthesis of the new DNA strands is carried out by an enzyme called DNA polymerase, which is not shown in the schematic.

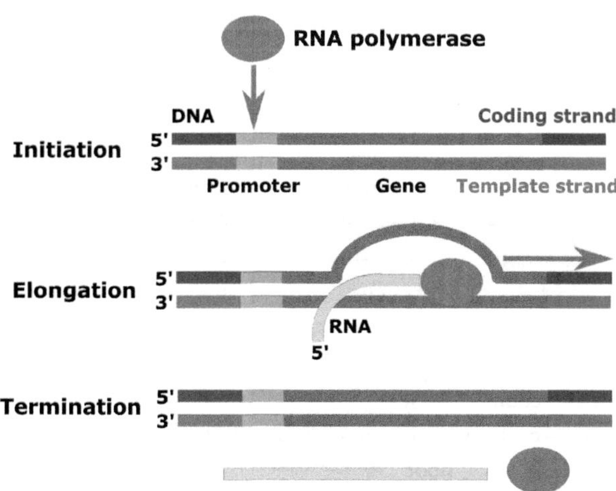

FIGURE 5.5
Schematic representation of the generation of RNA by the use of DNA as a template in a process called transcription. The process of making an RNA copy of a gene involves three main steps: initiation, elongation, and termination. Initiation involves the binding of RNA polymerase, the enzyme responsible for the copying, to the promoter region of the DNA. During elongation, the RNA polymerase traverses the template strand of the DNA and creates an RNA copy of the gene that is complementary to the template strand. The DNA double helix is unwound during this process. Termination entails the release of the RNA polymerase and the completed RNA copy of the gene from the DNA, and the restoration of the DNA double helix.

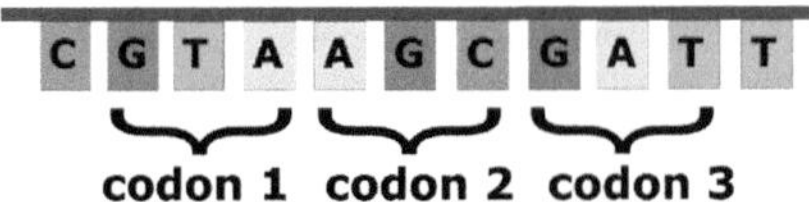

FIGURE 5.6
A codon is made up of three DNA or RNA building blocks. The schematic depicts three consecutive codons along a DNA strand.

components. But before this is covered, it is important to understand the nature of the genetic code: each triplet of bases in DNA (or RNA) represents a codon (Fig. 5.6) that encodes a particular amino acid. For example, in the case of RNA, the bases A, U, and G form the codon AUG, which encodes the amino acid methionine.

Since there are 20 amino acids (Fig. 5.7) and 64 possible codons, nearly all amino acids are encoded by more than one codon (Fig. 5.8). In fact, the codon frequency for each amino acid bears some relationship to the relative abundance of that amino acid in proteins: for example, the common amino acid serine is encoded by six different codons, whereas the rarer amino acid tryptophan is encoded by only one codon.

Following the transcription off a DNA template and further processing, the RNA

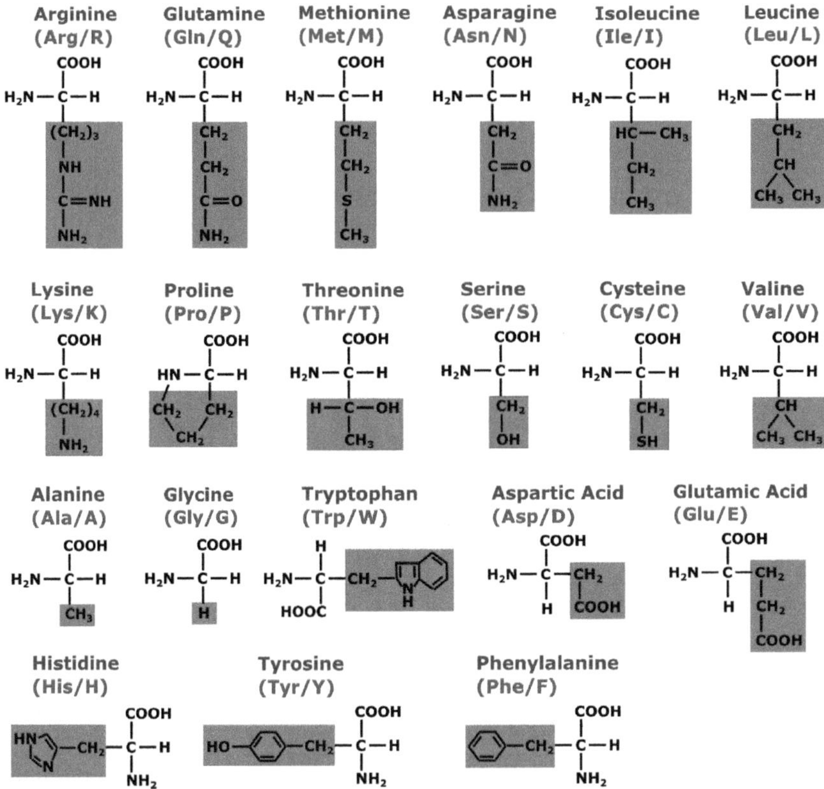

FIGURE 5.7
The chemical structures of the 20 common amino acids. The side chain (highlighted in orange) of each amino acid differentiates it from the others in terms of properties such as size and polarity. The abbreviated names (one- and three-letter) for each amino acid are shown in parentheses.

1st	2nd base											
base	U			C			A			G		
U	UUU	Phenylalanine	F	UCU	Serine	S	UAU	Tyrosine	Y	UGU	Cysteine	C
	UUC	Phenylalanine		UCC	Serine		UAC	Tyrosine		UGC	Cysteine	
	UUA	Leucine	L	UCA	Serine		UAA	Stop		UGA	Stop	
	UUG	Leucine		UCG	Serine		UAG	Stop		UGG	Tryptophan	W
C	CUU	Leucine	L	CCU	Proline	P	CAU	Histidine	H	CGU	Arginine	R
	CUC	Leucine		CCC	Proline		CAC	Histidine		CGC	Arginine	
	CUA	Leucine		CCA	Proline		CAA	Glutamine	Q	CGA	Arginine	
	CUG	Leucine		CCG	Proline		CAG	Glutamine		CGG	Arginine	
A	AUU	Isoleucine	I	ACU	Threonine	T	AAU	Asparagine	N	AGU	Serine	S
	AUC	Isoleucine		ACC	Threonine		AAC	Asparagine		AGC	Serine	
	AUA	Isoleucine		ACA	Threonine		AAA	Lysine	K	AGA	Arginine	R
	AUG	Methionine	M	ACG	Threonine		AAG	Lysine		AGG	Arginine	
G	GUU	Valine	V	GCU	Alanine	A	GAU	Aspartic acid	D	GGU	Glycine	G
	GUC	Valine		GCC	Alanine		GAC	Aspartic acid		GGC	Glycine	
	GUA	Valine		GCA	Alanine		GAA	Glutamic acid	E	GGA	Glycine	
	GUG	Valine		GCG	Alanine		GAG	Glutamic acid		GGG	Glycine	

FIGURE 5.8
The triplets of RNA bases (codons) that encode different amino acids.

molecule is exported from the nucleus and binds to large protein-RNA complexes in the cytoplasm called *ribosomes*. Ribosomes are small machines that play a central role in protein synthesis (Note 5.3). Although they are present in both prokaryotes and eukaryotes, their composition in these two different types of organisms differs slightly in size and properties. This difference has allowed the development of antibiotics that specifically target bacterial ribosomes without affecting the ribosomes of the infected person who is being treated. Ribosomes comprise two subunits. The small subunit binds the RNA molecule, whereas the larger subunit is the catalytic site where bonds between amino acids are formed to produce a protein whose amino acid sequence is specified by the sequence of codons in the RNA molecule. Specifically, individual codons in the RNA molecule are recognized by triplet RNA bases called anticodons in adaptor molecules called transfer RNAs (tRNAs). A tRNA molecule binds to a codon at one end via its anticodon and to the specific amino acid encoded by the codon at the other end (Fig. 5.9). Therefore, amino acids do not directly bind to the RNA molecule, but are inserted into a protein chain by tRNA adaptors. Protein synthesis thus proceeds as anticodons of tRNA molecules pair with the codons in the RNA molecule, and amino acids are added one by one to a growing protein chain as the ribosome moves along the RNA molecule (Fig. 5.9).

Addition of an amino acid to the growing protein chain involves an enzymatic reaction that connects two amino acids by a covalent bond. In the specific case of proteins, these connecting bonds are called peptide or amide bonds (Fig. 5.10). The enzymes that are involved in connecting amino acids to tRNA molecules are called synthetases and, like the enzymes that duplicate DNA, have proofreading capability, in this case for double-checking that the correct amino acid is enzymatically linked. This and other proofreading

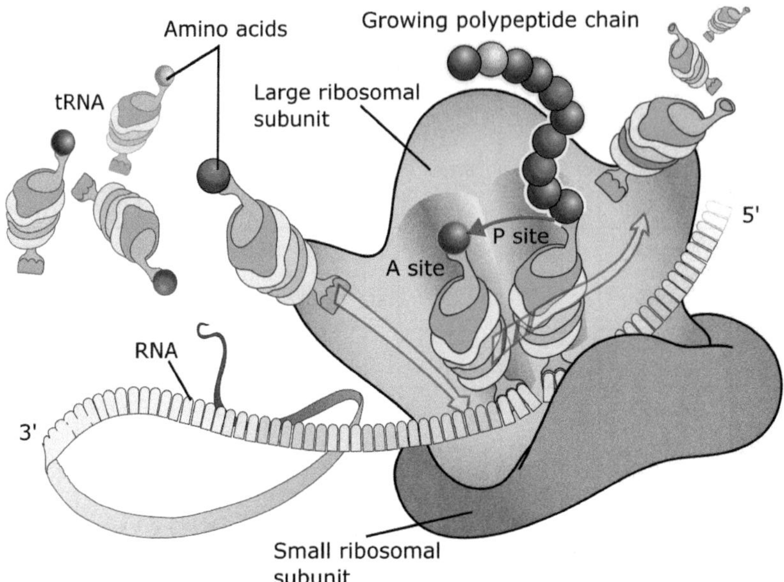

FIGURE 5.9
Schematic illustration of protein synthesis. In a process called translation, a ribosome moves along a strand of RNA (towards the 3′ end) and inserts amino acids into a growing protein chain using adaptor molecules called tRNAs. A tRNA that recognizes the RNA codon currently positioned in the A (aminoacyl) site of the ribosome binds to this codon via its anticodon. The amino acid attached to this tRNA is then joined to the growing protein chain held by the tRNA in the P (peptidyl) site of the ribosome. The growing protein chain is thereby transferred to the tRNA in the A site (indicated by red arrow), and as the ribosome moves, this tRNA is moved into the P site as the P site tRNA exits the ribosome. The protein synthesis continues in the same manner with the binding of a tRNA to the next codon in the A site, and terminates when a stop codon in the RNA is reached. [See Figure Credits.]

mechanisms help to ensure that the correct amino acid is added to the growing polypeptide chain. Polypeptide is a term used to refer to a sequence of amino acids linked by peptide bonds. Some proteins consist of a single polypeptide, while others consist of a complex of multiple polypeptides, as we will see below in the discussion of the quaternary structure of proteins.

The translation of an RNA template into an amino acid sequence begins and ends with a *start codon* and a *stop codon*, respectively. The codon for methionine, AUG, is the most common start codon. An example of a stop codon is UAA (Fig. 5.8). The ends of the synthesized protein are called the N-terminus and the C-terminus because the first amino acid in the sequence has an amino (NH_2) group and the last amino acid has a carboxyl (COOH) group (Fig. 5.10). These two groups are not involved in peptide bond formation and are the only amino and carboxyl groups in a protein that are not in the side chains (Section 5.4) of its amino acids. Once the full protein sequence has been generated, the attached ribosome recognizes the stop codon and dissociates from the RNA. The ribosome is usually reused by another RNA molecule in the cell. In fact, multiple ribosomes can load onto a single RNA molecule which gives rise to RNA-ribosome assemblies called polysomes that can be visualized using electron microscopy and resemble beads on a string.

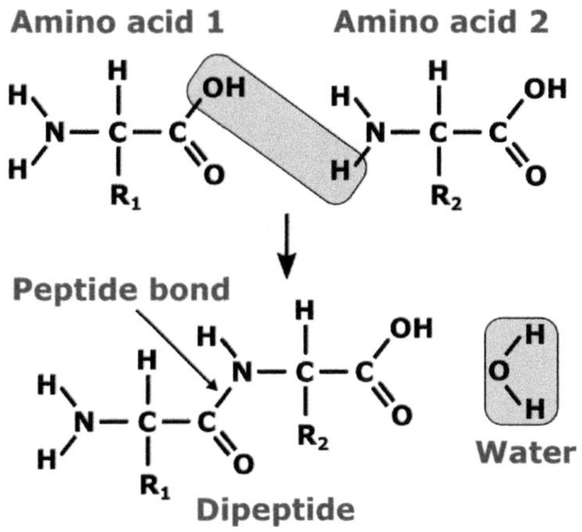

FIGURE 5.10
Peptide bond. The formation of the bond occurs between the carboxyl group of one amino acid and the amino group of the other, and results in the release of a water molecule. The two connected amino acids are called a dipeptide. The symbols R_1 and R_2 denote the side chains (i.e., R groups) of the two amino acids.

5.4 Structures of proteins

The amino acids that constitute proteins contain different side chains (also called R groups) (Fig. 5.7) which confer specific chemical properties. For example, some side chains are negatively charged (under physiological conditions), whereas others (lysine, arginine) are positively charged. Some amino acids are not charged and are highly hydrophobic (i.e., uncharged, or nonpolar), and are therefore not compatible with mixing with water.

The side chain of cysteine is unusual since it can bond covalently with another cysteine side chain to form a cystine bridge (Fig. 5.11). Cystine bridges can form both within a folded

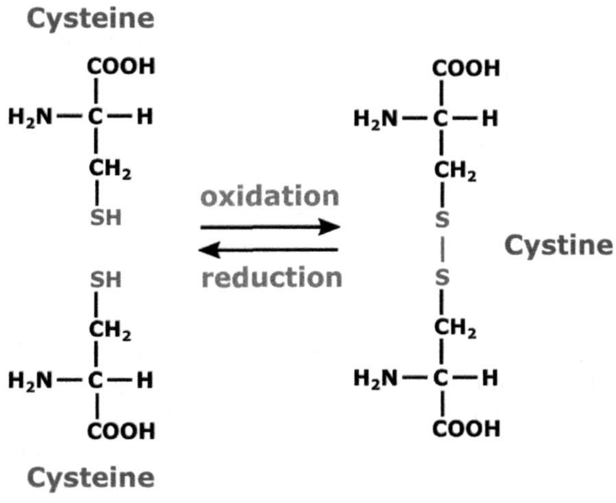

FIGURE 5.11
Cystine bridge created by the oxidation of two cysteine SH groups to form a disulfide bond.

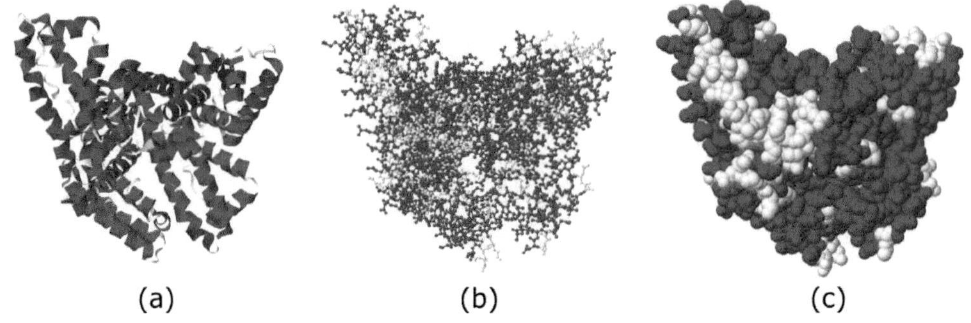

FIGURE 5.12
3D representations of protein structure. (a) The ribbon representation provides a view of the path and organization of the polypeptide backbone. (b) The ball-and-stick representation gives a molecular view of the protein by showing its atoms and the bonds between them as balls and sticks, respectively. (c) The space-filling representation shows the effective shape and dimensions of the protein by displaying its different atoms as proportionately sized balls that reflect the atoms' actual sizes. The protein used for the illustration is human serum albumin. [See Figure Credits.]

protein chain (intramolecular) or between different protein subunits (intermolecular), and they can greatly stabilize the overall folded state of a protein.

The arrangement of amino acids in space is determined by the 3D shape of the protein. Adopting the correct 3D shape is essential for the correct function of a protein. Indeed, some diseases are caused by alterations (or mutations) in proteins that result in improper or unstable folding. Proteins have many different functions in cells. For example, scaffold proteins serve as docking sites to bring multiple proteins that are involved in a particular signaling pathway together, and proteins of the enzyme class interconvert substances by inducing chemical changes. Another very important class of proteins is the receptor class. Proteins of this class carry out signaling and transport functions in cells. Indeed, receptors are of major interest to cell biologists since their location in cells and their dynamic behavior are directly relevant to their function.

How proteins fold into their 3D shape has attracted much attention over the past several decades. It is clear that for the majority of proteins, the structure into which they fold without assistance from other macromolecules in the cell is determined by their amino acid sequence. For some proteins, however, "helper" proteins called chaperones assist in their folding. The folded structure of a protein can be represented schematically in different ways, and three commonly used representations are shown in Fig. 5.12.

Proteins are classified into four levels of structure: primary, secondary, tertiary, and quaternary (Fig. 5.13). The primary structure represents the sequence in which the amino acids are covalently linked to each other and can be deduced from the DNA sequence that encodes the protein. In fact, the determination of a DNA sequence still remains much easier than that of a protein's amino acid sequence. The secondary structure represents local regions of ordered folding that result in several classes of structural elements: alpha helix (α-helix), beta sheet (β-sheet), and reverse turn. The propensity of a given region of a protein to form these secondary structural elements is again determined by the amino acid side chains (Fig. 5.7). For example, the amino acid proline is designated as a helix breaker and is never found in helices, but is found in reverse turns.

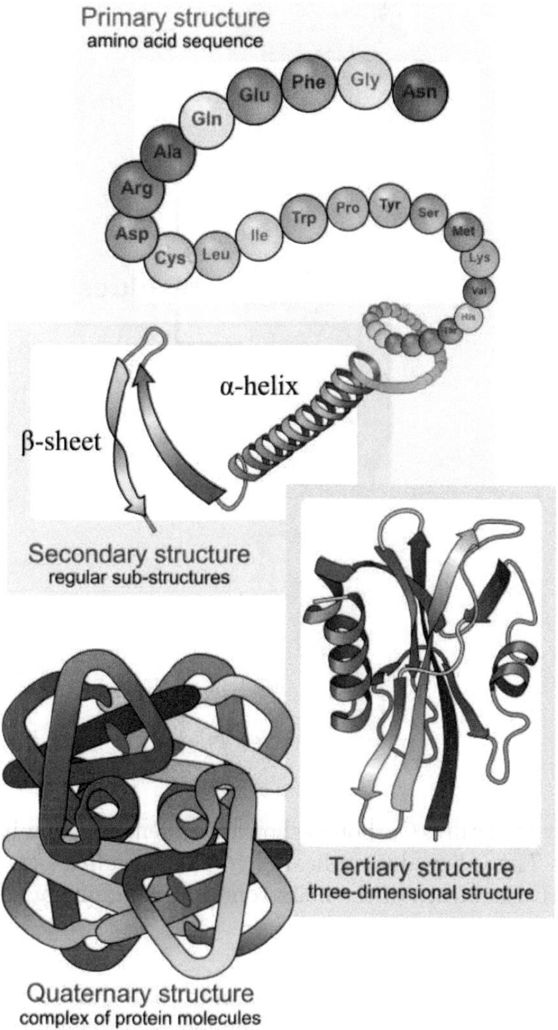

FIGURE 5.13
The four levels of protein structure. The primary structure is given by the sequence of amino acids. The secondary structure refers to local regions of ordered folding of the amino acid sequence, exemplified by the commonly occurring α-helix and β-sheet. The tertiary structure refers to the compact 3D structure that results from the folding of secondary structures upon one another. The quaternary structure refers to the number and arrangement of multiple protein molecules that together form a multi-subunit protein complex. [See Figure Credits.]

α-helices are the most common form of secondary structure in proteins, and have 3.6 amino acid residues per turn (Fig. 5.14). This means that amino acids with bulky side chains that are 3 or 4 residues apart in the primary sequence disfavor α-helix formation. The α-helix is stabilized by hydrogen bonds between the C=O group and the N-H group of amino acids that are four residues apart. Other helical structures (3 and 4.4 residues per turn) can also form in proteins, but these are not present for more than a few residues. α-helices can also be right-handed or left-handed (analogous to screws that turn in the clockwise and anticlockwise directions as they move away from the observer, respectively).

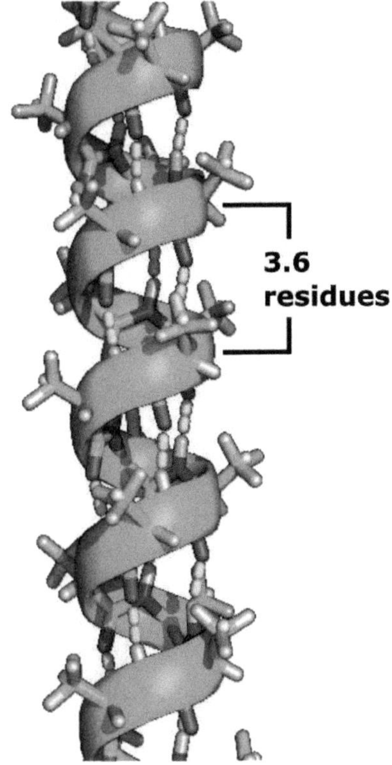

FIGURE 5.14
The α-helix secondary structure. Carbon is shown in green, nitrogen in blue, oxygen in red, and hydrogen in gray. The yellow dots indicate hydrogen bonding between amino acids that are four residues apart. Each turn of the helix contains 3.6 amino acid residues. [See Figure Credits.]

β-sheets are made up of two or more β-strands that can be parallel or antiparallel (Fig. 5.15). Amino acids in β-strands form extended conformations, and the side chains (R groups) of the residues project above and below the planes of the sheets. The strands associate with each other primarily through hydrogen bonding interactions between the C=O and N-H groups.

To form compact, stable structures, turns in the polypeptide chain are also necessary. These turns are called reverse turns if the C=O group of residue i forms a hydrogen bond with the N-H group of residue $i + 3$ (Fig. 5.16).

The polypeptide backbone has two angles that can vary due to rotational freedom (Fig. 5.17(a)). Specific ranges of these angles are allowed when a polypeptide folds into one of the different types of secondary structure, and this is illustrated by a Ramachandran plot (Fig. 5.17(b)). The original plot by Gopalasamudram Narayana Ramachandran was calculated before protein structures were available, and although in general the plot has held up to be consistent with the bond angles in solved structures of proteins, the agreement is not absolute.

How the secondary structures in proteins fold upon each other to form a compact structure is defined as the tertiary structure. Major steps have been made in understanding how individual amino acids contribute to the stability of the folded state of a protein by using an approach called site-directed mutagenesis (Note 5.4). This approach, pioneered by Sir Gregory Winter and Sir Alan Fersht, allows specific amino acids in proteins to be

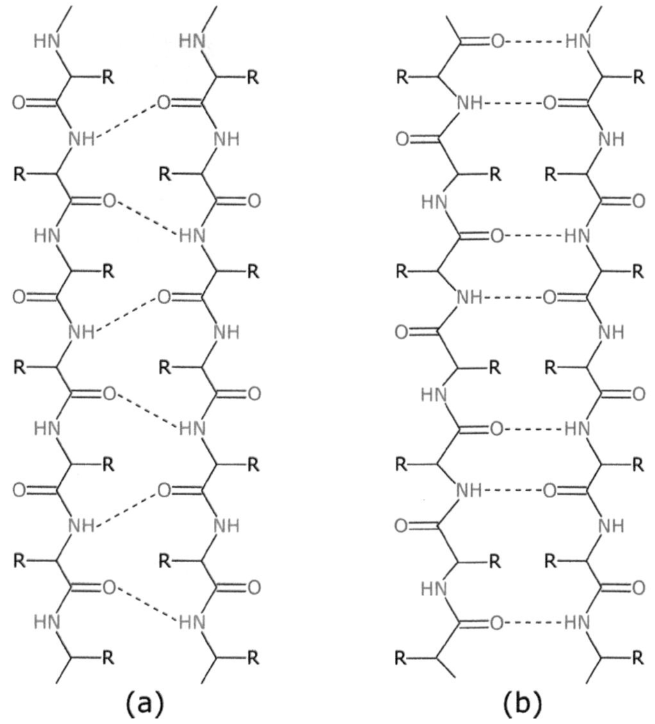

FIGURE 5.15
The β-sheet secondary structure. (a) β-sheet consisting of two parallel β-strands. (b) β-sheet consisting of two antiparallel β-strands. In both (a) and (b), the dotted lines indicate hydrogen bonding.

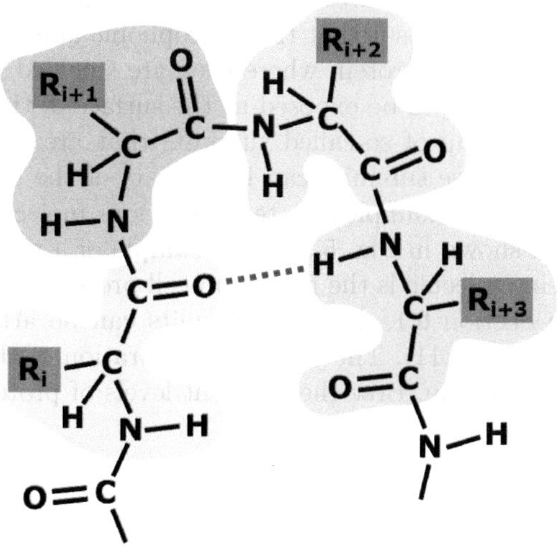

FIGURE 5.16
The reverse turn secondary structure. The blue dotted line indicates a hydrogen bond. The symbols R_i, R_{i+1}, R_{i+2}, and R_{i+3} denote the side chains of amino acid residues i, $i+1$, $i+2$, and $i+3$, respectively.

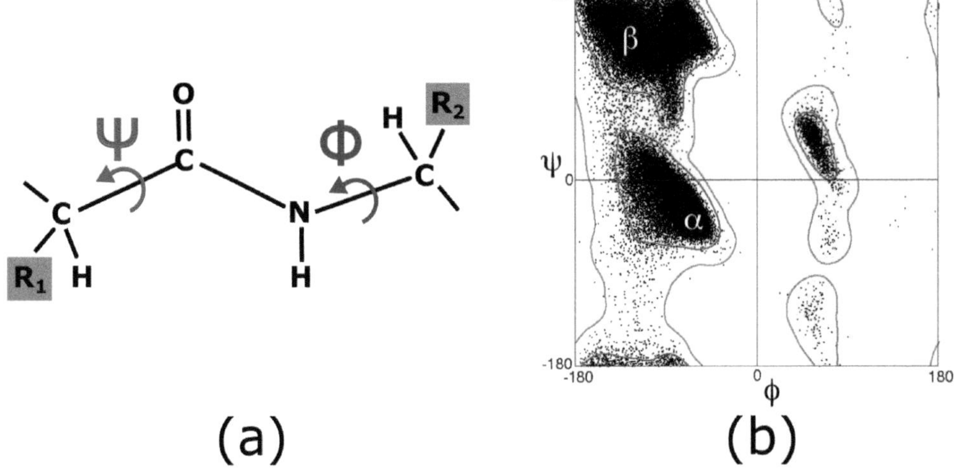

FIGURE 5.17
Polypeptide backbone angles and Ramachandran plot. (a) The angles ϕ and ψ of the polypeptide backbone. The symbols R_1 and R_2 denote the side chains of the two amino acid residues shown. (b) Ramachandran plot showing values for the ϕ and ψ angles that are compatible with α-helix or β-sheet formation. This particular plot was generated based on amino acid residues from many high-resolution protein crystal structures. [See Figure Credits.]

replaced. The effect of this replacement on the folded state and function of the protein can then be analyzed. This has led to important insight into how proteins fold and the role of specific amino acids in the function of a protein, but a complete discussion of this is beyond the scope of the current text. This work has shown that, particularly for larger proteins, transient folding intermediates exist on the folding pathway for a protein. In Fig. 5.18, the unfolded, intermediate, and folded states of a protein are depicted. An important feature for the folded state of most proteins is that the hydrophobic amino acids tend to be packed in the "hydrophobic core" of the protein where they are shielded from water. In contrast, polar and charged residues tend to be exposed at the surface of the folded protein.

Many proteins are made up of so-called subunits that are usually not covalently associated with each other. These subunits can be repeats of the same protein molecule or different protein molecules. An example of a tetramer (four molecules) of the same protein is the potassium channel shown in Fig. 5.19. An example of a tetramer comprising a pair each of a different protein molecule is the red blood cell protein hemoglobin. In some cases, as we shall see later in Section 6.1, different subunits can be attached to each other by covalent cystine bridges (Fig. 5.11). The overall configuration of the subunits is called the quaternary structure of a protein. How the different levels of protein structure are related is shown in Fig. 5.13.

5.5 Protein structure determination

While determining the sequence of a gene is relatively straightforward, the determination of the 3D structure of the protein that is encoded by the gene is typically much more

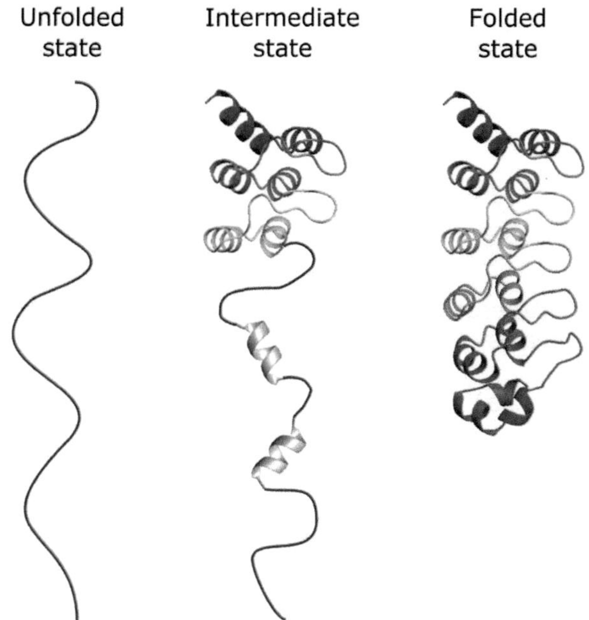

FIGURE 5.18
The unfolded, intermediate, and native states of a protein found in a prokaryotic microbe. The intermediate state consists of a partially folded version of the native state protein. [See Figure Credits.]

FIGURE 5.19
Structure of a potassium channel. A potassium channel is a tetramer consisting of four identical protein subunits, and its function is to transport potassium ions across cell membranes. Roderick MacKinnon was awarded the 2003 Nobel Prize in Chemistry for solving the 3D molecular structure of a potassium channel from *Streptomyces lividans* (pictured) and for determining the mechanism for the channel's selective transport of potassium ions. The purple dot in the center depicts a potassium ion. [See Figure Credits.]

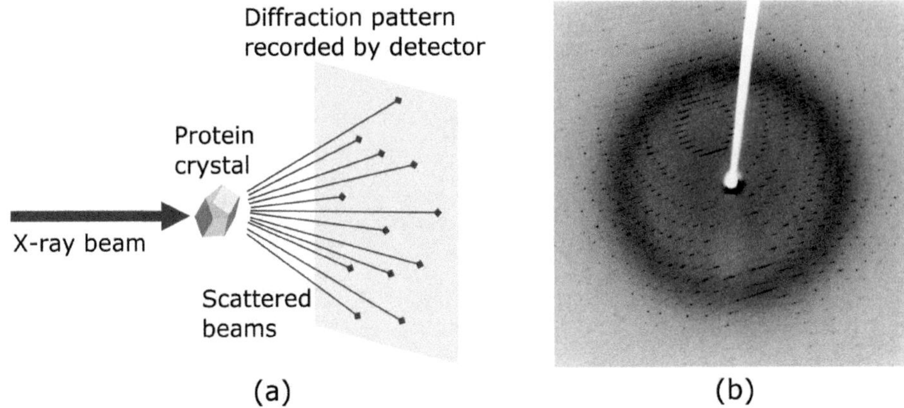

FIGURE 5.20
X-ray crystallography. (a) When an X-ray beam strikes a protein crystal, scattered beams are produced and captured by a detector, resulting in the recording of a diffraction pattern that contains information about the structure of the protein. (b) X-ray diffraction pattern of the SARS coronavirus main proteinase, 3CLpro. [See Figure Credits.]

difficult and time-consuming. The number of Nobel Prizes that have been awarded for the determination of protein structures is just one indication of the difficulty of this endeavor. For example, John Kendrew and Max Perutz received the 1962 Nobel Prize in Chemistry for solving the first X-ray structures of globular proteins, namely hemoglobin and myoglobin, both of which are transporters of oxygen. Roderick MacKinnon was awarded a Nobel Prize in Chemistry in 2003 for solving the 3D molecular structure of a potassium channel, a membrane protein that had eluded structural determination for many years (Fig. 5.19). The protein folding problem, i.e., the question of how the polypeptide chain encoded by a gene folds into its 3D structure, is one of the most important outstanding scientific questions. As we have seen, some basic principles are available to predict features of protein structures. However, there is no reliable way to deduce the 3D structure of a protein based on the genetic information, although theoretical approaches to predict this have shown major advances in recent years. Experimental techniques are therefore currently required to reveal the structure of a folded protein.

The primary techniques for protein structure determination are X-ray crystallography, nuclear magnetic resonance (NMR) spectroscopy, and a very recently developed method called cryo-electron microscopy (cryo-EM).

X-ray crystallography is based on crystallizing the protein of interest. The regular arrangement of the protein molecules in a crystal allows for the investigation of the protein structure using X-ray diffraction of the protein crystal. Based on the diffraction patterns produced by the crystal, the 3D spatial structure of the protein can be inferred. Figure 5.20(a) illustrates the generation of a diffraction pattern from the scattering of an X-ray beam by a protein crystal, and Fig. 5.20(b) shows the X-ray diffraction pattern of an enzyme found in the severe acute respiratory syndrome (SARS) coronavirus. A critical element in this approach is availability of crystals of the protein. Not all proteins crystallize easily and sophisticated protein chemistry might be necessary to obtain variants of the protein that are amenable to crystallization. Figure 5.21 shows a crystal of lysozyme, an enzyme that attacks bacterial cell walls.

The X-ray crystallographic structures of the extracellular regions of the two receptors (FcRn and transferrin receptor) that are used as model systems for subcellular trafficking in

FIGURE 5.21
Crystals of the protein lysozyme, stained with a blue dye. [See Figure Credits.]

this book have been solved, both as free receptors and bound to their ligands. Human transferrin receptor (Fig. 5.22) is an assembly of two proteins with three domains, with the helical domain forming the contacts between the two identical proteins. FcRn (Fig. 5.23) is an assembly of an α-chain and a smaller subunit called β_2-microglobulin that is non-covalently associated. FcRn is a member of the family of proteins called major histocompatibility complex (MHC) class I proteins, most of which play an important role in binding to foreign peptides from viruses and bacteria (i.e., pathogens) that are then recognized by the immune system to bring about clearance of the pathogen. A peptide is a short sequence of amino

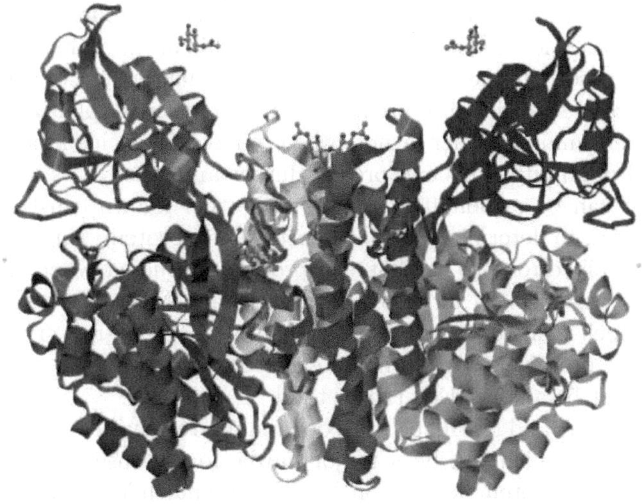

FIGURE 5.22
Structure of the ectodomain of the human transferrin receptor. The transferrin receptor is an assembly of two identical proteins, each with three domains. The two proteins are associated by interactions through their helical domains, shown in the center in yellow and red. The small ball-and-stick structures represent sugar molecules, and the green dots represent metal ions. [See Figure Credits.]

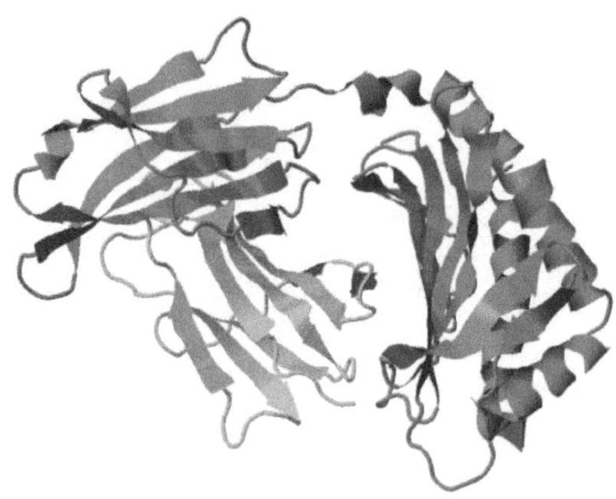

FIGURE 5.23
Structure of human FcRn. FcRn is a protein assembly consisting of the α-chain shown in orange and the smaller subunit β_2-microglobulin shown in green. [See Figure Credits.]

acids connected by peptide bonds, and can be derived from, for example, the degradation of proteins, including viral proteins in a virally infected cell.

An alternative technique for protein structure determination is NMR spectroscopy. Here the protein is produced with amino acids containing isotopes (usually ^{15}N or ^{13}C) with nuclear spins. The protein is then subjected to a very strong magnetic field that leads to the alignment of the magnetic spins in these amino acids with the strong field. The magnetic spins are then perturbed with the application of a weaker field that is orthogonal to the strong field (see Fig. 5.24(a) for a photograph of an NMR spectrometer). These perturbations of the magnetic spins can be detected and their decay to equilibrium can be observed. Complex perturbation experiments can be carried out, and their analysis can reveal important information about the spatial relationship of the amino acids with magnetic properties. This is based on the so-called nuclear Overhauser effect, which indicates the presence of two nuclear spins in close proximity to each other. Exploiting this effect, NMR spectroscopy can determine whether or not two specific amino acids in a polypeptide chain are in close spatial proximity. This can provide important constraints on the 3D structure of the protein that can be exploited, together with other known constraints, such as the Ramachandran angles, to determine the 3D structure of a protein. An important attractive feature about NMR spectroscopy is that the protein is studied when it is suspended in a liquid buffer (see Fig. 5.24(b) for a photograph of an NMR tube that is used to hold such a sample), which can be an important advantage over X-ray crystallography since crystallization of the protein is not required. A disadvantage of NMR spectroscopy is that larger proteins cannot be analyzed by the method.

X-ray crystallography and NMR spectroscopy require relatively large amounts of protein at high concentrations, which can be a major hurdle for the use of these techniques. By contrast, the amounts of protein needed for cryo-EM are much lower. Cryo-EM involves the use of a frozen protein sample that is bombarded with electrons. The pattern of scattered electrons on a detector can be used to determine the structure of the protein, avoiding the need to crystallize the protein. Although in the early stages of cryo-EM relatively low-resolution structures of proteins were obtained (leading to the branding of the technique as "blobology"), major developments have resulted in structures that approach the resolution generated using X-ray crystallography. The recognition of the technique as groundbreaking for structural biology was realized in 2017 when Jacques Dubochet, Joachim Frank, and

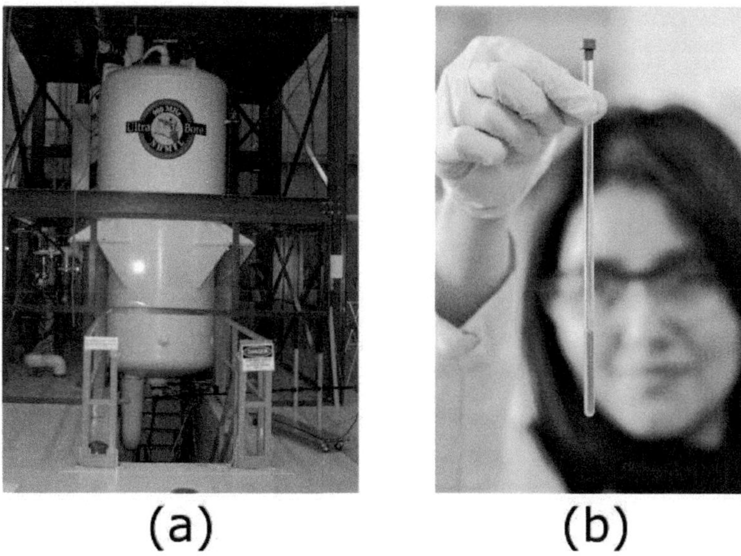

FIGURE 5.24
Nuclear magnetic resonance (NMR) spectroscopy. (a) An NMR spectrometer. (b) An NMR tube containing a liquid sample. [See Figure Credits.]

Richard Henderson were awarded the Nobel Prize in Chemistry for their pioneering research in this area.

6

Antibodies

Antibodies represent a major component of the immune system. They bind to large molecules such as peptides, proteins, and carbohydrates (compounds consisting of carbon, hydrogen, and oxygen atoms, including sugars). Molecules that an antibody can bind to are usually referred to as antigens. Typical examples of antigens are molecules that are foreign to the body to which the immune system elicits an immune response. There are, however, many other examples of antigens, such as proteins that are involved in autoimmune diseases and proteins that are of relevance in various applications of antibody technology. Antibodies can cause diseases such as arthritis by erroneously binding to components of the body such as proteins in the joints. The ability of antibodies to recognize specific molecules/cellular components (usually proteins) and to lead to their destruction has resulted in their use as agents for targeting antigens that are present on tumor cells. Antibodies that have been labeled with suitable reagents (e.g., radiolabels) can also be used to image tumors in the body. In addition, antibodies are widely used as laboratory reagents. For example, they are used in combination with (fluorescence or electron) microscopy to localize proteins or other macromolecules in cells. Their application in fluorescence microscopy is discussed at length in Section 9.4.1. Antibodies are also commonly used as detection agents in an array of diagnostic tests such as enzyme-linked immunosorbent assays (ELISAs) (Section 6.5.1). Indeed, widely available off-the-shelf pregnancy tests are based on antibody technology.

6.1 Structure of antibodies

Antibodies, or immunoglobulins, are Y-shaped molecules that are produced by B cells in the body and can be either expressed on the surface of the B cell or secreted as soluble molecules. In humans or mice, antibodies can be of several different classes: IgG, IgA, IgE, IgD, and IgM. Since soluble antibodies are the most relevant to the current text, these will be the focus of discussion. The overall structures of different antibody classes are shown in Fig. 6.1. IgM is usually found as a pentamer, by which we mean an assembly of five identical Y-shaped molecules. IgA, on the other hand, is usually found as a dimer, i.e., an assembly of two identical Y-shaped molecules. In contrast, the other antibody types, IgG, IgE, and IgD, are monomeric, meaning they are usually found as isolated molecules. For most applications in therapy, diagnosis, and laboratory research, antibodies of the IgG class are preferable. There are multiple reasons for this, including their longer persistence in the body and their ease of production relative to the other immunoglobulin classes. In humans, there are four subtypes of IgG — IgG1, IgG2, IgG3, and IgG4 — which have different functional activities.

The antibody molecule is modular in design, and comprises four polypeptide chains — two identical heavy chains and two identical light chains joined by cystine bridges, as illustrated in Fig. 6.2(a). (Note that the cystine bridges joining the heavy chain to the light chain, shown in Figs. 6.2(a) and 6.2(b), are omitted from all other similar depictions of antibodies in this book. In all cases, however, their presence is assumed.) Each chain

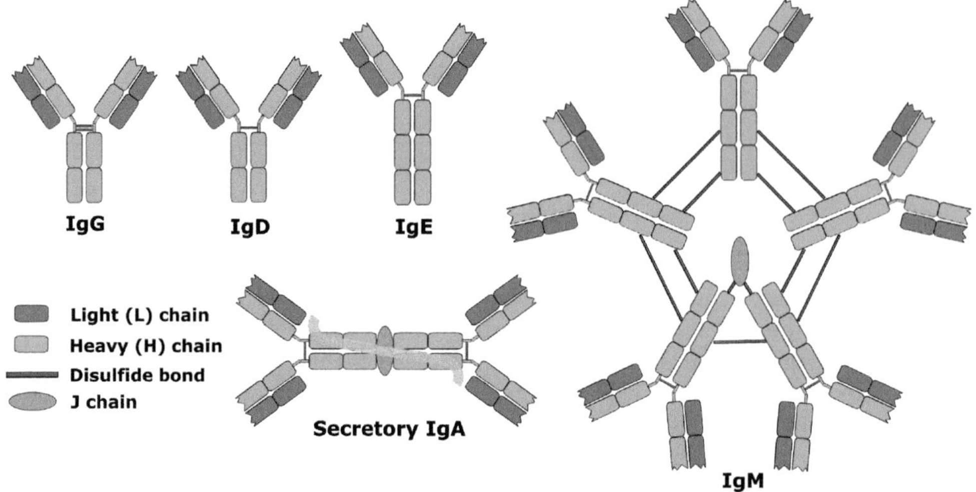

FIGURE 6.1
The five classes of antibodies. The yellow S-shaped structure in secretory IgA is known as the secretory component.

is made up of different immunoglobulin domains (Fig. 6.2(b)) connected to each other by flexible linkers. Each domain consists of two β-sheets that are pinned together centrally by a cystine bridge. Further dissection of the most widely used antibody, IgG, shows that it can be broadly divided into two parts: a pair of identical Fab (fragment antigen-binding) fragments and an Fc (fragment crystallizable) fragment (Fig. 6.2(b)). These regions can be made either by exposing an antibody to enzymes called proteases (papain or pepsin) that cut the heavy

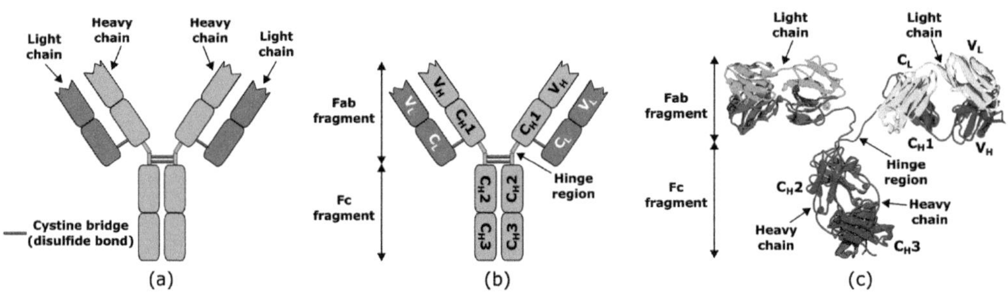

FIGURE 6.2
Structure of an antibody. (a) The polypeptide chains. An antibody comprises two identical heavy chains and two identical light chains, connected together by cystine bridges. (b) The fragments and the domains. An antibody is broadly divided into two parts: a pair of identical Fab fragments and an Fc fragment. Each of its four polypeptide chains is made up of different domains. A variable domain is indicated by the letter V, and a constant domain is indicated by the letter C. The subscript H or L specifies whether the domain is part of a heavy or light chain, respectively. The multiple constant domains of a heavy chain is further differentiated by a number. The V_H and V_L domains of a Fab fragment confer the specificity of the antibody for a particular antigen.(c) Ribbon representation of a mouse IgG2a antibody, with the two heavy chains shown in blue and red, the two light chains shown in green and yellow, and the domains and fragments labeled. [See Figure Credits.]

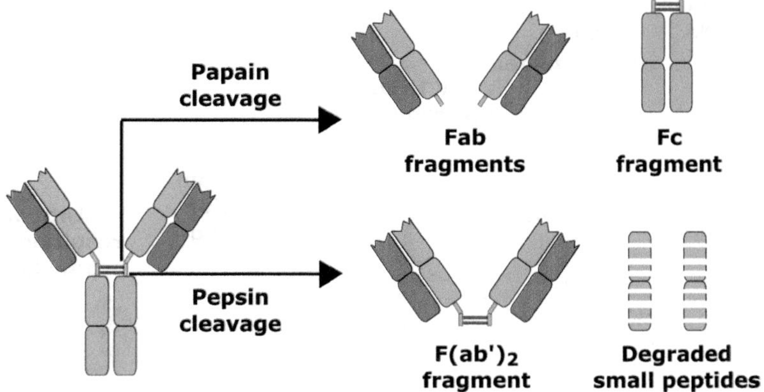

FIGURE 6.3
Cleavage of an IgG near its hinge region by the proteases papain and pepsin. Papain produces two Fab fragments and an Fc fragment. Pepsin produces an $F(ab')_2$ fragment and degraded parts of the Fc fragment.

chain polypeptide in the vicinity of the hinge region (Fig. 6.3), or by recombinant methods (Sections 7.1 and 7.2). Each Fab fragment contains the variable domains of the heavy and light chains (V_H and V_L, respectively) (Fig. 6.2(b)). These domains are responsible for conferring the specificity of the antibody for a particular antigen. In contrast with the variable domain sequences, the constant region sequences are essentially the same for all antibodies within a particular subclass (IgG1, IgG2, IgG3, or IgG4). In Fig. 6.2(c), a ribbon representation of a mouse IgG is shown with all the chains, fragments, and domains labeled.

6.2 Variable regions and binding activity

The question arises as to how antibodies can bind with high specificity to an almost unlimited number of different antigens. For this to be possible, they should have many different 3D shapes. Accordingly, an extraordinarily high number of different V region sequences, and a correspondingly high number of genes, would be necessary. Such a large number of genes would be hard to accommodate in the genetic material of a cell, and this therefore raised a fundamental question as to how it was achieved. Important insight into this question came from the Nobel Prize-winning work of Susumu Tonegawa and colleagues. This research led to the discovery that there are multiple genetic elements that encode antibody variable domains in cells. For example, for human heavy chain variable (V_H) domains, there are approximately 50 variable (V) gene segments, 30 diversity (D) gene segments, and 6 joining (J) gene segments. As the antibody-producing B cells develop, these gene segments undergo a process called recombination that combines one of the V gene segments with a D gene segment and a J gene segment to generate a VDJ gene segment that encodes the variable domain (Fig. 6.4). These rearrangements are specific to B cells which have the appropriate enzymatic machinery.

By combinatorial calculation, this gives rise to very large numbers of possible combinations. Superimposed on this, machinery in the cell gives rise to variable junctions at the V-D and D-J junctions, and further DNA alterations can be inserted into the genes by a

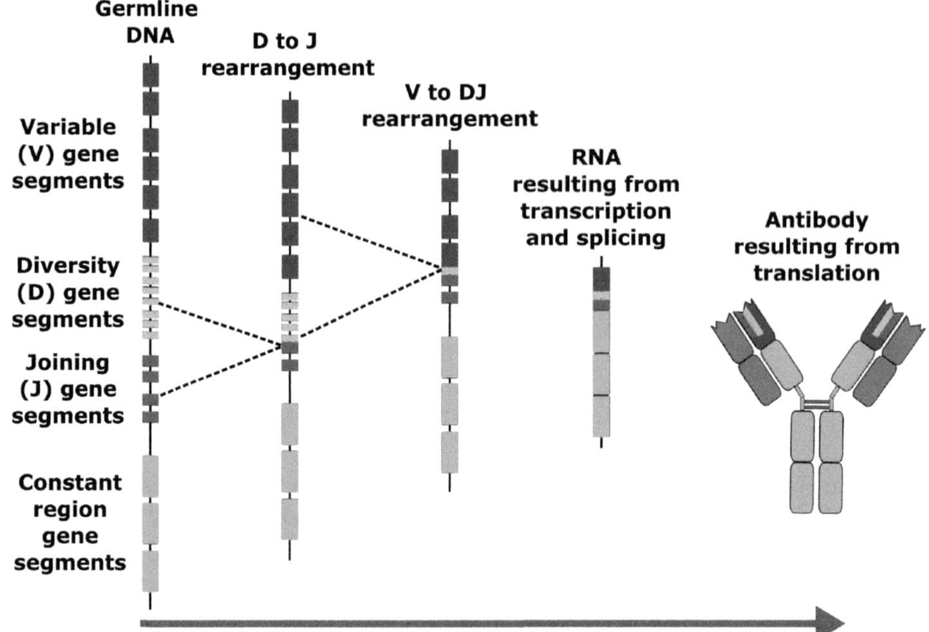

FIGURE 6.4
High-level schematic depicting V(D)J recombination of the gene segments that encode the variable domain of the antibody heavy chain in B cells. The process is illustrated starting from the germline DNA and ending with the generation of the antibody.

process called somatic mutation (Note 6.1). Similar processes produce light chain variable (V_L) domains, which have lower variability than heavy chain variable domains because their encoding gene segments do not include a D gene segment. Thus, each B cell produces an antibody that is distinct from the antibodies produced by other B cells. If the antibody binds to an antigen (e.g., of a pathogen) in the appropriate stimulatory environment, the B cell is activated. This activation can result in B cell expansion to larger numbers of B cells that all produce the same antibody in a process called *clonal expansion*. In addition, the activation can result in the B cells becoming *plasma cells* that secrete antibody into the extracellular space rather than displaying the antibody bound to their cell surface. These plasma cells are used to make hybridomas in the laboratory for the production of antibodies (Section 6.4.1).

The gene rearrangement process that occurs to produce the antibody repertoire is an unusual process, insofar as most cells do not rearrange their germline genes. Indeed, such a diversification process is not without a price. The aberrant activity of the rearrangement machinery in B cells can result in erroneous activation of genes that in turn can lead to tumorigenesis.

Importantly, the structure of the heavy and light chain variable domains (V_H and V_L domains, respectively) of antibodies is superbly designed as a framework to bind to different antigens. The β-sheets support three loops at the ends of each domain and these loops are the most diverse in sequence (Fig. 6.5). These loops are called complementarity-determining regions (CDRs). The third CDR for each variable domain is the most variable in both sequence and length since this is the region that is encoded by the V-D-J and V-J joining regions for the V_H and V_L domains, respectively. Significantly, this region of greatest sequence variation is located at the center of the antigen binding site.

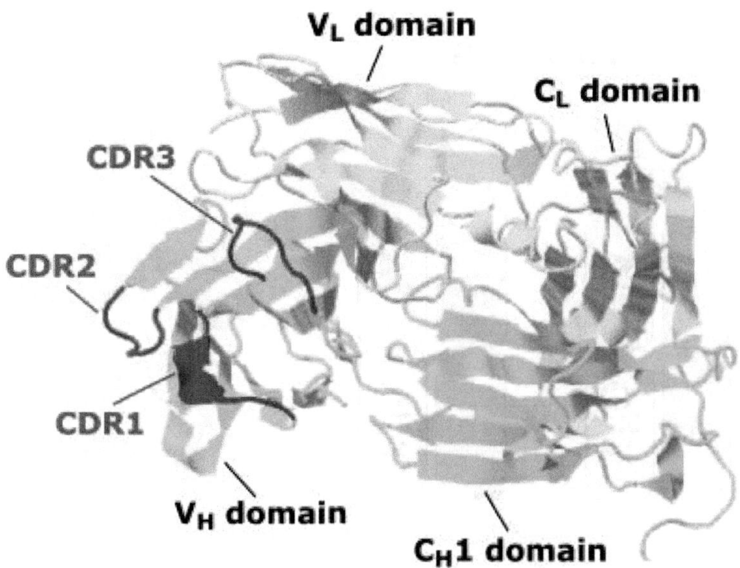

FIGURE 6.5
Complementarity-determining regions (CDRs) at the tips of the heavy chain variable domain of a mouse IgG2a antibody. The three CDRs are labeled 1, 2, and 3. [See Figure Credits.]

The antibody-antigen interaction is usually characterized by shape complementarity of the two interacting proteins. However, sometimes the shape of the antigen and/or antibody can remodel itself to optimize the binding interaction through a process called induced fit. Some antibodies have crevices into which small antigens can fit, whereas others have relatively flat surfaces or even protruding residues that are accommodated by a crevice in the antigen. This is illustrated in Fig. 6.6(a). To give a realistic example of the interaction between an antigen and an antibody, a representation of the enzyme hen egg white lysozyme in complex with the Fab fragment of an antibody is shown in Fig. 6.6(b). Antibody-antigen interactions are driven by electrostatic forces, hydrogen bonding, van der Waals interactions, and hydrophobic effects (Section 5.1). There are very rare examples where an antibody can bind covalently to antigen, although this represents a special case in which the antibody has been engineered to do this.

6.3 Constant regions

The constant regions of an antibody are, as the name suggests, conserved in sequence for each immunoglobulin type, although a very low level of sequence variation, called polymorphic variation, can occur. However, constant regions vary across different species. These differences in constant regions have two important consequences. First, when antibodies of mouse origin, for example, are delivered into humans, they are recognized as foreign and induce an immune response. This problem is discussed further in Section 7.3. Second, antibodies from different species can be used for the detection of two (or more) distinct proteins in cells for fluorescence microscopy (Section 9.4.1). These antibodies that bind to

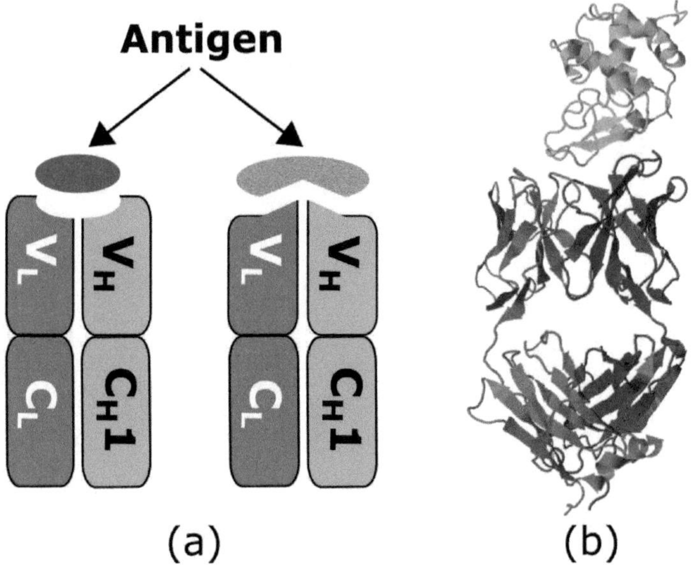

FIGURE 6.6
Complementarity of the two surfaces at the antibody-antigen interface. (a) Illustration of shape complementarity between an antigen and the Fab fragment of an antibody at the interface of interaction. (b) The enzyme hen egg white lysozyme (gray) in complex with the Fab fragment of a mouse antibody, shown in ribbon representation. The light and heavy chain components of the Fab fragment are shown in blue and red, respectively. [See Figure Credits.]

the different proteins can be detected using fluorescently labeled antibodies that interact with the different constant regions. Constant regions carry out multiple functions that differ from one immunoglobulin type to the next, as summarized below. Figure 6.7 summarizes a comparison between the different classes and subclasses of antibodies (Note 6.2).

IgGs are characterized by much longer in vivo half-lives relative to IgM, IgA, IgE, and IgD. This is because these antibodies bind to the Fc receptor, FcRn. This biological system is used as a model in several chapters of this book. The different classes of antibodies have different binding properties for receptors called FcγRs (IgG-specific), FcαRs (IgA-specific), and FcϵRs (IgE-specific). These receptors have distinct functional activities and are expressed on certain cell subsets. For example, FcγRs on macrophages and natural killer cells can interact with specific subclasses of IgG bound to a target such as a cancer cell. This process can result in cancer cell killing by either engulfment of the IgG-coated cell or the production of killing factors by the natural killer cell. Importantly, IgGs have very low affinities for binding to most of these FcγRs, so that FcγR-expressing cells only interact efficiently when the display of the IgG is multivalent (through so-called avidity effects). Proteins of the so-called complement cascade can also bind to IgG1, IgG3, and IgM and bring about a series of proteolytic steps that result in the destruction of the antibody-coated target.

It is important to note that the hinge region, which connects the Fab arms of an antibody to the Fc region, is highly flexible. This gives the Fc region flexibility once the antibody is bound to antigen that is present on, for example, cell surfaces. Such flexibility has been shown to be important for functional activity.

	IgG1	IgG2	IgG3	IgG4	IgM	IgA1	IgA2	IgD	IgE
Heavy chain	γ1	γ2	γ3	γ4	μ	α1	α2	δ	ε
Serum level (mg/mL)	9.0	3.0	1.0	0.5	1.5	3.0	0.5	0.03	$5x10^{-5}$
Half-life in serum (days)	21	20	7	21	10	6	6	3	2
Complement activation	++	+	+++	-	+++	-	-	-	-
Binding to Fc receptors	++	+	++	+	-	++	++	-	+++

FIGURE 6.7
Properties of human antibodies by class and subclass. In the bottom two rows, a larger number of plus signs indicates higher activity. A hyphen indicates negligible or no detectable activity.

6.4 Antibody production for laboratory and clinical use

6.4.1 The classical method: hybridoma technology

For many years, it has been known that the immunization of animals can be used to generate antibodies that bind to a specific antigen. For example, serum from horses that had been immunized with snake venom was used in the past to treat snake bites. In fact, even now the immunization of animals, such as mice or rabbits, with an antigen is a standard way to obtain antibodies against the antigen. This immunization results in "polyclonal" antibody responses, which are mixtures of different antibodies that bind to an antigen. The term polyclonal means that the antibodies in such a polyclonal antibody response exhibit possibly large differences in their sequences and therefore in their binding properties against the antigen. Further it has to be assumed that even animals of the same species will produce somewhat different repertoires of antibodies against the same antigen.

Being able to make large amounts of antibodies that have identical sequences is a critical step for the use of antibodies for both clinical and diagnostic/laboratory applications. The possibility to produce such "monoclonal" antibodies was realized by the development of *hybridoma technology* in the mid-1970s by César Milstein and Georges Köhler, which allows the production of almost unlimited quantities of a monoclonal antibody. This approach has revolutionized biology and medicine, and its significance was recognized by a Nobel Prize in 1984. In brief, hybridoma technology involves the generation of immortalized cell lines that can be cultivated in the laboratory to produce the same antibody. This is achieved by "fusing" the antibody-producing cells of an animal with a special cell line that grows indefinitely in culture (most cell lines do not do this, but die after several or less divisions) (Fig. 6.8). Monoclonal antibodies are typically made from rodents such as mice and rats, and hybridoma technology cannot be applied to humans since the spleen (a rich source of antibody-producing B cells) of the animal needs to be isolated. Usually the animal has been immunized (equivalent to vaccination) with the antigen of interest. This results in expanded numbers of B cells in the animal that produce antibodies recognizing this antigen with high affinity.

These fused cells are then "cloned" out, which means they are plated in wells of a plate at one cell per well and allowed to grow. Cells in a given well consequently produce the same antibody since each cell is a descendent of the single progenitor cell and the antibodies produced by a single B cell are all identical. Some of these clonal cell lines will produce antibodies in the culture supernatant that bind to the antigen of interest, and this can be

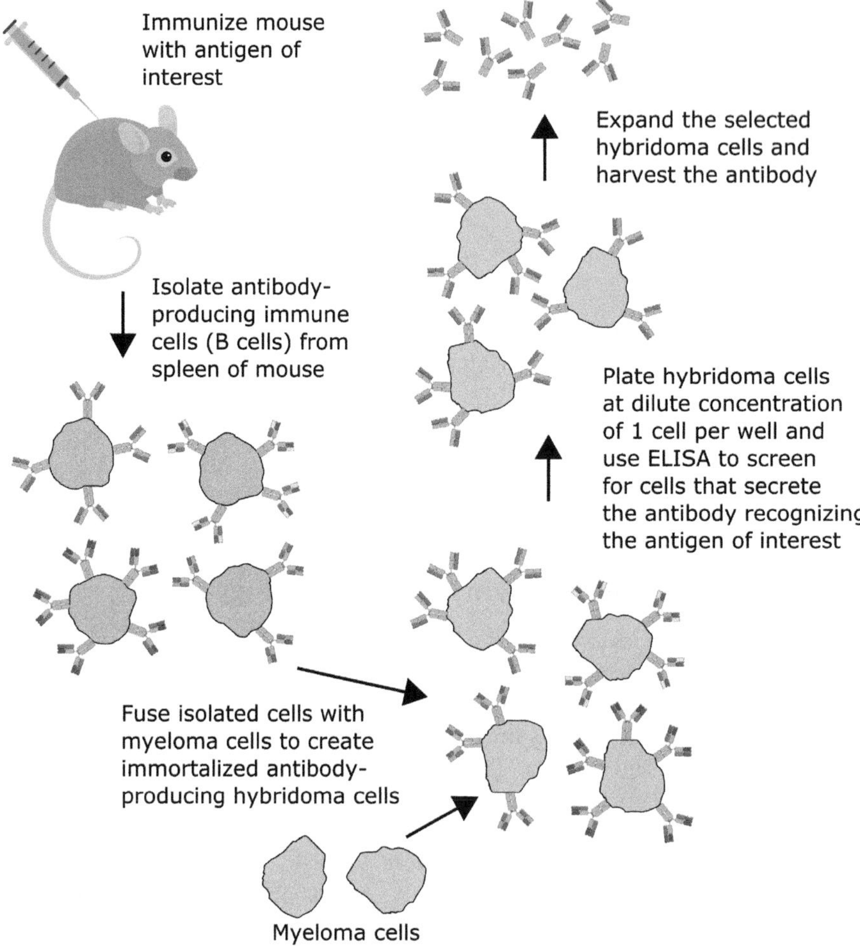
Immunize mouse
with antigen of
interest
Isolate antibody-
producing immune
cells (B cells) from
spleen of mouse
Fuse isolated cells with
myeloma cells to create
immortalized antibody-
producing hybridoma cells
Myeloma cells
Plate hybridoma cells
at dilute concentration
of 1 cell per well and
use ELISA to screen
for cells that secrete
the antibody recognizing
the antigen of interest
Expand the selected
hybridoma cells and
harvest the antibody

FIGURE 6.8
Main steps in monoclonal antibody production from mice using hybridoma technology.

screened for. The screening is usually done using an ELISA (Section 6.5.1), although other methods are possible.

Hybridoma technology opened the door to generating almost unlimited amounts of clonal antibodies that recognize a particular antigen. The immense value of this technology for research, diagnostics, and therapeutics was quickly realized (although despite the suggestion of César Milstein to the UK Medical Research Council to patent it, the technology was not believed at the time to be of potential commercial value! (Note 6.3)). However, although rodent antibodies are widely used in the laboratory and in diagnostic tests, they have limitations for use in humans. This limitation is due to the recognition of such rodent antibodies as foreign when delivered to humans. Such immunogenicity not only results in reduced levels of the antibody following delivery due to clearance by the human immune system, but can also lead to adverse events such as anaphylaxis and serum sickness. In addition, rodent antibodies have very short half-lives in humans due to their poor binding to human FcRn.

These problems, combined with the advent of appropriate molecular biology approaches, have led to the development of the field of antibody engineering that enables the production of monoclonal antibodies in "recombinant" form. "Recombinant" indicates that the antibody is generated by inserting the antibody gene(s) in an expression vector and producing the encoded antibody in cultured cells in the laboratory. These methods are described further in Chapter 7. As we will see there, one of the major achievements of the antibody engineering field is the isolation of antibodies that are suitable for use in therapy and diagnosis since they are very similar to human antibodies.

6.5 Diagnostic techniques using antibody detection methods

6.5.1 Enzyme-linked immunosorbent assay

Enzyme-linked immunosorbent assays (ELISAs) can be used to detect and quantitate proteins, small molecules, lipids, carbohydrate, or any other biological/chemical molecule in solution. The prerequisite for the use of this method is that there is a high affinity antibody, or polyclonal antibody, specific for the molecule that is being detected. ELISAs are a derivation of the radioimmunoassay (RIA) developed by Rosalind Yalow, for which she received the 1977 Nobel Prize in Physiology or Medicine. Nowadays, there are multiple ELISA formats. The specific format that is used for a given application depends on the nature of the molecule to be detected and the detection reagents available. Here we will consider two examples of commonly used formats.

Direct ELISA to detect an antibody specific for an antigen

The principal steps for the direct ELISA are shown in Fig. 6.9. The antigen is adsorbed to the plastic of 96-well plates. Typically proteins bind well to plastic due to hydrophobic interactions, whereas other molecules such as small chemicals may require specialized methods for capture. Sites on the plastic that are not bound by the protein are then "blocked" by exposure to blocking agents such as high concentrations of bovine serum albumin or a solution of reconstituted milk powder. Following blocking, the sample that putatively contains the antibody is added, in solution, at different dilutions to the wells of the plate. If a known amount of the same antibody is available, this can also be added to different wells of the 96-well plate at varying concentrations to generate a standard curve. This standard curve can be used to calculate the antibody concentration in the test sample. The antibody that binds to the immobilized antigen is detected using an enzyme-conjugated "secondary" antibody that recognizes the test antibody. For example, if the test antibody is of mouse IgG origin, then there are commercially available secondary antibodies that recognize all mouse IgGs, typically by binding to the constant regions of the mouse antibodies. This secondary antibody is chemically coupled to an enzyme (e.g., horseradish peroxidase) that converts a colorless substrate to a colored product. Hence the presence of color in the well, and the amount of colored product, following a defined reaction period can be determined spectrophotometrically in a multiwell plate reader.

Sandwich ELISAs for the detection of antigen

The requirement for a sandwich ELISA (Fig. 6.10) is that there are two antibodies (antibodies A and B) available that bind to distinct epitopes on the antigen that is to be detected. An epitope is a specific part of the antigen to which an antibody binds. Hence these two antibodies do not compete for binding to the antigen. Antibody A is adsorbed to the plastic of 96-well plates and unbound sites are blocked as for the direct ELISA. The

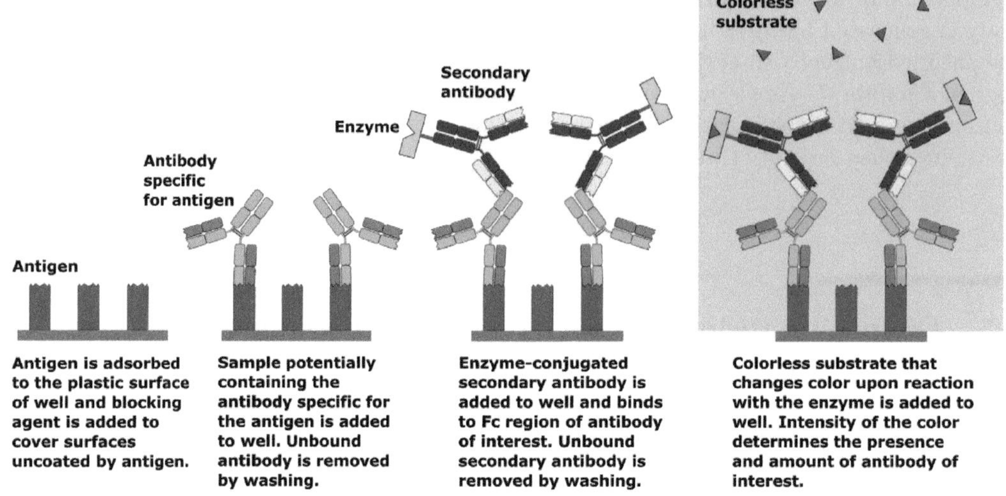

FIGURE 6.9
Principal steps of a direct ELISA for the detection of antibody specific for an antigen.

sample solution that putatively contains the antigen is then added, usually in dilutions, to the wells. The antigen is captured by the immobilized antibody A, and can be detected by adding antibody B. Antibody B can either be directly coupled to an enzyme that converts a colorless substrate to a colored product, or can be detected using an enzyme conjugated to a secondary antibody as for the direct ELISA. Analogously as for the direct ELISA, a standard curve can be generated from samples of known concentrations of the antigen and used to calculate the concentration of antigen in the test sample.

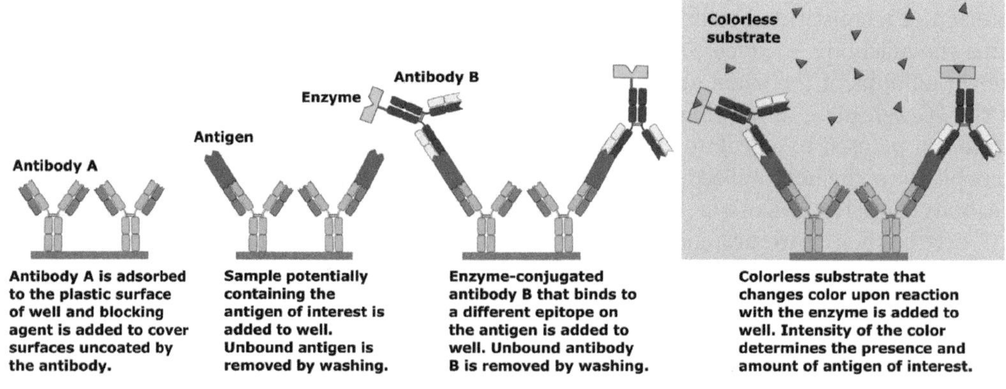

FIGURE 6.10
Principal steps of a sandwich ELISA for the detection of an antigen of interest.

6.5.2 Surface plasmon resonance for the quantitation of the affinity of an interaction

Surface plasmon resonance (SPR) is a physical phenomenon that forms the basis of a quantitative technique for the determination of the affinity of an interaction of two macromolecules. Typically, the technique is used to analyze protein-protein (or peptide) interactions, but can also be implemented to investigate the binding of other macromolecules to each other, e.g., lipids and carbohydrates. Here we will consider its application for the analysis of antibody-antigen interactions. In this case, the antibody (called "ligand") is immobilized on a sensor chip that comprises a gold surface with activated chemical groups for coupling proteins (Fig. 6.11). For example, the gold surface can be coated with carboxylate groups for coupling to exposed NH_2 groups on proteins via amine chemistry. Alternatively, other coupling methods can be used such as streptavidin-coated chips for coupling biotinylated antibodies. Streptavidin is a protein that binds biotin molecules with very high affinity, and the strong complex that it forms with biotin is commonly exploited for the specific attachment of a streptavidin-conjugated entity to a biotin-conjugated (i.e., biotinylated) entity (see Sections 8.7.1.2, 8.7.2, and 9.4.1 for more information and additional examples).

As illustrated further in Fig. 6.11, a beam of light is directed on the underside of the gold sensor chip at an *angle of incidence* that exceeds the *critical angle*. Total internal reflection of the beam is thus achieved, meaning that the beam is fully reflected at the gold surface (see Section 13.1.4 for a full discussion of total internal reflection). Even though the light beam is reflected, a special type of electromagnetic wave, called an *evanescent wave*, is generated that penetrates into the gold chip and is capable of exciting electrons in the chip. This excitation of electrons gives rise to the SPR phenomenon, and causes a drastic decrease in the intensity of the reflected light beam. SPR, however, occurs only when the light beam is reflected at a particular angle. This *angle of reflection* is referred to as the

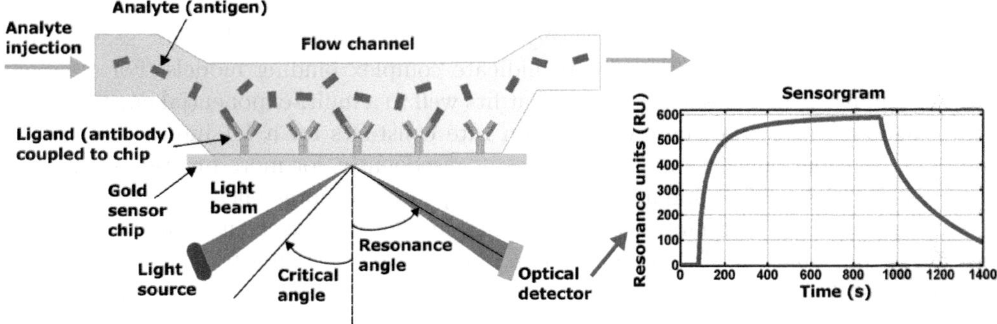

FIGURE 6.11
Using surface plasmon resonance to determine the affinity of an antibody for an antigen. The antibody is immobilized on the surface of a gold sensor chip, and the antigen is injected into a flow cell and allowed to bind to the immobilized antibody as it flows across the chip. The amount of antigen that binds to the antibody is then determined by changes in the resonance angle of the reflection of a beam of light that is directed on the bottom surface of the sensor chip at angles greater than the critical angle. The changes in the resonance angle are recorded over time by a detector and converted to resonance units. The resulting resonance unit versus time plot, which is referred to as a sensorgram, can then be used in conjunction with information such as the concentration of the injected antigen to calculate biologically significant quantities such as the equilibrium binding dissociation constant for the antibody-antigen interaction.

resonance angle, and its value is importantly sensitive to changes in the refractive index of the medium just above the coupling surface of the gold chip. As this refractive index depends on the amount of material that is bound to the chip, a sensor that detects changes in the resonance angle (by tracking the angle of incidence of the light beam that results in a reflected beam of minimal intensity) can determine the amount of antigen (called "analyte") that binds to the antibody immobilized on the coupling surface of the chip. Binding occurs as antigen is injected in a flow of solution across the chip, and changes in the resonance angle over time are recorded in resonance units, or RU, in an RU versus time plot called a *sensorgram*.

Hence, the affinity of the interaction between the antibody and antigen of interest can be measured as follows. When the antibody is coupled to the gold chip, the initial resonance angle can be quantitated as a value in RU. When antigen is then injected and binds to the antibody, the resonance angle changes, and a signal in RU is reported which is proportional to the amount of antigen bound. By performing injections with different concentrations of the antigen, a plot involving the antigen concentration and the amount of bound antigen can be generated based on the recorded sensorgrams. From this plot, equilibrium binding dissociation and association constants can be determined.

It is important to note that such SPR-based quantitation also allows the monitoring of the association and dissociation between antigen and antibody in real time. This can result in the extraction of association and dissociation rate constants from the data. However, these rate constants can be error-prone for several different reasons. First, so-called mass transport limitations can limit the rate of delivery of the analyte to the ligand binding sites, providing an upper limit to the association rate constant. Second, during the dissociation phase the analyte can rebind to ligand as it leaves the surface of the chip and encounters free ligand. This rebinding results in lower dissociation rate constants, and is enhanced by interactions that have fast association rate constants. Both mass transport limitations and rebinding can be reduced, but usually not eliminated, by increasing the flow rate for analyte injections and using low ligand coupling levels. Further, mass transport limitations are lower for smaller analytes. Collectively, these effects result in aberrant sensorgrams that can sometimes be misinterpreted to indicate complex binding models. For example, rebinding can result in a dissociation plot that fits well to a multi-exponential curve, leading to the extraction of two or more dissociation rate constants for a 1:1 interaction that in reality only includes one class of binding sites. Consequently, for more precise estimates of interaction kinetics, other approaches such as stopped-flow fluorescence are recommended. Nevertheless, SPR-based quantitation is a valuable and extensively used method for the high-throughput analysis of macromolecular interactions, particularly when comparative measurements of, for example, multiple mutated variants of a protein are sought.

7

Cloning of genes for protein expression

In the context of molecular biology, cloning is the isolation of a gene encoding a protein or a part of a protein and its amplification as identical copies. In this chapter, we present an overview of the approaches that can be used to modify and clone genes for the production of genetically engineered proteins. First, we will discuss how genes encoding the protein of interest are inserted into so-called expression constructs so that the corresponding protein can be produced in relatively large quantities. These approaches are now used, for example, in preference to hybridoma technology to make antibodies. Second, we will describe how genetic engineering can be used to modify proteins. Such modifications include the alteration of specific amino acids and/or the tagging of proteins with fluorescent proteins (Section 8.7.3). As discussed in other sections of this book, tagging proteins with fluorescent labels is an essential step for many fluorescence microscopy experiments.

7.1 Features of expression constructs

To produce a (modified) protein inside a cell, there are several prerequisites. The gene encoding the protein needs to be linked to a promoter to drive expression (for a detailed description of how the cellular machinery makes proteins from genes, see Section 5.3). The promoter-gene fusion needs to be in a form that can readily be duplicated, so that the DNA can be scaled up and transferred into cells that will express the protein (i.e., an expression host). Many copies of the DNA are needed to transfer the promoter-gene fusion into cells for expression, since the transfer process is very inefficient. Here we will only discuss the transfer of promoter-gene fusions as DNA into expression hosts, although other methods can be used, such as the delivery of the DNA packaged into viruses that infect the cells.

The scale-up of the promoter-gene fusion is typically accomplished by "cloning" the DNA segment into a *plasmid*. A plasmid is a closed circular piece of DNA that is commonly found in bacteria (Fig. 7.1). Plasmids are not part of the chromosomal DNA and were originally identified in bacteria as DNA segments that can replicate independently since they have their own origin of replication. They frequently carry genes that encode toxins and/or resistance to antibiotics and can be transferred from one bacterium to another. The origin of replication allows the plasmid to be replicated in bacterial cells so that it can be expanded. The plasmids that are currently used in molecular biology applications are engineered as described below. The presence of the antibiotic resistance gene enables the selection of bacterial cells that contain the plasmid, as plasmidless cells will die in the presence of antibiotic. Plasmids are usually generated and propagated in a variant of the bacterium *Escherichia coli (E. coli)* that is not pathogenic. The presence of the selectable marker is essential, since when bacterial cells are treated with a plasmid under conditions that result in uptake of the plasmid into the bacterial cells, the vast majority of cells will not take up any plasmid. Cells that contain the antibiotic resistance gene can therefore be selected from a mixed population of cells that do and do not have the plasmid. Plasmids

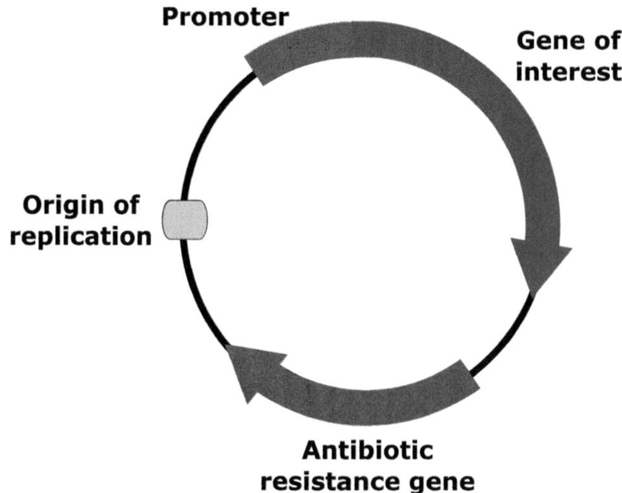

FIGURE 7.1
Schematic representation of a plasmid showing the origin of replication, the promoter, the gene of interest, and the antibiotic resistance gene.

can be transferred into a wide variety of cell types (e.g., mammalian, yeast, bacterial) for expression.

7.2 Methods for generating expression plasmids

In the following sections, we describe how molecular biology tools can be used to generate expression plasmids that can be used, in the context of this book, for the production of antibodies, proteins fused to fluorescent proteins, or modified proteins that can be site-specifically labeled with fluorophores. These sections are included for completeness, but are not essential for understanding the general principles. Before describing the detailed methods, we first discuss two important tools that are used in molecular biology, namely restriction enzymes and the polymerase chain reaction.

7.2.1 Restriction enzymes

A major step in molecular biology was made when bacterial enzymes called restriction enzymes were discovered. These enzymes make double-stranded breaks in DNA at specific sequences, and the breaks very often have overhangs, i.e., are not flush (Fig. 7.2). The realization that these enzymes can be used to cut up DNA sequences into defined segments, which could then subsequently be joined (ligated) together by an enzyme called DNA ligase, revolutionized molecular biology in the early 1980s. Many plasmids are designed to have multiple restriction sites adjacent to the promoter, which facilitates the insertion of genes of interest (tagged with restriction enzyme sites; see below) into these sites. The transcription of these genes into RNA by RNA polymerase is driven by the promoter that is already in the plasmid.

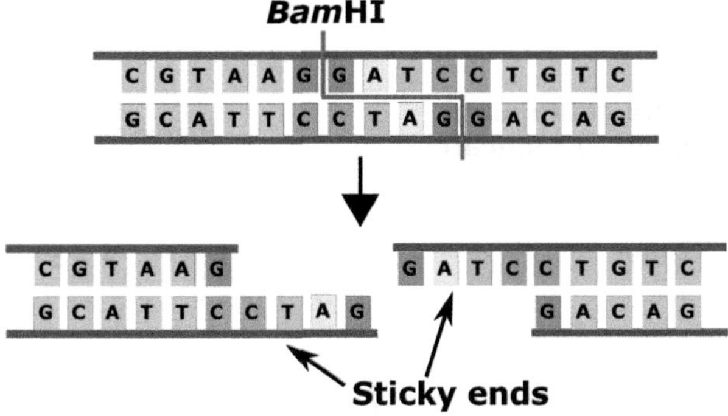

FIGURE 7.2
Restriction enzymes make double-stranded breaks in DNA at specific recognition sites. In the example shown, the restriction enzyme *Bam*HI recognizes the palindromic sequence GGATCC and makes a cut just after the first guanine on each strand. The overhangs in the resulting two pieces of DNA are often referred to as sticky ends.

7.2.2 Polymerase chain reaction

The *polymerase chain reaction* (PCR) is a widely used molecular biology technique that results in the amplification of specific genes or gene segments. The size of the fragment that can be amplified using this approach is in general limited to several thousand base pairs, and a requirement is that there is some knowledge of the sequence of the ends of the fragment that is to be amplified. This technique was invented by Kary Mullis in 1983, and is broadly used for multiple application areas in most molecular biology laboratories. The significance of the technique was recognized by the award of a Nobel Prize to Mullis in 1993.

The PCR involves the following steps (Fig. 7.3). First, relatively short single-stranded DNA molecules (around 20 to 30 base pairs), called "primers", which are complementary to the ends of the gene to be amplified, are synthesized by a DNA synthesis machine in the $5'$ to $3'$ direction. Second, the double strands of the so-called DNA template, which contains the gene (segment) to be isolated, are separated by heat (typically 94°C) and then cooled to around 50°C to 60°C in the presence of the oligonucleotide primers. This results in the "annealing" of the primers to the DNA template. Third, the annealed primers are extended at 72°C by a DNA polymerase that is heat-stable. In this regard, the isolation of a polymerase from the bacterium *Thermus aquaticus* (Taq polymerase), which resides in hot springs, represented a major breakthrough for the PCR technology, since all steps of the amplification cycle could be performed at relatively high temperature with retention of polymerase activity. Prior to the isolation of Taq polymerase, the DNA template/primer mix had to be cooled to much lower temperatures following the annealing step to allow the available heat-sensitive DNA polymerases to copy the DNA template, and new polymerase had to be added following each high temperature cycle. The PCR using Taq polymerase can be carried out by mixing the components in a single tube that is then placed in a thermocycler block that is programmed to run 30 or so repetitions of the second and third steps described above, although in its early stages it was performed by manually switching tubes between different water baths.

7.2.3 Details of approaches for generating expression plasmids

There are several steps that need to be carried out to generate a plasmid for the expression of a particular protein (protein X) using molecular biology techniques. These are described

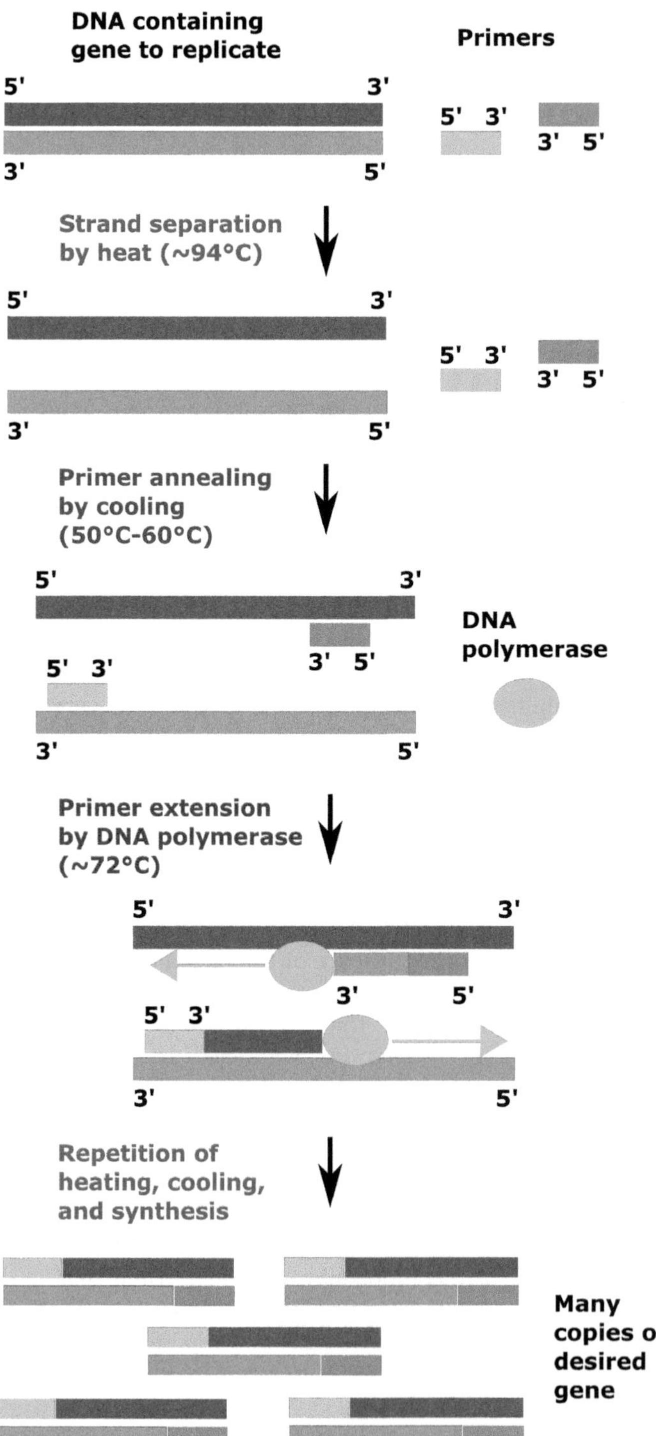

FIGURE 7.3
Schematic illustration of the steps of the polymerase chain reaction.

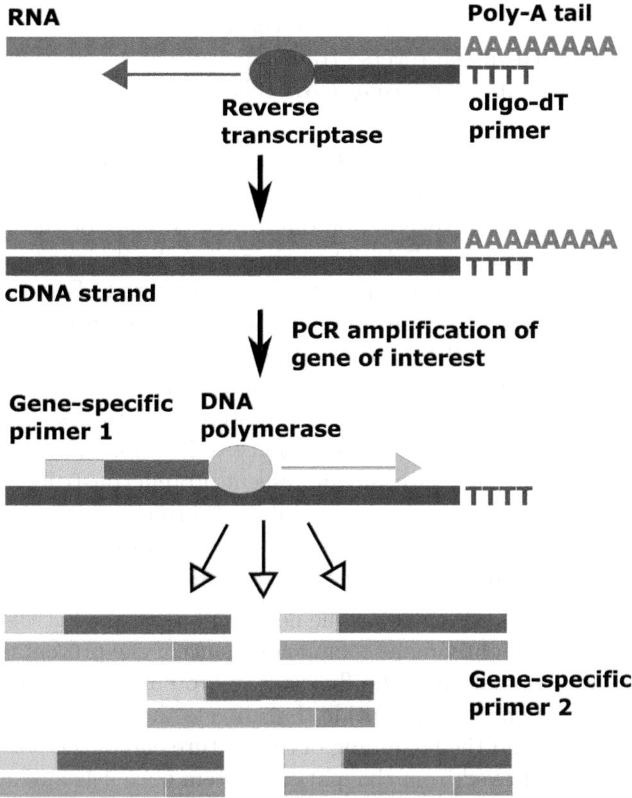

FIGURE 7.4
Reverse transcription PCR. First, RNA is reverse-transcribed using the enzyme reverse transcriptase to produce copy DNA, or cDNA for short. The cDNA is then used in the PCR (Fig. 7.3) as template with specific oligonucleotide primers, resulting in amplification of the gene of interest.

in further detail below, and involve the isolation of the gene, tailoring the gene so that it is inserted into a plasmid downstream of a promoter, and when appropriate, tailoring the gene so that it is either fused to a fluorescent protein gene (or other protein gene) and/or has mutated codons inserted for site-specific labeling (Section 8.7.1.1 and Note 5.4). In all cases, however, the cloning of a gene into a plasmid requires that multiple copies of the gene and the plasmid into which it is being inserted are available. The primary reason for this is that the transfer of the expression plasmid, once completed using the procedures described below, into *E. coli* is so inefficient that multiple copies are needed to obtain clones that have the required plasmid. Other reasons are that at all stages of the procedures, the pieces of DNA need to be readily detected and for this, many thousands of molecules are needed.

1. *Isolation of the gene from cellular RNA.* The gene encoding protein X is first isolated from cells that express the protein of interest. Assuming that the sequence is known, this is usually carried out using reverse transcription followed by the polymerase chain reaction (PCR). RNA is isolated from the protein-expressing cells and used as template by an enzyme called reverse transcriptase to make a copy of all genes that have been transcribed into RNA in the cells (Fig. 7.4). This can be achieved by annealing a synthetic DNA strand made of thymines (oligo-dT) that bind to the stretches of adenines (poly-A tail) present at the ends of all

mammalian RNAs (Fig. 7.4). As an alternative to using oligo-dT, gene-specific primers can be used in the reverse transcription reaction so that only the gene of interest is transcribed into *copy DNA* (cDNA). Multiple copies of cDNA are usually made.

2. *Modification and amplification of the DNA using the PCR.* The cDNA is then used in the PCR (see Section 7.2.2 for a description of this method) with DNA oligonucleotides that are specific for the 5′ and 3′ ends of the gene being targeted for amplification (Fig. 7.4). The primers can be designed in different ways depending upon the nature of the expression construct. For the expression of antibodies, the antibody gene is amplified with designed primers using the PCR and cloned into a plasmid that typically does not have any fluorescent protein gene in it (using the procedures described below). However, if the protein is to be expressed with a fluorescent protein at its C-terminus, this is usually carried out by designing the primer so that a restriction site (Section 7.2.1) is inserted at the 3′ end of the gene in-frame with the fluorescent protein construct. In-frame means that the protein translation machinery will read through the triplets of bases from the end of the gene of interest into the fluorescent protein gene. Consequently, protein X will be fused to the fluorescent protein gene.

   Sometimes the fluorescent protein gene needs to be attached at the N-terminus of the protein. This is slightly more complicated since if the protein is to be secreted or inserted into a membrane, the fluorescent protein gene needs to be inserted between the end of the leader peptide that directs the protein to the correct destination in the cell and the beginning of the mature end of the protein (the leader peptide is usually cleaved from the mature protein once it has reached the appropriate location). This step can be achieved using one of several different molecular biology approaches. One approach is to use the PCR to append restriction sites between the ends of the leader peptide and mature protein and to then append the same restriction sites (using the PCR) to the ends of the fluorescent protein gene.

   Analogous approaches can also be used to generate plasmids for the expression of proteins such as antibodies for site-specific labeling with biotin (Section 8.7.1.2) or fluorescent dyes (Section 8.7.1.3). In some cases, recombinant proteins are tagged with peptide tags to facilitate their purification (e.g., polyhistidine tags that allow binding to Ni^{2+}-NTA agarose beads that provide a solid matrix to capture the protein from solution). This can be achieved by appending the codons for the peptide tag using designed oligonucleotide primers and the PCR (Fig. 7.5).

3. *Ligation of the DNA into an expression plasmid.* Once the piece of DNA to be assembled into the full-length fusion protein has been amplified using the PCR with appropriate restriction site(s) appended, the gene needs to be ligated together to generate a continuous segment in a plasmid. In the current example for attaching a fluorescent protein to the C-terminus of protein X, the fluorescent protein is already in a plasmid vector that has appropriate restriction sites adjacent to the fluorescent protein gene. This plasmid and the PCR product are then digested with restriction enzymes (e.g., *Eco*RI, *Bam*HI) and ligated together using an enzyme called T4 DNA ligase (Fig. 7.6).

   An approach that is an alternative to the addition of restriction sites and restriction enzyme digestion is called TA cloning (Fig. 7.7). This is based on the activity of Taq polymerase (the DNA polymerase used in the PCR; see Section 7.2.2) to add an adenine that is not encoded by the template to the 3′ end of the DNA

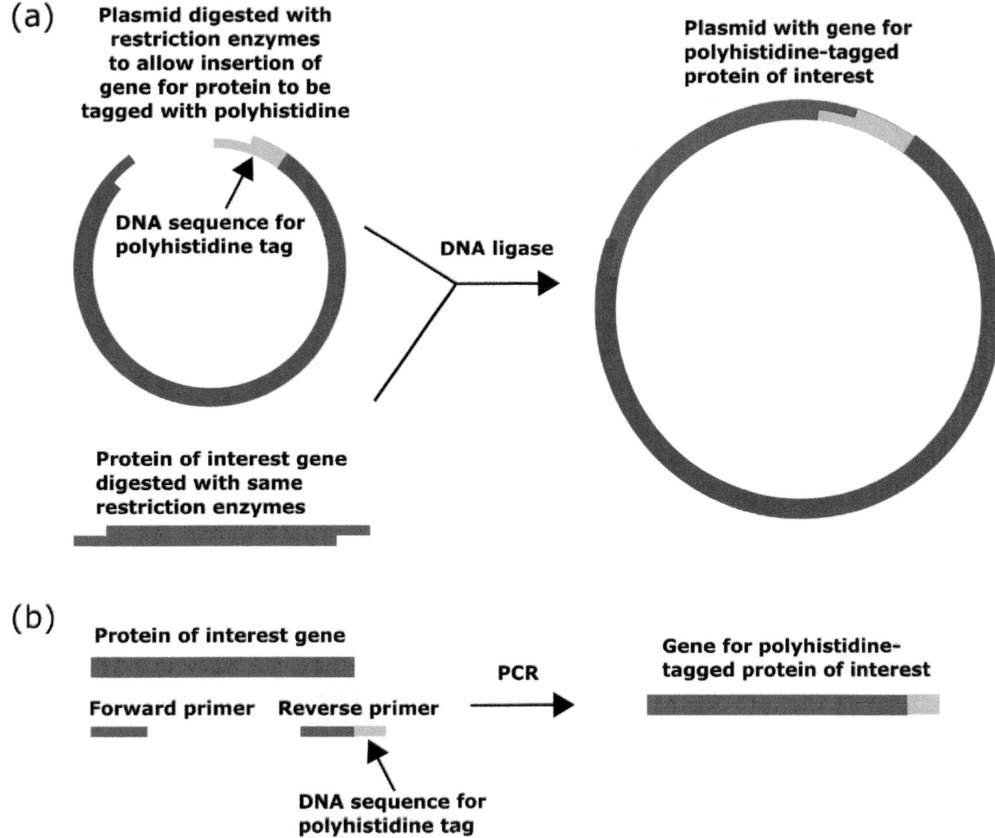

FIGURE 7.5
Two ways to add a polyhistidine tag DNA sequence to a gene of interest. The gene can either be (a) ligated in-frame into a plasmid that contains the sequence for a polyhistidine tag, or (b) extended with the polyhistidine tag sequence using the PCR with a primer that contains the sequence. In (b), the resulting PCR product is then ligated into a plasmid for expression of the tagged protein.

strand that is synthesized. Consequently, the PCR product has 3′ overhangs of adenine bases. These products can be ligated into vectors that have complementary, single-base thymine overhangs.

4. *Transferring the constructed plasmids into E. coli and selection of individual clones.* The so-called ligation mix that results from step 3 is then added to *E. coli* cells under conditions that force the cells to take up DNA (Fig. 7.8). The *E. coli* cells are selected on agar plates that contain the antibiotic for which the plasmid encodes resistance. Only plasmids that have been recircularized by ligase will confer antibiotic resistance and therefore colonies containing this can be selected. If the plasmid DNA is only digested with a single enzyme, there is also a step that is important to minimize the recircularization of the plasmid without the insertion of the protein X gene: before addition to the ligation mix, the plasmid is treated with an enzyme that modifies the ends of the linearized plasmid DNA so that they cannot ligate with themselves, but only with the "insert" DNA encoding protein X.

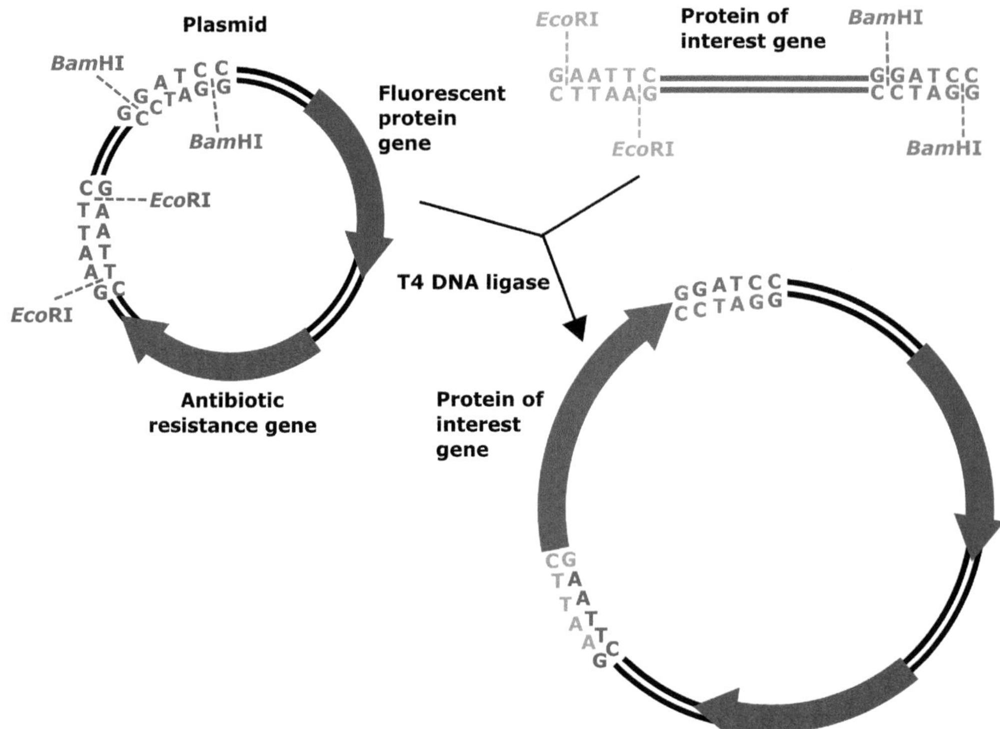

FIGURE 7.6
Insertion of the gene for a protein of interest into a plasmid. Digestion of both the gene and the plasmid with restriction enzymes *Eco*RI and *Bam*HI, followed by the ligation of the products of digestion using the enzyme T4 DNA ligase, yields the desired plasmid. The gene is inserted upstream of the gene for the fluorescent protein that will be attached to the protein of interest.

> During the selection process on agar plates, *E. coli* cells that are derived from a single (clonal) cell transformed by one of the plasmid molecules in the ligation mix grow up as colonies. The colonies that contain the plasmid with the correct insertion of the gene can be screened for and expanded in culture, and the DNA plasmid can be extracted for analysis by sequencing.

Once these steps have been completed, the plasmid is ready for use in subsequent experiments such as delivery into mammalian cells by a process called transfection (Section 7.2.4). Multiple parameters affect the efficiency of protein expression. First, some promoters are much more efficient at loading RNA polymerase for transcription than others, and promoter "strength" can therefore have a major impact. Second, proteins differ in intrinsic stability and less stable proteins do not fold as efficiently, leading to lower expression levels. Third, whether the codons encoding the amino acids are optimal for the particular expression host that is being used can affect the efficiency of translation by the ribosome. Specifically, the degeneracy of the genetic code (Fig. 5.8) means that most amino acids are encoded by more than one codon. A mammalian cell, for example, may have more tRNAs for one codon for a given amino acid relative to a bacterial cell, or even compared with a cell of a different mammalian species. Consequently biasing the codons in favor of the usage of the particular

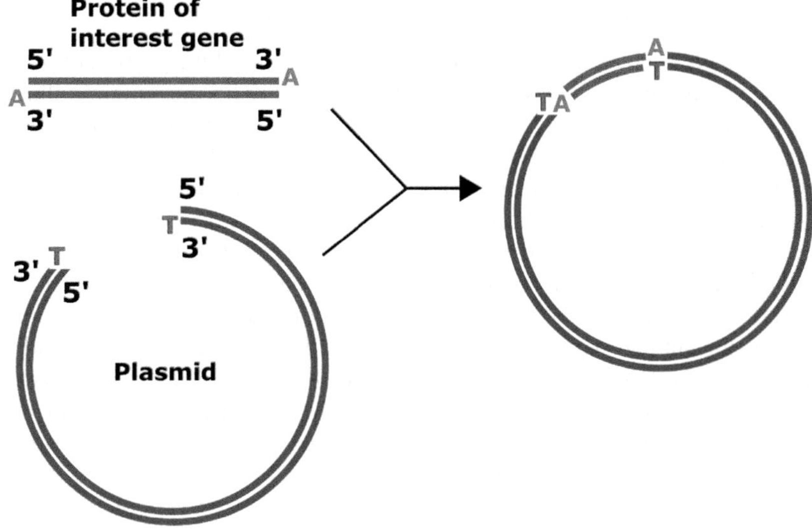

FIGURE 7.7
Constructing an expression plasmid by TA cloning. In this approach, an adenine is added to the 3′ ends of the gene for the protein of interest during the PCR. Similarly, a thymine is added to the 3′ ends of a blunt-end plasmid that has been cut with a restriction enzyme. The gene for the protein of interest is then inserted into the plasmid by the pairing of the adenine and thymine single-base overhangs.

expression host can be carried out, but this requires multiple additional modifications to the gene that will not be discussed further here.

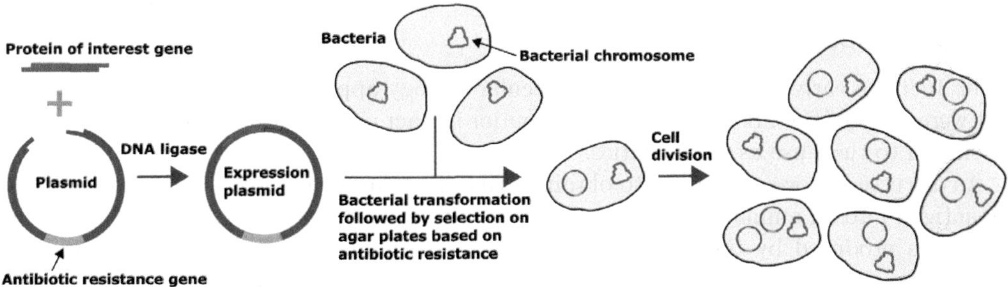

FIGURE 7.8
Cloning of a gene into an expression plasmid using bacterial cells. After ligation of the gene for the protein of interest into a plasmid, the resulting expression plasmid is introduced into bacterial cells. The bacterial cells then multiply via cell division to provide an unlimited source of the expression plasmid. Each bacterial cell usually has multiple copies of the expression plasmid. The expression plasmid can be isolated and used for transfection of mammalian or yeast cells, which then express the protein of interest. Alternatively, in some cases, the bacteria can be used themselves to express the protein of interest.

7.2.4 Transfection of mammalian cells for expression

Although other hosts, such as insect and yeast cells, can be used for protein expression, here we only discuss mammalian cell expression. This is primarily because mammalian hosts are used for the production of full-length antibodies for therapy (although antibody fragments are frequently produced in *E. coli*; see Section 7.3.3). In addition, the examples that we present in Chapters 2, 9, and 11 for fluorescence microscopy are images of mammalian cells that express proteins tagged with fluorescent proteins.

There are multiple ways to transfer plasmid constructs into mammalian cells. A commonly used approach is electroporation, which involves electrically shocking the cells in the presence of the DNA plasmid. All approaches to introduce plasmids into mammalian cells are relatively inefficient, necessitating the use of very large numbers of plasmid molecules (Section 7.1). Once the DNA plasmid is inside the cells, it can either stably insert into the chromosomal DNA ("stable transfectant") or be maintained for a limited number of rounds of division without recombination with the chromosomal DNA ("transient transfectant"). Transient transfections have the advantage that the cells can be analyzed within one to several days post-transfection and is frequently used for cells that cannot be maintained for many passages in culture. By contrast, stable transfectants are usually cloned out, a process that takes several weeks. These stable clones have the advantage that they represent a relatively homogenous population of cells that can be used, for example, as a stable cell line to express recombinant proteins, or in fluorescence microscopy experiments. However, there is a danger of using only a single cell line in imaging experiments: the cell line might be defective due to (over-)expression of the transfected protein and/or due to disruption of important functional genes by the insertion of the plasmid into the chromosome. For this reason, it is generally preferable to compare the properties of several different stable transfectants (Note 7.1).

7.3 Antibody engineering

In the 1980s to early 1990s, the use of molecular biology approaches (Sections 7.1 and 7.2) that were state-of-the-art at the time had a major impact on the antibody field, particularly in relation to their use as therapeutics.

As discussed above, a method to obtain antibodies against a specific antigen is to inject the antigen into an animal, such as a mouse, rabbit, or goat, and to extract the polyclonal antibodies produced by the immune system of the animal. B cells for use in hybridoma technology can be obtained from the spleen of the immunized animals. For diagnostic and other biosensing applications, antibodies of animal origin are usually sufficient. However, for therapeutic applications in humans, antibodies derived from an animal source pose a potentially serious problem. Animal antibodies have amino acid sequences, particularly in their constant regions, that are sufficiently different from the sequences of the corresponding human antibodies as to cause these antibodies to be recognized as foreign when injected into humans. Therefore the potential for an immune response exists, especially when repeated doses of the same antibody are used. This response not only reduces the dose of the therapeutic, but can also be dangerous for the patient. As a result, finding ways to obtain antibodies that are closer to those of human origin was a major priority. This priority could be addressed with modern molecular biology approaches. For example, the generation of cell-based systems for the large-scale production of antibodies that have been engineered so that they comprise human constant regions linked to rodent variable domains became

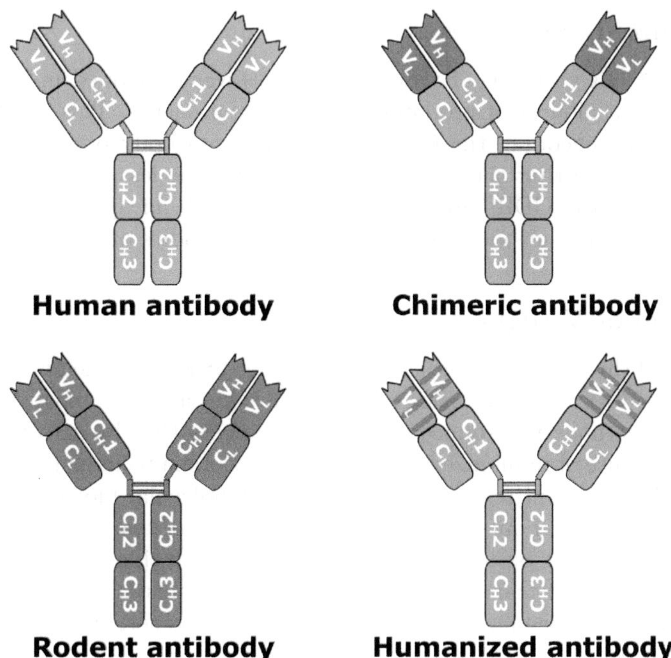

FIGURE 7.9
Chimeric and humanized antibodies. A chimeric antibody consists of rodent variable regions linked to human constant regions. A humanized antibody has the CDRs of a rodent antibody grafted onto a human antibody.

possible. In addition, technologies to convert, using molecular biology approaches, rodent antibodies into versions that more closely resemble human antibodies and are therefore less immunogenic were developed. The use of the PCR (Section 7.2.2) to isolate the genes that encode the antigen-binding sites, i.e., the V_H and V_L domains, of antibodies from antibody-producing cells, followed by the development of systems to select the antibodies that bind with highest affinity to a particular target, were developed as a pathway that bypassed rodent immunization and hybridoma generation for the isolation of antibodies.

7.3.1 Chimeric antibodies

Chimeric antibodies are molecules in which the variable domains of rodent (monoclonal) antibodies are connected, using molecular biology approaches, to the constant regions of human antibodies (Fig. 7.9). These antibodies are therefore approximately two-thirds human. This approach results in antibodies that are less immunogenic and have longer persistence in the body due to the interaction of the human constant region with human FcRn. Antibodies derived from rodents, on the other hand, do not bind well to this receptor. Indeed, antibodies of this class have been approved by the U.S. Food and Drug Administration for clinical use for the treatment of cancer and autoimmune diseases such as rheumatoid arthritis. This therefore represented a major step forward in the generation of antibodies that are suitable for use in humans.

7.3.2 Humanized antibodies

Humanization (or CDR grafting) of rodent antibodies was developed in the late 1980s by Sir Gregory Winter, a pioneer of protein engineering approaches. This involves grafting of the gene subsegments encoding the CDRs of a rodent antibody onto a human antibody framework, with the idea that the resultant antibody will be less immunogenic in humans (Fig. 7.9). Thus, only the CDRs are of rodent origin. Such an approach has been widely applied to convert rodent antibodies into humanized antibodies of clinical value.

7.3.3 Isolation of V regions

The development of the PCR (Section 7.2.2) in the late 1980s, combined with the realization that it could be used to isolate V region genes, resulted in a major expansion in the use of such recombinant methods to produce antibody fragments that bind to target antigen as an alternative to the use of immunized rodents and hybridoma technology. These fragments can then be linked to constant regions using methods of molecular biology. There are several steps that need to be carried out to isolate V regions with specific binding activities. First, the V region genes need to be isolated from antibody-producing cells. Second, the genes need to be transcribed and translated into functional proteins. Third, the V regions that bind to a chosen target antigen need to be isolated, and the genetic material encoding them needs to be sequenced.

The first step can be achieved by using the PCR. Specifically, for both heavy and light chain variable domains, the V region sequences are similar at the ends of the gene segments (i.e., V_H-D_H-J_H and V_L-J_L), whereas the sequences between these conserved regions can have quite high variability (giving each V region its particular binding specificity). Consequently, specific stretches of DNA (i.e., primers) can be designed that are complementary to the ends of the genes, and these can be used in the PCR to amplify the V region genes and generate libraries. The next step is to express the V regions in functional form so that they can bind to antigens. The heavy and light chain genes can be linked to each other so that they form a continuous polypeptide chain called a single-chain variable fragment (scFv). This scFv can be produced in *E. coli* as functional protein. However, the fundamental question arises as to how cells producing scFvs with the desired binding activity can be identified and isolated. This problem was overcome in the 1990s by using a process called phage display (Fig. 7.10 and Note 7.2).

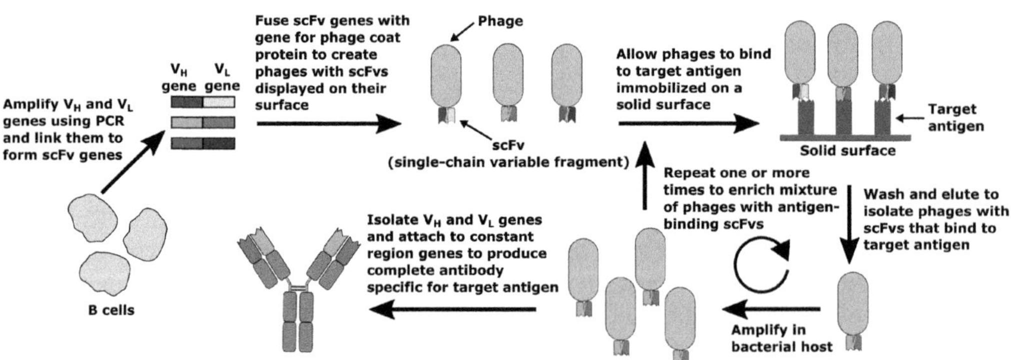

FIGURE 7.10
Main steps in the use of phage display to obtain and isolate genes encoding V regions that are specific for a target antigen.

Phages are like viruses, except they are propagated in bacterial cells. Phage particles can be made that have on their surface one type of scFv and within the phage particle, the genetic material that encodes this scFv is packaged. Using molecular biology approaches, large libraries of phage particles can be made that have different scFvs on their surface. Phage particles that have scFv that binds to the target antigen can be isolated as follows (Fig. 7.10). Phage particles are incubated with the target antigen, which is usually coated on a solid surface. Particles that do not bind to the antigen are removed by washing, while particles that bind to the antigen are eluted and amplified in a bacterial host. To obtain an enriched mixture of phage particles with scFvs that bind to the antigen, these steps can be repeated a number of times, using in each iteration the eluted and amplified particles from the previous iteration. This iterative procedure is often referred to as *panning*. After panning, the isolated phage particles can be propagated in *E. coli* and the genes encoding the binding scFvs isolated. Once these genes are isolated, they can be attached to constant region genes and produced as full-length antibody molecules (Section 7.2.3). Analogous approaches can also be used to make libraries of Fab fragments on the surface of phage. Alternative approaches, including the display of antibody libraries on the surface of yeast cells, can also be used to isolate antibody fragments with the desired binding properties. An important difference between yeast and phage is that yeast are larger in size. Consequently, yeast libraries displaying antibody fragments can be incubated with fluorescently labeled target antigens and sorted by flow cytometry (Note 7.3) based on their antigen-binding properties (Note 7.4).

The approach of V gene library generation has been used to produce therapeutic antibodies from human B cells. Consequently, the antibodies are completely human in sequence. The first of this class of antibodies to be approved by the U.S. Food and Drug Administration for clinical use is the blockbuster therapeutic, Humira. This antibody targets an inflammatory molecule called TNF-α, and was first approved in 2002 for the treatment of rheumatoid arthritis. Since then, it has been approved for the treatment of other diseases such as ulcerative colitis.

8

Principles of Fluorescence

As we have seen in the introductory part of this text, the labeling of molecules with fluorescent tags provides us with an exquisite tool to study specific molecules in their cellular context. Fluorescence is a physical phenomenon that occurs when photons are emitted by a fluorescent molecule, i.e., a fluorophore, as a consequence of the illumination of the molecule by light. This chapter will provide important background in physics and chemistry that is relevant for the understanding of fluorescence. Focus will also be given to the different types of fluorophores that are used in modern cellular microscopy.

8.1 Wave and particle description of light

A fascinating and sometimes confusing aspect of the electromagnetic theory of light is the duality between the wave description and the particle description of light. The concept of the wavelength of light is of course intimately related to the wave description of light. In Chapter 14, we will extensively study the wave nature of light to investigate the diffraction phenomena that are of importance in the image formation process of a microscope. Other aspects of light are best explained using the particle formulation of light that is based on photons as being the fundamental components of light. Photons are by their very nature quantized and are the basic notion of the quantum mechanical interpretation of light.

An important physical law is *Planck's relation*, which can be interpreted as a law that links the quantum interpretation of light to the wave interpretation, providing a relationship between the energy E of a photon and its wavelength λ, i.e.,

$$E = \frac{hc}{\lambda}.$$

Here c is the speed of light and h is *Planck's constant*, given by $h = 6.626 \cdot 10^{-34} J \cdot s$. Planck's relation states that the energy of a photon is inversely proportional to its associated wavelength. We therefore immediately have that lower wavelength light has higher energy than higher wavelength light. Specifically, blue light has higher energy than green light, which has higher energy than red light.

8.2 Jablonski diagram

A powerful tool in the description of the physical basis of fluorescence is the *Jablonski diagram* (Fig. 8.1). In this diagram, the energy levels of a fluorescent molecule are shown by horizontal lines. Two important states corresponding to the electron states are the ground state S_0 and the excited state S_1. The thinner lines in the Jablonski diagram indicate

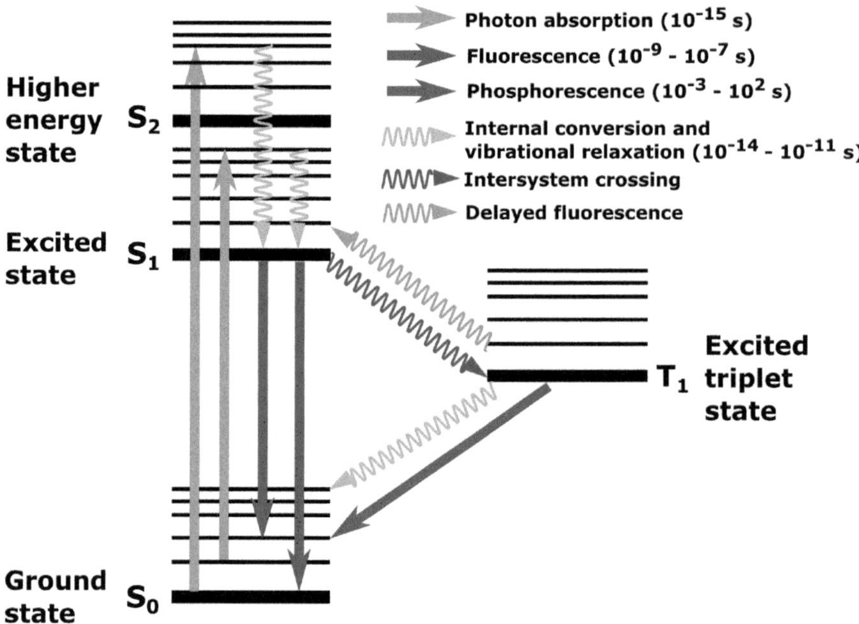

FIGURE 8.1
Jablonski diagram, showing various energy states and transitions between the states. Radiative transitions are depicted with straight arrows, and non-radiative transitions are depicted with squiggly arrows. The time scales on which some of the transitions occur are also given.

different vibrational states. Transitions between energy levels are shown by arrows. There are two important types of transitions. The radiative transitions involve the absorption or emission of a photon, whereas the non-radiative transitions involve other mechanisms such as the transfer of energy to the environment of the molecule. Of particular importance for us are the radiative transitions. The absorption of a photon results in a transition from the ground state to the excited state. Conversely, the emission of a photon results in a transition from the excited state to the ground state. In contrast to phosphorescence, the fluorescence emission process is almost instantaneous following the excitation of the molecule. The *fluorescence lifetime*, i.e., the time between the absorption of an excitation photon and the resulting emission of a photon is in the order of pico or nanoseconds. It is important to note that the emission process can also lead to additional changes in the energy levels, e.g., due to changes in the vibrational states, and can lead to dissipation of heat to the solvent. A third energy state that is sometimes of importance is the *triplet state*. Transitions from the excited state to the triplet state are non-radiative, i.e., the molecule appears dark. These transitions are important in localization-based super-resolution experiments, as discussed in Section 11.6.3.1.

8.3 Stokes shift

Since the energy released during the emission of a photon is typically no larger than the energy acquired during the absorption of a photon, the radiative energy associated with the emission process is less than the radiative energy associated with the absorption process. By Planck's relation, the wavelength of the emitted photon is therefore larger than the wavelength of the exciting photon. This phenomenon, called *Stokes shift*, is named after the Irish scientist George G. Stokes (1819–1903). The Stokes shift has very important implications for fluorescence microscopy, which we will see throughout this text.

There are situations, however, where the Stokes shift does not apply. Perhaps the most important of these exceptions can be found in two-photon excitation microscopy, where two relatively low-energy (i.e., long-wavelength) photons simultaneously excite the fluorescent molecule. The result is that the emitted photons are typically of shorter wavelengths than the exciting photons. In Section 10.5, two-photon excitation microscopy will be discussed in more detail.

8.4 Photobleaching

Given a steady source of appropriate exciting light, a fluorescent molecule can go through the absorption and emission cycle at a very rapid rate, thereby producing a steady stream of emitted photons. Unfortunately, most fluorescent substances cannot go through this cycle an infinite number of times. Similar to a light bulb, after a certain random number of cycles, the molecule stops emitting further photons. This event is referred to as *photobleaching*. Different types of fluorescent molecules have different expected times until photobleaching. For most experiments a microscopist would want to rely on the use a fluorescent molecule that is as *photostable* as possible, i.e., whose expected time to photobleaching is as long as possible. Figure 8.2 shows the results of the repeat imaging of an individual fluorescent molecule. The detected number of photons is relatively high and constant until the molecule photobleaches and the measured photon count drops in one step to the background level. If we image an ensemble of fluorescent molecules, the number of detected photons in a given time interval typically decreases exponentially with a certain *photobleaching rate*. An exam-

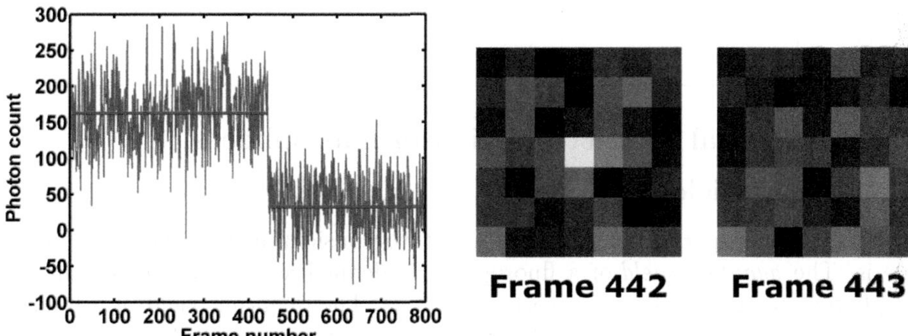

FIGURE 8.2
Photobleaching of a single ATTO 647N fluorescent dye molecule. The plot shows in blue the number of photons detected from the dye molecule (and a background component) over 800 images, each acquired with a 40-ms exposure. The red lines indicate the average photon counts per image before and after the molecule photobleaches. The last image before, and the first image after, the photobleaching event (frames 442 and 443, respectively) are shown next to the plot.

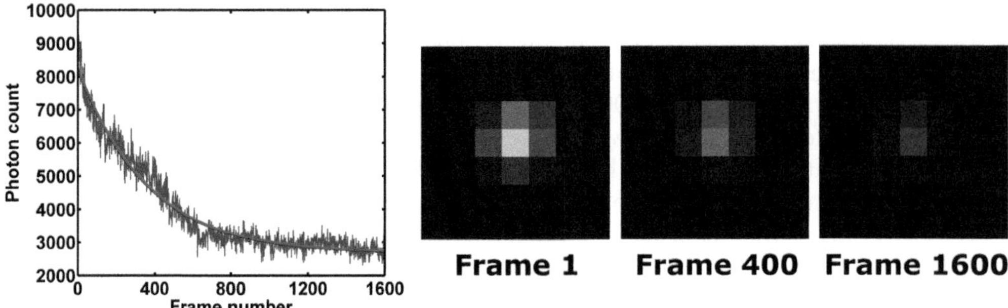

FIGURE 8.3
Photobleaching of a 100-nm TetraSpeck bead. The plot shows in blue the number of photons detected from the bead (and a background component) over 1600 images, each acquired with a 100-ms exposure. The red curve is an exponential decay function that has been fitted to the photon counts. The exponential decay constant (photobleaching rate) is 0.0277 s^{-1}. Frames 1, 400, and 1600 of the 1600-image sequence are shown next to the plot.

ple of this is given in Fig. 8.3, where the photobleaching curve of a fluorescent bead (which contains many individual dye molecules) is analyzed. The exponential decay behavior of an ensemble of fluorescent molecules is consistent with the fact that the time to photobleaching of each of the molecules has an exponential distribution. The relatively large number of molecules imaged in an ensemble obscures the drop in the emitted photons of the individual fluorescent molecules as they photobleach. The nature of the photobleaching process is not fully understood. However, it is clear that oxygen in the solvent can significantly worsen photobleaching. Therefore, so-called *antifade agents* or *oxygen scavengers* are often used to remove oxygen from the solvent. This is certainly a very appropriate approach for the preparation of fixed cell samples. In live cell samples the use of a scavenger is often problematic since oxygen is vital for the proper functioning of a living cell, and the oxygen scavenger may lead to toxicity and other detrimental effects on the cells.

8.5 Photophysical characterization of fluorophores

8.5.1 Quantum yield

When designing an experiment, it is important to consider how "bright" a fluorescent molecule is. The *quantum yield* of a fluorophore is defined as the ratio of the number of emitted photons to the number of absorbed photons. The quantum yield of a fluorophore therefore indicates how efficiently the fluorophore converts absorbed photons into emitted photons. The maximum quantum yield is 1. However, non-radiative transitions from the excited state of a fluorophore can lead to quantum yields that are significantly less than 1. The quantum yield depends on the specifics of an experiment, such as the environment of the fluorophore.

As examples, we consider the classical fluorescent dyes fluorescein isothiocyanate (FITC) and tetramethylrhodamine isothiocyanate (TRITC) (Fig. 8.4), which are not as frequently used today as they are supplanted by more photostable alternatives such as dyes from the Alexa Fluor family (Section 8.7.1). FITC has a quantum yield that varies between 0.3 and

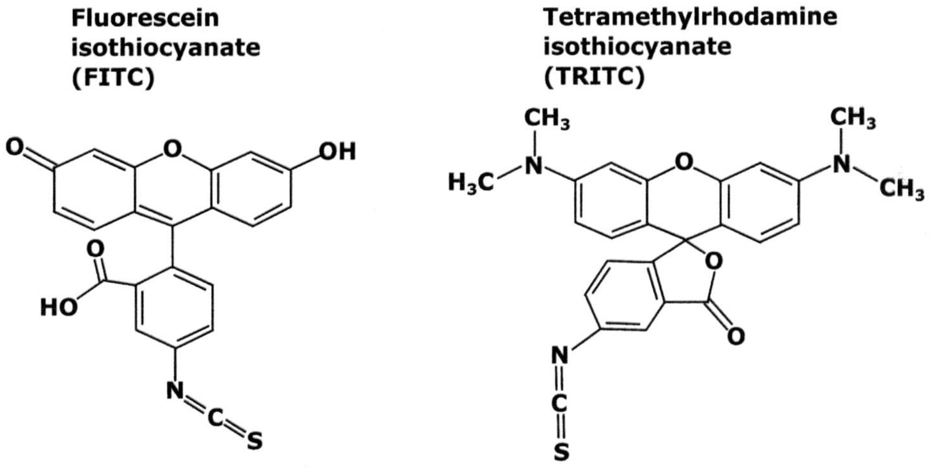

Fluorescein isothiocyanate (FITC)

Tetramethylrhodamine isothiocyanate (TRITC)

FIGURE 8.4
Chemical structures of the classical fluorescent dyes fluorescein isothiocyanate (FITC) and tetramethylrhodamine isothiocyanate (TRITC).

0.85, depending on the conditions under which the value is measured. TRITC is generally reported to have a quantum yield that is significantly lower than that of FITC.

8.5.2 Beer-Lambert law, effective absorption cross section and molar extinction coefficient

While it is important to know how efficiently a fluorophore translates absorbed photons into emitted ones, it is also important to know how well a fluorophore, or any molecule, can absorb photons. The *Beer-Lambert law* describes the intensity of a beam of light of a certain wavelength as it traverses through a liquid sample. As the beam traverses deep into the sample, its intensity decreases as its photons are absorbed or scattered by the molecules in the sample. The Beer-Lambert law is given by

$$I_1 = e^{-\sigma n_c l} I_0, \tag{8.1}$$

where l is the path length, i.e., the length of the light path in the sample. The symbol I_0 denotes the intensity of the light at the beginning of the light path, and the symbol I_1 denotes the intensity of the light at distance l into the sample. Besides the path length, the exponent depends on the *effective absorption cross section* σ and the *number concentration* n_c of the molecules in the sample, i.e., the number of molecules per cubic centimeter. The effective absorption cross section can be interpreted as the obstacle, described by a 2D area in square centimeters, that a molecule presents in the path of the beam of light. The Beer-Lambert law shows that due to light absorption by the molecules in the sample, the light intensity decreases exponentially as the light passes through the sample. Given a certain concentration of molecules n_c and the effective absorption cross section σ (i.e., the size of the obstacle that a molecule presents for the light beam), the amount of absorption increases, and therefore the light intensity decreases, with increasing depth l that the light beam traverses into the sample.

An equivalent way of looking at the Beer-Lambert law leads to an expression that is also often used. This expression is referred to as the *absorbance* of the sample, and it can be obtained by applying the logarithm with base 10 to the quotient I_1/I_0, i.e.,

$$-\log_{10}\left(\frac{I_1}{I_0}\right) = -\log_{10}(e^{-\sigma n_c l}) = \frac{-1}{\ln(10)}\ln(e^{-\sigma n_c l}) = \frac{1}{\ln(10)}\sigma n_c l = \epsilon C l, \tag{8.2}$$

where

$$\epsilon = \frac{\sigma}{\ln(10)} \cdot \frac{n_c}{C} = \frac{N_{Avo}}{1000 \cdot \ln(10)}\sigma \approx 2.615 \times 10^{20}\sigma$$

is the *molar extinction coefficient* and C is the molar concentration, i.e., the molecule concentration in moles per liter. Note that $C = n_c \cdot \frac{1000}{N_{Avo}}$, where $N_{Avo} \approx 6.022 \times 10^{23}$/mol is Avogadro's number, and therefore the unit of the molar extinction coefficient is area per mole (e.g., cm^2/mol), or equivalently, the inverse of the product of molar concentration and length (e.g., $M^{-1}\text{cm}^{-1}$). Equation (8.2) shows that given the same concentration and path length, a larger molar extinction coefficient means a higher absorbance, and therefore that the fluorophore can absorb photons more efficiently.

One of the reasons FITC and TRITC were very widely used was their ability to efficiently absorb photons. FITC and TRITC are characterized by relatively high molar extinction coefficients of 72000 $M^{-1}\text{cm}^{-1}$ and 107000 $M^{-1}\text{cm}^{-1}$, respectively.

We need to stress that the effective absorption cross section σ, and hence the molar extinction coefficient ϵ, are wavelength-dependent. This will be of relevance when we discuss the excitation spectrum of a fluorophore in Section 8.6.

8.5.3 Brightness of a fluorophore

In the prior subsections we discussed how effectively a fluorophore can absorb light and the efficiency by which a fluorophore can convert the absorbed photons into emitted photons. The first property is characterized by the effective absorption cross section or molar extinction coefficient, and the latter property is characterized by the quantum yield of the fluorophore. The *brightness* b of a fluorophore is then defined as the product of the molar extinction coefficient ϵ and the quantum yield Φ, i.e.,

$$b = \epsilon \cdot \Phi.$$

Clearly the brightness of a fluorophore depends on the excitation wavelength and the environment of the fluorophore, due to the dependency of ϵ and Φ on these experimental variables.

It would seem that the brightness of a fluorophore is the primary criterion for its choice in an imaging experiment. This is, however, not always the case as in some applications the photostability is more important. At least in some circumstances, a lower brightness of the fluorophore can be compensated for by increasing the power of the exciting light source.

8.6 Excitation and emission spectra

We have pointed out earlier the wavelength dependencies of properties of fluorophores such as the extinction coefficient. It is therefore important to understand how the photon emission properties of a fluorophore depend on the wavelength of the exciting light and on the wavelength at which the emission is observed. For this purpose, commonly two plots are considered, the *excitation spectrum* and the *emission spectrum*. The excitation spectrum is aimed at depicting the response of the fluorophore to different exciting wavelengths, whereas

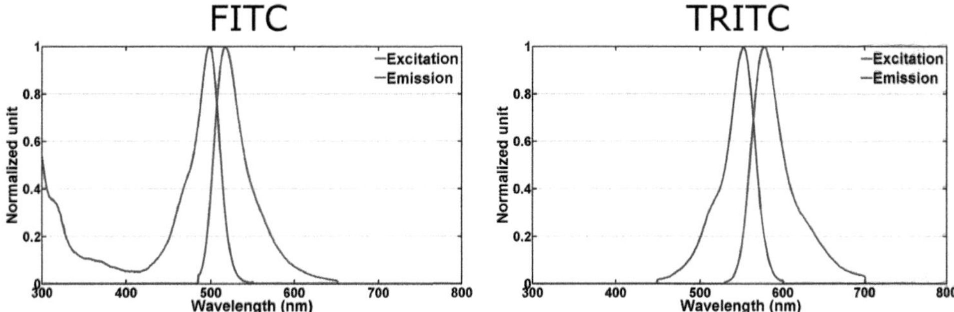

FIGURE 8.5
Excitation and emission spectra of fluorescein isothiocyanate (FITC) and tetramethylrhodamine isothiocyanate (TRITC). FITC has an excitation maximum of around 495 nm and an emission maximum of around 519 nm. TRITC has an excitation maximum of around 547 nm and an emission maximum of around 572 nm. [See Figure Credits.]

the emission spectrum is designed to show the dependence of the emitted photons on the wavelengths at which the emission is measured. The excitation spectrum is obtained by measuring the intensity of the emitted fluorescence at a specific wavelength, usually close to the emission maximum, and by varying the wavelength of the exciting light over a large interval. The emission spectrum is obtained by a complementary experiment, whereby the fluorophore is excited at a specific wavelength, usually close to the excitation maximum, and the intensity of the emitted light is measured over a large wavelength range. It is important for the interpretation of emission spectra that they are generally independent of the excitation wavelength for which they are measured. Figure 8.5 shows the excitation and emission spectra of FITC and TRITC. Note that as a consequence of the Stokes shift, the emission spectrum has nonzero intensities above the excitation wavelength. A symmetry can also often be observed between the emission and the excitation spectrum. This symmetry holds for many, although not all, fluorophores. It can be attributed to the symmetries in the Jablonski diagram for absorption and emission processes.

8.7 Fluorophores

Over the years, many fluorophores have been discovered or synthesized in the laboratory. There are three types of fluorophores that are currently used in cellular imaging. They are the classical chemical dyes, the semiconductor-based quantum dots, and the fluorescent proteins.

8.7.1 Chemical fluorescent dyes

Most chemical fluorescent dyes are small *organic*, or carbon-containing, compounds. Historically, small molecule fluorophores, such as fluorescein and rhodamine, have been widely used to label proteins for use in microscopy and other applications such as flow cytometry. Although still widely used in fixed cell imaging, where photobleaching can be reduced by the use of appropriate mounting media, the photo-instability of these fluorophores has severely limited their use in live cell imaging. The primary reason for this is that mounting medium

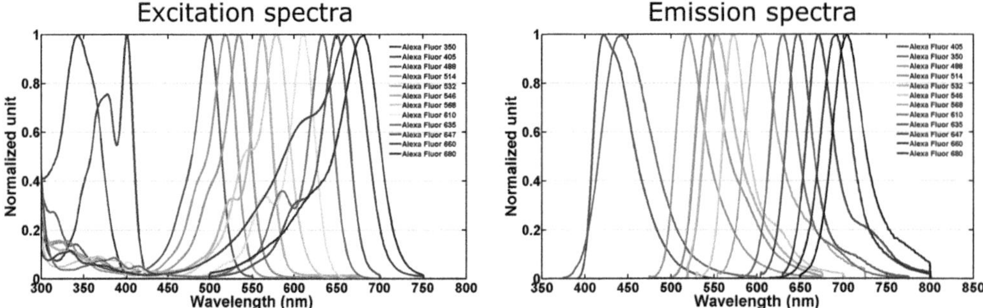

FIGURE 8.6
Excitation and emission spectra of 12 dyes from the Alexa Fluor family. Each dye is named according to the wavelength of its excitation peak. [See Figure Credits.]

can only be used with fixed (dead) cells since it is highly toxic to live cells. Consequently, for most cellular processes these fluorophores lose their fluorescence so rapidly that the proteins or other macromolecules of interest cannot be tracked for a sufficient time to generate informative data. These problems have prompted a major expansion in the development of fluorophores with increased photostability and higher photon output over the past decade or so. A class of such fluorophores is the Alexa Fluor dye series, or the more recently developed ATTO series. These fluorophores have a range of excitation and emission profiles (Fig. 8.6), facilitating their use in multicolor imaging experiments.

8.7.1.1 Labeling of proteins via cysteine or lysine residues

How are chemical fluorescent dyes such as Alexa Fluor or ATTO dyes incorporated in imaging experiments? One method is to use these dyes in chemically reactive forms that covalently attach to a protein by reacting with the side chains of cysteine (-SH) or lysine (-NH_2) residues in the protein (Fig. 8.7). This approach to labeling is carried out with a purified protein of interest that is subsequently taken up by the cells to be imaged, or with a purified protein (such as an antibody) that is then used to detect a protein of interest in fixed cells. Since a protein can have multiple cysteine and lysine residues, it will frequently have several labels attached, particularly if the coupling ratio of dye to protein is greater than one. Provided that label incorporation at multiple sites does not perturb the function of the protein, this is advantageous since it improves protein detection. However, for some applications such as those based on Förster resonance energy transfer (FRET) (Note 8.2), and for certain single molecule imaging experiments where one fluorophore per protein is a necessity, methods to label a protein at a specific location have been developed. Site-specific labeling can be achieved in some cases by mutating all except one cysteine residue in the protein to other amino acids such as serine or alanine, with the potential caveat that this can result in a misfolded protein. An analogous approach, however, cannot typically be taken for lysine since lysine is a very common amino acid in most proteins.

8.7.1.2 Labeling of proteins with fluorophore-conjugated streptavidin

An alternative approach for the labeling of proteins is to use the streptavidin-biotin system (Fig. 8.8). Streptavidin is a protein that has four identical subunits that can each bind with very high affinity to biotin. This protein is available with a large number of different fluorophores (e.g., Alexa Fluor dyes) coupled to it. Proteins of interest can be modified by a chemical reaction leading to the covalent coupling of biotin, a small chemical that binds

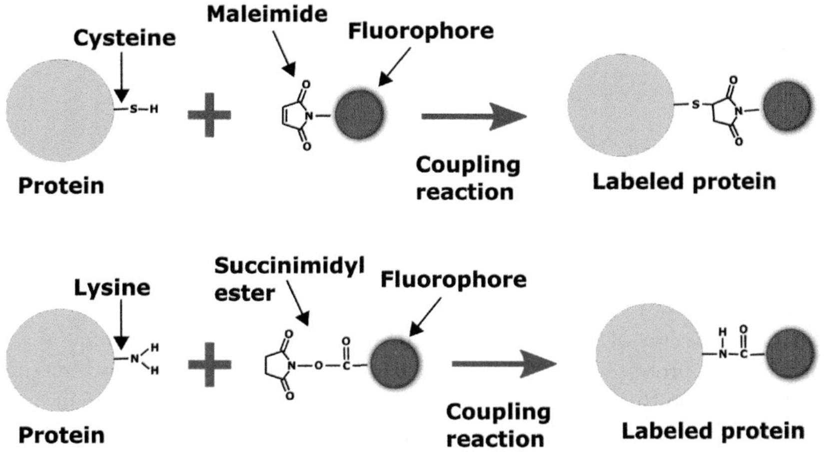

Cysteine
Maleimide
Fluorophore
Protein
Coupling reaction
Labeled protein

Lysine
Succinimidyl ester
Fluorophore
Protein
Coupling reaction
Labeled protein

FIGURE 8.7
Labeling of a protein via a cysteine or lysine residue. For the derivatization of cysteine, the fluorophore is commonly attached to the protein as a result of the reaction of a maleimide group with the thiol (-SH) group of a cysteine residue of the protein to form a stable carbon-sulfur bond. For the derivatization of lysine, the fluorophore is commonly linked to the protein by the reaction of a succinimidyl ester group with the amino (-NH_2) group of a lysine residue of the protein to form a stable amide bond.

with very high affinity to streptavidin. Biotin is covalently attached to the NH_2 groups on the lysine residues and N-terminus of the protein, usually resulting in the incorporation of multiple biotins at different sites in the protein and consequent heterogeneity following streptavidin binding. This problem can be circumvented by the modification of the protein

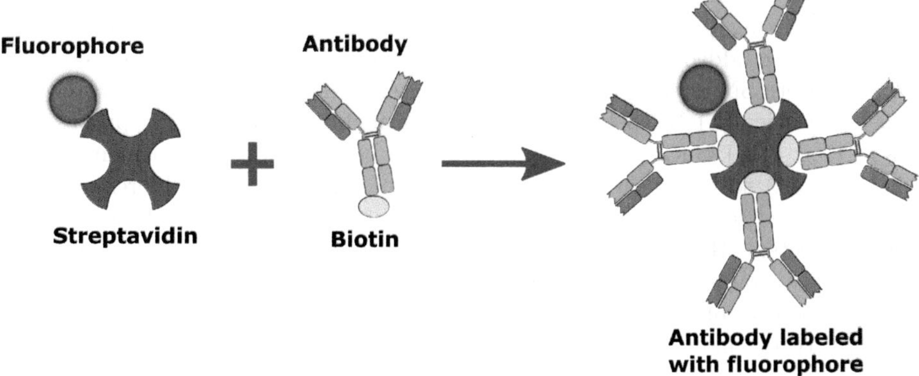

FIGURE 8.8
Streptavidin-biotin system for the specific association of two molecules. In the example shown, a fluorophore attached to streptavidin and a biotinlyated antibody are brought together by the strong interaction between streptavidin and biotin, resulting in the labeling of the antibody with the fluorophore. Note that while the example shows the maximum number of four biotinylated antibodies bound to a single streptavidin molecule, the stoichiometry of the complex can be changed by adjusting the concentrations of the biotinylated antibody and streptavidin.

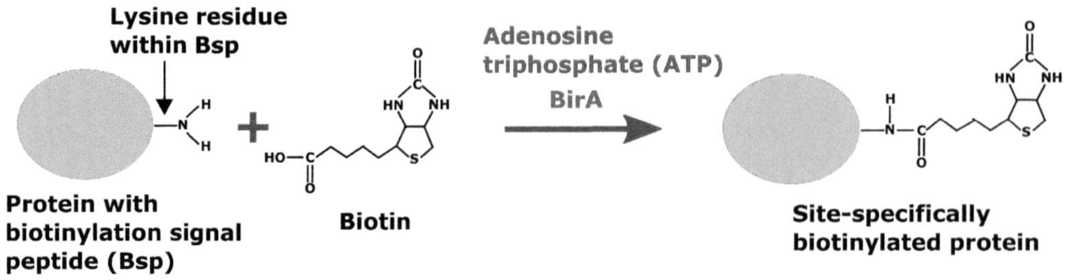

Lysine residue within Bsp

Protein with biotinylation signal peptide (Bsp)

Biotin

Adenosine triphosphate (ATP) BirA

Site-specifically biotinylated protein

FIGURE 8.9
Site-specific biotinylation of lysine residue within the biotinylation signal peptide (Bsp) tag on proteins. The protein with the Bsp tag is mixed with biotin, adenosine triphosphate (ATP), and the enzyme BirA. This reaction is usually carried out with purified recombinant proteins, but cells that express both recombinant protein (with Bsp tag) and BirA can also be used to produce site-specifically biotinylated protein.

gene with a short lysine-containing sequence that encodes a biotinylation signal peptide (Bsp). In the presence of biotin, an enzyme called BirA can be used to covalently link the biotin to this lysine residue, resulting in site-specific modification with one biotin per protein molecule (Fig. 8.9). As in the case of the direct labeling of a protein with a chemically reactive form of a dye, the streptavidin-biotin system for protein labeling is applied to a purified protein that is subsequently introduced into the cells to be imaged. The protein can either be introduced into the cells as labeled streptavidin-protein complexes, or the biotinylated protein can be incubated with the cells and subsequently detected with labeled streptavidin. Alternatively, streptavidin-protein complexes can be used to detect proteins of interest in fixed cells.

8.7.1.3 In situ labeling of proteins in cells using peptide tags

An approach for labeling proteins in situ in cells involves the engineering of a protein to add a peptide tag (e.g., "SNAP-tag" or "CLIP-tag") at the N- or C-terminus. This peptide tag can be added to the protein using the method of protein fusion discussed in Section 7.2.3. Both SNAP- and CLIP-tags are proteins that have enzymatic activity. These tags can be specifically labeled by the enzymatic attachment of synthetic probes that are fluorescent (Fig. 8.10). Such probes are designed to be able to cross cell membranes (i.e., are cell permeant) so that they enter the cytosol of cells. Once the fluorescent probe is covalently attached to the SNAP- or CLIP-tag, it remains inside the cell associated with the targeted protein. Hence proteins inside living cells can be labeled with this approach and subsequently tracked in real time. This approach is being continually improved upon by the generation of different tag sequences that can be labeled with distinct probes and thereby allow two or more different proteins in a cell to be tracked.

8.7.2 Quantum dots

Although great improvements have been made in their spectral properties and photostability, organic fluorophores still have a finite lifetime. This is particularly so for live cell imaging, where the photostabilizers that are present in fixed cell mounting medium often cannot be used due to toxicity. Over the past decade, the use of fluorescent particles called quantum dots (QDs), which are far more photostable than traditional organic dyes, has greatly expanded. The extended photostability of QDs enables imaging over significantly

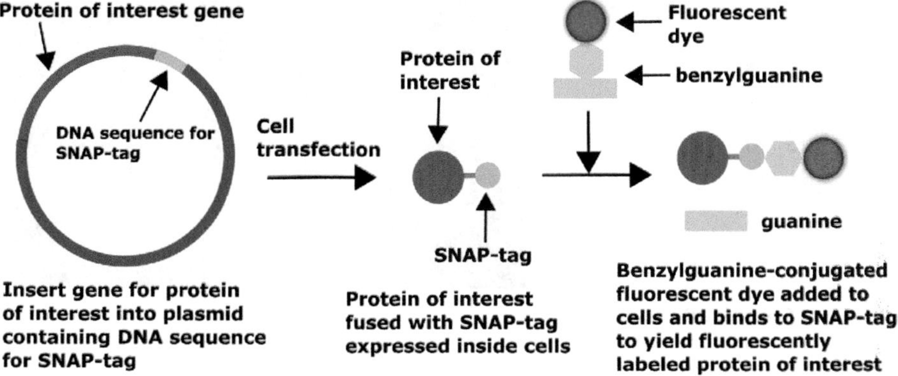

FIGURE 8.10
Schematic showing the labeling of proteins using SNAP-tags. Labeling of proteins using CLIP-tags occurs in an analogous fashion, but uses benzylcytosine-conjugated fluorescent dyes and results in the release of a cytosine molecule instead of a guanine molecule.

longer time periods, but the use of QDs is not without issues such as the intermittent nature of their fluorescence emission, often referred to as *blinking* (Note 8.3).

A QD is a semiconductor particle typically composed of a core enclosed by a shell. A common combination consists of a cadmium selenide (CdSe) core and a zinc sulfide (ZnS) shell (Fig. 8.11(a)). In a core-shell structure, the shell has the important function of improving the particle's quantum yield. The size of QDs is a determinant of their spectral properties. As QDs get bigger, the wavelength of their emission peak increases (Fig. 8.11(b)). Including the layer of chemically reactive groups added to its surface for the attachment of a protein of interest (Fig. 8.11(a) and Section 8.7.2.1), a QD typically has a diameter in the range of 10 to 20 nm. QDs emit fluorescence over a relatively narrow range of wavelengths, but their excitation can be achieved by illumination with light from a broad range of wavelengths (Fig. 8.12).

8.7.2.1 Labeling of proteins with quantum dots

QDs can be attached to proteins using a number of different approaches. For example, QDs coupled with chemically reactive carboxylate groups can be coupled to proteins by "amine" coupling chemistry through exposed NH_2 groups on the protein. These NH_2 groups are on lysine side chains, but can also include the N-terminal NH_2 group of the protein. Alternatively, biotin groups can be coupled to the amino groups of lysines and then the biotinylated proteins added to QDs that are coated with streptavidin.

However, both of the coupling methods described above can result in different orientations of the protein on the QD, due to the multiple lysines that can be derivatized. Both methods can also lead to clumping, or aggregation of the proteins and QDs, since one QD can bind to more than one protein molecule. An alternative strategy is therefore to genetically modify the protein with a Bsp tag followed by site-specific biotinylation (Fig. 8.9). The site-specifically biotinylated protein is then added to streptavidin-coated QDs.

The typical methods of protein labeling with QDs are therefore similar to those described in Sections 8.7.1.1 and 8.7.1.2 for protein labeling with chemical dyes, and are carried out with a purified protein of interest that is then taken up by cells as QD-protein complexes before they are imaged. Alternatively, QD-protein complexes can be used to detect proteins of interest in fixed cells.

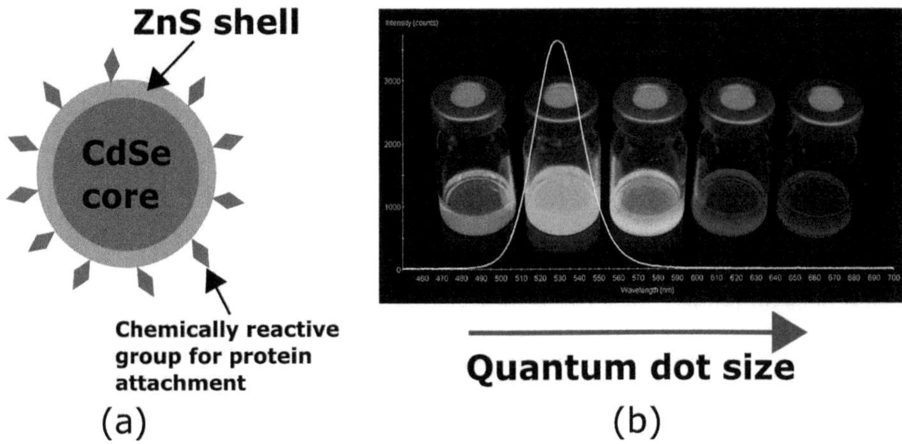

FIGURE 8.11
Quantum dots. (a) Structure of a QD with a cadmium selenide (CdSe) core and a zinc sulfide (ZnS) shell. (b) Solutions of CdSe QDs that emit photons of different wavelengths upon excitation by ultraviolet light. The wavelength of the emitted light increases with the size of the QD. [See Figure Credits.]

8.7.3 Fluorescent proteins

The limitation of using chemical dyes or QDs to label proteins is that, with the notable exception of using the relatively new approach involving peptide tags for the in situ labeling of proteins in cells (Section 8.7.1.3), chemical dyes and QDs cannot in general be used to label specific proteins amongst a mixture of proteins, such as within a cell. This traditional limitation has led to the use of protein engineering to tag proteins of interest, usually at

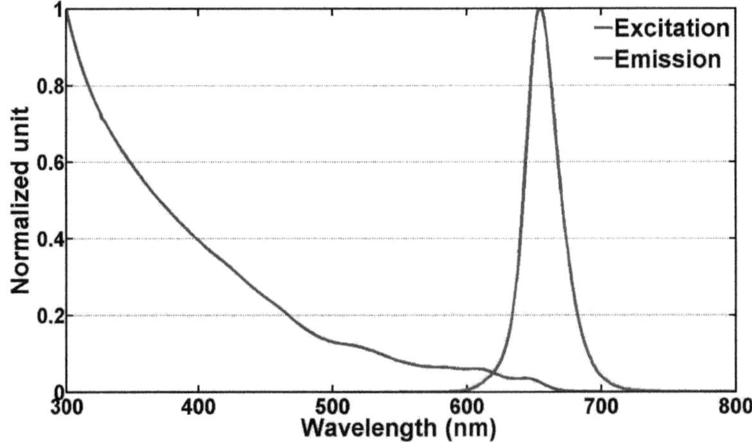

FIGURE 8.12
Excitation and emission spectra of Qdot 655, a QD available from Invitrogen Molecular Probes. Qdot 655 has a CdSe core and a ZnS shell, and emits fluorescence maximally at 655 nm. The plot shows that the excitation spectrum of a QD spans a broad range of wavelengths and increases with decreasing wavelength. [See Figure Credits.]

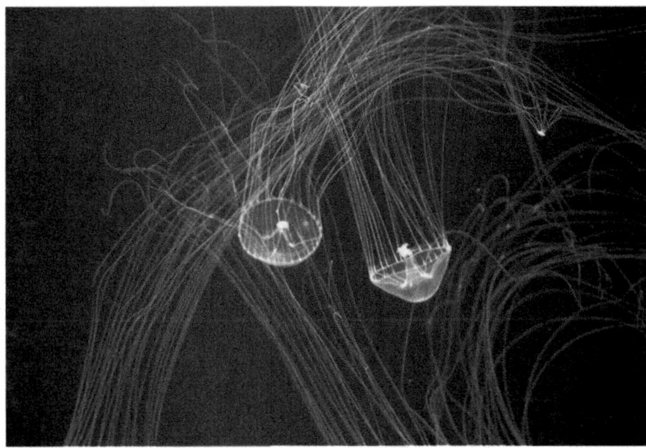

FIGURE 8.13
Jellyfish *Aequorea victoria*.

their N- or C-terminus, with a fluorescent protein. (See Section 7.2.3 for a discussion on the molecular approaches to protein fusion.) Green fluorescent protein (GFP) is the first fluorescent protein to be used in this approach and has been widely adopted as a fluorescent label of choice in many cell biological applications. GFP is a protein composed of 238 amino acids (26.9 kDa). It was isolated from the jellyfish *Aequorea victoria* (Fig. 8.13). Martin Chalfie, Osamu Shimomura, and Roger Y. Tsien were awarded the 2008 Nobel Prize for their discovery and development of GFP. GFP has an excitation peak at wavelength 395 nm and a minor one at 475 nm. Its emission peak is at 509 nm, which is in the lower green portion of the visible spectrum. GFP has a β-barrel structure, consisting of eleven β-strands that form a barrel and an α-helix containing the chromophore that runs through the center of the barrel (Fig. 8.14). It has five shorter α-helices that cap the ends of the barrel.

A significant improvement to the spectral properties of GFP was introduced in 1995 by Roger Tsien. By replacing a serine residue in wild-type GFP with a threonine residue, a variant of GFP was obtained that had its major excitation peak at 488 nm. The shift of the excitation peak from 395 nm to 488 nm was of particular practical importance as it made GFP compatible with standard filter sets that were readily available to fluorescence microscopists. This GFP variant and its derivatives, such as enhanced GFP (EGFP) which is obtained by the replacement of a second amino acid residue and is significantly brighter than wild-type GFP, have become standard in cellular fluorescence microscopy. Going forward, we will simply refer to GFP with major excitation peak at 488 nm as GFP (Note 8.4).

Following the development of GFP, it became clear that other fluorescent proteins with different and non-overlapping spectral properties would be desirable to allow two or more proteins to be tracked in the same cell. This led to the use of protein engineering to produce GFP variants that are spectrally distinct. The many GFP color mutants include the blue fluorescent protein (EBFP, EBFP2, Azurite, mKalama1), cyan fluorescent protein (ECFP, Cerulean, CyPet), and yellow fluorescent protein (YFP, Citrine, Venus, YPet) derivatives. These variants have different excitation and emission spectra, as shown in Fig. 8.15.

In the meantime, many other fluorescent proteins have been isolated, and in some cases, further improved by engineering. A prominent example is the family of red fluorescent proteins, which has expanded the available palette to include the red region of the visible spectrum. The first protein in this family was isolated from the reef coral *Discosoma* sp., and is called DsRed. DsRed has a β-barrel structure similar to that of GFP, and has maximum excitation and emission wavelengths of 558 nm and 583 nm, respectively. Despite the

FIGURE 8.14
Ribbon representation of the 3D structure of GFP. GFP has eleven β-strands that form a barrel, and an α-helix containing the fluorescence-emitting chromophore (shown in ball-and-stick representation) running through the center of the barrel. It also has five shorter α-helices that cap both ends of the barrel. [See Figure Credits.]

red fluorescence that it offers, however, DsRed has the undesirable property that it forms tetramers, raising the question of how significantly it can alter the localization and function of the protein of interest to which it is attached. To overcome this problem, a monomeric form of DsRed, called monomeric red fluorescent protein 1 (mRFP1), was obtained by making a substantial number of amino acid substitutions in DsRed. Besides its monomeric nature, mRFP1 has the advantage that its maximum excitation and emission wavelengths (584 nm and 607 nm, respectively) exceed those of DsRed, providing increased spectral separation from other fluorescent proteins. mRFP1 was developed in Roger Tsien's laboratory,

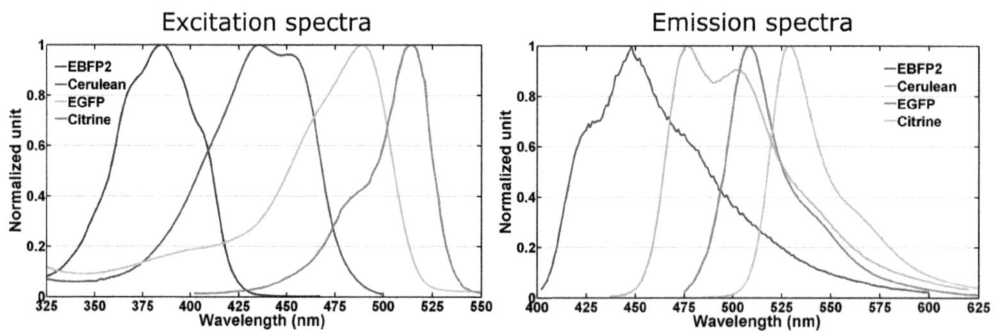

FIGURE 8.15
Excitation and emission spectra of four variants of GFP: enhanced blue fluorescent protein 2 (EBFP2), Cerulean, enhanced green fluorescent protein (EGFP), and Citrine. [See Figure Credits.]

which subsequently also engineered brighter and more photostable mRFP1 variants such as mCherry and mStrawberry (Note 8.5).

The choice of fluorescent protein will depend on the specifics of a particular experiment. For example, photostablity will be a key determinant for most live cell imaging applications. Moreover, the choice of fluorescent protein is sometimes limited by instability of the target protein following attachment of a particular fluorescent protein. This effect is not predictable, and under these circumstances, the optimal (least perturbing) fluorescent protein is determined experimentally.

8.7.4 Photoactivatable and photoswitchable fluorescent probes

Multiple fluorophores have been either developed or identified that are not constitutively fluorescent. Instead, by irradiation with light of particular wavelengths, they can be permanently switched from an "off" state in which they do not fluoresce to an "on" state in which they are capable of fluorescing, or they can be switched back and forth between these two states. A fluorophore belonging to the first category is referred to as photoactivatable, and a fluorophore in the second category is referred to as photoswitchable. Typically, a photoactivatable or photoswitchable fluorophore is switched to the "on" state by irradiation with blue or violet light.

Photoactivatable and photoswitchable fluorophores typically fall into two groups — fluorescent proteins and organic fluorophores. The latter group comprises conventional fluorophores, such as members of the Alexa Fluor and the ATTO dye families, which are photoswitchable when used in appropriate buffers (Note 8.6). A paradigmatic example, however, is an engineered form of wild-type GFP that can be photoactivated. This GFP mutant is named *photoactivatable GFP*, or PA-GFP, and is obtained by substituting a threonine residue in wild-type GFP with a histidine residue (Note 8.7). PA-GFP maintains wild-type GFP's major excitation peak at 395 nm, but has almost negligible absorbance in the region around the minor excitation peak at 475 nm. In this "off" state, PA-GFP effectively does not absorb 488-nm excitation light and therefore does not emit fluorescence. Upon photoactivation by irradiation with violet light (~400 nm), however, the absorbance in the region around 475 nm is dramatically increased, and the major excitation peak is shifted to approximately 500 nm. Subsequent excitation with 488-nm light then causes PA-GFP to emit fluorescence that is significantly more intense than that of wild-type GFP. Photoactivatable and photoswitchable fluorophores find significant application in localization-based super-resolution microscopy, which is discussed in Section 11.6.3.

8.7.5 Other labeling modalities

A large number of different labeled proteins or other macromolecules that are commercially available can be used in live cells to label specific pathways. For example, fluorescently labeled holotransferrin ("holo" indicates that it is bound to iron) labels the endosomal recycling pathway, whereas fluorescently labeled dextran, a polymer made of glucose molecules, can be used as a fluid phase marker that accumulates in lysosomes. Fluorescent low-density lipoprotein (LDL) also trafficks to lysosomes. Alternatively, the LysoTracker dyes, which only fluoresce at lysosomal pH, can be used to demarcate lysosomes in cells, although these dyes alter the pH within the acidic compartment and can therefore have the possible caveat that they may modulate trafficking pathways when used over prolonged periods. Other labels such as fura-2 and indo-1, whose emitted fluorescence intensity is dependent on the concentration of calcium ions in their environment, can be used as ratiometric calcium sensors to detect changes in calcium levels in cells. To quantify calcium levels, the fluorescence intensity of these dyes are measured at two different excitation or emission wavelengths, and the ratio of the two intensities is used to determine the concentration of calcium ions based on a calibration curve. In similar manner, the pH of compartments in cells can be measured by ratiometric imaging with pH-sensitive fluorophores such as fluorescein. For more details on ratiometric pH imaging, the reader is referred to Section 11.5.

9

Cells

Cells can be broadly classified into two groups: eukaryotic and prokaryotic. Eukaryotic cells are distinguished from prokaryotic ones by the presence of a nucleus and subcompartments called organelles. Bacteria and amoeba are representative examples of prokaryotic cells, whereas larger multicellular bodies, including animals, are made up of eukaryotic cells. The organization of eukaryotic cells will be considered here, as we look in some detail at the organelles of a cell. In addition, we will discuss proteins called receptors and the preparation of cell samples for fixed cell and live cell imaging. Throughout this chapter, we will use for illustration FcRn and transferrin receptor from our model systems.

9.1 Cellular structure

Within a larger animal, eukaryotic cells differentiate to become highly specialized tissues or organs such as liver or skin. Nevertheless, the overall structure of these cells is similar. Eukaryotic cells are bounded by a plasma membrane made up of a lipid bilayer and proteins (Fig. 9.1). The lipids in the plasma membrane have charged (hydrophilic) head groups with attached fatty acyl chains that are hydrophobic. Consequently in an aqueous environment, a bilayer structure with the head groups on the outside and the fatty acyl chains on the inside is energetically preferable. In general, all biological membranes are bilayers of this type, although the lipid composition can vary. The membrane proteins frequently have carbohydrate attached to them post-translationally (i.e., following translation) and are therefore glycoproteins. Membrane proteins fall into two broad categories. Those that are permanently attached to the membrane are called *integral membrane proteins*, and those that associate with the membrane only temporarily are called *peripheral membrane proteins*. Integral membrane proteins are usually *transmembrane proteins*, meaning that they span the entire lipid bilayer of the membrane. Some integral membrane proteins, however, are attached to the membrane only from one side. Peripheral membrane proteins associate with the membrane by attaching to integral membrane proteins, or by adhering to the periphery of the lipid bilayer. The interior of the cell is occupied by a fluid called the cytosol and many membrane-bound compartments called organelles. Collectively, the cytosol and all organelles except the nucleus are referred to as the cytoplasm. The organelles of a cell have distinct functions and can be distinguished from each other by the presence of specific proteins, although they will also share many proteins. The proteins that are specific to a given organelle can be used as markers for this compartment, provided that an antibody or other reagent that specifically binds to this protein is available.

The primary organelles and other components of a eukaryotic cell are illustrated in Fig. 9.2. The functions of the primary organelles can be summarized as follows.

Nucleus. The nucleus is bounded by a nuclear membrane with pores. This organelle contains the DNA which is replicated prior to cell division. The DNA is also transcribed into RNA in the nucleus. Following transcription, the RNA is modified by removal of intronic

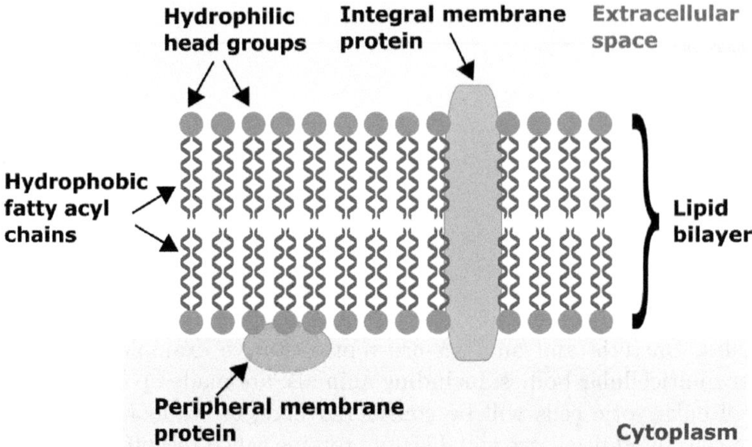

FIGURE 9.1
Plasma membrane of a eukaryotic cell. The plasma membrane comprises a lipid bilayer and proteins. In the lipid bilayer, the hydrophobic (lacking affinity for water) fatty acyl chains of the lipids are on the inside, and the hydrophilic (having affinity for water) head groups of the lipids are on the outside, with those of the outer leaflet facing the extracellular environment, and those of the inner leaflet facing the interior of the cell (i.e., the cytoplasm). An integral membrane protein is one that is permanently attached to the membrane, and a peripheral membrane protein is one that temporarily associates with either an integral membrane protein or the periphery of the lipid bilayer. The depicted integral membrane protein is one that spans the lipid bilayer and is therefore a transmembrane protein.

(noncoding) sequences and then exported to the cytoplasm through the nuclear pores. The proteins that are necessary to duplicate and transcribe DNA are present in the nucleus. In addition, the chromosomal DNA is highly compacted by association with proteins called histones.

Endoplasmic reticulum (ER). The ER is a reticular, highly interconnected mesh of membrane-bound compartments that are present in the cytoplasm. Proteins that are destined for membrane association (rather than soluble, cytosolic proteins) or secretion from the cell are synthesized on ribosomes that are attached to the ER, and the protein is therefore inserted into the ER membrane whilst it is being synthesized. This is called co-translational insertion. FcRn will be taken as an example. FcRn comprises a dimer of an α-chain and β_2-microglobulin. The α-chain is synthesized with a so-called signal peptide (also called a leader peptide) at its N-terminus, which directs this protein to be inserted into the ER membrane as a transmembrane protein (Fig. 9.3). As the signal peptide enters the ER, it is cleaved. The synthesis of the FcRn α-chain continues until a hydrophobic sequence ("transmembrane domain") is made that anchors the protein in the membrane. Protein synthesis continues so that the C-terminus of the protein is outside the ER (i.e., in the cytosol). Proteins with this orientation in the membrane (N-terminus inside the ER, C-terminus outside) are called type I transmembrane proteins. In addition, β_2-microglobulin is synthesized and inserted into the ER with a signal peptide. However, this protein does not have a transmembrane domain and is inserted into the ER as a soluble protein that can then associate with the FcRn α-chain.

By contrast with FcRn, the transferrin receptor has its C-terminus inside the ER and N-terminus outside following translation (Fig. 9.3) and is called a type II transmembrane

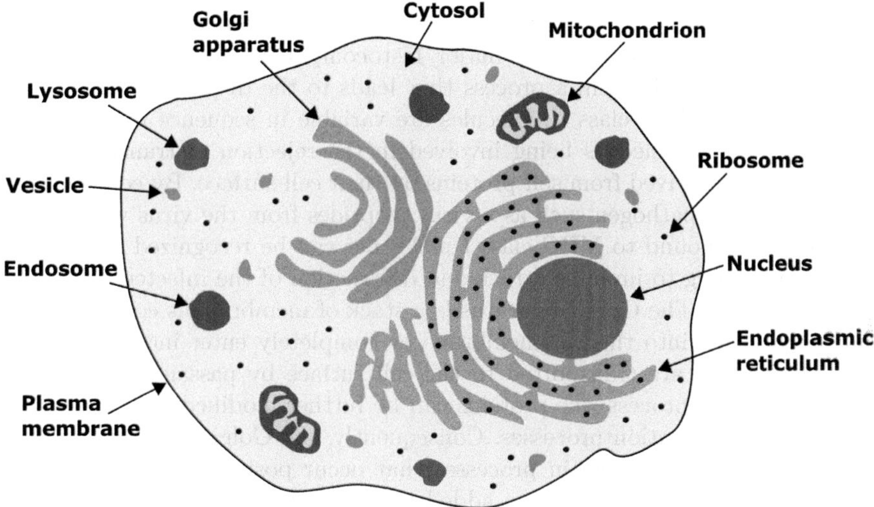

FIGURE 9.2
Primary components of a eukaryotic cell. A typical eukaryotic cell comprises the plasma membrane, the cytosol, and organelles that include the nucleus, mitochondria, the endoplasmic reticulum, the Golgi apparatus, lysosomes, endosomes, and vesicles. The ribosomes on which proteins are synthesized are also shown.

protein. In fact, this receptor is a complex of two equivalent type II membrane proteins. How this orientation is achieved is more complicated than for type I transmembrane proteins and will not be discussed further here.

In addition, peptides that are generated by the enzymatic chopping up, or proteolysis,

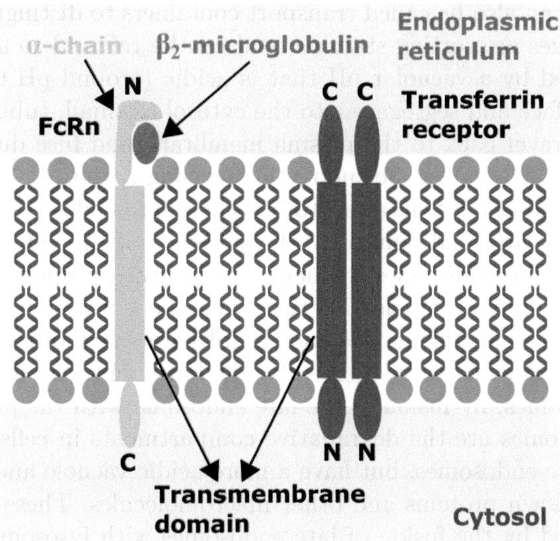

FIGURE 9.3
Orientation of FcRn and transferrin receptor in the lipid bilayer of the ER membrane. The letters N and C, respectively, denote the N-terminus and C-terminus of the α-chain of FcRn and each subunit of the transferrin receptor.

of proteins in the cytosol can enter the ER through specialized transport processes. These peptides can be loaded onto so-called major histocompatibility complex (MHC) class I molecules in the ER, resulting in a process that leads to the display of these peptides on the cell surface. (The MHC class I molecules are variable in sequence in different people, and were originally identified as being involved in the rejection of transplants.) Healthy cells display peptides derived from self-proteins on their cell surface. By contrast, if the cell becomes infected by a pathogen such as a virus, peptides from the virus will be displayed. These viral peptides bound to MHC class I molecules can be recognized as foreign by the immune system, leading to immune attack and destruction of the infected cells.

Golgi apparatus. The Golgi apparatus is a stack of membranous compartments. Proteins that are inserted into the ER membrane or completely enter into the ER (with no membrane association) are transported to the cell surface by passage through the Golgi apparatus. During this process, the proteins can be further modified by addition of carbohydrate and other maturation processes. Consequently, the Golgi plays an important role in the maturation of the protein in processes that occur post-translationally. In particular, the types of carbohydrate that are added to the protein can affect its function. For example, carbohydrate is covalently attached to the C_H2 domains of antibodies. The type of carbohydrate can affect the binding properties of antibodies of the IgG class to FcγRs (Section 6.3). Understanding how changes in carbohydrate affect the function of an IgG represents an area of active investigation, due to the widespread use of antibodies of this class as therapeutics.

Endosome. Endosomes are membrane-bound compartments that play a major role in the trafficking of proteins and other macromolecules within the cell. Any cell will form plasma membrane invaginations that bud into the cell. The membrane invaginations form small vesicles (around several hundreds of nanometers in diameter) that enter the cell and fuse with larger early endosomes (around 0.5–2 μm in diameter). An important class of these vesicles are called clathrin-coated vesicles, in which a protein called clathrin and associated proteins play an essential role in forming the bud. These buds are formed due to the ability of clathrin to generate cage-like structures. Sometimes these small vesicles are called endosomes, but they can also be called transport containers to distinguish them from endosomes. Early endosomes are sorting stations, and can be referred to as sorting endosomes. They are characterized by a vacuolar pH that is acidic (around pH 6.0). Membrane buds can form on their surface and segregate into the cytosol as small, tubulovesicular transport containers that can travel back to the plasma membrane and fuse during a process called exocytosis. This pathway is called the endocytic recycling pathway. As these recycling compartments form, the endosome "matures" to become a late endosome. Late endosomes can be distinguished from early endosomes by the presence of different protein markers and a lower vacuolar pH. Late endosomes frequently form invaginations into the compartment itself to generate intraluminal vesicles, or ILVs, i.e., the reverse of formation of recycling compartments. These compartments are called multivesicular bodies, or MVBs, and their contents are either transferred to lysosomes for degradation or, in some cases, released as microvesicles, or exosomes, by fusion of the late endosome with the plasma membrane.

Lysosome. Lysosomes are the degradative compartments in cells. They are similar in size to sorting and late endosomes, but have a more acidic vacuole and contain degradative enzymes that break down proteins and other macromolecules. These proteins and macromolecules are delivered by the fusion of late endosomes with lysosomes or the maturation of endosomes to form lysosomes. Most lysosomal enzymes require acidic pH to function properly. Thus, the lysosome is a "dead-end" for macromolecules that are internalized and do not get recycled from early/sorting endosomes. The breakdown of these macromolecules leads to nutrient supply to the cell, and lysosomal function is closely related to nutrient availability. In specialized cells called antigen-presenting cells, late endosomes and lysosomes are also the site at which peptides are loaded onto MHC class II proteins prior to delivery to the cell surface. The MHC class II proteins are, like MHC class I, involved in immune recognition since they bind peptides from pathogens that are subsequently recognized on the cell surface by T cells. However, these peptide-MHC class II complexes are, in general, formed from peptides that are derived from proteins that have been internalized into the cell, i.e., exogenous proteins. By contrast, MHC class I molecules are involved in binding to peptides generated from proteins that are usually synthesized intracellularly, such as viral proteins following viral infection (see section on endoplasmic reticulum above). A process called cross-presentation can, however, result in the generation of peptides from internalized proteins that are then loaded onto MHC class I molecules.

Mitochondrion. Mitochondria are highly specialized organelles that are the sites of production of energy, i.e., adenosine triphosphate (ATP) molecules, of the cell. The mitochondria are also the site of an important metabolic pathway called the citric acid cycle that generates a reduced metabolite called nicotinamide adenine dinucleotide hydrate (NADH). The NADH and succinate generated by the citric acid cycle are oxidized in a process called oxidative phosphorylation. Electrons produced during this process flow through a series of electron carrier proteins that are located in the inner mitochondrial membrane, resulting in the setting up of a proton gradient across this membrane. The backflow of these protons through a protein complex called ATP synthase provides the driving force for the production of ATP.

9.2 Receptors

Receptors are proteins that can respond to external molecules such as hormones, antigens, or neurotransmitters. These molecules are called ligands. Receptors can span the plasma membrane of the cell (e.g., transferrin receptor, growth factor receptors, and Fc receptors such as FcRn). Alternatively, some receptors, such as steroid hormone receptors (e.g., androgen receptor), are present in the cytosol of the cell. In most cases, plasma membrane receptors are the focus of microscopy experiments, and these will therefore be discussed further here.

Plasma membrane receptors serve several functions. Receptors such as the transferrin receptor transport their ligands into the cell to deliver nutrients or maintain homeostasis of the ligand in the body. On the other hand, growth factor receptors such as epidermal growth factor receptor (EGFR) regulate the growth and proliferation of the cell following ligand binding. Consequently, ligand concentrations can determine the proliferative state of cells. Similarly, the levels of hormones can be sensed by cells through cellular receptors. Thus, receptors confer the ability to respond to external stimuli on a cell. In many cases, ligand binding to the receptor will induce both the downstream signaling of the receptor and the internalization of the receptor so that it is degraded. This degradation serves to reduce receptor levels so that the cell is not overstimulated, and it represents an example of feedback downregulation.

The internalization and degradation of receptors can also be induced by the binding of an antibody, rather than ligand. This has attracted much interest as a way to inhibit growth factor-stimulated proliferation of tumors. In addition, receptor internalization provides a pathway to the interior of the cell which can be used to deliver toxic drugs. As a result, the study of the biology and subcellular trafficking of receptors that are the targets of antibody-based therapeutics is an area of active investigation.

9.3 Typical biological systems

9.3.1 Subcellular trafficking of the Fc receptor, FcRn

FcRn is an unusual receptor insofar as its natural ligand, IgG, binds very weakly through its Fc region at the near-neutral pH (pH 7.3–7.4) at which most cells are bathed, whereas binding is much stronger at acidic, endosomal pH. Consequently, IgG ligand is taken into cells by any process that involves the invagination of the plasma membrane into the cell. This pinching off of the membrane occurs during pinocytosis or other internalization pathways. Pinocytosis is the process whereby the invagination of the plasma membrane leads to the ingestion of extracellular fluid that contains proteins and other macromolecules by the cell. The IgG is transported into cells in the internalized vesicles and delivered to acidic sorting endosomes in which the pH is permissive for FcRn binding. FcRn-bound IgG is subsequently sorted into vesicular transporters that are delivered to the plasma membrane by exocytosis. The near-neutral extracellular pH for most cell types results in release of IgG. This FcRn-mediated sorting of IgG plays a central role in regulating IgG levels in the body. For example, if IgG levels increase above normal levels, FcRn becomes saturated and is no longer available to bind to IgG following IgG internalization. This unbound IgG is not sorted into the recycling pathway and instead enters late endosomes and lysosomes whereupon it is degraded.

Another important function of FcRn is to transport IgG molecules across cellular barriers, such as the gut epithelium, in a process called transcytosis. Indeed, FcRn is known as the neonatal Fc receptor due to its early identification in the 1980s as the transporter of IgG from mother's milk to the bloodstream of suckling rodents via gut epithelial cells. This IgG transport provides immunity during the postpartum phase (Note 9.1).

9.3.2 Subcellular trafficking of the transferrin receptor

All cells need iron for normal metabolic function. Iron is delivered in the form of ferric (Fe^{3+}):transferrin complexes ("holotransferrin") to cells. Holotransferrin binds to the transferrin receptor at the cell surface, is internalized and enters acidic sorting endosomes in cells. The acidic pH results in dissociation of the Fe^{3+} from transferrin to generate apotransferrin (i.e., iron-free transferrin). The released Fe^{3+} is reduced to ferrous iron (Fe^{2+}) and delivered into the cytosol by a transporter protein. The transferrin receptor:apotransferrin complex is subsequently recycled back to the plasma membrane in vesicular transporters that also contain FcRn:IgG complexes and fuse with the membrane by exocytosis. Importantly, the exocytic process leads to exposure of the transferrin receptor:apoptransferrin to the extracellular environment which is generally at near-neutral pH (pH 7.3–7.4), resulting in release of the apotransferrin. Following reloading of the apotransferrin with Fe^{3+}, the holotransferrin can bind at near-neutral pH to the transferrin receptor and undergo the cycle of internalization, Fe^{3+} release, and recycling.

9.4 Sample preparation

For the microscopic analysis of the distribution of one or more proteins in a cell, it is necessary to label the proteins with a fluorescent tag. The different kinds of labeling approaches that can be used to achieve this are discussed in Section 8.7. In the current section, we

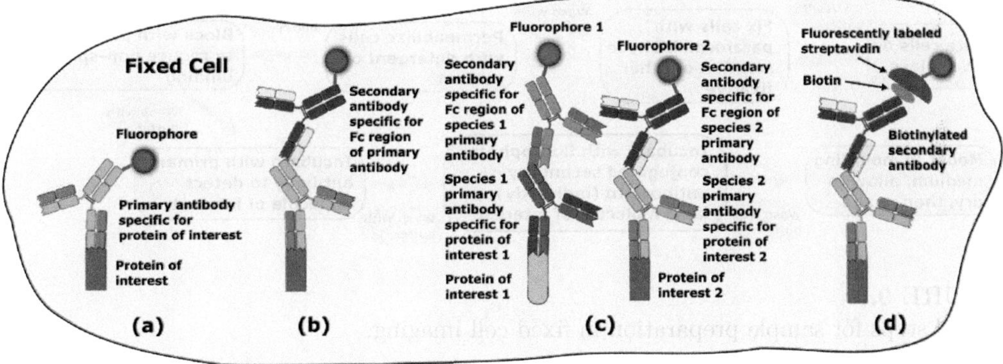

FIGURE 9.4
Common ways of labeling a protein of interest in a fixed cell. (a) Labeling is achieved with a fluorophore-conjugated primary antibody that is specific for the protein of interest. (b) Labeling is achieved with a fluorophore-conjugated secondary antibody that binds to the Fc region of the primary antibody. (c) Two different proteins of interest can be labeled using the approach of (b) when they can be detected with primary antibodies from different species. (d) Labeling can also be achieved with a biotinylated secondary antibody that binds to fluorophore-conjugated streptavidin.

present more detailed methods and several specific experiments that demonstrate how proteins can be labeled for the analysis of both fixed (dead) and live cells, using FcRn and the transferrin receptor as examples.

9.4.1 Labeling of proteins in fixed cells

Prior to the development of approaches for labeling specific proteins in live cells, detection of these macromolecules in fixed cells was the predominant method used in cell biology. A prerequisite for these experiments is almost invariably that an antibody specific for the protein of interest is available. This can be a monoclonal antibody (Section 6.4.1) or a polyclonal antibody mix. An antibody is, regardless of its source or route of generation, usually highly specific for the target protein so that it only binds to this protein and not to other macromolecules in the cell. Such an antibody can then be used to accurately localize a protein in a cell. These antibodies are called primary antibodies. For detection one can use either an antibody that is directly labeled with a fluorophore such as an Alexa Fluor dye (Fig. 9.4(a)), or a so-called labeled "secondary" antibody (Fig. 9.4(b)). Many labeled secondary antibodies are commercially available, and they specifically bind to the Fc region of the detection antibody. The Fc regions of antibodies from different species are sufficiently distinct to allow secondary antibodies that are specific for only one species to be generated. The use of a secondary antibody allows amplification of the fluorescent signal, since multiple secondary antibodies can bind to each primary antibody. In addition, multiple proteins can be detected in the same cell simultaneously if specific primary antibodies from different species are available and used in conjunction with secondary antibodies that only bind to primary antibodies from a particular species (Fig. 9.4(c)). Instead of a labeled secondary antibody, one can also use a secondary antibody that is conjugated to biotin (Fig. 9.4(d)). In this case, the fluorophore is attached to streptavidin, which binds specifically and strongly to biotin (Fig. 8.8). As is the case with the use of fluorophore-labeled secondary antibodies, this approach allows signal amplification. Specifically, multiple labeled streptavidin molecules

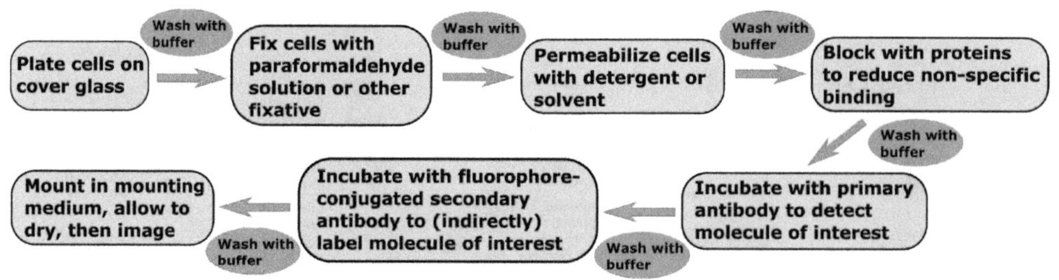

FIGURE 9.5
Typical steps for sample preparation in fixed cell imaging.

can bind to a biotinylated secondary antibody, which usually has more than one biotin attached to it.

9.4.2 Sample preparation for typical fixed cell experiments

Cells are first grown on cover glasses and then treated with chemicals such as paraformaldehyde solution that immobilize all macromolecules in the cells by extensive covalent cross-linking. This treatment is called fixation and kills the cells. The cells are then usually "permeabilized" with either a detergent such as saponin or a solvent such as methanol. The permeabilization step is required if proteins within the cell are to be detected. Without permeabilization, only proteins that are exposed on the plasma membrane of the cell will be accessible for binding to antibodies. Cells are then treated with a mixture of proteins in a blocking step that is designed to reduce nonspecific binding. This mixture usually contains proteins such as bovine serum albumin, which tend to bind to "sticky" or hydrophobic proteins. Following blocking, the cells are incubated with one or more detection antibodies, also referred to as primary antibodies. Bound primary antibodies are detected with one or more secondary antibodies. Each of these steps is followed by a washing step to remove excess fixation reagent, unbound primary antibody, etc.

An extremely important final step of fixed cell sample preparation is to treat the sample with mounting medium. The mounting medium not only allows the cover glass to be mounted on a glass slide to facilitate the subsequent microscopic imaging, but also has properties that reduce photobleaching during imaging. The specific properties will depend on the particular type of mounting medium, since some media are better than others for fluorophores with different spectral properties. A schematic summarizing the method described is shown in Fig. 9.5.

One type of fixed cell experiment that is frequently used for the initial characterization prior to live cell imaging is to analyze the distribution of a protein that is tagged with a fluorescent protein such as GFP. For example, to determine the localization of FcRn in cells, the cells are first transfected (Section 7.2.4) with an expression plasmid encoding FcRn-GFP. The cells are then fixed and permeabilized, so that antibodies that bind to markers of early or sorting endosomes and lysosomes (early endosome antigen 1, or EEA1, and lysosomal-associated membrane protein 1, or LAMP1, respectively) can enter the cells and bind to their intracellular targets. In Fig. 9.6, imaging data obtained with FcRn-GFP and either anti-EEA1 antibody or anti-LAMP1 antibody is shown. The anti-EEA1 and anti-LAMP1 antibodies used in this example are mouse antibodies and can therefore be detected with a fluorescently labeled secondary antibody that binds to mouse antibody constant regions. The data demonstrates that FcRn is primarily present on EEA1-positive endosomes, and cannot be detected in LAMP1-positive lysosomes.

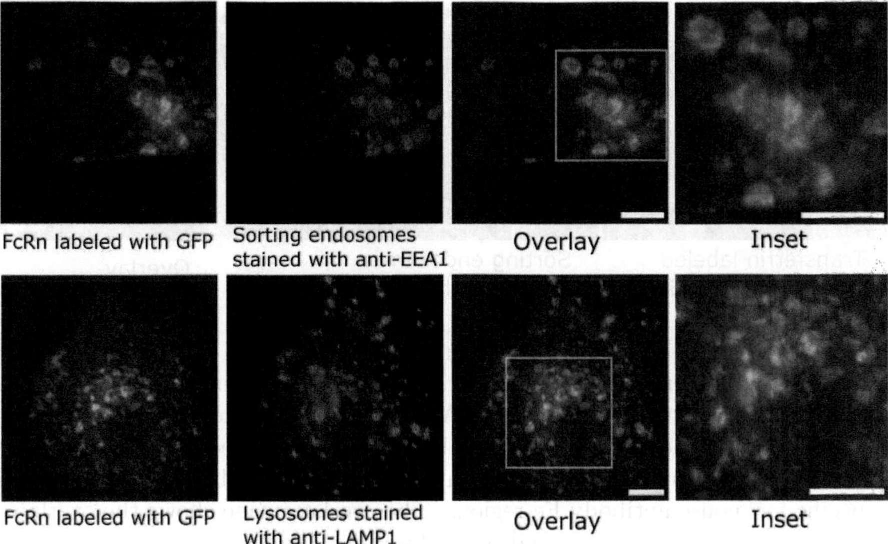

FIGURE 9.6
Fluorescence microscopy images showing the distribution of FcRn in HMEC-1 cells transfected with GFP-tagged FcRn. In the top row, FcRn labeled with GFP and sorting endosomes stained with mouse anti-EEA1 antibody are shown separately in green and red, respectively. In the overlay of the green and red images, and even more clearly in the inset from the overlay image, FcRn can be seen from the good agreement between the green and red signals to be localized in the sorting endosomes. In the bottom row, FcRn labeled with GFP is again shown in green, and lysosomes stained with mouse anti-LAMP1 antibody are shown in red. In this case, it can be seen from the poor overlap between the green and red signals in the overlay image that FcRn is not localized in lysosomes. In both cases, the fluorescence signal corresponding to the stained cellular compartments is obtained with an Alexa Fluor 568-labeled secondary antibody that is specific for mouse antibody Fc regions. Scale bars: 5 μm. [See Figure Credits.]

An example of a fixed cell experiment related to studying the transferrin receptor, in which cells are pre-treated with a fluorescently labeled protein prior to fixation and subsequently stained with an antibody, would be the use of labeled transferrin and an anti-EEA1 antibody. The cells are incubated with fluorescently labeled transferrin under physiological conditions (i.e., at 37°C in a controlled atmosphere). Following uptake into cells via binding to the transferrin receptor, transferrin accumulates in early endosomes, or sorting endosomes, from which it will be recycled out of the cells. Consequently, incubation of cells with labeled transferrin, followed by a short wash-out phase, results in localization of transferrin to sorting endosomes. In the example shown in Fig. 9.7, these sorting endosomes are detected using an antibody specific for EEA1 using the same approach as in Fig. 9.6. This experiment leads to the conclusion that the transferrin receptor (and its ligand, transferrin) traffick into sorting endosomes in cells.

9.4.3 Sample preparation for typical live cell imaging experiments

Live cell experiments differ from fixed cell experiments in several fundamental respects. First, the cells are plated in specialized dishes that usually have glass cover slips attached

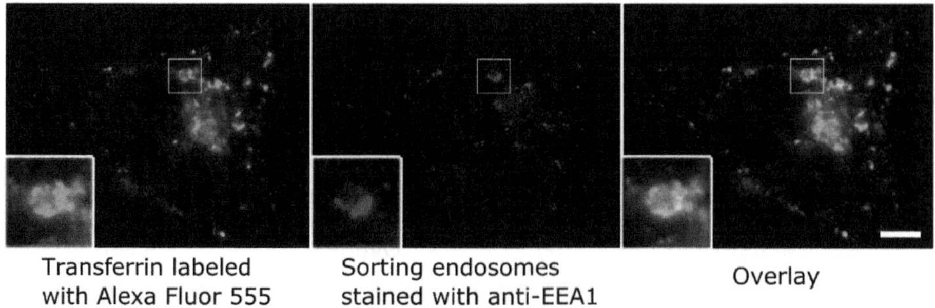

FIGURE 9.7
Fluorescence microscopy images showing the localization of transferrin on sorting endosomes in an HMEC-1 cell. Transferrin labeled with Alexa Fluor 555 is shown in green, and sorting endosomes stained with mouse anti-EEA1 antibody are shown in red. The fluorescence signal for the sorting endosomes is obtained with an Alexa Fluor 647-labeled secondary antibody that is specific for mouse antibody Fc regions. The overlay image shows the overlap of the fluorescence signals from transferrin and the endosomes. The inset shows a single endosome with localized transferrin. Scale bar: 6 μm.

to the bottom. This allows the cells to be maintained in their growth medium during the imaging experiment. An alternative, but less frequently used approach, is to illuminate the cells from the upper side using a dipping objective that can be submerged in the medium. Second, the cells usually express fluorescently labeled proteins. In the case of the use of fluorescent proteins such as GFP, this is frequently achieved by transfection (Section 7.2.4), in which an expression construct that links the gene encoding the protein of interest to the gene encoding the fluorescent protein is introduced into the cells. An alternative to the use of proteins that are genetically tagged with fluorescent proteins such as GFP is the use of SNAP or CLIP tags (Section 8.7.1.3). Nevertheless, the use of these tags also necessitates the transfection of an expression construct encoding the protein of interest with the SNAP or CLIP peptide sequence. Third, since mounting medium is not used in these experiments, photobleaching becomes a significant problem. This limitation can be reduced in several ways, including the use of low excitation light levels and more photostable fluorophores. For example, QDs (Section 8.7.2) are significantly more photostable than traditional organic dyes. Fourth, the cells need to be maintained under physiological conditions during the imaging experiments. This is frequently accomplished by placing the microscope in a temperature-controlled chamber (at 37°C for mammalian cells) with a regulated atmosphere (e.g., 5% carbon dioxide). Alternatively, the temperature of imaging dishes can be maintained at 37°C by using an objective warmer.

An example of a live cell imaging experiment in the context of our model systems is to compare the intracellular fate of two antibodies that do and do not bind to the Fc receptor, FcRn, because they have different Fc regions. For this, endothelial cells can be transfected with GFP-tagged FcRn and then incubated with Alexa Fluor-labeled antibodies. One of these antibodies is of the IgG1 subclass and binds with good affinity to FcRn. The second antibody is a mutated variant in which a histidine residue in the Fc region has been changed (mutated) to an alanine residue to eliminate binding to FcRn. Transfected cells are incubated with these antibodies, and after washing out excess antibodies, the cells are imaged at physiological temperature (37°C). The images shown in Fig. 9.8 show individual frames from typical movies that would be obtained in such an experiment. As expected, FcRn-GFP can be detected in circular sorting endosomes. The Alexa Fluor-

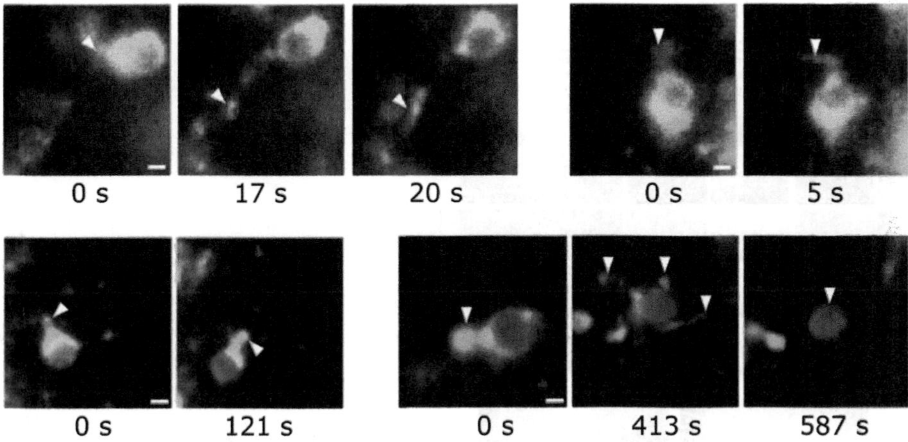

FIGURE 9.8
Live cell fluorescence microscopy images showing the sorting of IgGs by FcRn in endosomes of HMEC-1 cells transfected with GFP-tagged FcRn. Frames from movies of four sorting endosomes are shown with the time indicated below each frame. In each of the two sets of frames shown in the top row, GFP-labeled FcRn (green) and Alexa Fluor 546-labeled IgG1 (red) are seen to move away (see arrows) together in tubules from the sorting endosome. In each of the two sets of frames shown in the bottom row, the IgG1 (red) is of a mutated form that does not bind to FcRn (green), and therefore does not leave the sorting endosome in tubules containing FcRn (see arrows). Instead, the mutated IgG1 remains in the vacuole of the endosome. Note that at time point 587 s in the set of frames on the right, the sorting endosome no longer has a detectable level of FcRn, and only the mutated IgG1 is detected in the structure (see arrow). Scale bars: 1 μm. [See Figure Credits.]

labeled IgG1 is associated with FcRn-GFP on the periphery of these compartments, and can also be detected in tubulovesicular recycling compartments that are FcRn-positive and separate from the sorting endosomes. These recycling compartments will transport FcRn and bound IgG1 to the plasma membrane of the cell to undergo exocytosis. By contrast, the distribution of the mutated IgG1 is distinct. This antibody accumulates inside the sorting endosomes and does not overlap with FcRn. In addition, FcRn-positive tubulovesicular recycling compartments leave the sorting endosomes without detectable mutated IgG1. Consequently, the IgG1 mutant remains in the sorting endosome, which matures to become a late endosome. These late endosomes will subsequently fuse with lysosomes to deliver their contents, such as the IgG1 mutant, for degradation.

More recently, the use of QDs to label proteins in cells for live cell imaging has become very common. QDs can be coupled to target proteins in several different ways (Section 8.7.2). An experiment in which individual QD-labeled IgGs are tracked as they enter the cell and go into sorting endosomes is shown in Fig. 9.9. The IgG used in this experiment is engineered so that it binds to FcRn molecules on the plasma membrane. Cells were transfected with fluorescent protein-tagged FcRn and subsequently incubated with very low concentrations of IgG that had been site-specifically tagged with a biotin molecule (using the Bsp tag and BirA method; see Section 8.7.1.2) and linked to QDs coated with streptavidin (which binds tightly to biotin; see Section 8.7.2). Low concentrations of QD-IgG complexes were used to allow the tracking of single IgG molecules. The cells were imaged using multifocal plane microscopy (MUM) (Sections 10.2.2 and 11.7). In Fig. 9.9, a QD-labeled IgG molecule can be seen moving on the plasma membrane, followed by arrest of movement and internalization

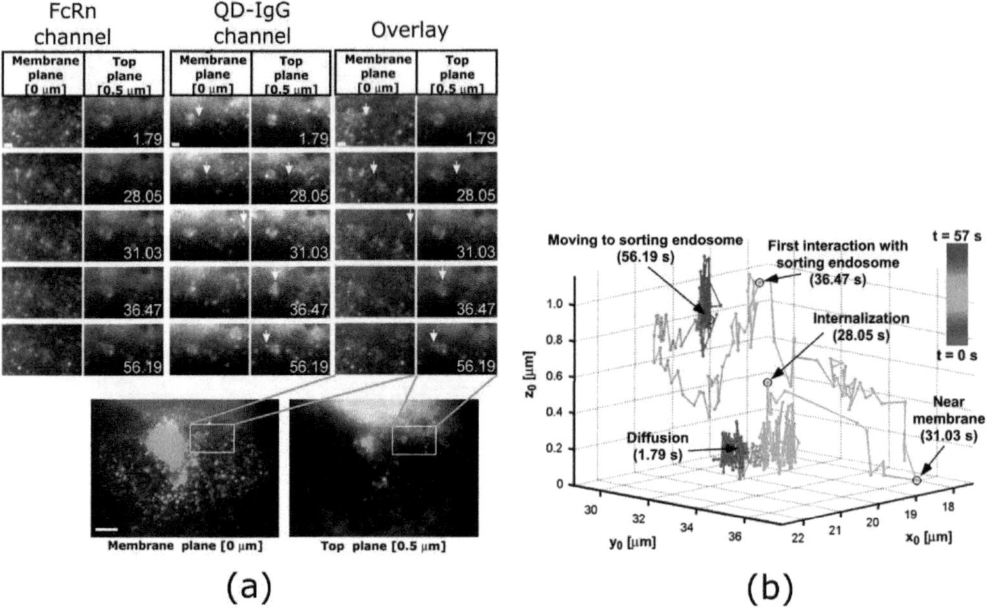

FIGURE 9.9
3D trajectory of a QD-labeled IgG molecule in an HMEC-1 cell transfected with fluorescent protein-tagged FcRn, imaged using multifocal plane microscopy. (a) Images of the fluorescent protein-tagged FcRn and QD-IgG channels, along with their overlay, acquired at time points (shown in seconds) corresponding to the five events of interest highlighted in (b). In each row, the pair of images of each channel were simultaneously acquired at a focal plane that coincides with the plasma membrane and a (top) focal plane that is 0.5 μm above the plasma membrane. Moreover, in the pair of overlay images, fluorescent protein-tagged FcRn is shown in green and QD-IgG is shown in red. The tracked QD-IgG is indicated by a white arrow. Scale bars: 1 μm. Below the rows of images, full-sized images of both focal planes are shown with a white box that indicates the region in the cell in which the trajectory of the QD-IgG is observed. The red haze seen in both focal planes is due to the presence of QD-IgGs in the imaging medium. Scale bar: 5 μm. (b) 3D trajectory of the tracked QD-IgG of (a) that is color-coded to indicate time. The color change from red to green to blue represents increasing time. Each of the five QD-IgG positions indicated by arrows corresponds to a row of images in (a). The QD-IgG is initially observed to randomly diffuse on the plasma membrane. Internalization of the QD-IgG into the cell is then seen as an abrupt change in its z position that indicates movement into the cell to a distance of 0.3 μm above the plasma membrane. After internalization, the QD-IgG is seen to move in a highly directed manner, taking an elaborate route to traffick deep inside the cell (0.8 μm above the plasma membrane) until it reaches a sorting endosome. The QD-IgG is then seen to briefly interact with the sorting endosome, loop around it, and after several repeated contacts, merge with the sorting endosome. [See Figure Credits.]

into the cell. This QD-labeled IgG then moves between the two focal planes that are being imaged before entering a sorting endosome that can be seen in the upper focal plane. The fluorescence image data is shown in panel (a) of the figure, and the trajectory of the QD-labeled IgG, determined using a localization algorithm (Section 19.6) for the analysis of MUM data, is presented in panel (b) of the figure (Note 9.3). For additional examples of live cell imaging using MUM, see Section 11.7.

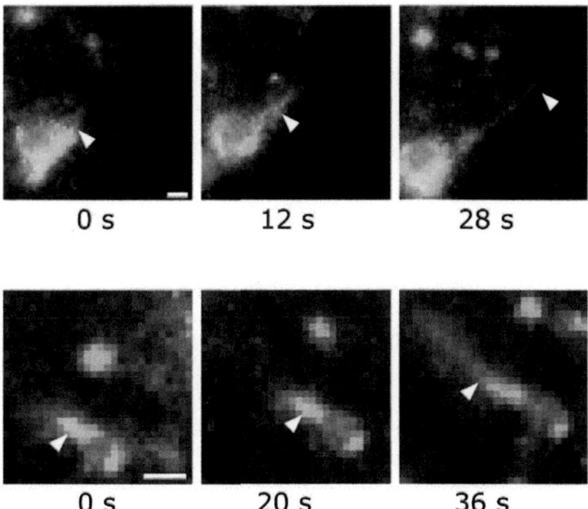

FIGURE 9.10
Live cell fluorescence microscopy images showing Alexa Fluor 546-labeled transferrin (red) and GFP-labeled FcRn (green) in tubular structures (see arrows) that extend from a sorting endosome. The top and bottom rows show frames from movies of two different HMEC-1 cells that have been transfected with GFP-tagged FcRn, with the time indicated below each frame. In each case, the interior of the sorting endosome is occupied by Alexa Fluor 647-labeled dextran (blue), which does not traffick with transferrin and FcRn. Scale bars: 1 μm. [See Figure Credits.]

In another example of a live cell imaging experiment, the question as to whether FcRn and transferrin (and transferrin receptor) separate from sorting endosomes into the same tubulovesicular recycling compartments was addressed. Cells were transfected with GFP-tagged FcRn and incubated with fluorescently labeled dextran and transferrin under physiological conditions. Dextran has no cellular receptor and is taken into cells by fluid phase processes (i.e., plasma membrane invaginations that encompass the extracellular medium) and accumulates in the vacuole of sorting endosomes. Following this treatment with fluorescently labeled dextran and transferrin, the cells were washed to remove the fluorescently labeled material that is outside the cells and imaged. In Fig. 9.10, the separation of transferrin and FcRn from sorting endosomes into tubulovesicular recycling compartments, without detectable levels of dextran, can be observed. This experiment shows that FcRn and transferrin (receptor) follow the same recycling pathway in cells.

Notes

Chapter 5

1. For additional information about the landmark work that solved the 3D structure of DNA, the reader is referred to [100, 124].
2. The 3′ end of a DNA strand refers to the third carbon atom in the sugar ring attached to a hydroxyl group, whereas the 5′ end corresponds to the fifth carbon atom attached to a phosphate group.
3. The 3D structure of a ribosome has been solved and is described in more detail in [12, 95, 108].
4. Codons in a gene can be changed, or mutated, using site-directed mutagenesis. This technique involves the alteration of one or more bases in one or more specific codons of a gene with the goal of changing the targeted amino acid(s) that are encoded by these codons. Multiple experimental approaches have been developed over the past several decades to achieve this (e.g., [17]).
5. The structure of the protein 3CLpro is described in [4].

Chapter 6

1. For more details of the process of somatic mutation, see [35, 129, 120].
2. For further details concerning the behavior of antibody (sub)classes, see [71, 18].
3. For a historical account of the failure to patent Milstein and Köhler's hybridoma technology, see [65].

Chapter 7

1. For more details concerning expression plasmids and protein production, see [125].
2. The 2018 Nobel Prize in Chemistry (half-share with Frances Arnold) was awarded to Sir Gregory Winter and George P. Smith for their demonstration that peptides and antibody fragments could be displayed on the surface of phage, enabling the selection of peptides or antibody fragments with specific binding activities. This technology has resulted in the isolation of therapeutic antibodies that are fully human in sequence.
3. Flow cytometry (Fig. A) involves the labeling of cells or other particles with fluorescently tagged antibodies or other proteins that potentially bind to cellular components. After labeling, the cells are analyzed using a flow cytometer, in which they are passed through an excitation light beam as individual cells. The fluorescence emitted by each cell is detected using a detector such as a photomultiplier tube (Section 12.2). This technique allows the levels of fluorescence associated with each cell in a population to be analyzed, and very large numbers of cells can be characterized per minute. In addition, by using lasers with different excitation wavelengths and emission filters (Section 10.7) with different passbands, multiple distinct fluorescent labels can be studied at the same time.

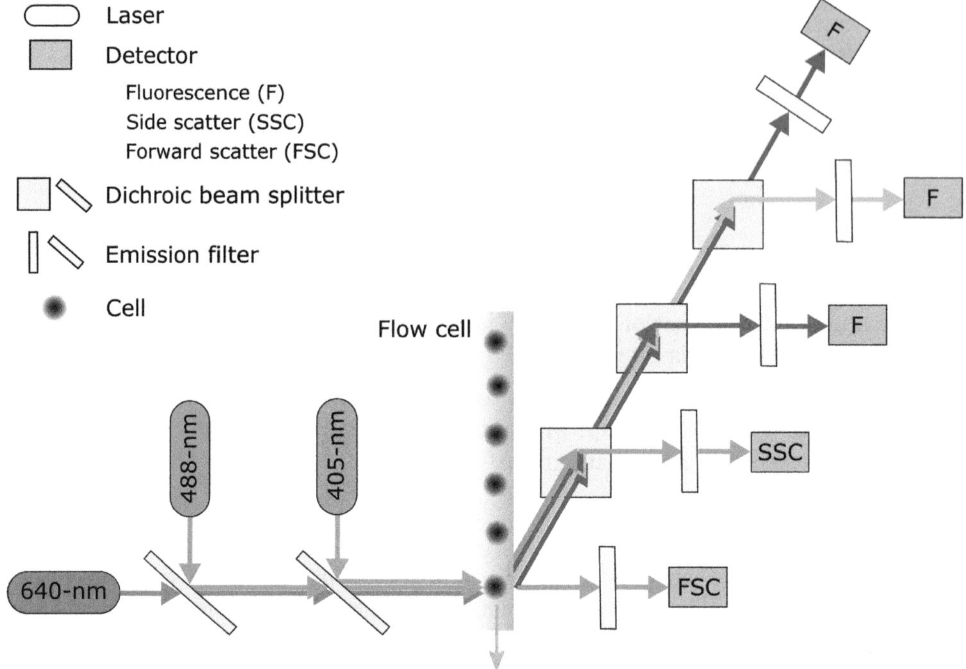

FIGURE A
Flow cytometry. In this technique, cells in a liquid stream contained in a flow cell are individually passed through an excitation light beam, allowing the fluorescence levels associated with each cell to be analyzed individually. This schematic illustrates the use of lasers of different wavelengths and an appropriate set of dichroic beam splitters, emission filters, and detectors to analyze the levels of fluorescence associated with the distinct fluorophores with which a cell may be labeled. Also shown in the schematic is the use of two additional detectors to record excitation light (typically from the 488-nm laser) that is scattered by a cell. The forward scatter (FSC) detector measures the intensity of light that is scattered along the direction of the laser beam, while the side scatter (SSC) detector, which is positioned in a direction that is orthogonal to the laser beam, measures the intensity of light that is scattered in directions outside of the laser beam path. The FSC signal provides a measure of the size of a cell as its intensity is proportional to cell size, and the SSC signal is an indicator of the internal complexity of a cell, as cells with a more granular morphology are expected to produce more side-scattered light. Whereas the fluorescence and SSC signals are commonly detected using photomultiplier tubes, the typically strong FSC signal is commonly detected with a photodiode which is less sensitive but also less expensive.

Most flow cytometers, however, do not give spatial information concerning the location of the fluorescence within the cell.

As shown in Fig. A, a flow cytometer can also be used to analyze the size and morphological complexity of each cell through the measurement of the intensity of excitation light that is scattered by the cell. Moreover, not illustrated in the figure is the ability of some flow cytometers to physically sort the cells into different populations based on their fluorescence and light-scattering characteristics. Often referred to as a cell sorter, such a flow cytometer assigns an electrical charge to each cell as determined by its measured optical properties, and uses electromagnets to divert the cell into the appropriate container.

4. For more details concerning antibody display and selection technologies, see [59].

Chapter 8

1. For more on the types and spectral properties of fluorophores used in fluorescence applications, see, e.g., [61].

2. FRET refers to the physical phenomenon in which a fluorophore that absorbs excitation light transfers the excitation energy to a nearby fluorophore instead of emitting fluorescence. The fluorophore that transfers the excitation energy is referred to as the donor, and the nearby fluorophore on the receiving end of the transfer is referred to as the acceptor. As the donor and acceptor are typically chosen to have well-separated excitation spectra and well-separated emission spectra, the absence and occurrence of FRET can be detected as fluorescence from the donor and the acceptor, respectively.

   The probability that an excited donor molecule transitions back to the ground state via energy transfer to an acceptor molecule instead of fluorescence emission is termed the FRET efficiency. Due to the fact that the FRET efficiency is proportional to the inverse of the sixth power of the distance that separates the donor and the acceptor, FRET typically occurs only when the donor and acceptor are no more than a few nanometers apart. FRET is therefore a valuable tool for probing nanometer-scale distances between macromolecules or within a macromolecule. FRET-based spectroscopy techniques, for example, are commonly employed for the investigation of a protein's conformational states. By attaching a donor and an acceptor fluorophore to the protein of interest, and by simultaneously but separately measuring the fluorescence emitted by both the donor and the acceptor over time, the FRET efficiency can be estimated for each detected protein molecule as the ratio of the number of photons detected from the acceptor to the total number of photons detected from both donor and acceptor. A common analysis is then to infer the distribution of the different conformations of the protein of interest from a histogram of the estimated FRET efficiencies. Assuming, for example, that the protein is labeled in such a way that a small distance between the donor and acceptor molecules indicates a folded protein structure and a larger distance indicates an unfolded structure, two prominent peaks might be observed in the histogram — one that corresponds to a high FRET efficiency and represents the population of folded protein molecules, and one that corresponds to a lower FRET efficiency and represents the population of unfolded protein molecules.

   FRET-based spectroscopy is commonly implemented using a confocal microscope (Section 10.4) or, in the case of a widefield microscope, a TIRFM setup (Section 11.4). In more advanced applications that require more sophisticated experimental setups, data besides fluorescence intensities can be collected that, for example, relate to the fluorescence lifetime of the donor fluorophore or the rotational mobility of the donor and acceptor fluorophores. For much more in-depth information relating to FRET and FRET-based spectroscopy techniques, including schemes involving more than two fluorophores, guidelines for donor and acceptor fluorophore selection and sample preparation, and details regarding various types and methods of data analysis, we refer the reader to the literature (e.g., [101, 98]).

3. For a comprehensive discussion of the properties of QDs and the advantages and disadvantages of using QDs as fluorescent labels in biological imaging, see, e.g., [122]. Despite their extended photostability relative to that of traditional fluorophores, QDs have been reported to photobleach. In addition, the wavelength of their fluorescence emission has been observed to gradually decrease over time, a phenomenon that could amount to an effective reduction or loss of detectable signal in the wavelength window of an emission filter. For investigations of the photobleaching and "blue shift" properties of QDs, see, e.g., [117, 32, 6].
4. The first demonstration that GFP can be expressed in prokaryotic and eukaryotic cells, and that it can therefore be used for protein localization in living cells, was reported in [22]. The development of the GFP variant that has its major excitation peak at 488 nm was reported in [48].
5. Details relating to the isolation of DsRed, the development of mRFP1, and the engineering of mRFP1 variants can be found in [67], [20], and [102], respectively.
6. For details relating to the use of conventional organic dyes as photoswitchable fluorophores, see [47, 116].
7. The development of PA-GFP is described in [78].

Chapter 9

1. For more details concerning FcRn and IgG trafficking in cells, see [23].
2. The fixed cell experiments presented in Section 9.4.2 that investigate the distribution of FcRn and the live cell experiments presented in Section 9.4.3 that study the sorting of IgGs by FcRn and the recycling of FcRn and transferrin are described and discussed in detail in [74].
3. For further details of the MUM experiment described in Section 9.4.3 for the 3D tracking of QD-labeled IgG, see [88].

Exercises

Chapter 8

1. Given that the time to photobleaching of each fluorescent molecule in an ensemble has an exponential probability distribution, show that the photobleaching of the ensemble is described by an exponential decay function.
2. Determine the effective absorption cross sections for the dyes FITC and TRITC, assuming their molar extinction coefficients are as given in Section 8.5.2.
3. The absorbance of a liquid sample of FITC was found to be 0.558. Given that the path length used for the determination was 2 cm, and assuming that the molar extinction coefficient for FITC is as given in Section 8.5.2, what was the molar concentration of FITC in the liquid sample?
4. For a path length of 3 cm and a molar extinction coefficient for TRITC as given in Section 8.5.2, calculate the molar concentration of TRITC in a liquid sample that will result in the absorption of 65% of the light received by the sample.
5. The absorbance of a liquid sample of a fluorophore was determined to be 0.425. What percentage of the light received by the sample was absorbed by the sample?
6. Fluorophores A and B are excited at the same wavelength under identical experimental conditions. Fluorophore A has a molar extinction coefficient of 68000 $M^{-1}\text{cm}^{-1}$ and a quantum yield of 0.77, whereas fluorophore B has a molar extinction coefficient of 75000 $M^{-1}\text{cm}^{-1}$ and a quantum yield of 0.69. Which fluorophore has the greater brightness?

Part III

Optics and Microscopy

Overview

Having introduced the ideas from cell biology and chemistry that are necessary for our treatment of cellular microscopy, we are now in a position to discuss microscopy instruments and present the corresponding optical principles. In the prior part, we have also addressed the preparation of samples for microscopy experiments. Here we will be in a position to bring all these threads together and examine in some detail different prototypical microscopy experiments. Again we will be able to illustrate many of the key ideas within the context of exploring our example biological systems, i.e., the elucidation of the role of the receptor FcRn in the transport of antibody molecules and the transport of iron by means of the transferrin system. We will also discuss the detectors commonly used for image acquisition in microscopy, with particular emphasis given to topics relevant to the analysis of the data they produce. Additionally, we will take a fairly close look at aspects of image formation in a microscope according to geometrical optics and diffraction theory.

10

Microscope Designs

In Sections 3.2 and 3.3, we encountered the basic layout of microscopes for both transmitted light microscopy and fluorescence microscopy. In the current chapter, we will begin by discussing the light path of a widefield microscope, the light path of a confocal microscope, and configurations for imaging in three dimensions and for capturing multiple colors. Besides allowing for a discussion of various microscopy modalities, this will set the stage for a more in-depth presentation of the individual components of a microscope and for further theoretical considerations of the optical principles behind image formation in a microscope.

10.1 Light path for widefield fluorescence microscopy

We have already encountered the basics of the light path for an epifluorescence microscope in Section 3.3 of the introductory part to this book. Importantly, the excitation light reaches the sample through the objective, and the light emitted by the fluorescent molecules in the sample is also collected by the objective (Fig. 3.7). This necessitates a critical element in the light path, the dichroic beam splitter, which allows the excitation light to be reflected into the back aperture of the objective. (The back aperture refers to the opening through which light enters and exits the rear side of the objective.) At the same time, the dichroic beam splitter allows the emitted light to pass through to reach the detector or the eyes of the microscopist. This important optical element will be discussed in detail in Section 10.7.

The excitation light, be it through a lamp, laser, or LED (Section 3.5.1), reaches the sample through the objective after being reflected by the dichroic beam splitter. It is often also important that the light passes through an excitation filter to ensure that only the desired wavelengths of light reach the sample. The light emitted by the fluorescent labels in the sample is collected by the objective, passed through the dichroic beam splitter, and then focused onto the detector. An emission filter is often necessary to ensure that only light within the wavelength range that matches the emission wavelength of the fluorophore reaches the detector, and that scattered light, sample autofluorescence, and other sources of background light are suppressed. Scattered light refers to any light, including excitation light, any emitted fluorescence, and ambient light, that is scattered by the surfaces of the optical elements in the microscope or by the sample. Sample autofluorescence is the fluorescence emitted by light-absorbing molecules that naturally occur in the sample.

10.1.1 Infinity-corrected light path

Older microscopes have a design of the light path such that the objective focuses the image directly onto the detector or, if the sample is observed by eye, onto an intermediate image plane that is then observed by the microscopist through an eyepiece (Fig. 10.1). This design has several problems as changing the position of the objective or inserting additional optical

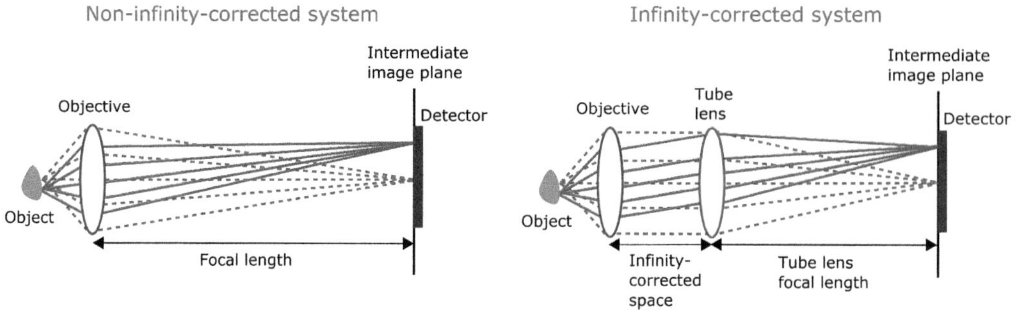

FIGURE 10.1
Comparison of a non-infinity-corrected and an infinity-corrected optical system. The paths of two sets of rays originating from two different points in the imaged object are shown in each system. In the non-infinity-corrected system, the rays in each set are focused by the objective directly onto the intermediate image plane where the detector is positioned. In the infinity-corrected system, they are mapped to parallel rays by the objective, which are then focused by the tube lens onto the intermediate image plane.

elements in the light path between the objective and the detector can severely impact the optical system, for example through changing its focus level.

To address these problems, modern microscopes use infinity-corrected optics. In this design, the objective is in fact split up into two parts. An infinity-corrected objective focuses the sample to infinity. This means that rays emanating from a point in focus of the objective are mapped to parallel rays by the objective. These parallel rays are then focused onto the detector or intermediate image plane by a focusing lens called the *tube lens*. The space between the objective and the tube lens is referred to as the *infinity-corrected space*. The focal length of the tube lens varies somewhat depending on the manufacturer, with Zeiss using a focal length of 165 mm, Olympus using a focal length of 180 mm, and Nikon and Leica using a focal length of 200 mm. It is also important to note that some manufacturers, such as Zeiss, also use the tube lens for color correction purposes.

10.2 Imaging in three dimensions

For a specific position of the objective with respect to the sample, a specific focal plane in the sample will be imaged. Many samples, however, have an appreciable thickness that requires the imaging of different focal planes in order to obtain images of the parts of the sample that are outside the current focal plane. The imaging of different focal planes within a sample can be achieved by sequentially changing the focus position of the microscope, or by simultaneously acquiring images of distinct focal planes using a non-conventional microscope configuration.

10.2.1 Focus control and acquisition of z-stacks

The traditional method for imaging different focal planes within a sample involves changing the focus position of the optical system. This can be done by changing the position of the objective with respect to the sample, or by changing the position of the sample with respect to the objective. The first option is the one typically used on inverted microscopes. Changing the objective position can either be done using a manual focus knob on most systems, or with motorized focus systems that use mechanical mechanisms or piezo devices. For the second option, a stage is used that allows the sample to be moved. Mechanical and piezo-based stages are available that differ in capability and cost. With these devices, increments in the low tens of nanometers are reliably achievable.

Using the coordinate system that is typically applied in microscopy, whereby the xy planes lie parallel to the focal plane and the z-axis coincides with the optical axis, a set of acquisitions with different focus positions as specified along the optical axis is referred to as a *z-stack*.

10.2.2 Multifocal plane microscopy

The use of mechanical or piezo devices to change the focus position of the optical system is particularly suitable for the 3D imaging of a fixed cell sample. For the 3D imaging of a live cell sample, however, this method is often inadequate as intracellular events of interest typically involve objects such as vesicles and proteins that move at a much faster rate than the speed at which the focus position can be changed. Important events that occur concurrently at different depths within the cell sample can therefore be missed while the current focal plane is being imaged or while the focus position of the optical system is being changed. An alternative 3D imaging approach that addresses this important problem is to simultaneously image different focal planes within the cell sample.

The simultaneous imaging of different focal planes, or multifocal plane microscopy (MUM), can be implemented by partitioning the fluorescence collected from the sample between detectors that are positioned at different distances from the tube lens. The different positions of the detectors with respect to the tube lens correspond to planes in the image space that have unique conjugate planes in the object space, thereby allowing the imaging of a distinct focal plane within the sample by each detector. An illustration of a four-plane MUM setup is shown in Fig. 10.2.

10.3 Imaging of multiple colors

When a single fluorescently labeled protein is to be imaged in a sample, a standard microscope system configuration with a single detector will typically suffice. Very often, however, multiple proteins that are distinguished by fluorophores of different colors need to be imaged in a sample (see, for example, the experiments described in Chapters 2 and 9.) In such cases, a single-detector system that images one color at a time may also be used, but will typically involve additional software-controlled hardware devices that control the excitation light that illuminates the sample and the fluorescence that reaches the detector during the acquisition of a given image. In such a single-detector system for two-color imaging, for example, the high-level illustration of Fig. 10.3(a) shows that a software-controlled mechanism is used to ensure that at any given time, the sample is illuminated by a range of wavelengths that are appropriate for the excitation of one of the labeled proteins of interest. By synchronizing the selection of excitation wavelengths with the exposure of the detector, a sequence of images is acquired that consists of alternating images, or sets of images, of the two proteins.

An obvious limitation of imaging one color at a time is that none of the labeled proteins of interest can be continuously imaged in time. Furthermore, the need to operate additional hardware components will often mean a slower overall image acquisition. A solution that

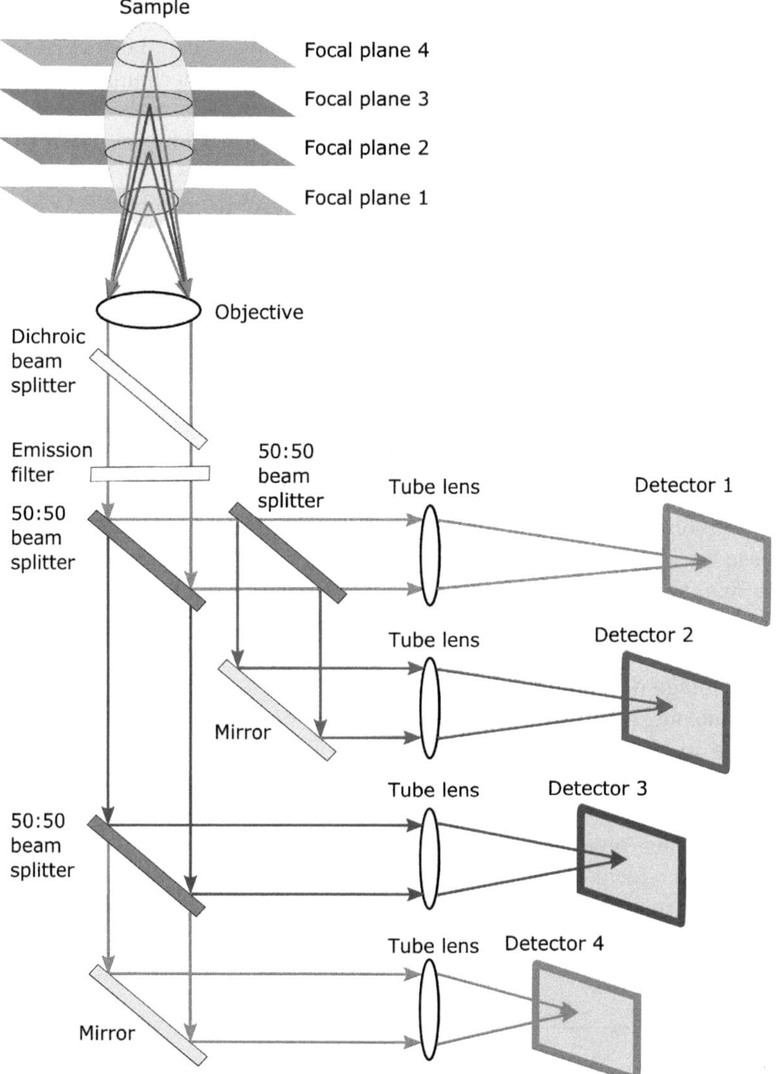

FIGURE 10.2
Four-plane multifocal plane microscopy (MUM) emission light path. The simultaneous imaging of four different focal planes within a sample can be realized with four detectors that are positioned at different distances from the tube lens. A 50:50 beam splitter is used to reflect 50% of the collected fluorescence towards one pair of detectors and transmit the other 50% towards the other pair of detectors. An additional 50:50 beam splitter is then used in each of the two resulting light paths to evenly divide the fluorescence between the two detectors. The four different colors are used to distinguish the light paths corresponding to the four focal planes, and do not represent fluorescence of different colors.

overcomes these issues is the simultaneous imaging of the multiple colors by one or multiple detectors. A high-level illustration of this approach is given in Fig. 10.3(b) for the two-color scenario. The schematic depicts the simultaneous illumination of the sample by two wavelength ranges, made possible by an appropriate optical component. This allows the simultaneous detection of the fluorescence from the two labeled proteins of interest by two

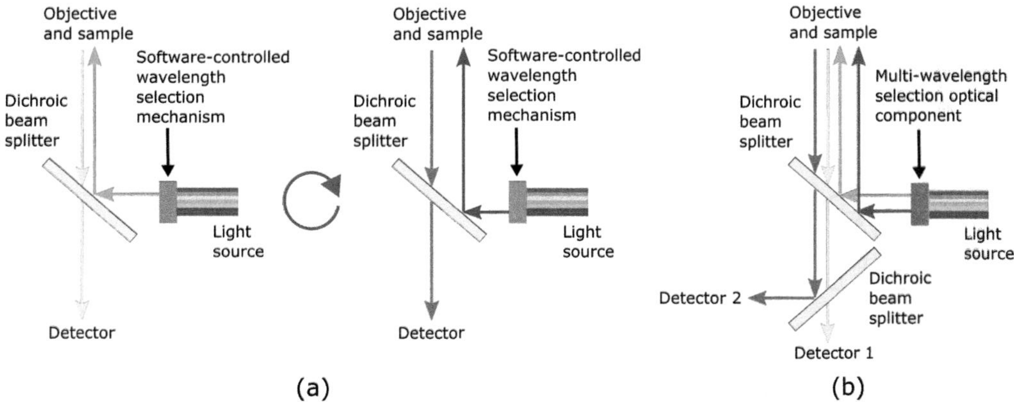

FIGURE 10.3
Sequential and simultaneous two-color imaging. (a) Sequential imaging configuration. A software-controlled mechanism is used to select the wavelength range for the excitation of each of the two differently colored fluorophores, ensuring that the sample is illuminated with only one wavelength range at any given time. Excitation by each wavelength range is synchronized with the exposure of the detector, so that a sequence of images is acquired that comprises alternating images, or sets of images, of the two fluorophores. (b) Simultaneous imaging configuration. The sample is illuminated simultaneously with both wavelength ranges and the fluorescence emitted by the two fluorophores are separately but simultaneously detected by two detectors. The simultaneous illumination is made possible by an optical component that selects both excitation wavelength ranges.

detectors, each dedicated to the imaging of one of the proteins. It is also possible to record both images side-by-side onto a single detector by replacing the second dichroic beam splitter on the light path with a commercially available device called an image splitter. An image splitter contains a dichroic beam splitter, but also has additional optical components that align the two colors separated by the dichroic beam splitter side-by-side without combining them.

In both Figs. 10.3(a) and 10.3(b), optical filters and beam splitters are largely omitted that are crucial for the imaging configurations shown. Optical filters and beam splitters will be discussed in detail in Section 10.7, and multi-color imaging will be revisited in Section 10.7.2 with particular emphasis placed on the required optical filters and beam splitters.

10.4 Light path for confocal microscopy

Confocal microscopy is an optical imaging technique that uses point illumination and a spatial pinhole to eliminate out-of-focus light in the light paths for the emitted fluorescence.

In a confocal microscope, light rays from the excitation light source, which is usually a laser, are focused by a primary focusing lens through a pinhole, and then passed through a secondary lens which produces parallel light rays (see Fig. 10.4, which also shows the light path for a widefield fluorescence microscope for comparison). A *pinhole* is a simple optical device, an opaque disk with a small hole at the center. The parallel rays are then reflected by a dichroic beam splitter and collected by an objective which focuses them onto

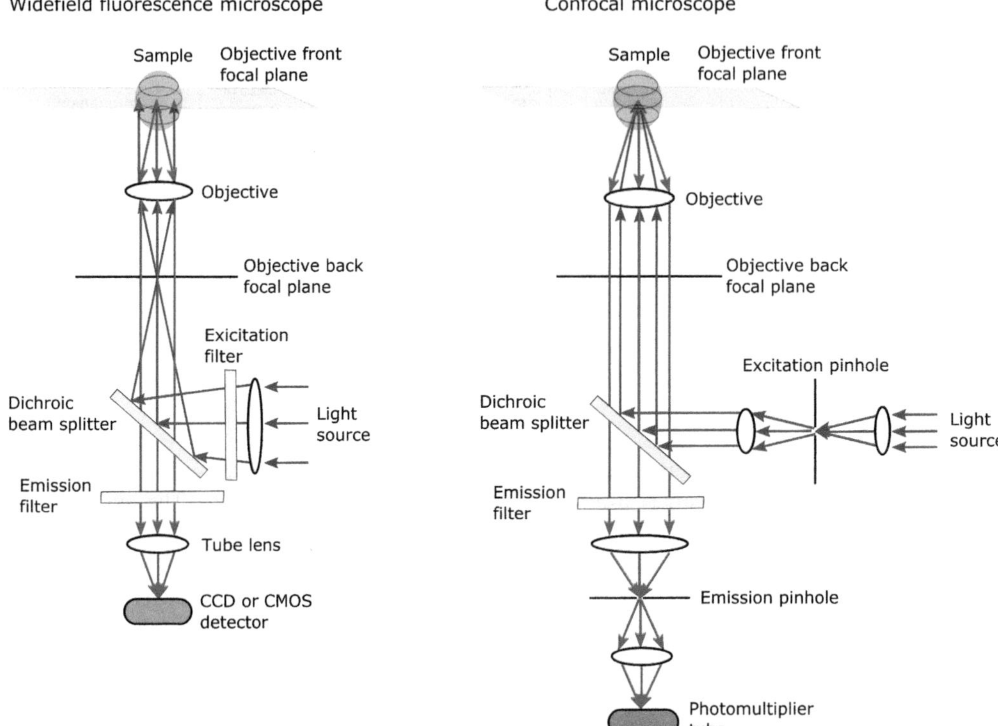

FIGURE 10.4
Comparison of the light paths for a widefield fluorescence microscope and a confocal microscope. Note that in the widefield configuration, the emission light path is shown for only fluorescence that emanates from a single point in the front focal plane. In reality, however, the widefield illumination means that there will also be fluorescence emanating from other points in the front focal plane as well as points outside of the front focal plane.

the sample. Therefore, the excitation in a confocal microscope produces a focused spot of light in the sample. Through a scanning system, this focus spot is moved incrementally throughout the sample. Figure 10.5 depicts such a system, by which a focal plane in the sample can be systematically scanned. The image is obtained by measuring the emitted fluorescence from each of the spot locations throughout the sample. Therefore, the emission optical system is designed to obtain the fluorescence signal from the illuminated focus spot in the sample. As the excitation beam not only illuminates the focus spot of interest, but also other parts of the sample, if only to a significantly lesser extent, the emission optics is designed to only collect the fluorescence from the illuminated focus spot and to reject any emitted fluorescence from outside of this volume. This is achieved by the use of a second pinhole. The light emitted from the sample is collected by the objective and then focused by a focusing lens onto the emission pinhole. The light that passes through the pinhole is then focused again onto a detector, typically a point detector such as a photomultiplier tube (Section 12.2). The fundamental role of this pinhole is that light from outside the illuminated volume will be blocked from reaching the detector (Fig. 10.6). It is this rejection of emitted light from out-of-focus locations that is one of the main advantages of the confocal imaging approach.

In a widefield configuration, the image detector can image a large number, possibly millions, of pixels at a time. In contrast, in a confocal configuration, the image is scanned one

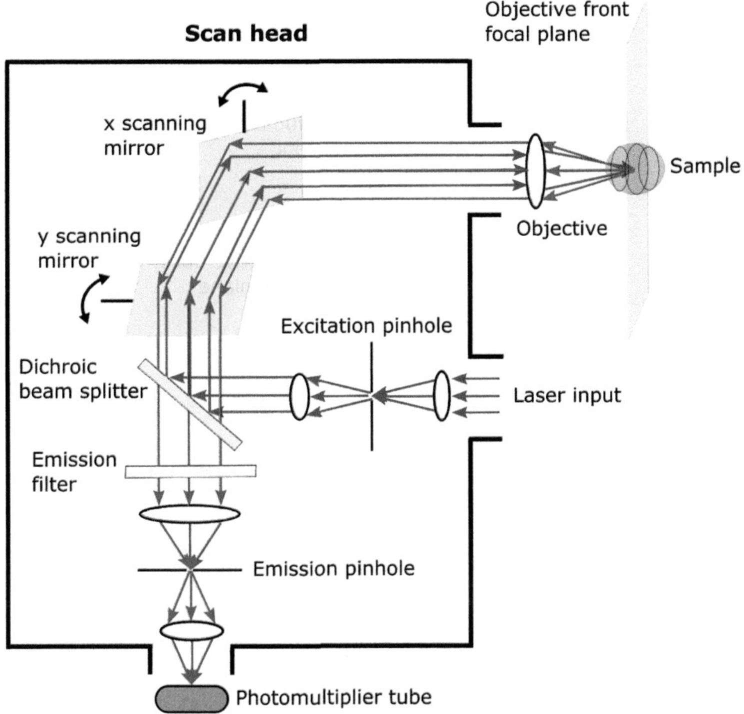

FIGURE 10.5
Use of mirrors to scan the excitation laser beam. In a confocal laser scanning microscope, mirrors are used to scan the excitation laser beam across a focal plane in the sample. One mirror changes the position of the laser focal spot in the x direction, and the other changes the position in the y direction. The parts of a confocal system that are typically housed in a scan head are enclosed in thick lines. A scan head has a port for attachment to a microscope, and often also supports ports for connections to a laser and a photomultiplier tube.

point at a time. A confocal microscope therefore needs to have a very fast scanning system to be able to acquire images at the same rate as a widefield microscope. This necessitates a very short acquisition time for each position of the focusing system and therefore requires a high photon flux. The scanning system and a significant part of the excitation and detection system of a confocal microscope are often contained in a *scan head* (Fig. 10.5).

An approach to improve on the acquisition speed of a confocal microscope is to parallelize the pinhole optics by using a system that employs multiple pinholes simultaneously. The *spinning disk*, or *Nipkow disk*, provides a system that implements such an approach (Fig. 10.7).

10.5 Two-photon excitation microscopy

Similar to confocal microscopy, two-photon excitation microscopy is an imaging modality that improves image quality by minimizing the amount of out-of-focus light that is detected. In a two-photon excitation modality the wavelength of the excitation light is usually in the near-infrared range. The fluorophore of interest, which normally requires the absorption of

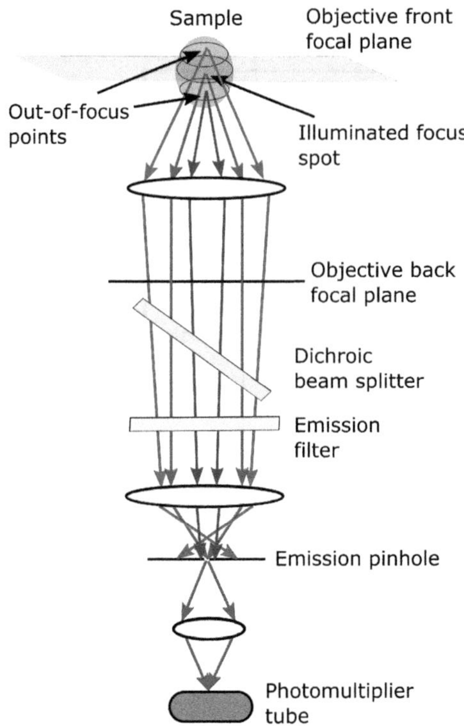

FIGURE 10.6
Rejection of fluorescence from outside the illuminated volume by the emission pinhole of a confocal microscope. The diagram shows that fluorescence originating from the illuminated volume is able to pass through the emission pinhole and reach the detector, but fluorescence originating from a plane above and a plane below the focal plane is blocked by the emission pinhole.

only one photon with a suitable wavelength in the visible or near-ultraviolet range, can therefore only be excited by the near-simultaneous absorption of two photons of the relatively low-energy excitation light. Due to the very low probability of two photons being absorbed by a fluorophore at nearly the same time, the near-infrared excitation light must be applied at a high photon density. Two-photon excitation microscopy is therefore typically implemented with an excitation laser that produces extremely short pulses of near-infrared light at very high power. The focus spot of this excitation laser is moved throughout the imaged sample using a scanning system, much like in the case of confocal microscopy. However, unlike in the confocal technique, no pinhole is necessary for the rejection of out-of-focus fluorescence in the emission path. This is because two-photon excitation effectively generates fluorescence only at the focal plane, by virtue of the fact that only in the illuminated focus spot is the excitation photon density high enough for the simultaneous absorption of two photons by a fluorophore to occur with substantial frequency. The rare occurrence of two-photon absorption outside the focal plane, together with the fact that light of longer wavelengths experiences less scattering in the sample than light of shorter wavelengths, also means that compared to confocal microscopy, more excitation light reaches the focal plane in two-photon excitation microscopy. This is an especially important advantage when it comes to the imaging of focal planes deeper within the sample. Naturally, the effective absence of two-photon absorption outside the focal plane also translates to reduced photobleaching of fluorophores outside the focal plane.

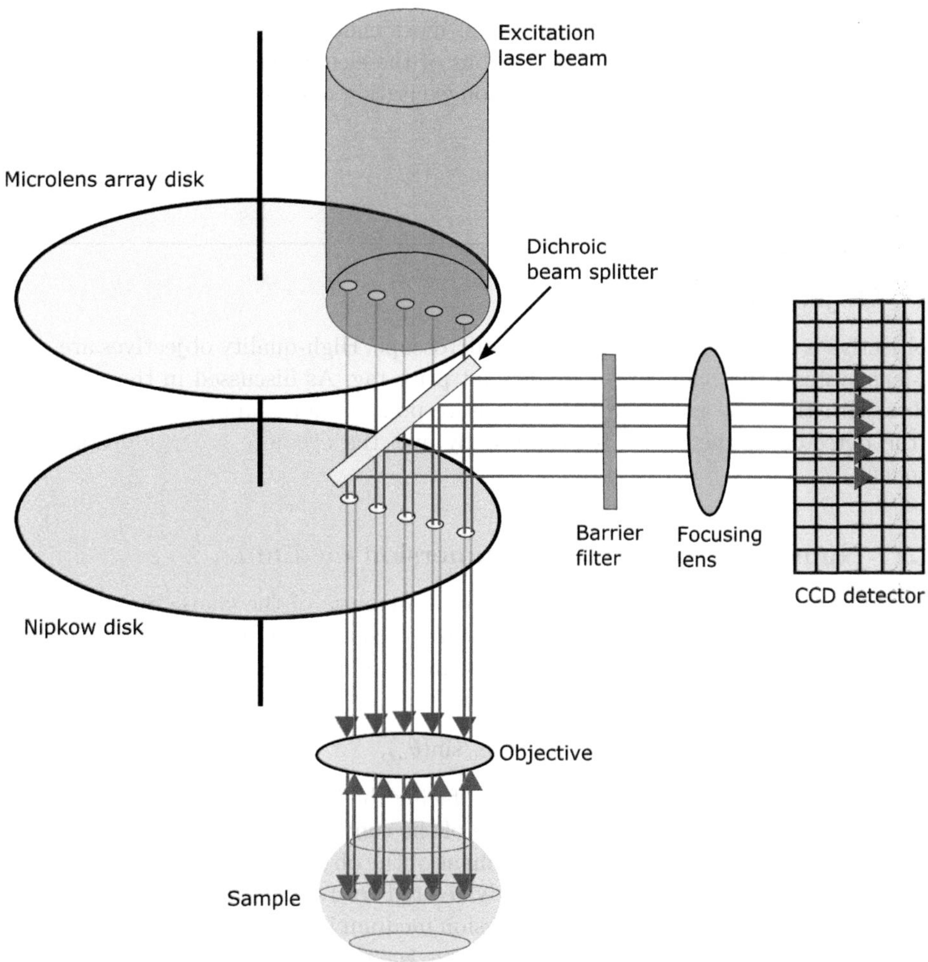

FIGURE 10.7
Use of a Nipkow disk to speed up image acquisition. In a spinning disk confocal microscope, a Nipkow disk, which contains pinholes arranged in spirals, allows the scanning of the excitation laser beam across a focal plane in the sample using multiple pinholes at a time. The fluorescence collected from the multiple laser focal spots in the sample is passed through a barrier filter and then focused onto a CCD-based detector. The purpose of the barrier filter is to prevent excitation light reflected by the disk from reaching the detector. The implementation shown was invented by Yokogawa Electric Corporation. It employs an additional spinning disk that contains a microlens for each pinhole in the Nipkow disk. The microlenses significantly improve the sensitivity of the technique by increasing the amount of excitation light that is focused into the pinholes of the Nipkow disk.

Use of long-wavelength excitation light in two-photon excitation microscopy provides the further advantage of reducing phototoxicity to the cell sample. Phototoxicity refers to cell damage caused by irradiation with light, and is primarily attributed to the production of reactive oxygen species which cause damage to the cell through the oxidation of biological macromolecules such as DNA and proteins. Reactive oxygen species are produced by the reaction of excited molecules with oxygen, and can therefore result from the absorption of excitation light by the fluorophores of interest and by light-absorbing molecules that are endogenous to the cell. Since in a typical cell most endogenous molecules are excited by light in the ultraviolet, violet, and blue regions of the electromagnetic spectrum, the long-wavelength excitation light used in two-photon excitation microscopy produces significantly lower levels of phototoxicity.

10.6 Objectives

The objective is the optical centerpiece of a microscope. High-quality objectives are complex multi-lens systems that can carry a significant price tag. As discussed in the prior section, together with the tube lens, the objective is responsible for mapping the imaged object onto the detector. If the image is observed by eye, the eyepiece is an additional optical component.

10.6.1 Numerical aperture and immersion medium

The magnification of the optical system is the most basic of the characteristics of an objective. However, for modern research-grade microscopes, it is possibly one of the least important ones. The quality of the acquired image is more importantly reflected in the numerical aperture of the objective, defined as

$$n_a := n_o \sin(\theta_o), \tag{10.1}$$

where n_o is the refractive index of the medium facing the front element of the objective, and θ_o is the half angle that describes the opening of the cone defined by a point in focus on the optical axis and the aperture of the front element of the objective (Fig. 10.8). The numerical aperture of an objective is therefore directly related to its light-gathering capability. The inclusion of the refractive index of the immersion medium in the definition is justified by the considerations in Section 13.1.5, where we analyze the path that an extreme ray takes when traveling from the objective to the sample. This path includes a segment corresponding to the layer of immersion medium that is traversed by the ray upon exiting the objective. Figure 10.8 illustrates how an objective with a larger numerical aperture can capture more of the light that is radially emitted from a fluorescent object than an objective with a smaller numerical aperture. As we will see in Chapter 14, the numerical aperture is also an optical characteristic that is of major significance in the diffraction description of image formation by a microscope.

Already in Sections 4.1 and 4.2 in the introductory part of the book, the numerical aperture played a major role in the description of the point spread function of a microscope and in the formulation of Rayleigh's resolution criterion. Since the sine function has values between 0 and 1, the numerical aperture can attain values between 0 and n_o. This shows that the numerical aperture is crucially dependent on the refractive index of the immersion medium, i.e., the medium facing the front element of the objective. As the refractive index of air is 1, the maximum numerical aperture for an air objective is 1. For water immersion objectives, it is 1.33, i.e., the refractive index of water. In terms of objectives designed for oil immersion, the maximum numerical aperture for the most typical oil immersion objective on the market is 1.515, but some specialty objectives with higher maximum numerical apertures are also available.

It should be pointed out that the refractive index of the immersion medium also needs to be matched to the refractive index of the cover glass that is being used, with the exception of dipping objectives that are designed to work without cover glass.

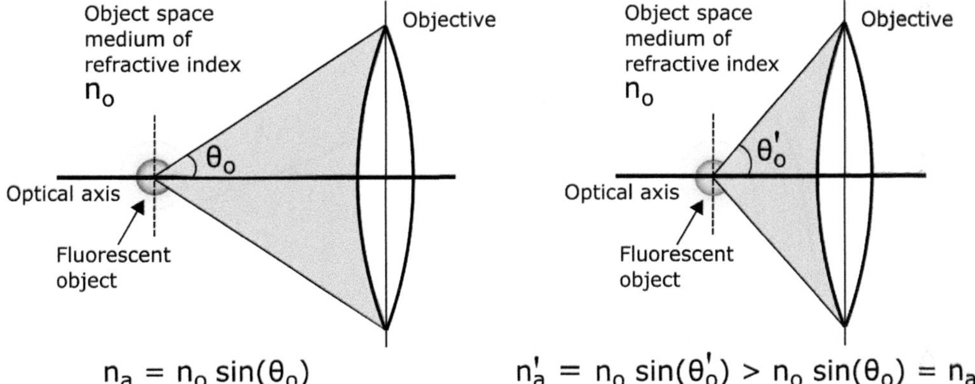

FIGURE 10.8
Numerical aperture. The sketch on the left shows that the numerical aperture n_a of an objective is given by the product of the refractive index n_o of the objective immersion medium and the sine of the half angle θ_o of the maximum cone of light defined by the aperture of the front element of the objective. The sketch on the right shows that an objective with a larger numerical aperture n'_a can capture more light from a fluorescent object by virtue of having a maximum cone with a larger half angle θ'_o.

10.6.2 Corrections

A major reason for the complexity of high-quality objectives is that they are expected to produce "perfect" images in a number of regards that cannot be achieved with simple optical elements. The specific aspects are referred to as *corrections*. An important correction requires objectives to produce a field of view that lies in a flat image plane. Such objectives are usually referred to as objectives with *planar* or *plan* corrections. For colocalization experiments it is very important to know how the objective behaves when different wavelengths are imaged through the objective. If two fluorescent labels coincide in the sample, it is important that they also coincide in the image. However, simple lenses have focusing properties that are wavelength-dependent. Therefore *color corrections*, often denoted by the term *apochromatic* or *apo*, are necessary to guarantee that different labels that coincide in the object also coincide in the image. There are objectives with different levels of color correction on the market. They vary based on the range of wavelengths that are corrected, and it is important to make sure that an objective has indeed the appropriate color correction for the planned experiment. The problem for an experimentalist is that the manufacturers, as with other optical characteristics of objectives, do not publish the precise criteria that they employ to assess the level of correction. This is of particular concern for applications requiring highly accurate measurements, where uncertainties regarding the optical properties of the microscope can call into question various assumptions that underpin a particular experiment. In addition, it is often unclear to what extent corrections are valid for out-of-focus imaging applications.

10.6.3 Transmission efficiency

For high-sensitivity experiments it is very important that the optical system absorbs as few photons as possible. The *transmission efficiency* of an objective, i.e., the percentage of the light that passes through the objective, can vary significantly from objective to objective. Figure 10.9 shows an example of a transmission curve. Clearly the transmission efficiency of

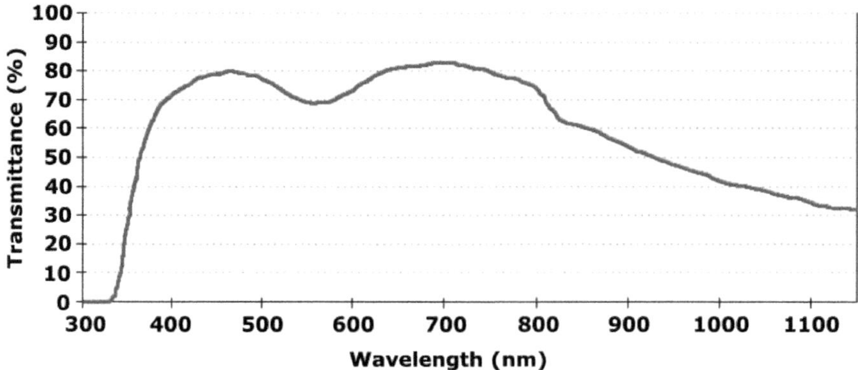

FIGURE 10.9
Transmission curve for the Zeiss Plan-Apochromat 100×/1.4 oil objective of Fig. 3.12. [See Figure Credits.]

an objective is wavelength-dependent and depends on the type of glass elements that have been used in the construction of the objective. Especially when experiments are carried out at the extreme ends of the visible spectrum, care needs to be taken to make sure that the objective has sufficient transmission at the wavelengths of interest. For example, ratiometric calcium imaging with fura-2 requires, for one of the fluorescence intensity measurements needed to form the desired ratio, excitation at a wavelength of around 340 nm. Many objectives, however, have too low a transmission efficiency in that wavelength neighborhood, such that excitation of the dye at around 340 nm becomes very difficult if not impossible. The objective of Fig. 10.9, for instance, would not be a particularly good choice given its very low transmission efficiency at 340 nm. Instead, one might consider using objectives made with special glasses such as quartz, which have significantly higher transmission efficiencies in the near-ultraviolet range of wavelengths.

As a general rule, the higher the level of corrections that the objective achieves, the lower the transmission efficiency. For example, if only one wavelength range is to be imaged by the objective, it would often be advantageous to use an objective with the same numerical aperture but without color correction, as the color correction typically reduces the transmission efficiency of the objective and is not required for the particular experiment.

10.7 Optical filters

We have seen in Section 10.1 that the optical light path of a fluorescence microscope crucially depends on the use of optical filters and dichroic beam splitters. In this section, we will discuss important aspects of the design of these optical components. An optical filter is an optical element through which light is passed in a wavelength-dependent manner, i.e., light of certain wavelengths can pass through the optical filter, whereas light of other wavelengths is blocked. The manufacture of filters is complex and not all designs that might be desired can be achieved in practice. For example, multiple steep rises and falls of the spectrum are difficult to achieve.

An appropriate choice of optical filters is important to carry out highly sensitive experiments. The role of the *excitation filter*, i.e., the filter that is in the excitation pathway,

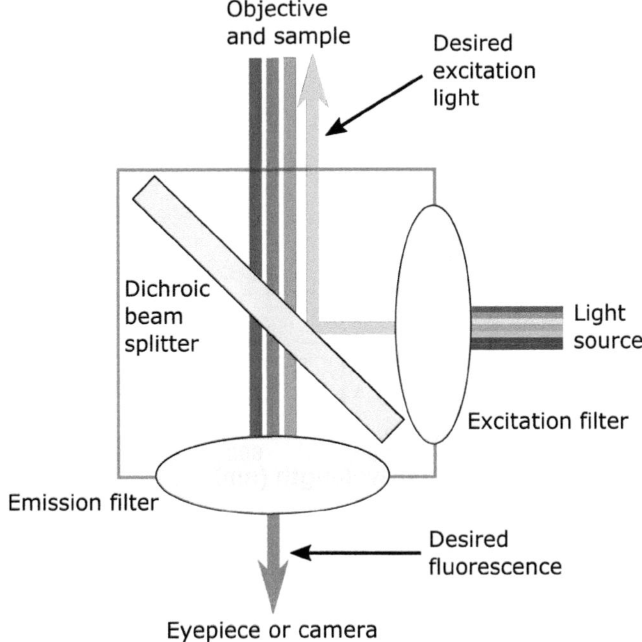

FIGURE 10.10
Filtering of light by the excitation filter, emission filter, and dichroic beam splitter in a fluorescence microscope. The square outline depicts the filter cube in which these three optical components often reside.

is to ensure that only the intended wavelengths reach the fluorescently labeled sample (Fig. 10.10). It is therefore critical that wavelengths outside the particular excitation wavelength range are suitably suppressed. A region of wavelengths in which a filter is designed to allow light to pass through is known as a *passband* of the filter, whereas a wavelength range in which light is blocked from passing through the filter is known as a *rejection band*. The choice of passband in an excitation filter will critically depend on the light source. For lasers that only have one wavelength, often only a very narrow "clean-up" filter is used to remove impurities in the spectrum of the laser. For weaker light sources, the passband of the filter cannot be too narrow as otherwise the light intensity might not be sufficient for appropriate excitation of the fluorescently labeled sample.

The role of the *emission filter*, i.e., the filter that is in the emission pathway, is to selectively allow the signals from the excited fluorophore of interest to pass through to the detector, and to reject all other signals, such as scattered excitation light, light emitted by other fluorophores in the sample, and ambient light (Fig. 10.10). The optical design of this filter is not straightforward due to some critical tradeoffs. In order to let as much as possible of the desired signal pass through the filter, a very wide spectral window would seem most appropriate. This is particularly relevant since the fluorescence signal of interest is often very weak. On the other hand, noise signals, such as scattered light and fluorescent signals from other fluorophores in a cellular sample, are often so large that they would drown out any of the actual signal of interest. Therefore, keeping the range of wavelengths small at which light passes through the filter might be of importance to reduce the inclusion of noise sources in the acquired data. The design of the emission filter is also intimately related to the design of the excitation filter. Any excitation light that is passed through the excitation filter typically has very high power and is prone to have a strong influence on the emission

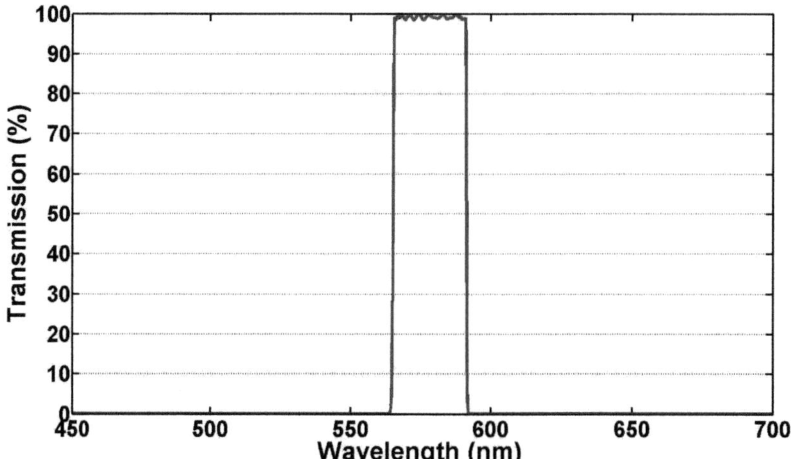

FIGURE 10.11
Transmission spectrum of a filter manufactured by Semrock (Part # FF01-578/21). [See Figure Credits.]

light path due to reflection and scattering of the light. It is therefore necessary that the emission filter is designed so that it does not allow light in the passband of the excitation filter to pass through.

Figure 10.11 shows the spectral characteristics of a particular optical filter. This *filter spectrum* shows the amount of light as a percentage that is transmitted by the filter as a function of wavelength. The part of the spectrum where the curve is close to 100% implies that most of the light at those particular wavelengths will pass through the filter. On the other hand, in the other regions of the spectrum, transmission of nearly 0% indicates that most of the light is blocked at those wavelengths.

We can see in Fig. 10.10 that there is a central element in the light path of a fluorescence microscope at which the excitation and emission light paths meet and separate. The *dichroic beam splitter* has the complex task of reflecting the excitation light into the objective while letting the emitted light pass through to the emission filter and the detector in the opposing light path. The performance of a dichroic beam splitter is characterized by a spectral graph similar to that of an optical filter. The spectral graph for a dichroic beam splitter shows the amount of light that is transmitted as a function of wavelength. Light that is not transmitted will be reflected rather than simply blocked as in the case of an optical filter. Therefore, with a dichroic beam splitter, we speak of passbands and *reflection bands*. For most designs, the reflection property depends crucially on the dichroic beam splitter being precisely oriented at 45° with respect to the light path.

It is clear from the above descriptions that the excitation filter, the emission filter, and the dichroic beam splitter need to be designed as a unit. In fact, in many standard microscopes they are placed together in a *filter cube*. An example of a filter cube is shown in Fig. 10.12(a), and an example of a filter set that could be placed in such a cube is shown in Fig. 10.12(b).

The fraction of the amount of light that is transmitted at a particular wavelength λ is measured by the *optical density* (OD_λ), which is defined as the negative of the logarithm to the base 10 of the ratio of the intensity I of the transmitted light to the intensity I_0 of the light that enters the optical element, i.e., $OD_\lambda = -\log_{10}(I/I_0)$. Note that the ratio I/I_0 is simply the fraction of light that is transmitted at wavelength λ, and is typically expressed

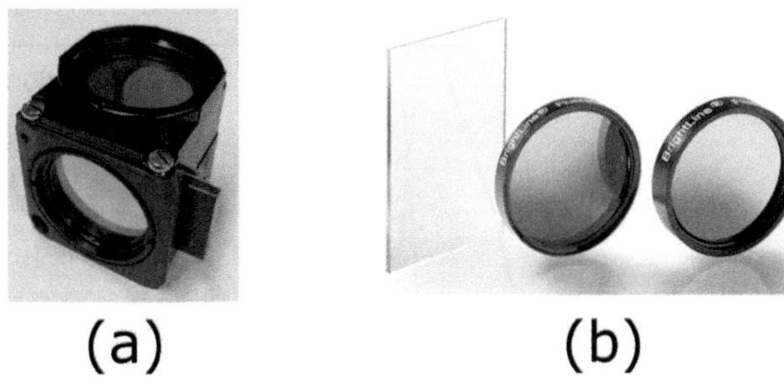

FIGURE 10.12
Filter cube and filter set. (a) A filter cube for a Zeiss microscope, manufactured by Zeiss. (b) A filter set manufactured by Semrock. The round components are the excitation and emission filters, the rectangular component is the dichroic beam splitter. [See Figure Credits.]

as a percentage transmission as shown in the filter spectrum of Fig. 10.11. A typical design of the blocking area of an excitation or emission filter has an optical density specification of greater than 5. For the filter of Fig. 10.11, for example, the rejection bands within the visible region have, for the most part, an optical density value of greater than 6, meaning that less than 0.0001% of light is transmitted at wavelengths within those bands.

10.7.1 Example: a filter set for a GFP-labeled protein

In Chapters 2 and 9, we discussed the problem of the localization of proteins of interest in a cell, specifically using FcRn and the transferrin receptor in our examples. The approach to visualize a receptor in a cell is to label the receptor with a fluorescent molecule such as GFP (Section 8.7.3). In order to image the labeled protein in a cell with a fluorescence microscope, we need to use an appropriate combination of optical filters and dichroic beam splitter.

To arrive at a proper choice of a filter and dichroic combination, often called a *filter set*, it is important to first consider the excitation and emission spectra of the fluorophore that is to be imaged. In Fig. 10.13, we see that the excitation spectrum of GFP has a maximum near 488 nm. An excitation wavelength near 488 nm is therefore appropriate, and there are a number of lasers on the market that have a laser line at 488 nm. When such a laser is used, a clean-up filter with a narrow passband centered at the 488-nm wavelength will often suffice as an excitation filter. If the excitation is to be done with an arc lamp instead, then we need to choose a filter with a wide enough passband so that sufficient light can be delivered to the sample. For the excitation filter in the filter set of Fig. 10.14, the transmission is above 90% for the wavelength range of 451.5 nm to 486.5 nm.

We see from the emission spectrum of GFP that its peak of emission is around 510 nm. This means that it would be desirable to have an emission filter whose passband includes this peak and as much as possible of the emission spectrum of GFP in order to maximize the detection of the emitted light from the fluorophore. The emission filter in Fig. 10.14, for example, has a suitable transmission profile with transmission above 90% in the range of 505.5 nm to 544.5 nm.

To complete the filter set, a dichroic beam splitter is necessary. We know that the dichroic needs to reflect the excitation light in the passband of the excitation filter and pass through the wavelengths in the passband of the emission filter. This provides the design requirements for the dichroic beam splitter. We can see from its spectral graph in Fig. 10.14 that the

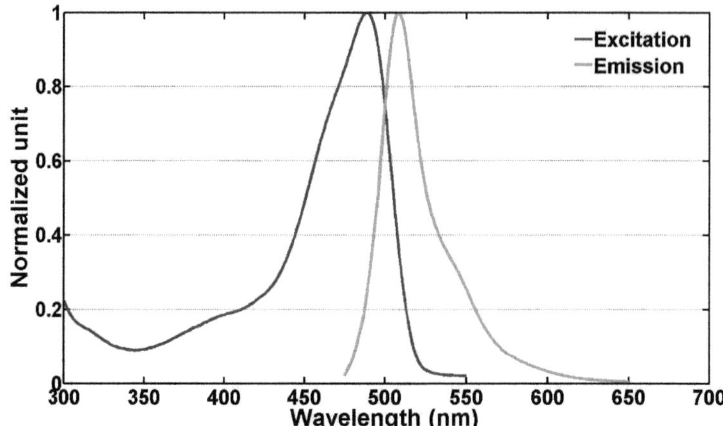

FIGURE 10.13
Excitation and emission spectra of green fluorescent protein (GFP). [See Figure Credits.]

spectrum just manages to rise fast enough to achieve this specification. A major challenge in the design of filter sets is that, especially in the case of dichroic beam splitters, arbitrarily fast-rising profiles are not possible. So it is often the dichroic beam splitter that determines the design of the excitation and emission filters. In the filter set of Fig. 10.14, the passbands of the excitation and emission filters are chosen to be as close as possible to the transition part of the dichroic beam splitter so that as much as possible of the excitation light near the transition is delivered to the sample, and as much as possible of the fluorescence near the transition is captured from the fluorophore.

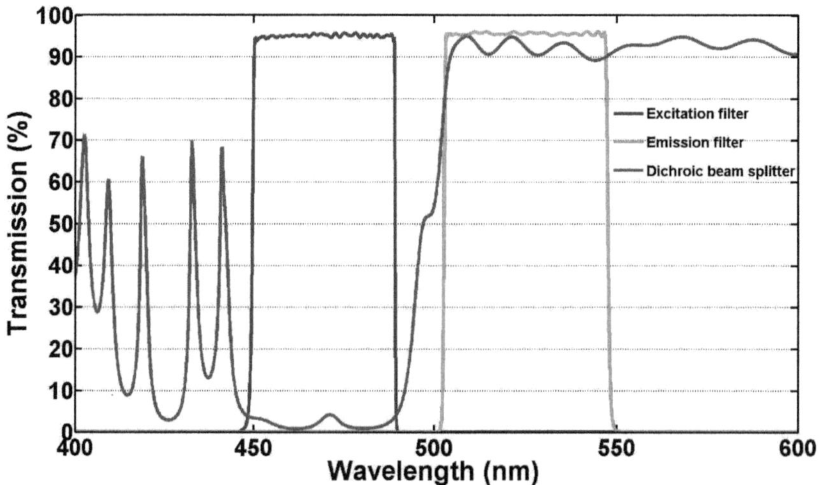

FIGURE 10.14
Transmission spectra of the excitation filter, emission filter, and dichroic beam splitter of a filter set that is optimized for the imaging of GFP-labeled samples. This particular filter set is manufactured by Semrock (Part # GFP-A-Basic-000). [See Figure Credits.]

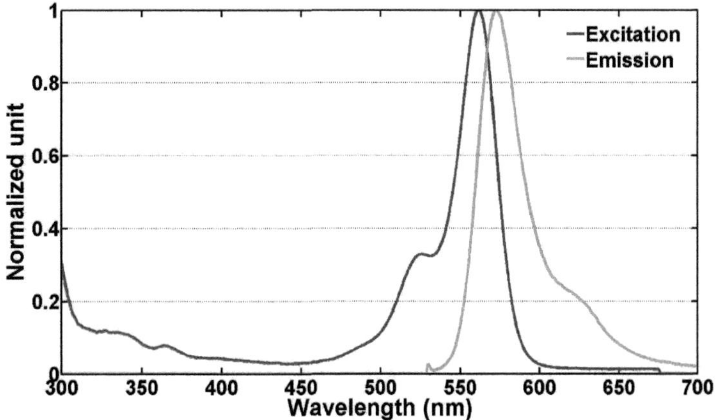

FIGURE 10.15
Excitation and emission spectra of Alexa Fluor 546. [See Figure Credits.]

The question very naturally arises as to why the passband of the emission filter is restricted to a relatively small range. Why would the upper limit of the passband of the emission filter not be chosen to be much higher to capture as much fluorescence as possible from the fluorophore? In fact it could and in many cases should be chosen to be much higher. The filter set of Fig. 10.14 is designed for a situation where a fluorophore like TRITC (Fig. 8.4) or Alexa Fluor 546 (Fig. 10.15) that emits maximally at a wavelength above 550 nm might be in the sample in addition to GFP. If no other fluorophore and no significant autofluorescence are present in the sample, then the emission passband should be chosen to be as large as possible. We will discuss the imaging of samples with multiple fluorophores in Section 10.7.2.

What about the rejection bands? For the excitation filter of Fig. 10.14, for a large part of the wavelength range the optical density is above 5, i.e., the transmission is less than 0.001%. The same is true for the emission filter in the filter set. The high optical density values are desirable, as they indicate the rejection bands are effective at preventing excitation light of unwanted wavelengths from reaching the sample, and fluorescence of unwanted wavelengths from reaching the detector.

It is interesting to see that much better performing filter sets are available than the one we have just discussed. At a higher price, the filter set in Fig. 10.16 is available also for the imaging of GFP. Here the passbands of the excitation and emission filters are characterized by a significantly higher transmission. At the same time, the rejection bands of these filters still have similar specifications as those of the earlier filter set. The dichroic beam splitter here also provides improved performance relative to that of the earlier filter set, as it has a passband of higher transmission, a reflection band of lower transmission, and a transition that is steeper. The result is a filter set that will allow much more light to be collected from the sample. This can be of great importance for high sensitivity experiments such as single molecule experiments.

10.7.2 Imaging of multiple fluorophores

When two different fluorescently labeled proteins are to be imaged in a sample, filter design has to take this into account. If the two proteins are to be imaged sequentially with one detector in a widefield microscope, then a software-controlled device is typically used to

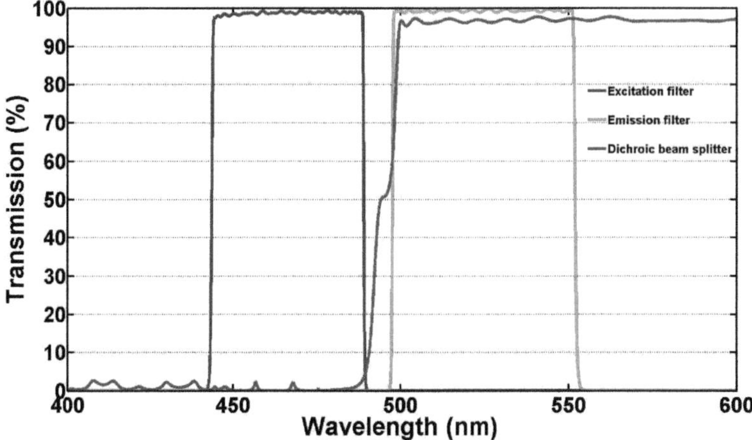

FIGURE 10.16
Transmission spectra of the components of a filter set for GFP imaging that has higher transmission bands and a steeper dichroic transition than the filter set of Fig. 10.14. This particular filter set is manufactured by Semrock (Part # GFP-4050B-000). [See Figure Credits.]

change the optical filters so that for a given image, only one of the fluorophores is excited and its fluorescence emission properly captured. One possibility is to use a filter cube turret (Fig. 10.17) to switch the filter sets for the two fluorophores in and out of the light path as needed. Given that the two fluorophores used are GFP and Alexa Fluor 546, for example, the filter cube turret would be used to switch between a filter cube containing the filter set of either Fig. 10.14 or Fig. 10.16 for GFP, and a filter cube containing the filter set of Fig. 10.18 for Alexa Fluor 546. Since the turning of a filter cube turret is a relatively slow process, this approach is most appropriate for the imaging of fixed cells or relatively slow events in live cells.

To image faster events in a live cell sample sequentially with one detector, one approach is to use a filter wheel (Fig. 10.19) to switch between the filters for the excitation of the two

FIGURE 10.17
Filter cube turret. The photograph on the left shows a filter cube turret that holds up to six filter cubes. The photograph on the right shows the same filter cube turret positioned under the objective of an inverted microscope.

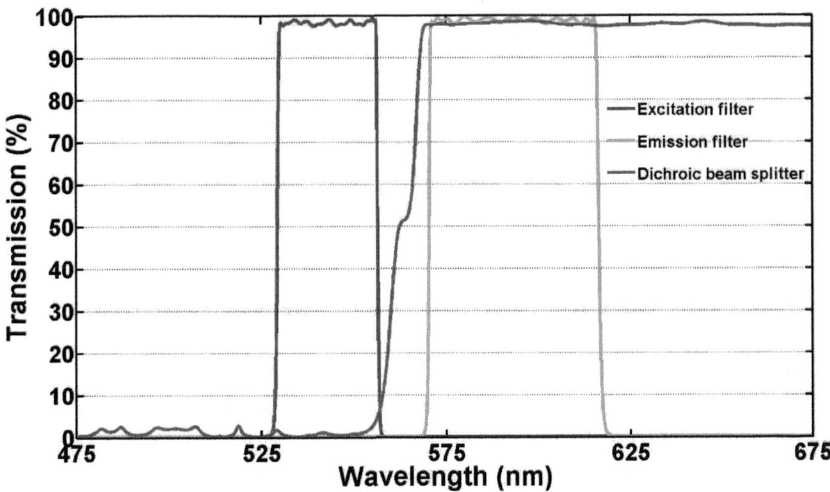

FIGURE 10.18
Transmission spectra of the excitation filter, emission filter, and dichroic beam splitter of a filter set that is appropriate for the imaging of Alexa Fluor 546-labeled samples. This particular filter set is manufactured by Semrock (Part # TRITC-B-000). [See Figure Credits.]

fluorophores (see Fig. 10.3(a) for a high-level sketch of this type of approach). The filter wheel is generally faster than a filter cube turret, but a disadvantage of this approach is that the same dichroic beam splitter is used for the imaging of both fluorophores, and it may not be the ideal choice for one or both of the fluorophores. This type of configuration will typically also require the use of a filter wheel in front of the detector to switch between the filters for the fluorescence emitted by the two fluorophores. When this approach is implemented with lasers for the excitation of the two fluorophores, each laser is passed through an appropriate clean-up filter. To repeatedly switch between the two lasers, shutters (Fig. 10.20) capable of blocking laser beams can be used to control which laser beam is passed through to the sample. Alternatively, lasers that can be modulated between an on and an off state can be used.

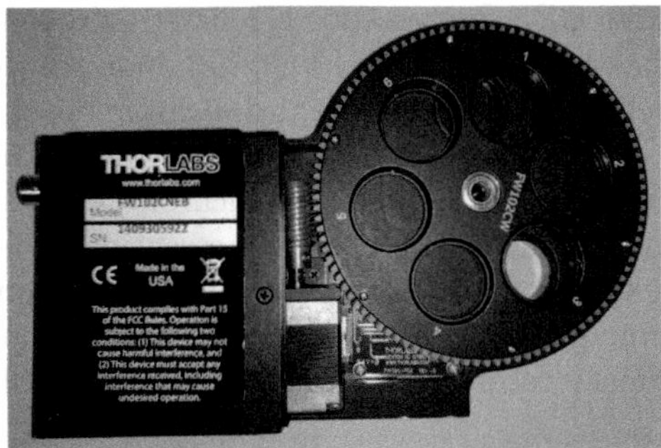

FIGURE 10.19
Filter wheel that holds up to six filters. Software commands can be used to turn the wheel so that the desired filter is placed in the light path as needed. This particular filter wheel is manufactured by Thorlabs (Part # FW102CW).

FIGURE 10.20
Shutter for blocking a laser beam. Software commands can be used to close or open the aperture in the center of the device. When the aperture is closed (left image), the laser beam is blocked. When it is open (right image), the laser beam is passed through. This particular shutter is manufactured by Vincent Associates. [See Figure Credits.]

If the two fluorescently labeled proteins exhibit very fast dynamics, it may be necessary to image both fluorophores simultaneously. In this case, a more complex optical filter and dichroic beam splitter configuration is needed. Different approaches are again possible, but the common theme is to simultaneously excite the two fluorophores in the sample and to use a dichroic beam splitter combination that allows the separation of the emitted fluorescence by color (Fig. 10.3(b)). When the light source is a lamp, an excitation filter with two passbands is needed to simultaneously pass through the wavelength ranges for the excitation of the two fluorophores. When the light source comprises two lasers, a dichroic beam splitter can instead be used to combine them. A configuration that uses lasers is illustrated in Fig. 10.21, which uses a sample containing a GFP-labeled protein and an Alexa Fluor 546-labeled protein as an example. The illustration expands on the schematic of Fig. 10.3(b), and shows that a second and a third dichroic beam splitter are used. The second beam splitter reflects the combined beam onto the sample, so that both proteins are excited at the same time. It also transmits the fluorescence emitted by both GFP and Alexa Fluor 546. It should be noted that this beam splitter is special in that it has four passbands and four reflection bands (Fig. 10.22). The two lasers are reflected onto the sample by two different reflection bands, and the fluorescence from GFP and Alexa Fluor 546 are transmitted by two different passbands. The passband for GFP fluorescence is located between the reflection bands for the two lasers, while the passband for Alexa Fluor 546 fluorescence is located immediately after the second of these reflection bands. The third dichroic beam splitter reflects the GFP fluorescence towards one detector, and transmits Alexa Fluor 546 fluorescence towards a second detector. Note that by replacing the first dichroic beam splitter and the clean-up filters with an excitation filter with two passbands, essentially the same configuration can also be used with lamp excitation. Note also that while the configuration of Fig. 10.21 uses one detector for the imaging of each fluorophore, it is also possible, as mentioned in Section 10.3, to simultaneously record the images of both fluorophores side-by-side onto a single detector.

We should also mention that while a multi-band dichroic beam splitter such as the second beam splitter in this example has the important advantage of enabling simultaneous multi-color imaging, its optical properties will often be suboptimal to those of its single-band alternatives because of the more complex design specifications that it has to meet. For the particular beam splitter used in this example, an obvious drawback is that it does not pass through as much of the GFP fluorescence to the detector. This can be seen by comparing the

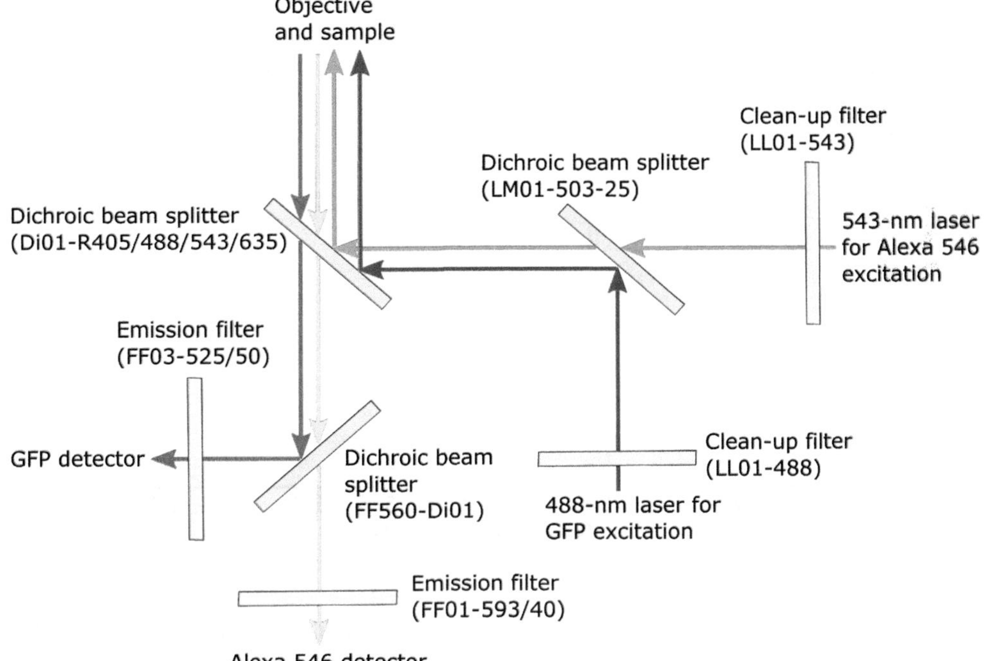

FIGURE 10.21
Optical filter configuration for the simultaneous imaging of GFP and Alexa Fluor 546. This particular configuration uses a 488-nm laser for the excitation of GFP and a 543-nm laser for the excitation of Alexa Fluor 546, each of which is passed through a clean-up filter. The two laser beams are combined with a dichroic beam splitter that reflects the 488-nm laser beam and transmits the 543-nm laser beam. The combined laser beam is directed toward a second dichroic beam splitter that reflects light of wavelengths 488 nm and 543 nm towards the objective and sample. At the same time, this second dichroic beam splitter transmits light of wavelengths at and near the maximum emission wavelengths of GFP (510 nm) and Alexa Fluor 546 (572 nm). The transmitted fluorescence emission is then separated by a third dichroic beam splitter, which reflects light of wavelengths below 560 nm towards the emission filter and detector for GFP, and transmits light of wavelengths above 560 nm towards the emission filter and detector for Alexa Fluor 546. The part numbers given in parentheses indicate filters and dichroic beam splitters manufactured by Semrock that can be used to implement this configuration.

overlap between the passband of the GFP emission filter with the corresponding passband of the multi-band dichroic beam splitter in Fig. 10.22 and the passband of the single-band dichroic beam splitter in Fig. 10.16.

10.8 Transmitted light microscopy

The main focus of this book is on fluorescence microscopy due to its relevance for molecular cell biology experiments. However, as already mentioned in Chapter 3, transmitted light microscopy does have important applications. We therefore very briefly discuss here a couple

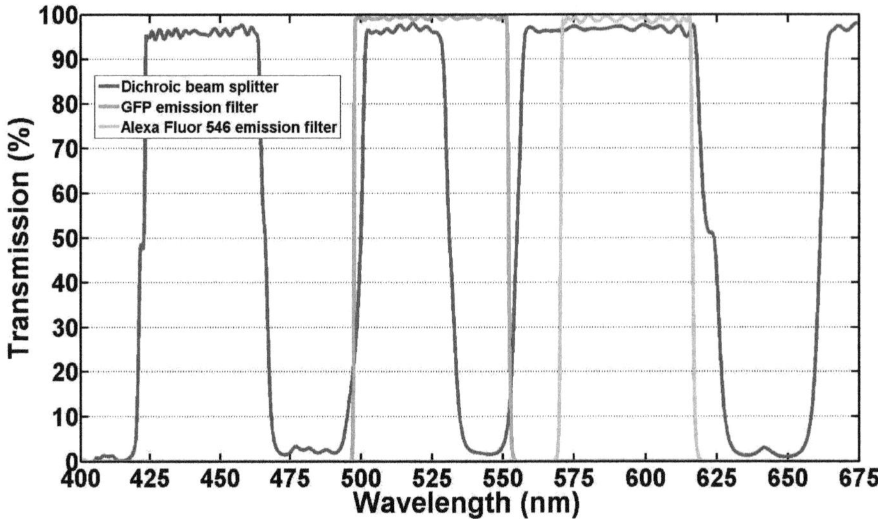

FIGURE 10.22
Transmission spectrum of a multi-band dichroic beam splitter (Semrock Part # Di01-R405/488/543/635) suitable for the imaging of GFP and Alexa Fluor 546. The spectra of the GFP emission filter (Semrock Part # FF03-525/50) from the filter set of Fig. 10.16 and the Alexa Fluor 546 emission filter (Semrock Part # FF01-593/40) from the filter set of Fig. 10.18 are also shown. [See Figure Credits.]

of important approaches to transmitted light microscopy, namely phase-contrast and differential interference contrast (DIC) microscopy. These approaches provide solutions to the main problem associated with the transmitted light microscopy of cellular samples, which is that cells are to a large extent translucent, and therefore almost "invisible" in the sense that some structural details are not readily observable as variations in the intensity of the detected light. As discussed in more detail in Section 14.1, we can associate a phase to light. This phase is often affected by the cellular sample, and in particular, different cellular compartments can have different effects on the phase of the transmitted light. Phase-contrast and DIC microscopy are examples of techniques that are based on revealing changes in the phase of the transmitted light, and that consequently enable the visualization of the cellular structures that give rise to them.

Phase-contrast microscopy reveals structural details in a cell that are not visible when observed using a simple transmitted light microscope by converting phase differences in light that passes through the sample into corresponding variations of intensity in the acquired image. When a sample is illuminated, some of the light passes through the sample undeviated while the remainder is scattered by the various cellular structures. The undeviated light does not interact with the sample and therefore contains no information about its structural details. On the other hand, the scattered light captures structural information in the form of a phase shift that results from its traversal through the various structures. The scattered light is typically significantly weaker in intensity compared to the undeviated light, and its phase is shifted by approximately $-90°$ relative to that of the undeviated light. The undeviated and scattered light combine through interference (Section 14.1.3) to form the image that is observed. In the absence of any contrast-enhancing optics, interference of the undeviated and the scattered light, given the nature of their intensity and phase differences, produces an image with an intensity that is not significantly different from the intensity of its undeviated light component. The image therefore shows relatively little intensity contrast between the cellular structures which scatter the illuminating light and the parts of the sample through which the illuminating light passes undeviated. Naturally, relatively few structural details can be seen in such an image. In a phase-contrast microscope, a condenser annulus in the illumination light path is used in conjunction with a phase plate that is built into the objective to spatially separate the undeviated and the scattered light collected by the objective. This separation allows the manipulation of only the undeviated light by the phase plate, which includes a phase shift of $-90°$ to correct for the difference in phase relative to the scattered light. The phase correction enables the undeviated and the scattered light to constructively interfere, thereby increasing the image contrast by raising the intensity of the areas that contain the cellular structures above the intensity of the areas that do not. Importantly, the phase plate also substantially reduces the intensity of the undeviated light so that the scattered light becomes a significant contributor to the intensity of the image. These phase and intensity manipulations by the phase plate therefore convert unobservable phase differences between the undeviated and the scattered light into observable variations in intensity that reveal the structural details of the sample in a high-contrast image (Note 10.1).

DIC microscopy is a technique that also produces images with higher contrast than simple transmitted light microscopy. The DIC technique works by converting gradients in the optical path length of the sample into differences in intensity in the acquired image. Optical path length (Section 14.1.1.5) is defined as the product of the distance traversed by a light wave and the refractive index of the medium through which it propagates. Since a cell contains structures of varying dimensions and refractive indices, light waves that propagate through different parts of the cell will traverse different optical path lengths which contain information about the structures. Exploiting the idea that structural details can be detected as differences in the optical path lengths traversed by two closely spaced light waves, the technique propagates pairs of illuminating waves through the sample and then combines each pair to obtain information about the difference in the optical path lengths traversed by the two waves. Specifically, a linear polarizer and a Nomarski prism are used to split the illuminating light into pairs of closely spaced waves that are passed through the sample. The two waves in each pair will propagate through the sample in parallel, but will traverse different optical path lengths provided that they pass through areas of differing refractive indices and/or thicknesses. After traveling through the objective, the two waves in each pair are recombined by a second Nomarski prism to produce a wave that effectively encodes the difference between the optical path lengths traversed by the two waves. Finally, a component of this combined wave is passed through a second linear polarizer so that the two constituent waves can undergo interference to contribute intensity to the acquired image that is proportional to the difference in the optical path lengths they traversed. Gradients in the optical path length are thus transformed into intensity variations that enable the visualization of the structural details of the sample. It is important to note that the second polarizer also eliminates background light by blocking waves that originated from wave pairs that traversed identical optical path lengths (i.e., that did not encounter optical path differences due to the sample). In practice, the polarizer and prism on the illumination side are inserted into the condenser which focuses light onto the sample, and the polarizer and prism on the detection side are introduced into the light path in a slider that is inserted below the objective (assuming that an inverted microscope is used) (Note 10.2).

11

Microscopy Experiments

Microscopy is a very versatile technique. Microscopes can be equipped in many different configurations, allowing for a large variety of experiments to be carried out. Here we want to present a number of key experiments to illustrate examples of standard microscope configurations and some of the most commonly performed experiments. The types of experiments covered include localization and association experiments, total internal reflection fluorescence microscopy experiments, single molecule microscopy experiments, and multifocal plane microscopy experiments.

11.1 Fixed cell experiments

As discussed in introductory Sections 2.2 and 2.3, the most basic experiments in cellular microscopy are localization and association experiments, where the location and association of labeled molecules are investigated. These types of experiments are typically carried out using fixed cells as the dynamic processes in the cell are not relevant for these types of studies. In fact, carrying out localization and association experiments in live cells could be problematic, as the movement of the objects of interest during the acquisition can cause problems in the interpretation of the data.

The microscope that is used for the imaging of fixed cells, especially association experiments, is typically one with a device that allows for the efficient changing of filter cubes, often a *filter cube turret* under the objective (Fig. 10.17). This enables the imaging of different fluorophores with little effort needed to change the filter set. As we are discussing the imaging of fixed cells there is no demand on the speed of these filter changes. Of course, for the imaging of a single fluorophore in a localization experiment, such a filter-changing device is not necessary.

11.1.1 Localization of FcRn

One of the two biological problems that we introduced in Part I addresses the question of the role that the receptor FcRn plays in the transport of IgG molecules. We describe here the most fundamental experiments designed to help identify where FcRn is located within a cell. To this end, we conduct the following individual experiments.

We begin by imaging labeled FcRn in a human endothelial cell. The model proposed for the function of FcRn postulates that the endothelial cells that line the blood vessels play a role in the functioning of FcRn. FcRn is labeled with GFP by fusion of the FcRn gene with this fluorescent protein gene. Human endothelial cells are cultured and transfected with FcRn-GFP using the procedure described in Section 7.2. They are grown on cover glasses, fixed, and mounted on slides with mounting medium that also contains anti-photobleaching reagents.

For the imaging we use a standard microscope (Fig. 3.7) with a filter set that is designed to image GFP-labeled samples, as was discussed in Section 10.7.1. A high magnification and high numerical aperture objective, such as a 100×/1.4 Plan-Apochromat oil objective (Fig. 3.12), is appropriate. A standard CCD or CMOS camera (Section 12.3) is typically used for such experiments. Each of the leftmost images in Fig. 9.6 shows the result of an acquisition carried out as we have described. What can, at first, be disconcerting to a newcomer to fluorescence microscopy is that the contours of the cell are not easily visible. This is due, in this case, to there not being any appreciable level of the labeled protein on the plasma membrane of the cell. An additional image acquired in transmitted light mode, however, can be used to visualize the outline of the cell.

The two leftmost images in Fig. 9.6 show that FcRn is seen in ring-like structures with a diameter of around 1 μm and other smaller structures. A control experiment with an unlabeled sample needs to be conducted to verify that the observed structures are in fact due to the labeled FcRn, and to exclude the possibility that they are due to autofluorescence or other contaminants.

11.1.2 Association experiments with FcRn, EEA1, LAMP1, and transferrin

The above localization experiment provides information about the basic geometry of FcRn-positive structures in a cell. However, from this basic localization experiment we cannot obtain any clear information regarding the identity or the function of the FcRn-positive structures. The standard approach to obtaining answers to these questions is to carry out association experiments. In these experiments, the molecules of interest are imaged at the same time as molecules that label known structures. In our situation, because of what we know about FcRn, we might suspect that FcRn is present in early endosomes. A standard label for early endosomes is early endosome antigen 1 (EEA1). Antibodies are available that recognize this marker. Therefore the FcRn-GFP-transfected cells used for the localization experiment discussed in Section 11.1.1 can be fixed, and EEA1 can be labeled using the antibody-based labeling strategies discussed in Section 9.4.1.

A critical consideration, however, is which fluorophore should be used for the labeling of the anti-EEA1 antibodies. As our cellular sample is already transfected with FcRn-GFP, we need to use a fluorophore that is spectrally separated from GFP so that, with minimal "cross contamination", we can assign an acquired image either to FcRn or to EEA1. In this case we have chosen Alexa Fluor 568. The plots in Fig. 11.1 show that there is only a small amount of overlap in the excitation and emission spectra for the two fluorophores. For the sequential imaging of GFP and Alexa Fluor 568, a suitable filter set for GFP is shown in Fig. 10.14 (and an even better performing alternative is shown in Fig. 10.16), and an appropriate filter set for Alexa Fluor 568 is shown in Fig. 11.2. If we wanted to have even further separation between the spectral responses, we could in fact use a dye like Cy5, which has even less spectral overlap with GFP (Fig. 11.3). However, as exemplified by the curves in Fig. 12.8 for the Hamamatsu ORCA-ER CCD camera and the Andor Zyla 5.5 CMOS camera, some standard detectors can have significantly reduced quantum efficiency in the emission wavelength range of Cy5, i.e., above 650 nm. This might be one reason for preferring a fluorophore with a lower emission wavelength.

The actual experiment is carried out using a microscope as described for the localization experiment. Two exposures are acquired, one with the GFP filter set in the light path, the other with the Alexa Fluor 568 filter set in the light path. As we use grayscale cameras, we obtain two grayscale images — one for the FcRn-GFP signal, the other for the EEA1-Alexa Fluor 568 signal. An example grayscale FcRn-GFP image is again shown as the top left image in Fig. 9.6, pseudo-colored in green. An example grayscale image of EEA1-

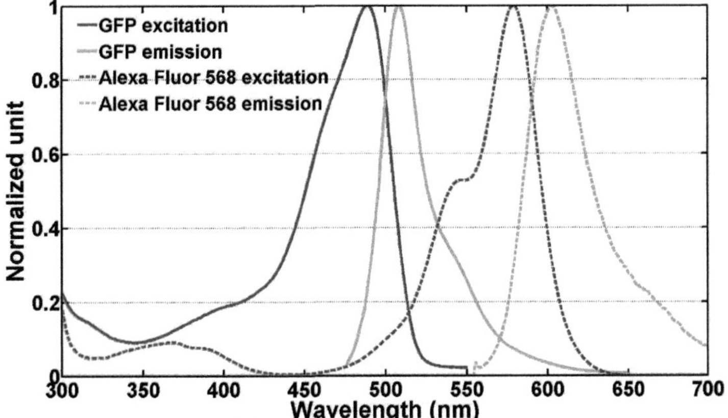

FIGURE 11.1
Excitation and emission spectra of GFP and Alexa Fluor 568. [See Figure Credits.]

Alexa Fluor 568 in the same field of view is shown to the right of the FcRn-GFP image, pseudo-colored in red. To better demonstrate the association of FcRn with EEA1, these two grayscale images are further displayed together as a single overlay image, shown as the third image in the top row of Fig. 9.6. This is done by simply embedding the two grayscale images into an RGB image format to form a "color" image. Here we can clearly see significant areas of overlap between the two signals (see also the inset in the top row of Fig. 9.6 for a zoomed-in version of the overlay image), indicating that FcRn-GFP is located on early endosomes.

To make sure that there are no significant bleed-through or cross contamination problems, control experiments need to be conducted. This can be done, for example, by carrying out the above described experiments with samples that are only labeled with FcRn-GFP and verifying that no significant signals are observed in the acquisition of the image cor-

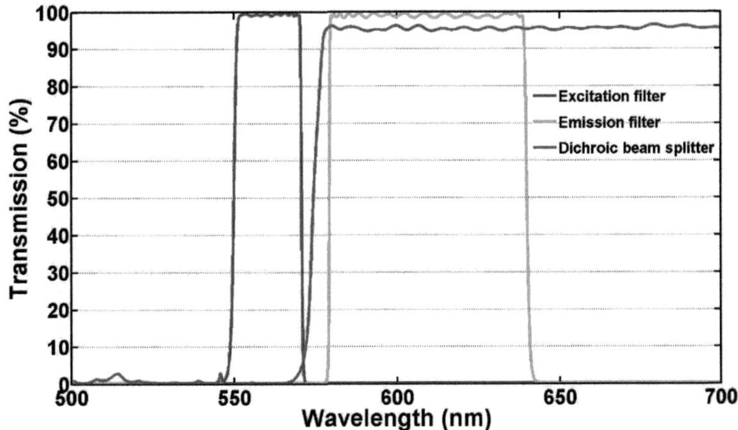

FIGURE 11.2
Transmission spectra of the excitation filter, emission filter, and dichroic beam splitter of a filter set that is optimized for the imaging of Alexa Fluor 568-labeled samples. This particular filter set is manufactured by Semrock (Part # LF561-B-000). [See Figure Credits.]

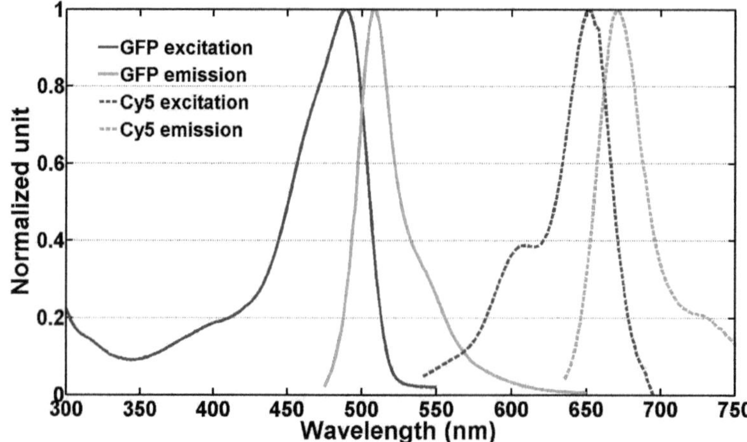

FIGURE 11.3
Excitation and emission spectra of GFP and Cy5. [See Figure Credits.]

responding to the Alexa Fluor 568 filter set. Similarly, we want to make sure that for a sample in which only Alexa Fluor 568 is present, there is no significant signal in the channel corresponding to the GFP filter set.

A similar association experiment can be carried out to investigate the presence of FcRn in lysosomes. Analogous to the use of anti-EEA1 antibody for identifying early endosomes, antibodies that bind to lysosomal-associated membrane protein 1 (LAMP1), a standard lysosomal marker, can be used to identify lysosomes. Using Alexa Fluor 568 to label the anti-LAMP1 antibody, acquisition under the same experimental setting for the imaging of FcRn and EEA1 yields images exemplified by those shown in the bottom row of Fig. 9.6. The FcRn-GFP grayscale image is again peudo-colored in green, while the LAMP1-Alexa Fluor 568 grayscale image is pseudo-colored in red. In this case, the overlay image and its inset show no appreciable overlap between the FcRn-GFP and LAMP1-Alexa Fluor 568 signals, indicating that there is no significant presence of FcRn in lysosomes. We will continue this discussion of the association of FcRn with lysosomes in Section 11.5, where we will make these last statements more precise.

The biological question presented in Section 2.5 addresses the question of how iron is transported by transferrin. Since the transferrin receptor is known to be responsible for the transport of transferrin within cells, it is instructive to compare the intracellular location of FcRn with that of the transferrin receptor. In Fig. 2.8, results are shown for an association experiment that was carried out by pre-pulsing FcRn-GFP-transfected HMEC-1 cells with Alexa Fluor 555-labeled transferrin. In the overlay image, we see a significant association between the FcRn and transferrin signals, as both signals are present on structures that we had identified as early endosomes, as well as on smaller "spots" which are likely smaller transport vesicles in the cell. The FcRn pathway and the transferrin transport pathway therefore show some overlap.

11.1.3 Pulse-chase verification of fate of mutated IgG

Fixed cell experiments are not only conducted to determine the location of proteins in cells (Section 11.1.1) and to correlate proteins with intracellular structures (Section 11.1.2). They are also carried out to obtain insights into the cumulative, relatively long-duration dynamic

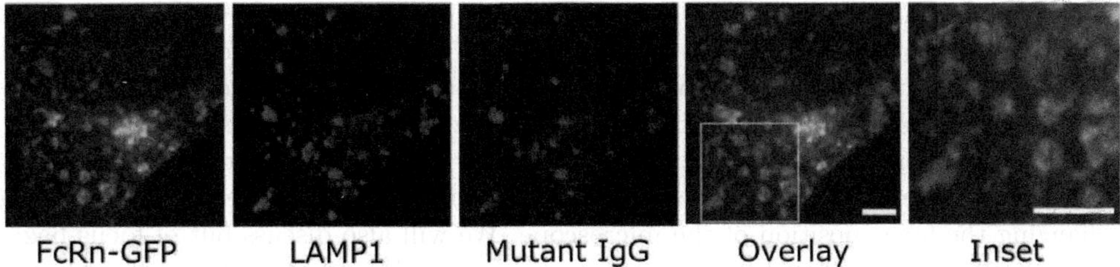

FIGURE 11.4
Verification of fate of mutated IgG with a pulse-chase experiment. Images shown are those of an FcRn-GFP-transfected HMEC-1 cell that was pulsed for one hour with Alexa 647-labeled mutated IgG that does not bind to FcRn, and chased for another hour under physiological conditions. The image of LAMP1-positive compartments is obtained with an Alexa Fluor 568-labeled secondary antibody that binds to an anti-LAMP1 primary antibody. The overlay of the FcRn-GFP, LAMP1, and mutated IgG images demonstrates that the mutated IgG is destined for degradation in lysosomes. The inset shows the region enclosed by the white box in the overlay image. Scale bars: 5 μm. [See Figure Credits.]

developments in a cell. In this type of fixed cell experiment, commonly referred to as *pulse-chase experiments*, cells are first incubated with, and allowed to take up, a fluorescently labeled molecule of interest for a given time duration. Following this pulse period, the cells are chased for a specified time in medium in which they continue to function. To study the intracellular fate of the molecules of interest after the chase period, the cells are then fixed, stained with appropriate markers for compartments of interest, and imaged.

To verify the fate of the IgG in Fig. 9.8 that is mutated so that it does not bind to FcRn, we can conduct the following pulse-chase experiment. We first pulse FcRn-GFP-transfected human endothelial cells with Alexa Fluor 647-labeled mutated IgG for one hour. We then wash the cells to remove excess mutated IgG, and chase them in medium without mutated IgG for another hour. The cells are then fixed and stained with anti-LAMP1 antibody to identify the lysosomes, as we expect the mutated IgG to not be recycled by FcRn and to eventually be degraded in lysosomes.

The fixed cells are then imaged, and Fig. 11.4 provides an example of the results that have been obtained in such an experiment. The figure shows three images of a cell acquired in succession using three different filter sets. An image of FcRn-GFP-positive recycling compartments is shown pseudo-colored in green, an image of LAMP1-positive degradative compartments (visualized with an Alexa Fluor 568-labeled secondary antibody) is shown pseudo-colored in red, and an image of the Alexa Fluor 647-labeled mutated IgG is shown pseudo-colored in blue. An overlay of these three images clearly shows that the mutated IgG introduced during the pulse period has ended up in the lysosomes, and not the recycling compartments, of the cell. This result confirms our expectation that the mutated IgG is destined to be degraded rather than recycled.

11.2 Imaging a 3D sample

In the experiments discussed so far, we have implicitly assumed that we can obtain all of the relevant information by taking one image of a cell from the given focal plane. If this is not the case due to the thickness of the sample, we might be missing important information from parts of the sample that are too far from the focal plane. For thick samples we therefore need to acquire images at different focal positions within the sample in order to obtain more complete information. In this section, we will briefly discuss the experimental design considerations associated with the traditional 3D imaging approach introduced in Section 10.2.1, whereby different focal planes within the sample are sequentially acquired by changing the focus position of the microscope. We will also discuss out-of-focus haze, a common problem encountered in the imaging of 3D samples, due to which relatively weak signals of interest in or near the plane of focus are obscured by strong signals from parts of the sample that are farther away along the optical axis. For the 3D imaging technique introduced in Section 10.2.2, by which different focal planes within the sample are imaged simultaneously rather than sequentially, we defer the discussion of experimental design considerations to Section 11.7, where we also provide two detailed examples of experiments carried out using the technique.

11.2.1 Acquisition of z-stacks

One way to address the problem that a sample is too thick to be represented with a single image is to acquire a z-stack of images (Section 10.2.1), i.e., images at different z positions, or focal levels, throughout the cell. For an example, Fig. 11.5 shows a number of images taken in a z-stack of a human prostate cancer cell with fluorescently labeled plasma membrane, nucleus, and vesicles called exosomes.

In the design of a z-stack experiment, decisions have to be made as to how many images should be taken and at which focal positions. Ideally one would want to say that as many images as possible should be taken at focal positions separated by spacings that are as small as possible. However, there are many reasons why this might not be a good approach. Foremost amongst them is that fluorescent labels photobleach (Section 8.4). This means that often there is a severe limit on the number of images that can be taken before the fluorescence signal is diminished to the point where good images can no longer be acquired. The number of images that is to be taken would also depend importantly on the purpose of the experiment. Z-stacks acquired for the study of image analysis approaches, such as deconvolution (Chapter 21), typically require more images with smaller focal spacings than z-stacks that are acquired for the visual localization of, for example, the distribution of large organelles in a cell.

11.2.2 Out-of-focus haze

In standard fluorescence microscopy experiments, the sample is uniformly illuminated with the excitation light. This means that all fluorescently labeled parts of the sample will emit fluorescent light. Consider Fig. 11.6(a), where we show a simplified representation of a cell with three fluorescently labeled organelles. If we take an image such that organelles A and B are in focus, we will see focused images of organelles A and B (Fig. 11.6(b)). In addition, we will observe a diffuse and hazy image of organelle C. The appearance of the out-of-focus image is due to the image formation process along the optical axis, which we will investigate in detail in Chapter 14.

This aspect of the image formation process has significant implications for the imaging of a 3D sample. Specifically, small or weakly labeled parts of a sample can easily be obscured by the emission from strongly labeled parts of the sample that are located at other positions along the optical axis. In fact, various techniques have been developed to address this problem, at least partially. For example, confocal imaging has advantages in that the pinhole structure can exclude out-of-focus light (Section 10.4). Another example is two-photon

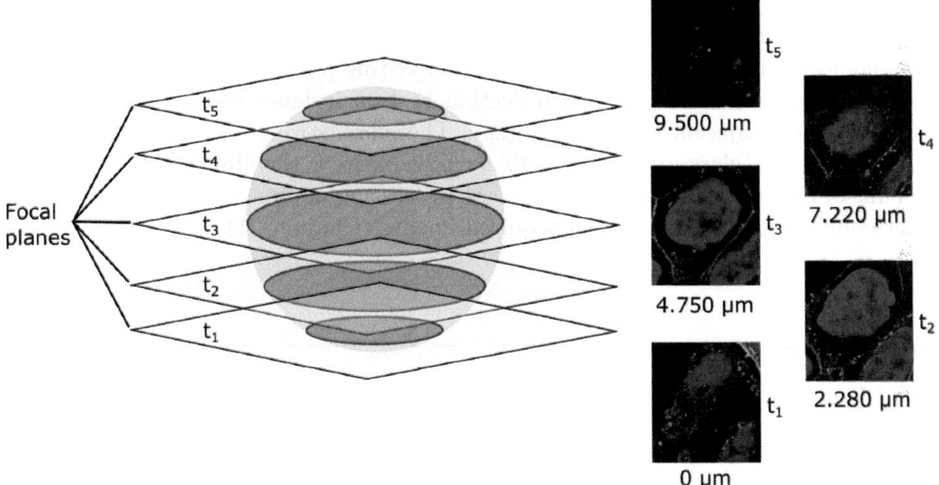

FIGURE 11.5
Z-stack schematic and example. The schematic shows a z-stack to comprise a set of images acquired sequentially at different focal positions within the sample. The example shows five images from a z-stack acquisition of a human prostate cancer cell. In both the schematic and the example, the symbols t_1 through t_5 are time points that indicate the order of image acquisition. In the example, the bottom image and the top image were acquired at focal positions corresponding to the bottom and top of the cell, respectively. Shown below each image is the precise focal position at which that image was acquired. The plasma membrane of the cell, displayed in green, was stained with Alexa Fluor 488-conjugated cholera toxin B subunit (CTxB), a protein which binds to the lipids in the membrane. The cell nucleus, displayed in blue, was stained with Hoechst dye, which binds to DNA. Vesicles called exosomes, displayed in red, were detected with a mouse antibody that was specific for CD63, a protein marker for intracellular vesicles. The fluorescence signal associated with the exosomes was obtained with an Alexa Fluor 555-conjugated secondary antibody that was specific for mouse antibody Fc regions. Scale bar: 4 μm.

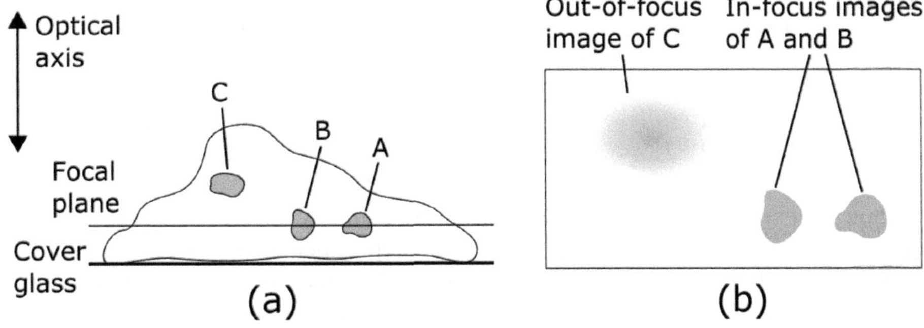

FIGURE 11.6
Out-of-focus haze. (a) A cell contains fluorescently labeled organelles A, B, and C, of which A and B are located at the focal plane. (b) Organelles A and B are in-focus in the acquired image, while the out-of-focus organelle C appears as a diffuse and hazy mass.

excitation microscopy (Section 10.5), in which flurorescence emission is effectively limited to the focus spot of the excitation laser by the requirement that a fluorophore is excited only by the near-simultaneous absorption of two exciting photons. Total internal reflection microscopy, which will be introduced in Section 11.4, is designed to only illuminate a thin layer of the sample adjacent to the cover glass. Therefore, strongly labeled parts of the cell away from the cover glass are not, or only weakly, excited, thereby significantly reducing the amount of out-of-focus light. Lastly, image processing approaches have been developed to remove out-of-focus light from the acquired images (Chapter 21).

11.3 Live cell experiments

When living cells are imaged, the experiments that are carried out can be significantly more complex than fixed cell experiments. The purpose of a live cell experiment is typically to observe the dynamics in the sample. The problem then is to match the image acquisition speed to the dynamics that are to be observed. If the dynamics are sufficiently slow, an experimental setting described in the prior section on fixed cell imaging (Section 11.1) can be used. In this case, many aspects of the experimental design are very similar to those described there. Of course, an important point is that the sample has to be imaged at different time points so that dynamics can be observed that are of interest. This will again lead to considerations such as the necessity to avoid severe photobleaching problems.

The reason that a typical fixed cell configuration may not suffice for capturing fast dynamic events is the need to image samples labeled with different fluorophores at a high acquisition rate. In the fixed cell configuration, images of the different fluorophores are sequentially acquired. This sequential imaging strategy is time-consuming, and is made worse by the usually significant time that it takes to switch the different filter sets in and out of the light path.

There are different approaches to the imaging of fast dynamics in live cells. All aim at reducing the time-consuming parts of the acquisition process in a standard fixed cell microscope configuration. A basic approach is to keep the flow of the sequential acquisition as shown in Fig. 10.3(a), but to use instrumentation that permits faster changes between the imaging settings for the different fluorophores (see, for example, the discussion of the use of filter wheels in Section 10.7.2). Even further improvements in speed, however, can be made by effectively parallelizing the acquisition process to allow the simultaneous capturing of images corresponding to the different fluorophores. This can be done by employing multiple detectors, one for each fluorophore, as illustrated in Fig. 10.3(b). This approach is more expensive to implement and, importantly, requires the use of a more sophisticated combination of optical filters (see the discussion and the simultaneous dual-color imaging example in Section 10.7.2). However, it represents the most time-efficient way to carry out live cell imaging.

The choice of exposure time is always a serious issue in any fluorescence microscopy imaging experiment. In the case of live cell microscopy experiments, this is particularly critical as there is a tradeoff between the need to minimize photobleaching of the fluorophores and the need to obtain images with sufficient signal from the fluorophores to produce meaningful images. A further variable is the chosen magnification and, related to this, the pixel size of the detector. The smaller the magnification for a given pixel size, the higher the number of collected photons per pixel, but this higher photon count is gained at the expense of potentially losing significant details in the image due to the relatively coarse pixelation.

Further problems in live cell microscopy that should not be underestimated are related to the necessity to image live cells at 37°C rather than at room temperature. This requires the use of a heating system. Heating enclosures for the microscope, or heating systems for the objective, are often used. A concern with all these systems is significant sample drift due to the temperature gradients in the system. This can make frequent refocusing a necessity, which can be problematic for the analysis of the data.

11.3.1 Example: FcRn-mediated IgG trafficking

We can perform live cell microscopy to continue the study of the role of FcRn in human endothelial cells. To investigate how IgG molecules are trafficked by FcRn, we carry out a pulse-chase experiment, in which FcRn-GFP-transfected human endothelial cells are first incubated with fluorescently labeled IgG molecules for one hour. Immediately following the pulse period, the cells are washed and then imaged at 37°C with a single camera at a rate of two frames, one of FcRn and one of IgG, every 1.2 s, and with an exposure time of 500 ms per frame.

The considerations for the choice of fluorophore for the labeling of the IgG molecules are similar to those we encountered when we discussed association experiments, but include two important additional issues to account for. In contrast to fixed cell microscopy experiments, photobleaching becomes a very important issue as the total exposure time for the sample is typically much longer in live cell experiments. Furthermore, reagents that reduce photobleaching, such as antioxidants, can mostly not be used in live cell experiments as they can interfere with cellular processes. For our current example, the relatively photostable Alexa Fluor 546 is chosen as the label for IgG. The second issue relates to the choice of dichroic beam splitter and emission filter. Due to the the need to image the dynamics of both FcRn-GFP and the fluorescently labeled IgG molecules at a fast rate, only one fixed dichroic beam splitter is used in the light path, rather than having two that need to be switched in and out of the light path as in the association experiments discussed in Section 11.1.2. As FcRn and IgG are distinguished by differently colored fluorophores that are maximally excited by light in two different wavelength ranges, this single dichroic beam splitter must have reflection bands that direct both ranges of excitation wavelengths toward the sample, and passbands that transmit fluorescence in both emission wavelength ranges of the two fluorophores. Given that a sequential imaging approach is used to capture images of FcRn and IgG in alternating fashion on a single detector, one possible solution for the fluorescence detection is to use an emission filter for each fluorophore and to switch them in and out of the light path using a filter wheel that is positioned directly in front of the detector. For our current example, however, we instead use a single dual-band emission filter that has passbands for transmitting both emission wavelength ranges. This single-filter solution has the advantage of eliminating the time needed to turn and stabilize the filter wheel after the acquisition of each image.

In terms of sample illumination, a 488-nm laser is used to excite FcRn-GFP and a 543-nm laser is used to excite Alexa Fluor 546-labeled IgG. Shutters (Fig. 10.20) are used to repeatedly switch between the two lasers, alternately allowing only one of the laser beams to reach the sample during the acquisition of each image.

The transmission spectrum of a multi-band dichroic beam splitter that is appropriate for the imaging of GFP and Alexa Fluor 546 is shown in Fig. 10.22. The multi-band dichroic beam splitter is again important because of the speed advantage that it confers on a sequential multi-color image acquisition. As mentioned in Section 10.7.2, however, compared with single-band dichroic beam splitters, less ideal optical properties can often be expected for multi-band dichroic beam splitters due to the more complex design requirements needed to support the excitation and emission light paths for two or more fluorophores. As previously pointed out, for example, use of the multi-band beam splitter of Fig. 10.22 can be expected to result in the detection of a weaker GFP signal because the beam splitter does not capture the emission wavelength range of GFP as fully.

Examples of images of the trafficking of Alexa Fluor 546-labeled IgG by FcRn-GFP in live HMEC-1 cells, acquired in an experiment as described, are shown in Fig. 9.8. In both sets of images shown in the top row, the FcRn and the IgG channels are overlaid in green and red, respectively. In both cases, there is clear overlap between FcRn and IgG as they are both transported from a sorting endosome in a tubule. These results provide visual evidence for the transport of IgG by FcRn.

In a similar experiment where wild-type IgG molecules were replaced with mutated IgG molecules that no longer bind to FcRn, live cell imaging reveals a very different behavior for IgG. The two sets of images in the bottom row of Fig. 9.8 show that the mutated IgG molecules primarily localize in the vacuoles of the FcRn-positive sorting endosomes. In contrast to the experiment with wild-type IgG, FcRn-positive tubules that leave the endosomes do not contain mutated IgGs. This difference in transport behavior corroborates the understanding that FcRn is indeed responsible for the transport of IgGs in cells. Sorting endosomes, in particular, appear to play a central role in the transport of IgG molecules in FcRn-expressing cells.

11.4 Total internal reflection fluorescence microscopy (TIRFM)

A persistent problem when imaging cells is the out-of-focus fluorescence that can provide very significant background to the part of the cell that we want to image (Section 11.2.2). Such out-of-focus fluorescence can have detrimental effects on image quality, in particular by overwhelming low signals. Therefore, finding ways to selectively excite the part of the cells that we seek to image has been a high priority. Unfortunately, this is generally not a straightforward task. One particular technique, *total internal reflection fluorescence microscopy* (TIRFM), solves this problem very well for one particular situation. With this approach a very thin layer of the sample, just above the cover glass, can be illuminated. In fact, the illumination level decreases exponentially with increasing distance from the cover glass. The result is that in most circumstances only the cellular membrane adjacent to the cover glass is illuminated. This therefore provides an excellent imaging approach to investigate dynamic processes on the plasma membrane with significantly less background signal due to fluorescently labeled structures within the cell (Note 11.2).

The principle behind TIRFM is total internal reflection, which is discussed in Section 13.1.4. If rays of light travel through a medium of a certain refractive index n_1 and encounter, at a planar interface, another medium with refractive index n_2 such that the angle of incidence θ_1 is larger than the critical angle given by $\theta_c = \arcsin\left(\frac{n_2}{n_1}\right)$, then the incident beam will be reflected and not refracted. The definition of the critical angle already implies that total internal reflection is only possible when $n_2 < n_1$, i.e., when the refractive index of the second medium is lower than the refractive index of the first medium.

11.4.1 Objective-based total internal reflection fluorescence microscopy

In the most popular implementation of TIRFM, the illuminating beam is delivered to the sample through the objective. The illumination light therefore travels through the objective, the immersion medium, and the cover glass to the sample (Fig. 11.7). The interface at which

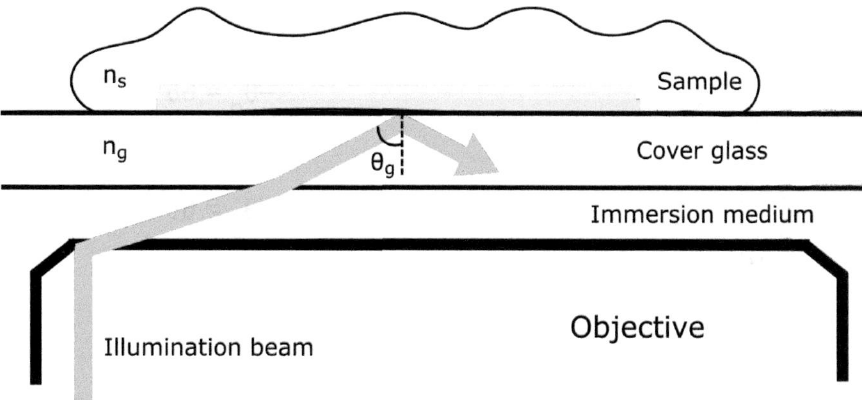

FIGURE 11.7
Objective-based total internal reflection fluorescence microscopy. The illumination beam travels through the objective, the immersion medium, and the cover glass, and impacts the cover glass-sample interface at an angle of incidence θ_g that is larger than the critical angle. The beam is reflected back into the cover glass, but an evanescent wave is generated that illuminates only a thin layer of the sample adjacent to the cover glass. The symbols n_g and n_s denote the refractive indices of the cover glass and the sample, respectively.

total internal reflection of the illuminating beam has to be achieved is the interface between the cover glass and the sample.

A living cell sample has to be cultured in a dish in an appropriate cellular medium or buffer. These liquids will have a refractive index close to that of water, i.e., close to $n_s = n_2 = 1.33$. The refractive index of the cover glass is close to $n_g = n_1 = 1.52$. For this situation, the condition on the possibility of total internal reflection is satisfied as $\frac{n_s}{n_g} = 0.875 < 1$. In fact, the critical angle for this example is $\theta_c = 61°$. To be able to carry out TIRFM, an illumination beam needs to be delivered from the objective that has an incident angle θ_g at the cover glass-sample interface that is larger than this critical angle.

On the opposing side of the cover glass, an evanescent wave will be formed which will illuminate only a thin layer of the sample just above the cover glass. Specifically, this evanescent wave will have an intensity I that decreases exponentially with the distance z from the cover glass as

$$I = e^{-\frac{z}{d}}, \tag{11.1}$$

where

$$d = \frac{\lambda}{4\pi}\left(n_g^2 \sin^2(\theta_g) - n_s^2\right)^{-\frac{1}{2}},$$

with λ the wavelength of the illuminating light. This shows that in the case where the refractive indices are as given above, if the illuminating light has an incidence angle of $\theta_g = 65°$ and a wavelength of $\lambda = 488$ nm, the strength of the evanescent wave decreases to half of its value at the cover glass-sample interface at a depth of $z = 75$ nm into the sample, and to 10% of its value at the cover glass-sample interface at $z = 249$ nm into the sample (Fig. 11.8). Equation (11.1) further shows that to achieve illumination of an even thinner layer of the sample for the same wavelength of the illuminating light and the same glass and sample refractive indices, one can increase the incidence angle of the illuminating light at the cover glass-sample interface. For example, from Fig. 11.8, we see that by increasing the incidence angle by 5° to $\theta_g = 70°$, the evanescent wave intensity decreases to the 50% and 10% levels at the shallower depths of $z = 52$ nm and $z = 172$ nm into the sample,

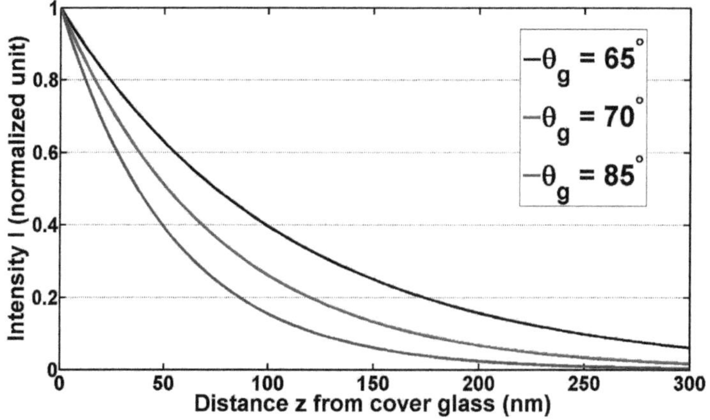

FIGURE 11.8
Exponential decay of evanescent wave strength in TIRFM. For three different angles of incidence θ_g of the illuminating light at the cover glass-sample interface, curves are shown for the intensity I of the evanescent wave as a function of the distance z from the interface. The wavelength of the illuminating light is $\lambda = 488$ nm, and the refractive indices of the cover glass and the sample are $n_g = 1.52$ and $n_s = 1.33$, respectively.

respectively. The figure also shows that increasing the angle further to $\theta_g = 85°$ reduces those depths even more to $z = 37$ nm and $z = 124$ nm.

A practical illustration of total internal reflection illumination with different angles of incidence is given in Fig. 11.9. The images shown are of two fixed human prostate cancer cells stained with Alexa Fluor 555-conjugated cholera toxin B subunit, which binds to lipids in the plasma membrane. A visual comparison shows that with increasing angle of incidence, out-of-focus fluorescence from the interior of the cell is reduced as the penetration of the resulting evanescent wave is reduced, and Alexa Fluor 555-labeled structures on the plasma membrane can therefore be seen with increased clarity. An image of the cells acquired with widefield illumination is also shown. Compared with the TIRFM image acquired with the largest angle of incidence, we can clearly see from this widefield image that the signals

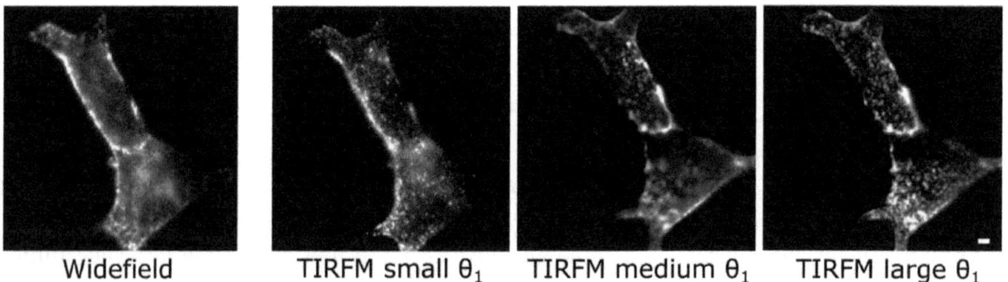

FIGURE 11.9
Plasma membrane imaged using widefield illumination and total internal reflection illumination with three different incidence angles θ_g. The image is of two adjacent fixed human prostate cancer cells. The plasma membrane is stained with Alexa Fluor 555-conjugated cholera toxin B subunit. Scale bar: 2.5 μm.

from the Alexa Fluor 555-labeled structures on the plasma membrane are almost entirely overwhelmed by out-of-focus haze from the cell interior.

11.4.2 Exocytosis imaged by total internal reflection fluorescence microscopy

Exocytosis is the fundamental cellular process by which intracellular vesicles and tubules merge with the plasma membrane and discharge their contents into the extracellular space. This is a means of moving cargo from within the cell to its exterior. In addition, this is an important process by which receptors are transported from the interior of the cell to the plasma membrane. Imaging these vesicles and tubules is particularly challenging in live cell microscopy as they often move very rapidly and are typically very weakly labeled. Using conventional illumination approaches will illuminate many structures within the cell that produce out-of-focus fluorescence (Section 11.2.2) that can easily obscure the exocytic processes.

Here we illustrate TIRFM by continuing our discussion of FcRn-mediated IgG transport. We use an imaging configuration where we simultaneously image FcRn-GFP and Alexa Fluor 546-labeled IgG. Specifically, GFP and Alexa Fluor 546 are simultaneously excited in TIRFM mode with a 488-nm laser and a 543-nm laser, respectively. Moreover, as illustrated in Fig. 10.21, the simultaneous illumination by the two lasers and the separate but simultaneous detection of GFP and Alexa Fluor 546 fluorescence are realized by using a multi-band dichroic beam splitter in a two-camera configuration. Note that if we are only interested in investigating exocytosis of either IgG or FcRn, we could use a one-camera system equipped with a single-band dichroic beam splitter. Here, however, we are interested in determining whether or not IgG and FcRn exocytic events are associated with one another, and therefore require a more complex experimental setup.

To image FcRn-mediated IgG exocytosis, we pulse FcRn-GFP-transfected HMEC-1 cells with Alexa Fluor 546-labeled IgG and follow that with an exchange of medium. The medium exchange is designed to remove fluorescently labeled IgGs from the medium that would otherwise contribute to the background signal, despite the use of total internal reflection excitation. Image acquisition using the described setup then produces data such as that presented in Fig. 11.10. The figure shows a time sequence of images displaying the overlay of the GFP and Alexa Fluor 546 channels, which were separately but simultaneously acquired by two grayscale cameras. In this sequence, an exocytic compartment approaching the plasma membrane is observed as an increase in the intensities of both GFP and Alexa Fluor 546 at the exocytic site delineated by the white box. A characteristic "puff" is then seen, in both color channels, as the exocytic compartment merges with the plasma membrane at 1.5 s and releases IgG molecules into the extracellular space. The "puff" marks the peak in the intensities of both channels at the exocytic site, following which the diffusion of FcRn into the plasma membrane and of IgG into the extracellular space are observed as a decrease in the fluorescence intensity. The accompanying fluorescence intensity plot provides a quantitative description of these same events. Importantly, as these events coincide in the FcRn and IgG channels, we can conclude that IgG exocytosis is mediated by FcRn.

11.5 pH measurement and ratiometric imaging

The fluorescence properties of a molecule often depend on its environment. Particularly relevant is the dependence of the excitation and emission spectra on the pH of the environment,

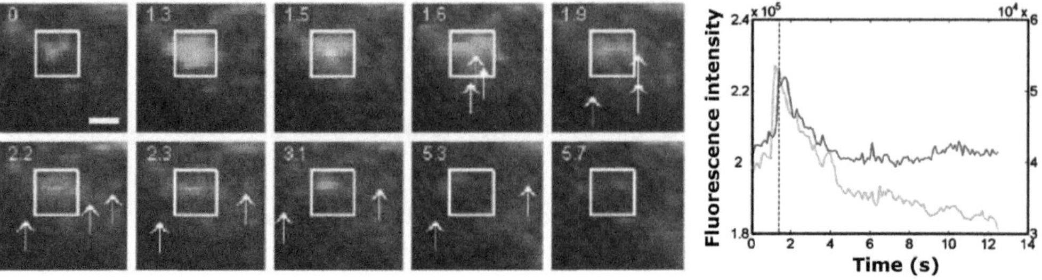

FIGURE 11.10
FcRn-mediated exocytosis of IgG in an HMEC-1 cell. Images were acquired with a dual-color TIRFM setup whereby FcRn-GFP and Alexa Fluor 546-labeled IgG were separately but simultaneously imaged with two cameras. The image sequence, with time indicated in seconds, displays the grayscale images from the two cameras in overlay, with FcRn pseudo-colored in green and IgG pseudo-colored in red. In each image, the white box delineates the exocytic site, and when present, a white arrow indicates an individual IgG molecule. The accompanying plot shows, as a function of time, the GFP (green) and the Alexa Fluor 546 (red) fluorescence intensities at the exocytic site. The vertical dashed line in the plot marks the time point (1.5 s) at which IgG molecules are released into the extracellular space. The intensities are given by the sum of the values of the pixels comprising the exocytic site. The left axis shows the intensity for GFP, and the right axis shows the intensity for Alexa Fluor 546. Scale bar: 1 μm. [See Figure Credits.]

as the pH of the cellular environment can vary from about 7.2 in the cytosol to 6.0–6.5 in early endosomes and 4.5–5.0 in lysosomes. Therefore, if during an experiment the fluorescence level of a probe changes, it is important to ensure that it is not due to changes in the pH of the environment. For example, in a live cell experiment, GFP-tagged protein might be detected in one subcellular compartment, such as an early endosome, but not in another, for example a lysosome. Examining GFP emission as a function of pH (Fig. 11.11) shows that the emission at lysosomal pH is significantly lower than the emission at endosomal pH. Therefore the protein might in fact be present in the lysosomes with the fluorescence being too low for detection.

This leads us to revisit the association experiment discussed in Section 11.1.2, involving FcRn and lysosomes. The experiment described there used GFP as a label for FcRn. Therefore, considering the significantly reduced GFP emission at lysosomal pH, it is important to examine the experiment in more detail. The labeling of FcRn with GFP was done by attaching GFP to the C-terminus of FcRn. Since FcRn is a type I transmembrane protein (Fig. 9.3), this means that the expression of FcRn-GFP on membranes is such that GFP is exposed to the cytosol. The result of the experiment thus indicates that there is no significant presence of FcRn on the lysosomal membrane. However, based on this experiment we cannot say much in terms of the presence of FcRn in the lumen of lysosomes. The absence of GFP signal from the lysosomal lumen could be due to FcRn not being there, or it could be due to the GFP signal being undetectable because of the low lysosomal pH. In fact, in experiments using the pH-insensitive mRFP1 as a label for FcRn, it has been shown that FcRn can be detected in the lumen of lysosomes. However, since lysosomes are the compartments responsible for breaking down proteins and other macromolecules, it is not completely surprising to find evidence of any protein in lysosomes.

The pH dependence of fluorophores can be used to our advantage, as we can use it to develop experiments to determine the pH of the environment of the fluorophore. If we

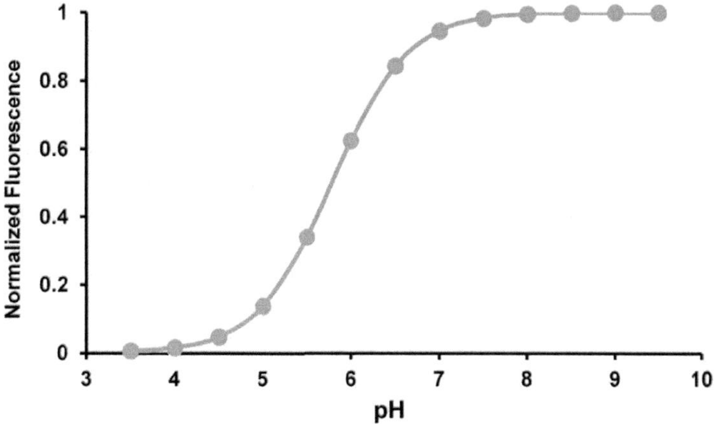

FIGURE 11.11
Dependence of EGFP fluorescence emission on the pH. The plot shows the fluorescence intensity of EGFP suspended in buffers with pH values ranging from 3.5 to 9.5. The fluorescence intensity weakens with decreasing pH. [See Figure Credits.]

consider the pH dependence of the excitation spectrum of fluorescein (Fig. 11.12(a)), we see that the spectrum exhibits a significant dependence on the pH at wavelengths near 490 nm. Therefore, the pH of the environment can be determined by measuring the fluorescence intensity that results from illumination with 490-nm light. The problem with this simple approach is that a change in the fluorescence intensity can be due to something other than a change in the pH. The photobleaching of the fluorophore or a change in the concentration of the fluorophore, for example, can reduce the measured fluorescence intensity even though the pH has not changed. An approach to overcome this problem is to carry out ratiometric

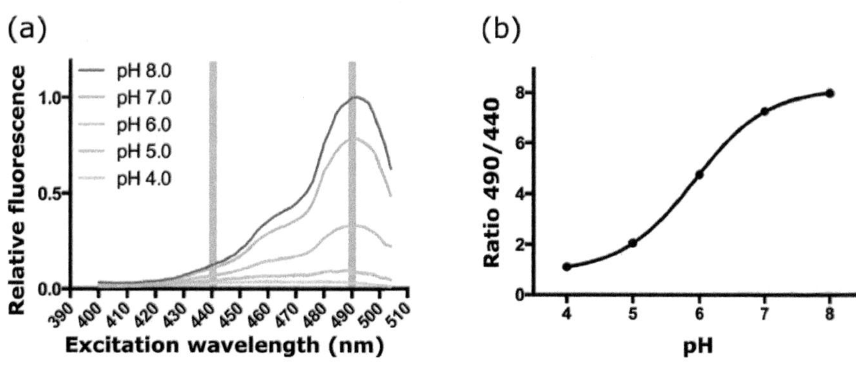

FIGURE 11.12
Dual-excitation ratiometric imaging with fluorescein. (a) pH-dependent excitation spectra of fluorescein, with emission measured at 520 nm. The spectra were acquired using fluorescein-labeled dextran dissolved in calibration buffers of known pH values. (Dextran, which trafficks to lysosomes after cellular uptake, is used here as the experimental goal is to measure lysosomal pH.) The gray bars indicate the excitation wavelengths (490 nm and 440 nm) to be used for the ratiometric imaging. (b) Calibration curve generated from the spectra of (a) by calculating the ratio of the fluorescence intensities corresponding to excitation wavelengths of 490 nm and 440 nm. [See Figure Credits.]

imaging in which the fluorescence intensity is measured sequentially at two different excitation wavelengths, and the pH is determined by the ratio of the two measured intensities. The use of a ratio corrects for events such as photobleaching and changes in fluorophore concentration, as these events affect the fluorescence intensity in the same way regardless of the excitation wavelength. By taking the ratio of fluorescence intensities measured in controlled environments of known pH values, a calibration curve can be generated that maps intensity ratios to pH values. Such a curve can then be used in an experiment to assign a measured intensity ratio to a specific pH value.

In choosing the two excitation wavelengths to use, it is important to ensure that the resulting fluorescence intensity ratios provide a good indicator of the pH value. A commonly employed approach is to select the first excitation wavelength to be one at which the fluorescence emission is very sensitive to pH changes, and the second excitation wavelength to be one at which the fluorescence emission is minimally sensitive to pH changes. In this way, fluorescence intensity ratios are obtained that can be used to effectively discern pH values over a useful range. In the case of fluorescein, we see from Fig. 11.12(a) that we can choose 490 nm to be the first wavelength and 440 nm to be the second wavelength based on the significant and minimal sensitivity, respectively, of the fluorescence emission to the pH at these excitation wavelengths. In Fig. 11.12(b), a calibration curve is shown for fluorescein that gives the relationship between the pH and the fluorescence intensity ratio obtained using the 490-nm and 440-nm excitation wavelengths (Note 11.5). It is worth noting that unlike fluorescein, there are fluorophores, such as the wild-type GFP mutant ratiometric pHluorin, whose excitation spectrum is characterized by high pH sensitivity in two different wavelength ranges, in such a way that as the pH increases, the fluorescence intensity increases upon excitation with light in one wavelength range, but decreases upon excitation with light in the other wavelength range. With such fluorophores, the two excitation wavelengths are chosen near the peaks of the two wavelength ranges (Note 11.6).

What we have described above is the dual-excitation approach to ratiometric imaging, in which the sample is excited at two wavelengths and the fluorescence emission is measured at one wavelength. Provided that a suitable fluorophore is used, the dual-emission approach, based on the same principles but looking instead at the pH-dependent emission spectrum of the fluorophore, can also be employed in which the sample is excited at one wavelength and the emission is measured at two wavelengths.

In the calculation of the ratio of fluorescence intensities, care needs to be taken to adjust the intensities for background fluorescence that is not associated with the fluorescence of interest, as otherwise significant errors can occur (Exercise 11.1).

11.6 Single molecule microscopy

11.6.1 Bulk versus single molecule experiments

When we consider the images produced by a fluorescence microscopy experiment, such as those of GFP-labeled FcRn and Alexa Fluor-labeled endosomal and lysosomal proteins EEA1 and LAMP1 in Fig. 9.6, we need to realize that what we see is the result of emitted light from many labeled molecules. We therefore refer to this type of imaging as a *bulk microscopy experiment*. However, when we motivate the use of fluorescence microscopy techniques, such as in Section 3.3, we think of fluorescent molecules that are used to label individual specific molecules in the cell (e.g., Fig. 9.4). The reason why we cannot see the individual molecules in images such as those in Fig. 9.6 is twofold. First, in classical fluo-

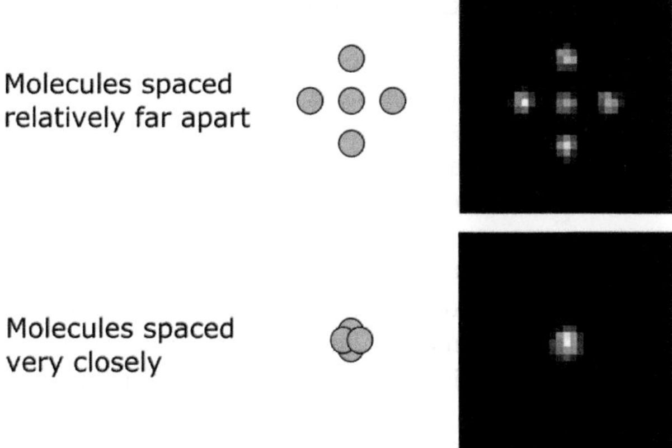

FIGURE 11.13
Illustration of the crowding problem in single molecule microscopy. When individual molecules are spaced sufficiently far apart, they can be distinguished as separate diffraction spots in the acquired image. When they are spaced very closely, they become difficult to distinguish as their diffraction spots overlap significantly in the acquired image. In the simulated image for the uncrowded scenario, each of the four outer molecules is located 1 μm away from the center molecule. In the simulated image for the crowded scenario, each of the four outer molecules is located only 200 nm away from the center molecule.

rescence microscopy the signal produced by an individual fluorescent molecule is typically too low to be detected above the background of scattered photons and the noise level of the camera. With modern, highly sensitive and low-noise cameras this reason is becoming less important. In fact, the development of highly sensitive cameras that allow for the visualization of individual molecules was one of the major advances that made possible the advent of *single molecule microscopy*.

The second reason is much more fundamental and has to do with the inherent limitations of optical microscopes. We know from our basic discussion concerning the resolution of an optical microscope (Section 4.2) that it is difficult to distinguish point sources that are spaced too closely together. Here we can roughly think of a distance of several hundred nanometers as the cutoff below which distinguishing individual point sources becomes difficult. In a cellular context, we can expect many of the types of molecules that we want to image to be spaced much more closely than this. So from an optical point of view these molecules are "crowded" together. We speak of the *crowding problem* to refer to the situation in which the molecules are spaced too closely to be easily distinguishable when imaged through an optical microscope. If we manage to prepare our sample in such a way that the fluorescently labeled molecules are spaced "far" apart, then we can hope to image individual molecules. These two scenarios are illustrated in Fig. 11.13. The approach of preparing the sample in a way that minimizes the crowding problem is often used in single molecule tracking experiments (Section 11.6.2), where in many cases it is sufficient to label a few representative molecules, rather than all of the molecules of the specific type that is being studied (Fig. 11.14). Localization-based super-resolution microscopy is another important approach that addresses the crowding problem to obtain high-resolution structural information. Here, a complete image is obtained by iteratively imaging sparse subsets of the molecules that make up the full sample. This method will be discussed in more detail in Section 11.6.3.

11.6.2 Single molecule tracking experiments

Single molecule tracking experiments aim to capture the trajectory of one molecule, for example a receptor or ligand of a receptor, as a function of time. If the movement of interest is in a cell, this cell would usually be a live cell, and we are therefore in the realm of live cell microscopy. As discussed above, in order to be able to observe a single molecule, the experimental setting needs to be such that it can be observed without incurring the crowding problem.

Photobleaching of the fluorescent label is one of the most serious problems for single molecule tracking. Essentially all currently available fluorescent molecules bleach so rapidly in aqueous solutions that, under normal conditions, tracking is impossible for more than a few seconds. This can be improved in some situations by attaching multiple fluorophores to the tracked molecule, but it nevertheless poses a fundamental problem. One approach to overcome this problem is to use QDs (Section 8.7.2) as they can maintain fluorescent properties for extended periods of time. Their relatively large size and more complex labeling process can, however, pose problems of their own.

In Section 11.4, we discussed TIRFM for the analysis of dynamic events at or close to the plasma membrane. In particular, we showed in Section 11.4.2 how TIRFM can be used to study exocytosis at the plasma membrane. The experimental setup for a single molecule TIRFM tracking experiment is no different from that of a standard TIRFM experiment. Special consideration needs to be given to having a highly sensitive camera with the lowest possible noise levels, especially for the readout noise. The camera also needs to support the frame rate that is required to obtain satisfactory tracking information. Further considerations for the choice of a camera will be given in Chapter 12. Great care needs to be taken with the choice of camera type. For example, an EMCCD camera (Section 12.3.3) should be used only when it provides better performance than a CCD or CMOS camera, i.e., when the photon count per pixel is very low (Sections 17.4 and 18.6.2). The dual-color TIRFM experiment described in Section 11.4.2 utilizes cameras that satisfy the sensitivity and frame rate requirements for capturing the trajectories of individual molecules. Examples of trajectories of IgG molecules immediately following an exocytic event, captured using the dual-color setup, are shown in Fig. 11.15. The figure also provides examples of trajectories of individual FcRn molecules imaged in a similar, but single-color, TIRFM experiment. Each of the

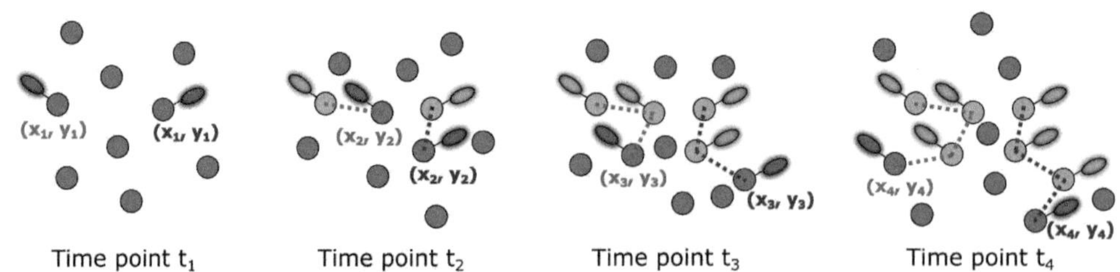

FIGURE 11.14
Single molecule tracking facilitated by the labeling of a small subset of the molecules of interest. The schematic shows the tracking of two fluorescently labeled molecules over four sequentially acquired images. The fact that most of the molecules of interest are not labeled simplifies the task of determining the movement of each of the two labeled molecules from image to image and the task of accurately estimating the positional coordinates of each of the two molecules from each image. In the illustration, the gray-colored molecules indicate the positions of the two labeled molecules in previous images, and the dashed lines and positional coordinates specify the trajectories of the two molecules.

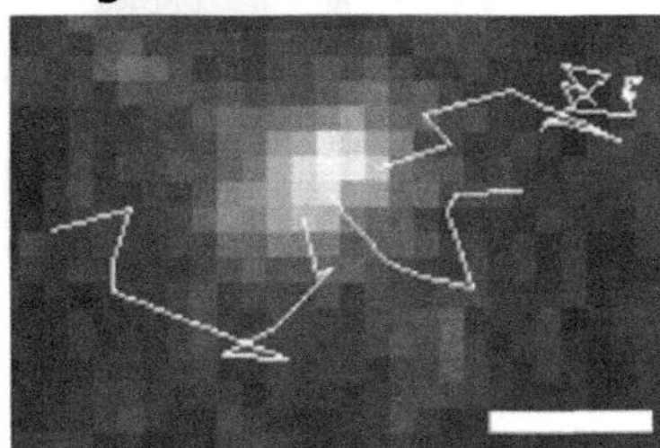

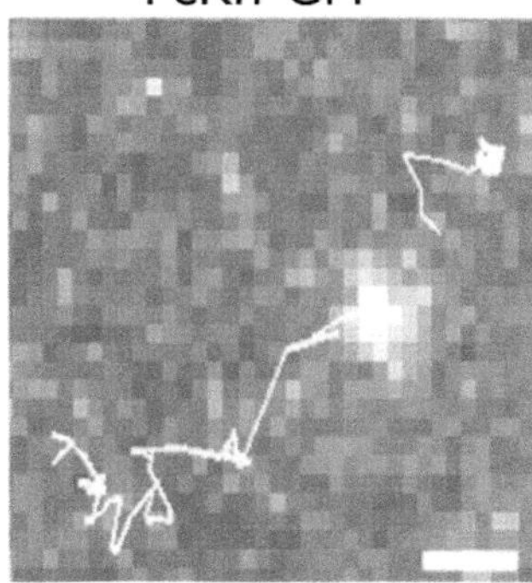

FIGURE 11.15
2D trajectories of individual IgG (left image) and FcRn (right image) molecules immediately following exocytosis, imaged using TIRFM. The IgG molecules are labeled with Alexa Fluor 546, and the FcRn molecules are labeled with GFP. Each image shows the trajectories of three individual molecules, superimposed on the image capturing the moment when the exocytic compartment merges with the plasma membrane. Each trajectory is determined by localizing the molecule in consecutive images. Scale bars: 1 μm. [See Figure Credits.]

trajectories shown was obtained by localizing the molecule in consecutive images using a model-based approach that estimates the molecule's positional coordinates via the fitting of a model (such as the Airy profile of Eq. (4.1)) to a region of the image that contains the molecule. Model-based localization is a commonly used approach for determining the trajectories of single molecules. The 3D trajectories of the QD-labeled molecules in Fig. 9.9 and the example of Section 11.7.2, for example, were also obtained using this approach. Model-based localization and, more generally, model-based parameter estimation, will be discussed in detail in Chapter 16.

11.6.3 Localization-based super-resolution microscopy

Localization-based super-resolution microscopy exploits an important property of some fluorophores in order to solve the crowding problem that is responsible for the limitations in resolution (Note 11.8). This key property is *stochastic excitation*. Given appropriate experimental conditions, such as specific excitation schemes and/or buffer conditions, some fluorophores exhibit stochastic excitation. By this we mean that the fluorophore will either be in an "off" state or in an "on" state. In the former case, the fluorophore resides in a stable *dark state* in which it cannot be excited and therefore cannot emit fluorescence. In the latter case, the fluorophore is in an activated state in which it can absorb excitation light and potentially emit fluorescence as a result. The amount of time that such a fluorophore spends in the "on" and "off" states is stochastic, but the use of appropriate experimental conditions allows the tuning of the proportion of molecules in the "on" state versus molecules in the "off"state to ensure that the small number of molecules that are emitting fluorescence at any given point in time are sufficiently sparsely located so that the crowding problem is overcome.

Localization-based super-resolution microscopy experiments are based on the premise that a large number of images are taken of the sample that is subjected to stochastic excitation conditions. In each of the images, the fluorescence from only a small number of isolated single fluorophore molecules is expected to be detected. The locations of these

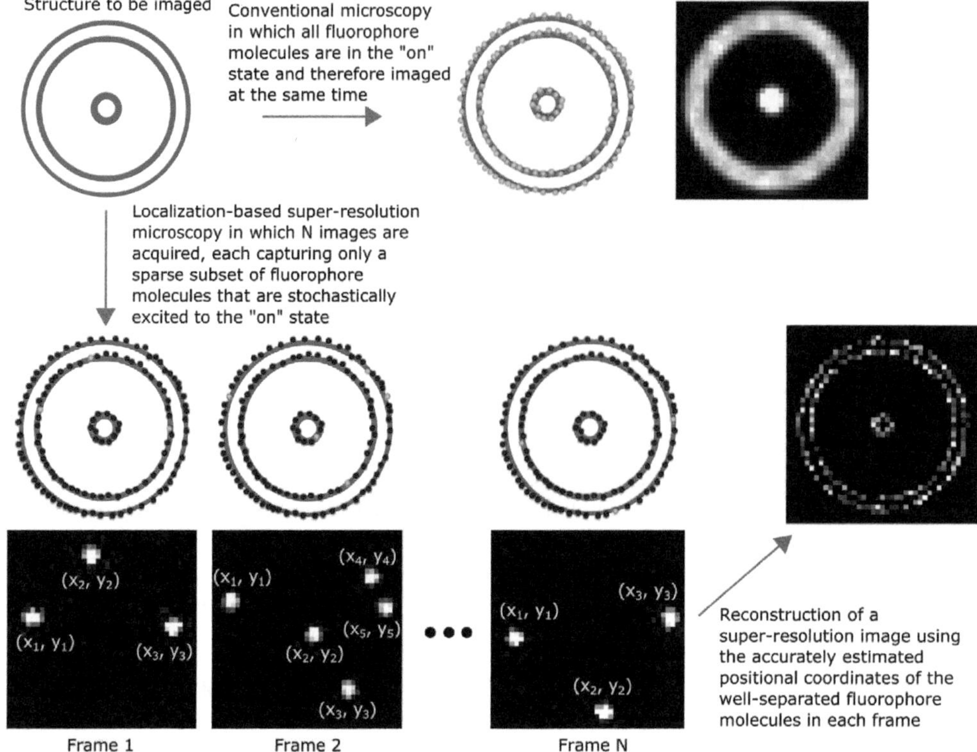

FIGURE 11.16
Conventional versus localization-based super-resolution microscopy. In conventional microscopy, all fluorophore molecules that label a structure essentially fluoresce at the same time. Therefore, a low-resolution image is obtained of closely spaced structures as well as small structures due to the significant overlap of the diffraction spots corresponding to all of the fluorophore molecules that label these structures (in other words, due to the crowding problem). In the example shown, the simulated image shows that the two outer rings of the imaged three-ring sample are not easily distinguishable, and that the innermost ring is seen as a blob. In localization-based super-resolution microscopy, only a sparse subset of the fluorophore molecules emit fluorescence in a given image. Therefore, by acquiring many such images and accurately determining the positional coordinates of the well-separated diffraction spots in each image, a high-resolution image of the sample can be constructed using the estimated positional coordinates. The simulated image shows that, using this technique, the two outer rings of the three-ring sample can be distinguished, and the innermost ring can be seen as a ring instead of a blob.

isolated molecules are estimated from the images, and a high-resolution image of the sample is obtained using these estimated locations. The "quality" of the high-resolution image depends crucially on the accuracy with which the fluorophore molecules are localized. This topic will be analyzed in great detail in Chapter 17. As illustrated in Fig. 11.16, this method allows one to resolve closely spaced structures and to more clearly visualize small structures that would otherwise be misrepresented by the imaging of all fluorophore molecules at the same time.

Although localization-based super-resolution microscopy can be said to overcome the classical resolution limit, we need to be careful what we mean by overcoming the resolution limit, as nothing is done to change the inherent optical properties of the microscope. What is done is to cleverly manipulate the sample and devise an experiment so that the crowding problem is addressed. We therefore are able to obtain high-resolution images with an instrument whose optical properties are unchanged.

There are two main variations of localization-based super-resolution microscopy which differ by the type of stochastically excited fluorophore that is used. In one variation, the fluorophores used are fluorescent proteins that are photoactivatable or photoswitchable (Section 8.7.4). In some cases, photoconvertible fluorescent proteins whose emission can be changed from one wavelength to another by irradiation with violet light are also used. Use of fluorescent proteins has the advantage of enabling the in situ labeling of proteins of interest in cells by way of genetic engineering (Section 7.2.3). The other variation of localization-based super-resolution microscopy makes use of photoswitchable organic fluorophores (Section 8.7.4), such as Cy5 (Section 11.1.2) and members of the Alexa Fluor and ATTO dye series. Use of organic dyes is advantageous as they are generally brighter and more photostable than fluorescent proteins (Note 11.9).

A basic localization-based super-resolution microscopy setup that images a sample labeled with only one fluorophore typically uses two lasers — a usually blue or violet activation laser that, as mentioned in Section 8.7.4, switches the photoactivatable or photoswitchable fluorophore to the "on" state, and an excitation laser that causes a fluorophore in the "on" state to potentially emit fluorescence. To maintain an environment in which only a small subset of well-isolated fluorophore molecules are emitting fluorescence at any given time throughout an experiment, illumination of the sample with the two lasers needs to be administered at appropriate intensity levels and with suitable time durations. It may be the case, for example, that a short pulse of the activation laser needs to be applied at regular intervals, and that the intensity or duration of the pulse needs to be increased gradually as the population of non-photobleached fluorophores decreases with time, in order to maintain a relatively constant density of fluorophore molecules in the "on" state over the course of an experiment (Note 11.10).

In the case of the organic fluorophore-based implementation of the technique, a special imaging buffer is needed, in addition to the appropriate application of the activation and excitation lasers, to achieve the required photoswitching of conventional dyes. This is discussed next as we look in some detail at the photophysics of the stochastic excitation of organic dyes. For insights into the photochemical bases underlying the control of fluorescence emission by photoactivatable and photoswitchable fluorescent proteins, which typically involve light-induced changes to the ionization state or conformation of the protein's chromophore, we refer the reader to the literature (Note 11.11).

11.6.3.1 Photophysics of the stochastic excitation of organic fluorophores

Upon absorption of excitation light and entering the excited state, a photoswitchable organic fluorophore can either return to the ground state by emitting fluorescence or transition to the triplet state via intersystem crossing. While in the triplet state, the fluorophore can transition non-radiatively back to the ground state via internal conversion and become available again for light-induced excitation to the excited state. Alternatively, transition back to the ground state can be radiative, in which energy is released in the form of phosphorescence. These processes are depicted in the Jablonski diagram of Fig. 8.1.

Not shown in Fig. 8.1 is the fact that from the triplet state, transition back to the ground state is facilitated by the organic fluorophore reacting with oxygen. For the type of fluorophores and samples typically imaged in localization-based super-resolution experiments, this transition occurs practically exclusively by the non-radiative internal conversion process rather than by the emission of phosphorescence. Furthermore, not depicted in Fig. 8.1 is an

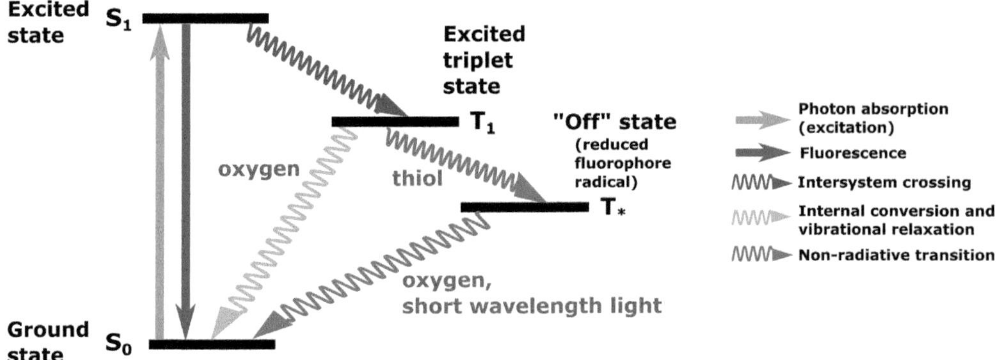

FIGURE 11.17
Photophysics of localization-based super-resolution microscopy with photoswitchable organic fluorophores. Absorption of excitation light causes the fluorophore to transition from the ground state to the excited state. From the excited state, the fluorophore can either return to the ground state by emitting fluorescence, or it can transition non-radiatively to the triplet state by intersystem crossing. Once in the triplet state, the fluorophore can return to the ground state by reaction with oxygen, or it can be switched "off" by reaction with a reducing agent (e.g., a thiol). In the "off" state, the fluorophore exists in a stable, non-fluorescent form as a reduced radical. The fluorophore can be brought out of the "off" state and returned to the ground state by reaction with oxygen. This restoration to the ground state is facilitated by the absorption of light of a shorter wavelength than the excitation light.

alternative path wherein the fluorophore reacts with a reducing agent such as a thiol (an organic compound that contains a carbon-bonded -SH group) to form a reduced fluorophore radical. The reduced fluorophore radical is stable, and in the context of localization-based super-resolution microscopy, importantly accounts for the "off" state of the fluorophore in which the fluorophore cannot absorb excitation light and emit fluorescence. Reaction of the reduced fluorophore radical with oxygen restores the fluorophore to the ground state, a process that is facilitated by absorption of light of a shorter wavelength than that of the fluorophore excitation light. These transitions, which revolve around the triplet state, are summarized in Fig. 11.17 with a Jablonski diagram.

From the above descriptions and Fig. 11.17, it is clear that while a reducing agent is crucial for switching fluorophores to the "off" state, oxygen plays the critical role of switching them to the "on" state by returning them to the ground state from which they can absorb excitation light and potentially emit fluorescence. A successful organic fluorophore-based implementation of a localization-based super-resolution experiment is thus highly dependent on the use of a suitable imaging buffer to ensure that in a given image, only relatively few fluorophore molecules are fluorescing while the majority of the remaining fluorophore molecules are in the "off" state. Such an imaging buffer typically contains a reducing agent such as the thiol β-mercaptoethylamine and an oxygen scavenging system that depletes oxygen. A commonly used oxygen scavenging system comprises the enzymes catalase and glucose oxidase.

In a typical experiment that uses photoswitchable organic fluorophores, the shorter wavelength light that facilitates the restoration of a reduced fluorophore radical to the ground state for subsequent excitation is provided by the blue or violet activation laser

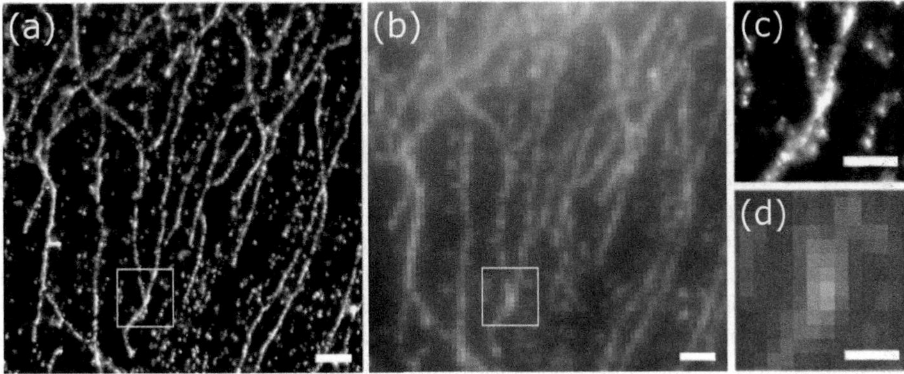

FIGURE 11.18
Microtubules in a BSC-1 monkey kidney cell, imaged with localization-based super-resolution microscopy and widefield microscopy. (a) High-resolution image reconstructed using super-resolution microscopy. (b) Widefield fluorescence image of the same field of view. (c) Region enclosed by yellow box in (a). (d) Region enclosed by yellow box in (b). The microtubules are stained with anti-tubulin primary antibody and labeled with Alexa Fluor 647-conjugated secondary antibody. Scale bars: 2 μm in (a) and (b), 1 μm in (c) and (d).

mentioned in Section 11.6.3. The excitation laser then provides the necessary light for effecting both the emission of fluorescence and the transition to the triplet state (Note 11.12).

11.6.4 A localization-based super-resolution experiment

A standard example that illustrates the power of localization-based super-resolution microscopy is the imaging of microtubules in cells. Such tubules form strands that help to maintain the structure of a cell and also function as tracks for the transport of intracellular cargoes. Figure 11.18(a) shows a high-resolution image of microtubules in a BSC-1 cell (a monkey kidney cell), reconstructed from the images produced by an experiment that uses the organic fluorophore Alexa Fluor 647. In this experiment, tubulin, the building block protein from which a microtubule is made, is stained with mouse anti-tubulin antibody and then labeled with Alexa Fluor 647-conjugated goat anti-mouse secondary antibody. The high-resolution image is generated using the location coordinates of individual Alexa Fluor 647 molecules, which are sparsely distributed in each image of the super-resolution data set. For comparison, a conventional widefield fluorescence image of the same microtubules is shown in Fig. 11.18(b). One can see that the high-resolution image provides a significantly clearer view of the microtubule strands. In the region enclosed by the insets of Figs. 11.18(c) and 11.18(d), for example, the presence of two microtubule strands is clearly revealed by the high-resolution image, but is obscured in the widefield image.

11.7 Multifocal plane microscopy

In Sections 10.2.1 and 11.2.1, we discussed the use of focusing devices to image parts of the sample that are far from the focal plane. With these devices, cells are imaged one focal plane at a time, as illustrated both schematically and by example in Fig. 11.5. This sequential imaging strategy has the major disadvantage that it is slow. Using the successive scanning technique to capture fast cellular dynamics is therefore often not possible. It is not a surprise that relatively few detailed investigations have been carried out to study subcellular trafficking behaviors in three dimensions. In contrast, there is a large number of very detailed investigations of dynamic processes on the plasma membrane, as these processes can be imaged without changing the focus of the microscope, as seen in Section 11.4.

To address this problem, *multifocal plane microscopy* (MUM) can be used. This imaging modality is based on imaging different focal planes simultaneously. The basic optical principle behind this approach is to split the emission light path with beam splitters. Each of the split beams is then focused with a tube lens onto a detector, and the positioning of the detector with respect to the tube lens determines the focal plane for the given light path. An illustration of this principle is given in Fig. 10.2.

In the design of a MUM instrument configuration, two important and related decisions need to be made. One, the number of focal planes to image needs to be determined. Two, the spacings between the focal planes need to be chosen. In order for an experimentalist to make these decisions, the purpose of the experiment needs to be very well understood. For example, if very precise structural information is required, then a very fine sampling, i.e., a relatively large number of focal planes separated by relatively small plane spacings, is required to finely span the structure of interest along the optical axis. On the other hand, if in a live cell experiment the itinerary of a subcellular organelle, vesicle, or tubule is to be analyzed, then relatively few focal planes with relatively large plane spacings are typically sufficient to coarsely span the expected axial range of movement to be recorded. For such tracking applications, the most important requirement is that *connectivity* is guaranteed. By connectivity we mean that when an object loses visibility in one focal plane, it becomes visible in a neighboring focal plane so that it can be continuously tracked. Connectivity does not only depend on the spacings between the focal planes, but clearly also on the size and brightness of the object that is to be tracked. Obviously, a very bright and large object will be visible above the background in a distant focal plane, whereas for a very dim and small object this will not be the case.

In terms of the number of focal planes to use, another important consideration is that the collected photons are distributed amongst the different focal planes. Therefore, assuming an equal distribution of photons, adding more focal planes reduces the number of photons that are detected in each plane, and accordingly reduces the image quality in each plane. For weakly labeled samples or those that are prone to photobleaching, it is therefore important to use the smallest number of focal planes to achieve the imaging goal.

From a quantitative perspective, the number of focal planes used and the spacings between the planes also play an important role in determining how well one can estimate, from the images produced by a MUM setup, the location of an object in the axial, or z, direction. The ability to axially pinpoint an object is crucial, for example, in a tracking application in which one needs to generate an accurate 3D itinerary of an object of interest. Axial localization, however, is challenging because of the optical microscope's poor depth discrimination capability, which refers to the microscope's inability to yield accurate information about the axial position of an object whenever the object is located in or near the focal plane. Provided that an appropriate number of suitably spaced focal planes is used, a MUM setup can overcome the depth discrimination problem. It does so by providing simultaneous views of an object from multiple focal planes which, when taken in aggregate, contain more information about the axial position of the object than just a single view of the object from the standard focal plane. With a MUM setup, it transpires that with fewer focal planes and relatively large plane spacings such as 500 nm, very accurate axial location estimates can be obtained.

For a simple illustration of the importance of choosing suitable plane spacings, we take a two-focal plane MUM configuration and show that by using a relatively wide spacing of 500 nm instead of a spacing of 100 nm, a molecule that is near-focus with respect to one of the focal planes can be localized on the optical axis with a significantly smaller standard deviation. In Fig. 11.19, an example is given in which a molecule is located just 30 nm above the first focal plane of the MUM setup. For a plane spacing of 100 nm, we see that the image of the molecule from the second focal plane is a relatively sharp image that is similar to the image from the first focal plane. In contrast, for the wider plane spacing of 500 nm, the image from the second focal plane is significantly more out-of-focus. Used in conjunction with the image from the first focal plane, however, the more out-of-focus image actually allows the estimation of the molecule's axial coordinate $z_0 = 30$ nm with a standard deviation δ_{z_0} that is nearly three times smaller than what can be expected if the sharper image from the 100-nm plane spacing is used instead. Intuitively, the superior performance provided by the wider spacing is due to the fact that two very different axial views of the molecule naturally yield more information about its axial location than two similar axial views. Indeed, the fact that two focal planes separated by the relatively small distance of 100 nm produce similar images of the molecule is a manifestation of the depth discrimination problem, and is directly related to the concept of *depth of field*, which quantifies how far away an object can be from the focal plane before the quality of its image changes appreciably (Section 14.5.5.2). Note that the δ_{z_0} values shown in Fig. 11.19 are in fact Fisher information-based lower bounds on the standard deviation with which the molecule's z_0 coordinate can be estimated. A general treatment of the theory of the Fisher information matrix is given in Chapter 17, while details concerning the more specific calculation of the lower bound δ_{z_0} can be found in Chapter 19. Note also that the optical microscope's depth discrimination capability is also investigated more thoroughly in Chapter 19 using the lower bound δ_{z_0}.

11.7.1 Focal plane spacing and magnification

Geometrical optical considerations can give us an understanding of the basic principles for obtaining the desired focal plane spacings. Using Eq. (13.13) in Section 13.3.2, which deals with the magnification in the axial direction, we see that if a detector is moved a distance of $\delta_i \approx 3.3$ mm towards the tube lens when the lateral magnification is $M = 100$ and the object space refractive index is $n_o = 1.518$, the conjugate plane, or the focal plane of the imaging system, is moved a distance of $\delta_o = 500$ nm further from the objective. This gives a relatively simple way to set up a multifocal plane microscope, starting with a standard fluorescence microscope. To obtain a two-plane configuration, for example, the emission light path of the microscope can be split between two camera ports. The detector that images the design focal plane is attached via a standard coupler to one port, while the other detector is attached to the second port via a custom coupler that positions it 3.3 mm closer to the tube lens. In this way, the latter detector will image a focal plane in the sample that is 500 nm above the design focal plane. It is important to emphasize that, as per Eq. (13.13), the spacing of the focal planes depends not only on the relative positions of the detectors, but also on the lateral magnification and the object space refractive index.

This approach to MUM leads to different lateral magnifications in the non-design focal planes. Using Eq. (13.14) in Section 13.3.3, and assuming the distance Δ_{i1} to be well approximated by a tube lens focal length of 160 mm, the lateral magnification M_2 of the focal plane that is 500 nm above the design focal plane in the above example, for instance, will be slightly smaller at 97.9. For some applications, such small changes in the lateral magnification are not very significant and could be ignored. However, for precise analyses such as those required in many single molecule applications, these changes should not be overlooked.

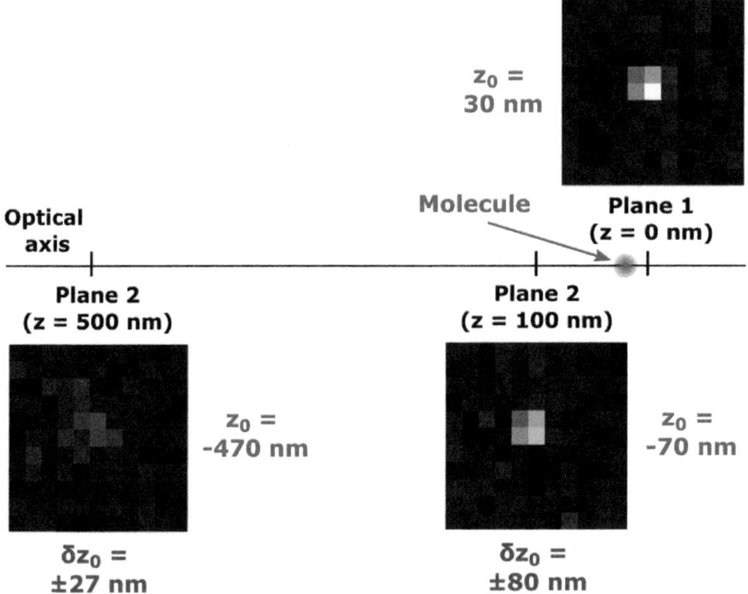

FIGURE 11.19
Effect of plane spacing on the axial localization performance of a two-plane MUM setup. Compared with a 100-nm plane spacing, a 500-nm plane spacing allows the axial coordinate z_0 of a molecule that is near-focus with respect to focal plane 1 to be estimated, using the pair of images of focal planes 1 and 2, with a standard deviation δ_{z_0} that is, in this specific example, nearly three times better (i.e., smaller). A representative image of the molecule is shown for focal plane 1, and for focal plane 2 corresponding to each spacing. Each image is simulated with the same 3D point spread function model, but with respect to a distinct focal plane. The value of z_0 with respect to the given focal plane is indicated in red. The value of δ_{z_0} for each spacing is shown in blue. For the simulation of the images, it is assumed that photons of wavelength 520 nm are collected from the molecule using an objective with a magnification of 100, a numerical aperture of 1.4, and an immersion medium of refractive index 1.518, and are split equally between two detectors that image the two focal planes.

11.7.2 Transferrin trafficking in epithelial cells

Monolayers of cells such as the epithelial cells that form a barrier between the bloodstream and cerebrospinal fluid have a thickness that is well above what can be visualized in one focal plane with a high numerical aperture and high magnification objective. A four-plane MUM setup as depicted in Fig. 10.2, in which the collected fluorescence is divided equally between the four detectors corresponding to the four focal planes, is therefore used to investigate the 3D transport of transferrin in a rodent epithelial cell monolayer. Using the designations in Fig. 10.2, focal plane 1 is positioned approximately 2 to 2.5 μm above the cover glass, focal plane 2 is positioned 2.18 μm above focal plane 1, focal plane 3 is positioned 2.4 μm above focal plane 2, and focal plane 4 is positioned 2 μm above focal plane 3. These plane spacings make possible the imaging of 3D trafficking pathways over a large axial range of about 10 μm.

This MUM setup is used to image not only the transferrin molecules, which we label with Qdot 655, but by changing the excitation laser, also the Alexa Fluor 488-labeled plasma membrane of the cells (i.e., via Alexa Fluor 488-conjugated cholera toxin B subunit

(CTxB), which binds to lipids in the cell membrane). A 635-nm laser is used to excite transferrin-Qdot 655, while a 488-nm laser is used to excite Alexa Fluor 488-CTxB. Note that the 635-nm wavelength falls within the weak tail of the broad excitation spectrum of Qdot 655 (Fig. 8.12). It is nevertheless used to image transferrin-Qdot 655 because shorter excitation wavelengths will result in higher phototoxicity (Section 10.5) to the cells. Continuous imaging of transferrin-Qdot 655 is carried out with an exposure time of 75 ms at 10 frames/s by each of the four detectors, and is only interrupted by the acquisition of one image of the Alexa Fluor 488-labeled plasma membrane once every 25 or 50 frames. Shutters (Fig. 10.20) are used to switch between the two lasers as needed, and on the emission side, an emission filter with multiple passbands is used that transmits the fluorescence from both Qdot 655 and Alexa Fluor 488.

An example of the 3D trajectory of a transferrin molecule that is acquired with such an experiment is shown in Fig. 11.20. In part (a) of the figure, images acquired from the four focal planes at a given time point in the trajectory are shown. The images are arranged according to focal plane depth, with the bottom image corresponding to focal plane 1. Each image shown is an overlay of a grayscale image of the transferrin-Qdot 655 channel and a segmented image of the Alexa Fluor 488-CTxB channel (i.e., the plasma membrane channel) rendered in green. For visualization, the tracked transferrin-Qdot 655 molecule is highlighted in blue and indicated by a red arrow. From a time sequence of such images from the four focal planes, a 3D trajectory comprising the 3D positional coordinates of the transferrin-Qdot 655 molecule at each time point is determined using a single molecule localization algorithm designed for MUM data (Section 19.6). This trajectory spans a time duration of nearly 29 s, and its xy-plane projection and a full 3D rendering are shown in parts (b) and (c) of Fig. 11.20, color-coded to indicate time. These visualizations show that the transferrin-Qdot 655 molecule is initially seen inside one cell of an epithelial cell monolayer. It then moves toward the lateral plasma membrane, undergoes exocytosis, and is quickly endocytosed, or internalized, into the adjacent cell. Importantly, this trajectory shows that transferrin can move at rapid rates and traverse long distances in the axial (i.e., the z) direction, thereby justifying the use of MUM for the imaging of these dynamic processes.

11.7.3 Imaging the pathway preceding exocytosis

In Section 11.4.2, imaging of exocytic events was discussed. It was seen that TIRFM is an important tool for decreasing the background fluorescence so that these events can be detected at the plasma membrane. However, if we want to image, at the same time, the events deeper in the cell that precede the exocytic events, TIRFM alone would not suffice as the very thin layer of illumination that it provides just above the cover glass would not allow the delivery of sufficient excitation light to the fluorescently labeled molecules of interest from deeper within the cell. To image these events, it is necessary to use widefield illumination. Therefore, if we want to image both an exocytic event and its preceding events, we need a modality that supports simultaneous total internal reflection and widefield illumination. Furthermore, it should be noted that using widefield illumination simultaneously with total internal reflection illumination would make the latter pointless if the same excitation wavelength is used. However, if different excitation wavelength bands are used, then widefield and total internal reflection illumination can be used at the same time. We therefore also require an experimental design that uses a dually labeled sample. In this way, one label will be excited with widefield illumination and will be observable throughout the cell, whereas the second label will only be excited with total internal reflection illumination and will therefore only be visible in proximity to the cover glass.

In order to study IgG exocytosis as well as the events preceding IgG exocytosis, we can

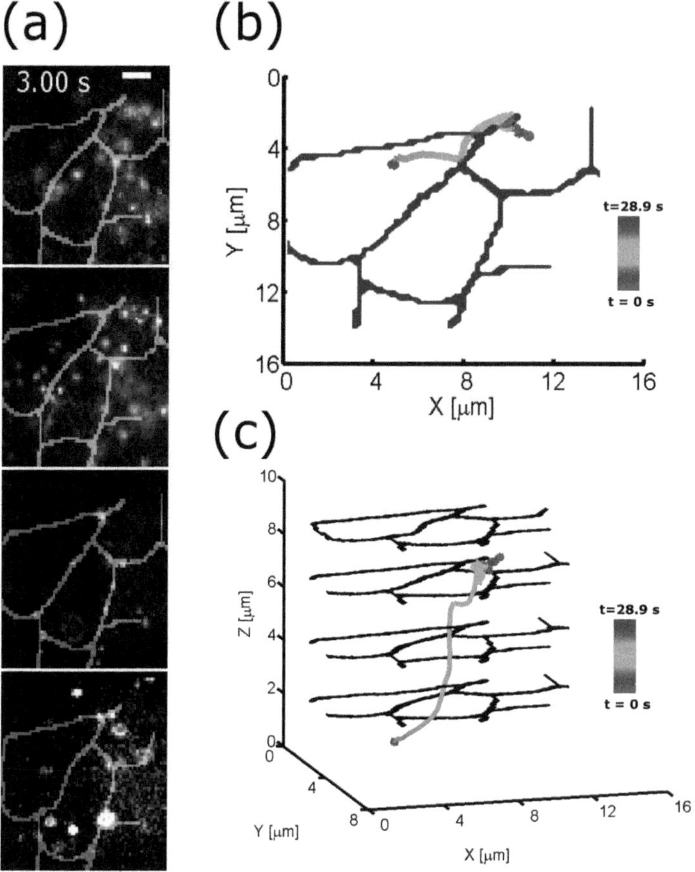

FIGURE 11.20
Transport of transferrin in a thick epithelial cell monolayer as imaged by a four-plane MUM setup. (a) Images acquired from the four focal planes at a given time point in the trajectory of a transferrin-Qdot 655 molecule. The images are arranged according to focal plane depth, with the bottom image corresponding to the plane closest to the cover glass. Each image is an overlay of a grayscale image of the transferrin-Qdot 655 channel and a segmented image of the plasma membrane channel (Alexa Fluor 488-CTxB) rendered in green. The transferrin-Qdot 655 molecule that is tracked is highlighted in blue and indicated by a red arrow. Scale bar: 5 μm. (b) The xy-plane projection of the 3D trajectory of the transferrin-Qdot 655 molecule highlighted in (a). The trajectory is determined using a single molecule localization algorithm for MUM data, and is color-coded to indicate time. From the xy-plane projection, the molecule is clearly seen to be transported from one epithelial cell to another. (c) The 3D trajectory of the transferrin-Qdot 655 molecule highlighted in (a). From this 3D rendering of the trajectory, the molecule is seen to traverse a large distance along the axial (i.e., the z) direction. [See Figure Credits.]

use a MUM setup as sketched in Fig. 11.21. Here a 100× TIRFM objective with a numerical aperture of 1.45 is used. A 488-nm laser line provides the total internal reflection illumination, and a 543-nm laser provides the widefield illumination. HMEC-1 cells are transfected with both pHluorin-FcRn and FcRn-mRFP1, so that pHluorin-FcRn in proximity to the plasma membrane can be imaged in TIRFM mode using the 488-nm laser line, while FcRn-

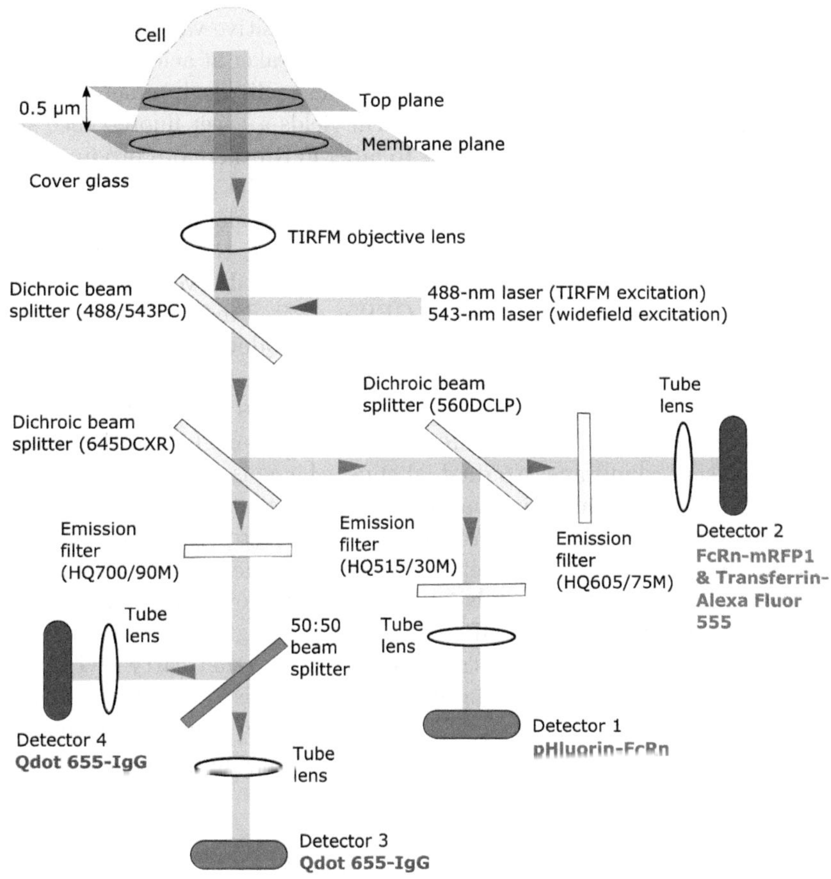

FIGURE 11.21
Two-plane MUM setup with simultaneous total internal reflection and widefield illumination that enables the imaging of both the FcRn-mediated exocytosis of IgG at the plasma membrane and the events deeper in the cell that precede the exocytic event at the membrane. One focal plane is referred to as the membrane plane, as it is positioned at the plasma membrane located adjacent to the cover glass. The other focal plane is called the top plane, as it is positioned a distance of 0.5 μm above the membrane plane. pHluroin-FcRn in proximity to the membrane plane is imaged using total internal reflection illumination with a 488-nm laser, while FcRn-mRFP1 and Alexa Fluor 555-labeled transferrin are imaged using widefield illumination with a 543-nm laser. Qdot 655-labeled IgG is imaged using both lasers, as Qdot 655 has a broad excitation spectrum. Using a combination of appropriate dichroic beam splitters and emission filters, the collected fluorescence is divided between four detectors. By their positioning with respect to the tube lens, detectors 1 and 3 image the membrane plane, while detectors 2 and 4 image the top plane. Detectors 1 and 3 capture the fluorescence from pHluorin-FcRn and Qdot 655-IgG, respectively, that are in proximity to the membrane plane. On the other hand, detectors 2 and 4 capture the fluorescence from deeper in the cell. Specifically, FcRn-mRFP1 and transferrin-Alexa Fluor 555 signals are captured by detector 2, and Qdot 655-IgG signals are captured by detector 4. The part numbers given in parentheses indicate emission filters and dichroic beam splitters manufactured by Chroma Technology that can be used to implement this configuration.

mRFP1 in the interior of the cell can be imaged in widefield mode using the 543-nm laser line. Here the pHluorin used is called ecliptic pHluorin which, similar to the previously mentioned ratiometric pHluorin (Section 11.5), is a pH-sensitive variant of wild-type GFP. Unlike ratiometric pHluorin, however, its fluorescence is weaker at acidic pH than at neutral pH when excited with 488-nm light (Note 11.6). In the acidic environment of intracellular compartments such as endosomes, ecliptic pHluorin yields a lower fluorescence signal than GFP. Use of ecliptic pHluorin instead of GFP to label FcRn therefore produces less out-of-focus haze (Section 11.2.2) on the plasma membrane, thereby allowing pHluorin-FcRn in proximity to the plasma membrane to be imaged in TIRFM mode without its signal being drowned in haze. This is a critical design choice, as otherwise the strong fluorescence from FcRn-GFP-positive endosomes from the cell interior could easily overwhelm the FcRn-GFP signal near the plasma membrane, even when GFP is excited in TIRFM mode.

The HMEC-1 cells are imaged in medium containing Qdot 655-labeled IgG and Alexa Fluor 555-labeled transferrin. The QD label for IgG is chosen for its brightness and photostability (Section 8.7.2), and is excited by both laser lines due to its broad excitation spectrum (Fig. 8.12). As in the case of FcRn, we want to image IgG in the cell interior as well as near the plasma membrane. No dual labeling scheme is necessary for IgG, however, because Qdot 655-IgG is introduced in quantities that are too small to generate out-of-focus haze of appreciable consequence. Transferrin has been shown to localize to the same sorting endosomes as FcRn (Fig. 2.8). It is therefore labeled with Alexa Fluor 555, which is more photostable than mRFP1 and has an emission spectrum that overlaps with that of mRFP1, and added as a supplement to FcRn-mRFP1 so that sorting endosomes in the cell interior, which are sources of FcRn-positive transport containers that carry IgG to the plasma membrane, can be visualized for longer durations of time. This is used to overcome the fact that mRFP1 photobleaches relatively quickly and, therefore, if not supplemented with a more photostable fluorophore, will not suffice to ensure that endosomes remain sufficiently detectable over the long periods of image acquisition that are often needed to capture the events of interest.

Figure 11.21 further shows that the MUM instrument employs two focal planes and four detectors. With the focal plane referred to as the membrane plane, TIRFM images are acquired of the plasma membrane adjacent to the cover glass. The other focal plane is positioned about 500 nm above the membrane plane and is used to image dynamic events deeper in the cell. This focal plane is referred to simply as the top plane. A dichroic beam splitter is used that reflects the 488-nm and 573-nm laser beams towards the sample for the simultaneous total internal reflection and widefield illumination, and at the same time allows the resulting fluorescence from the various proteins of interest to pass through. This fluorescence is then separated by color, by way of a second dichroic beam splitter that reflects light of wavelengths below $\sim$645 nm and transmits light of wavelengths above $\sim$645 nm. The former set of wavelengths includes the fluorescence emitted by pHluorin-FcRn, FcRn-mRFP1, and transferrin-Alexa Fluor 555, and is partitioned by yet another dichroic beam splitter. Wavelengths below $\sim$560 nm are reflected by this third dichroic beam splitter and passed through an emission filter for pHluorin fluorescence. Detector 1, which is positioned from the tube lens such that it images the membrane plane, therefore captures the fluorescence due to pHluorin-FcRn in proximity to the plasma membrane. On the other hand, wavelengths above $\sim$560 nm are transmitted by the third dichroic beam splitter and passed through an emission filter for mRFP1 and Alexa Fluor 555 fluorescence. Detector 2, which is positioned closer to the tube lens so that it images the top plane, thus captures the fluorescence due to FcRn-mRFP1 and transferrin-Alexa Fluor 555 from the cell interior.

The set of wavelengths above $\sim$645 nm that is transmitted by the second dichroic beam splitter contains the fluorescence emitted by Qdot 655-IgG, and it is accordingly passed

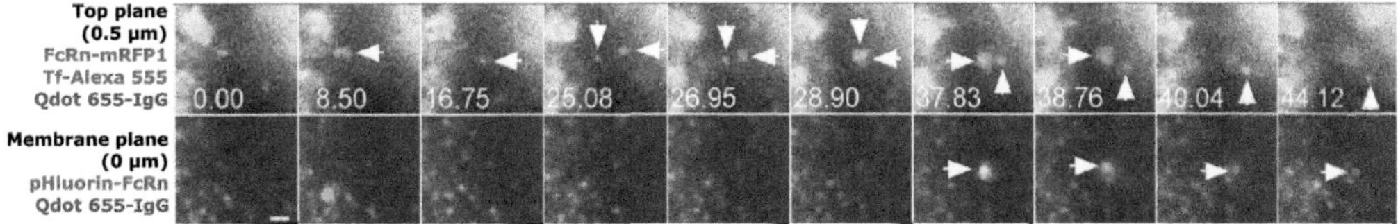

FIGURE 11.22
Time sequence of images of an HMEC-1 cell acquired using the two-plane MUM setup depicted in Fig. 11.21, showing FcRn-mediated IgG exocytosis and the events preceding exocytosis. The top row shows images of the top plane, each consisting of an FcRn-mRFP1/transferrin-Alexa Fluor 555 (Tf-Alexa 555) image acquired by detector 2 and a Qdot 655-IgG image acquired by detector 4 overlaid in green and red, respectively. The bottom row shows images of the membrane plane, each consisting of an pHluorin-FcRn image acquired by detector 1 and a Qdot 655-IgG image acquired by detector 3 overlaid in green and red, respectively. Events of interest are highlighted in blue, and the time at which each set of four corresponding images are acquired is shown in seconds in the overlay images of the top plane. The image sequence shows, on the top plane, an FcRn/transferrin-positive transport container with IgG (leftward arrow) to leave a sorting endosome at 8.50 s, and another FcRn/transferrin-positive transport container with IgG (downward arrow) to come into the same vicinity at 25.08 s. The two containers come close to each other before one of them exocytoses at 38.76 s, releasing IgG (rightward arrow) on the membrane plane. The remaining transport container (upward arrow) is then seen to move away from the area of observation in the top plane. Scale bar: 1 μm. [See Figure Credits.]

through an emission filter for Qdot 655 fluorescence. As we are interested in detecting the Qdot 655-IgG signal from both the cell interior and the plasma membrane, the fluorescence transmitted by the emission filter is divided equally by a 50:50 beam splitter between detectors 3 and 4, which are positioned exactly the same distances from the tube lens as detectors 1 and 2, respectively. Therefore, detector 3 captures the fluorescence due to Qdot 655-IgG near the plasma membrane, and detector 4 captures the the fluorescence due to Qdot 655-IgG from the cell interior.

The results of an experiment using the described MUM setup are shown in Fig. 11.22. Images of the top plane are displayed as overlays, at corresponding time points, of the FcRn-mRFP1/transferrin-Alexa Fluor 555 images (green) acquired by detector 2 and the Qdot 655-IgG images (red) acquired by detector 4. Similarly, images of the membrane plane are displayed as overlays of the pHluorin-FcRn images (green) acquired by detector 1 and the Qdot 655-IgG images (red) acquired by detector 3. Highlighted in blue are the events of interest, which begin in the top plane with an FcRn/transferrin-positive transport container with IgG (leftward arrow) leaving a sorting endosome at 8.50 s. It is then joined by another FcRn/transferrin-positive transport container with IgG (downward arrow) at 25.08 s in the same vicinity, and the two containers approach each other before one of them exocytoses at 38.76 s, releasing IgG (rightward arrow) on the membrane plane. The remaining transport container (upward arrow) then moves away from the area of observation in the top plane.

12

Detectors

Effective means of recording the images produced by a modern microscope are essential for recordkeeping and, even more importantly, for the analysis of experimental data. Technological advances over the past decades in detector development have played a crucial role in advancing the capabilities of modern microscopy. A wide variety of detectors with differing specifications are currently available. The specifics of the microscopy experiment dictate which detector should be used to allow for the best execution of the experiment.

In this chapter, we will introduce some of the most important detector types and discuss their important properties. There are essentially two classes of detectors, the image detectors and the point detectors. Whereas image detectors acquire a full image at each acquisition time point, point detectors collect only one data value at each point in time. While image detectors can provide excellent spatial resolution, they are typically relatively slow. In contrast, the lack of spatial information in a point detector is made up by extremely fast acquisition speeds combined with outstanding sensitivity. Unfortunately, to date no single detector can achieve all three properties (i.e., high spatial resolution, high temporal resolution, and high sensitivity) at the same time. We will first give an introduction to the basics of photon detection and image generation by a detector. We will then provide a brief discussion of point detectors before devoting the remainder of the chapter to image detectors.

12.1 Photoelectric effect

The physical basis for the detectors considered in this chapter is the *photoelectric effect*. This effect describes the conversion of photons into electrons as the photons impact a photon-sensitive material. Heinrich Hertz was the first to observe this conversion in 1887. It was Albert Einstein who received the Nobel Prize in Physics in 1921 for his 1905 publication (Note 4.3) in which he provided an explanation of the photoelectric effect in terms of *quanta of light*, i.e., photons. The photoelectric effect is the key to both image and point detectors, since it explains that photons that impact a suitable material can be converted into electrons (so-called *photoelectrons*). Once this conversion has taken place, the amount of converted photoelectrons can be measured using suitable electronic circuitry.

Light detectors or *photodetectors* are devices designed with materials that exhibit the photoelectric effect, combined with electronic circuitry that converts the acquired electrical charge into a digital signal. This conversion process is divided into two important steps. First, the photoelectrons need to be read out from the detector into the electronic circuitry. Then, as a second step, the resulting electronic signal is digitized using an *analog-to-digital converter* (A/D), such that a measurement of the amount of light that impacts a photosensitive element of the detector is represented by a digital signal.

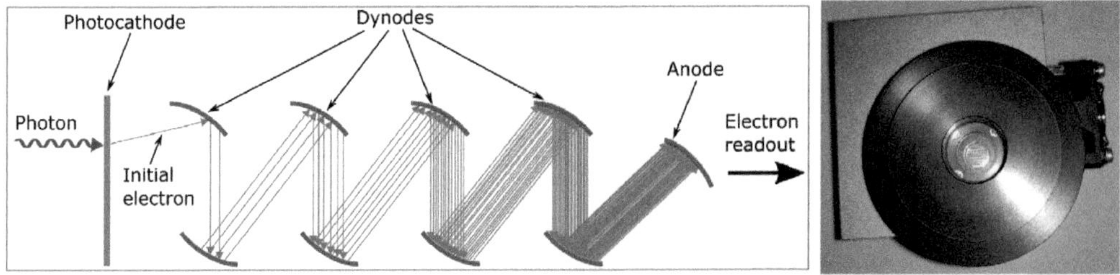

FIGURE 12.1
Photomultiplier tube. The schematic illustrates the detection of a photon by a photomultiplier tube. An incident photon impacts a photocathode, resulting in the generation of an initial electron by the photoelectric effect. The electron then enters a multiplier consisting of a series of electrodes called dynodes. At each dynode, a number of secondary electrons are generated per electron that impacts the dynode, and all electrons are then accelerated towards the next dynode where the same process occurs. This cascade results in the multiplication of a single initial electron into a large number of electrons at the last dynode (called the anode), from which they are read out by the detector. The photograph shows a detector that uses a photomultiplier tube. This particular detector (PMC-100-1) is commercially available from Becker & Hickl GmbH.

12.2 Point detectors

In imaging modalities such as the confocal microscope (Section 10.4), an image is formed by scanning a small spot of illumination through the sample and sequentially collecting the fluorescence emitted from each spot location. For these modalities, only the size of the measured signal is of relevance, and a detector that does not produce spatial information is adequate. What is critical, however, is that the detector needs to be able to rapidly acquire the data as the illumination spot is scanned through the sample. A point detector satisfies these requirements and is commonly used in scanning-based imaging modalities. The two main categories of point detectors, *photomultiplier tubes* (PMTs) (Fig. 12.1) and *avalanche photodiodes*, both employ signal multiplication to significantly amplify the detected signal. These technologies can also be used for single photon detection. In this context, the superb time resolution allows a very precise determination of the photon arrival time, which is often important for applications such as fluorescence lifetime measurements. As with image detectors (Section 12.3), the specific properties of the point detectors, such as spectral response, vary depending on the particular materials that are used. When using these detectors, it is important to keep in mind limitations that could affect their use in particular applications. For example, for the measurement of photon arrival times, the *dead time* of the detector is of relevance. In photon counting mode, after a photon has been registered, the detector cannot register a second photon during a short recovering time, i.e., the dead time.

12.3 Image detectors

An *image detector* is typically made up of a 2D array of photosensitive elements called pixels, each of which can be used to measure, independently from any of the other elements

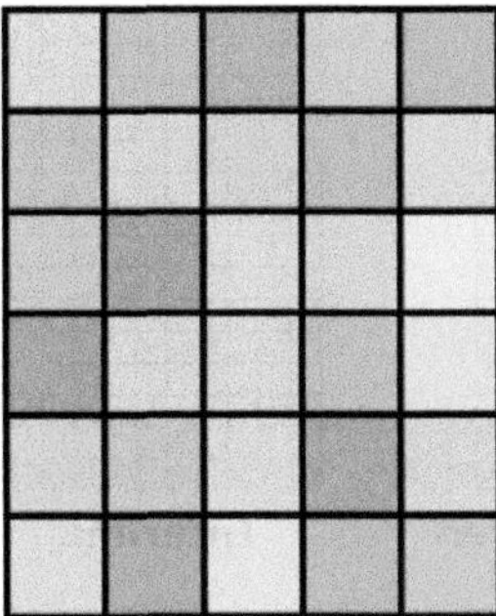

FIGURE 12.2
Sketch of an image detector. An image detector comprises a 2D array of photosensitive elements called pixels. The different amounts of light measured in the pixels are represented by different shades of warm colors, with lighter shades indicating more detected light.

of the detector, the amount of light that impacts it (Fig. 12.2). In this way the measured data provides a pixelated image of the light that impacts the detector array.

Image detectors are a standard tool for widefield cellular microscopy. A suitable detector has to satisfy a number of important properties to be useful in cellular microscopy. With *linearity of the response*, we mean that the acquired signal in the exposed image is proportional to the light that impacted the detector. This is an important criterion for a quantitative analysis of the acquired data. As with any electronic device, light detectors are not without noise sources that can corrupt the acquired data. *Low noise levels* are important since the signals that we wish to detect are often very low. Low light imaging, such as cellular microscopy, necessitates that these noise sources are kept to an absolute minimum so that they do not overwhelm the acquired signal. A third important property is the *uniformity of the response*, the desirable quality that the pixels comprising a detector produce signals of the same intensity level when illuminated with the same level of light.

The image produced by an image detector is an array of digital numbers. The most typical image formats that are relevant to us are grayscale images and color images. A grayscale image is a rectangular matrix whose entries contain positive values that stand for the measured signal in the corresponding pixel of the detector. The image is called grayscale as it does not contain any color or frequency information.

As an image can contain a million or more entries, storage space is an important issue. Therefore the entries of an image matrix are very rarely stored as floating point numbers. Instead, they are stored as unsigned integers. Consumer and lower-grade scientific cameras often use 8-bit encoding. However, the $2^8 = 256$ values are usually not sufficient for the dynamic range that we need for many scientific purposes. Therefore, 16-bit storage is typically used, although the signal in a pixel may only be digitized as a 12-bit or 14-bit value.

Color images are often stored in RGB format, which stands for "Red Green Blue". This is based on the decomposition of any color into its three constituent primary colors red, green, and blue. From a data storage point of view, an RGB image is nothing but a 3D array made up of three grayscale images of equal size. We should note that there are other color decompositions that lead to different but, from our point of view, equivalent image formats.

The various detectors that we will describe in subsequent sections have readout electronics associated with them that digitizes the registered optical data and converts it to a digital image. The digital image is then available for subsequent analysis and storage on a computational device.

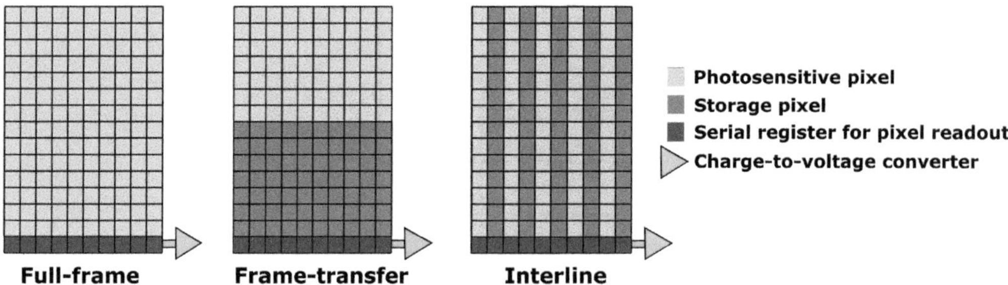

FIGURE 12.3
The three main architectures of a CCD chip.

12.3.1 Charge-coupled device (CCD) detectors

The *charge-coupled device (CCD) detector* is a detector that is frequently used for microscopy experiments due to its low noise levels, the linearity of its response, and the uniformity of the response over its pixels.

In a CCD detector the charges from all pixels are read out by the same electronic circuitry. The fact that a CCD detector only has one shared readout circuitry means that the charges from the different pixels on the detector need to be shifted to the readout circuitry. Different architectures have been implemented for CCD detectors that allow for the shifting of the charges to the readout circuitry. The main architectures are the *full-frame* architecture, the *frame-transfer* architecture, and the *interline* architecture, which are illustrated in Fig. 12.3. Common to all these architectures is the *serial register* that is responsible for the readout operation of the imaging chip.

In the full-frame architecture, all pixels are exposed to light and at the end of the exposure the pixels are shifted row by row into the serial register, where one row of pixels is read out at a time. In this way, one row after another will be read out in succession (Fig. 12.4). The problem with this approach is that the pixels remain exposed during the entire readout process. As the readout process needs to be relatively slow in order to reduce the readout noise level, the readout process can take a considerable amount of time. As the rows of pixels are being shifted one by one towards the serial register, continued exposure of the pixels can lead to significant smearing, unless a shutter external to the chip is used to block the incoming light during the readout process. Mechanical shutters can be slow and add significantly to the cost of the camera. Their integration with the camera also poses problems, as a shutter needs to be coordinated with the image readout process. The mechanical shutter approach is particularly problematic for the acquisition of images in fast succession. An additional disadvantage of the full-frame architecture is that the start of the exposure of a new image can only occur after the current image is completely read out. This means that slower acquisition frame rates can be expected compared to the other architectures.

The frame-transfer architecture is designed to address the problems with the full-frame design. In this architecture, the pixels in half of the chip are covered and are not photosensitive. This part of the chip acts as a storage area, while the other half of the chip functions as the actual image sensor. After exposure on the imaging array the accumulated electric charges are very quickly transferred into the storage area. This can typically be done without significant deterioration of the signal in well under a millisecond, i.e., much faster than the actual readout process. The pixels in the storage area are then shifted to the serial register for readout. This approach significantly reduces the exposure of the image during the readout process. The expense is that in this architecture half of the pixels, i.e., those in

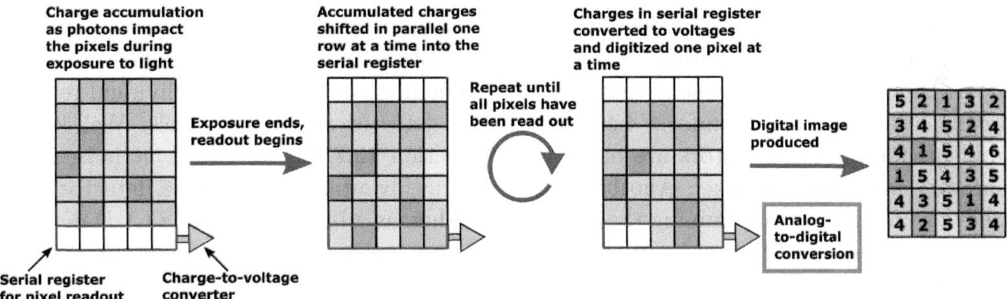

FIGURE 12.4
High-level illustration of the operation of a CCD detector. The schematic gives an overview of the process by which photons detected in the pixels of a CCD detector are converted to the digital values of an image.

the storage area, are lost to acquisition. It is important to note, however, that during the readout of an image, a new image can at the same time be exposed in the image area of the detector to speed up the rate of acquisition.

In the interline architecture, columns of photosensitive pixels are interlaced with columns of non-sensitive pixels for storage of the pixel charges during the readout process. After an image has been exposed the charges in each pixel column are shifted to the storage pixels in the neighboring storage column. The charges in the storage pixels are then read out through the serial register. Since it takes only a one-pixel shift to move charges to a neighboring storage pixel, the transfer occurs very rapidly and virtually eliminates the problem of smearing. An important advantage of this design is therefore similar to that of the frame-transfer design, i.e., the image pixels are protected from exposure during the readout process. A problem with this design, however, is that within the image area there are the columns of storage pixels. Since those pixels are not light-sensitive, the sensitivity, or *quantum efficiency* (Section 12.6), of the detector is diminished. To address this problem, in some detectors a microlens is used over each photosensitive pixel to guide to the pixel light that would otherwise impact the photo-insensitive areas adjacent to the pixel. In this fashion a fairly high quantum efficiency can be achieved. There is, however, still a loss of spatial resolution since the columns comprising the storage pixels are not represented in the image that is read out.

12.3.2 Complementary metal-oxide-semiconductor (CMOS) detectors

The rate of image acquisition by a CCD detector is relatively slow due to the sequential readout of the pixels that comprise an image. Furthermore, a limitation on the readout speed is imposed in order to guarantee a low readout noise level. Interest therefore persists in the use of *complementary metal-oxide-semiconductor (CMOS) detectors*, an alternative that can provide substantially faster readout times.

The main difference between a CCD and a CMOS detector is in the architecture of the readout process. In contrast to a CCD detector, the charge accumulated in each pixel of a CMOS detector is processed via the pixel's own readout circuitry. This parallelization eliminates the time that is required for the shifting of charges to a common readout circuitry and the subsequent pixel-by-pixel processing, and therefore very significantly improves the readout speed. A further advantage of CMOS detectors is that, at least in high-end versions of the detector, the readout noise level is even lower than that of CCD detectors.

While a CMOS detector's readout architecture confers the major advantage of much faster rates of image acquisition, it is also responsible for some important disadvantages. A particularly important issue is that it is very difficult to calibrate the individual readout circuitries to produce a uniform behavior of the detector across its pixels. Well-known problems for CMOS detectors are offset levels (Section 12.7.1.1) and readout noise levels that vary from pixel to pixel. Another important disadvantage of a CMOS detector is its behavior under hardware binning (Section 12.6). Whereas a binned pixel in a CCD detector has readout noise of the same variance as the readout noise of a regular pixel, an $N \times N$-binned pixel in a CMOS detector has readout noise with a variance that is equal to the sum of the variances of the readout noise in the N^2 component pixels. As a result, from a readout noise point of view, there is no advantage to hardware binning over software binning (Exercise 12.3) in the case of a CMOS detector.

12.3.3 Electron-multiplying charge-coupled device (EMCCD) detectors

CCD and CMOS detectors are very powerful. In Chapter 17, we will quantitatively analyze their capabilities in regards to estimating the location of a single molecule, and will show that for a relatively large number of acquired photons, their performance can be close to ideal. Where problems can arise with these detectors, however, is when the readout noise is significant in relation to the signal that is acquired. This is intuitively clear, as it is then difficult, if not impossible, to distinguish the signal that originates from the object of interest from the noise that is added by the detector electronics. *Electron-multiplying CCD (EMCCD) detectors* have been introduced to address this problem (Note 12.1). The basic idea of these detectors is that the signal detected from the imaged object is amplified before it is read out, so that readout noise can no longer overwhelm the now amplified signal. While this is a very powerful concept, these detectors also have problems that arise from the nature of the amplification process. Specifically, the amplification process is itself a random process, which means that the amplified signal cannot be precisely computed from the initial signal. It can only be determined with a very large prediction error. This means that the analysis of an image acquired with an EMCCD camera is not straightforward. As we will show in Sections 17.4 and 18.6.2, due to the stochastic nature of the amplification process, use of EMCCD detectors can in fact be detrimental in situations other than extreme low-light circumstances.

Signal amplification in an EMCCD detector occurs in a multiplication register that is situated immediately after the serial register (Fig. 12.5). The electrons in each pixel are sequentially shifted into the multiplication register, which consists of typically several hundred stages. In each stage, a voltage is applied that causes each electron to produce, with a relatively small probability (typically 0.01 to 0.02), a secondary electron through a physical phenomenon called impact ionization. All electrons that exit a stage enter the next, in which more secondary electrons are generated with the same probability by impact ionization. Due to the large number of stages in the multiplication register, there is a significant probability that the overall cascade produces a substantial increase in the number of electrons, despite the low probability of secondary electron generation in each stage. An amplified signal can therefore be expected that is significantly larger than the readout noise. In an EMCCD detector, the probability of secondary electron generation can be specified indirectly, but more intuitively, through a parameter called the *electron multiplication (EM) gain*. The value of the EM gain typically represents the average number of electrons that will result from the multiplication of a single initial electron.

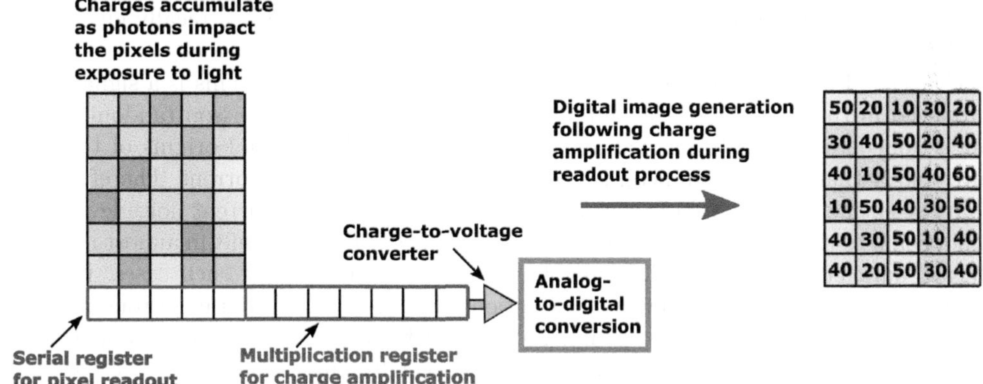

FIGURE 12.5
EMCCD detector. The operation of an EMCCD detector is essentially as described in Fig. 12.4 for a CCD detector, except the charges accumulated in the pixels during exposure to light are stochastically amplified as they pass through a multiplication register during the readout process. The image produced by an EMCCD detector can be significantly brighter as a result.

12.4 Randomness of photon detection and detector noise sources

Understanding the noise sources in an image detector is critical for the setup of a microscopy experiment and the proper analysis of the acquired data. Before looking at these noise sources, however, we first consider the photon signal that is detected in the pixels of a detector, and to which the detector will introduce noise of its own. The emission of photons by the imaged sample is intrinsically stochastic in time, such that repeat measurements of arriving photons by a detector pixel under unvarying conditions (i.e., same photon arrival rate, same exposure time, same photon wavelength, etc.) will not result in the same number of detected photons. This random fluctuation in the number of detected photons from repeat measurements of light of constant intensity is considered as noise by some, and is often referred to as *shot noise*. As we have seen in Fig. 4.6 and the associated discussion, the naturally fluctuating number of detected photons is well represented by a Poisson random variable. In this book, we do not characterize the stochastic fluctuations of the photons associated with an object of interest as noise, as it is a natural aspect of the signal. Therefore, instead of using the terminology "shot noise", we simply consider the photon signal to be stochastic, and model it with a Poisson random variable. The terminology "noise" will only be used to refer to noise due to sources other than the object of interest. These sources include the Poisson-distributed background component, which adds photons that originate from anything other than the object of interest, and the detector itself as we will discuss next.

Detectors are imperfect devices that introduce noise to the photon signals that they detect. Two detector noise sources are of particular significance as they are common to the image detector types discussed above. The first is the detector's readout electronics, which adds noise to the detected signal when the signal is read out. Readout noise is typically modeled as additive Gaussian noise (recall Fig. 4.7 and the associated discussion), and its variance depends crucially on the readout speed. For a given detector, it is typically the case that the higher the readout speed, the greater the variance (i.e., the greater the noise level). In manufacturer documentation for cameras, the noise level is typically given as the standard deviation of the Gaussian noise.

The second important detector noise source is the *dark current*. This is a signal that is introduced in the circuitry of the chip and is often modeled as a Poisson random variable that is added to the photon signal detected in a pixel. The thermal origins of this noise source imply that cooling significantly reduces the level of the dark current. Therefore high-quality scientific cameras are often significantly cooled. The dark current noise level is also proportional to the exposure time. Short exposure times will often result in almost negligible levels of dark current, and cooling might not even be necessary in such cases. However, cooling is essential in the very low-light situations that are encountered in astronomy where very long exposure times are necessary. For typical fluorescence microscopy applications, the dark current for scientific-grade cameras is, especially after cooling, often so low as to be negligible.

An important source of stochasticity that is unique to the EMCCD detector, which some may characterize as a noise source, is its signal amplification process. As discussed in Section 12.3.3, the multiplication of electrons by the multiplication register of an EMCCD detector is random in nature, meaning that the amplified electron count in a pixel is not simply a deterministic multiple of the initial Poisson-distributed number of photoelectrons in the pixel. Instead, the stochasticity of the multiplication leads to an amplified electron count that is represented by a random variable with a relatively complicated probability distribution. This probability distribution is typically determined by modeling the signal amplification as a branching process, and its precise expression depends on how the probability of secondary electron generation in each stage of the multiplication register is modeled. We will consider branching processes, and in particular branching processes for which the probability of offspring (i.e., secondary electron) generation is given by the geometric probability distribution, in some detail in Section 15.2.4 and Online Appendix Section G.1. Due to its stochastic nature, depending on circumstances signal amplification can be either beneficial or deleterious from the perspective of image data analysis, as we will see in Sections 17.4 and 18.6.2.

12.5 Grayscale and color cameras

In many low-light fluorescence microscopy experiments, sensitivity is crucial. Therefore the most sensitive detector should be used. For reasons that we will see shortly, this is typically a *grayscale detector*, i.e., a detector that produces an image that has no color information. Color detectors can be obtained by applying color-sensitive coatings to the pixels of a standard, i.e., grayscale, detector. In Fig. 12.6, the so-called *Bayer pattern* is shown in which the pixels produce red, green and blue signals. After acquisition, an interpolation algorithm is used to construct an image in RGB format based on the values of neighboring pixels. It can be easily seen here that this approach produces low sensitivity images since in each pixel light that does not match the color of the pixel is filtered out. Such designs are therefore typically not used in low-light fluorescence microscopy applications. A significantly more expensive alternative solution addresses this problem by employing three grayscale detectors (Fig. 12.6). A prism is used to split the incoming light into the three color components (red, blue, and green) and direct each of the three components to one of the three detectors. Each of these three detectors produces one image each, and the three images are combined to produce an RGB image. This design has significantly better sensitivity than the one based on the Bayer-patterned detector, as no light is filtered out due to its frequency content.

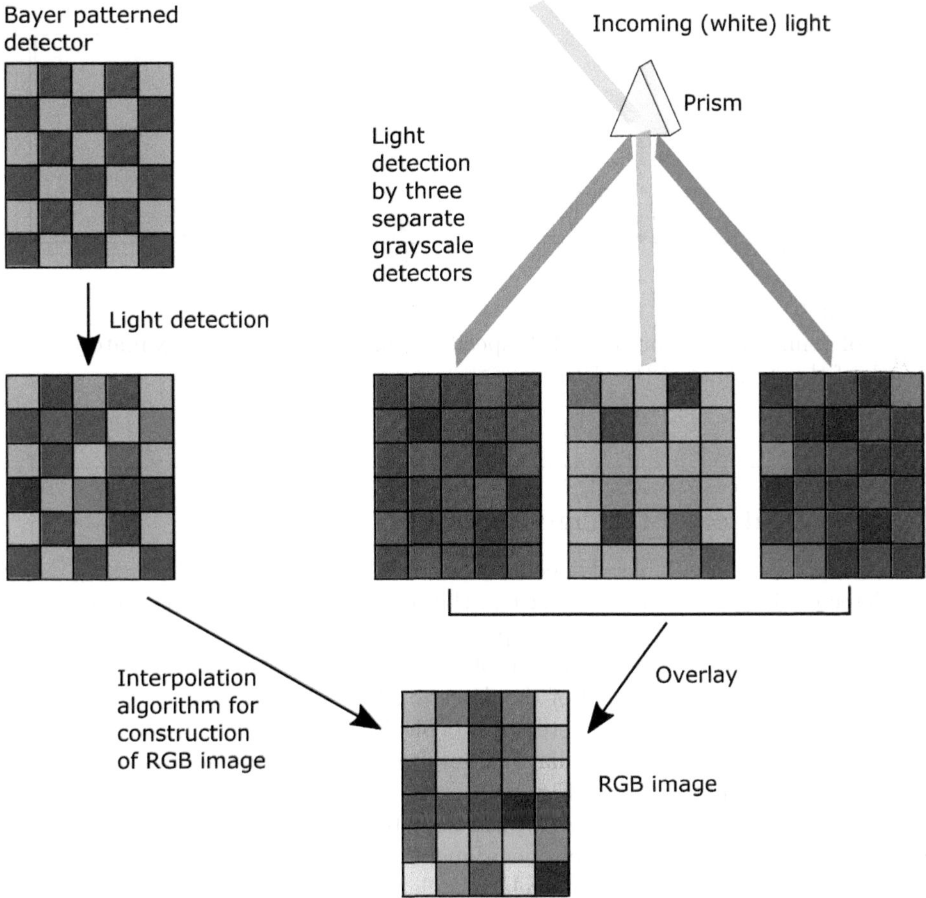

FIGURE 12.6
Two implementations of color detectors. In one approach, color filters arranged according to the Bayer pattern are applied to the pixels of a grayscale detector such that each pixel detects only red, green, or blue light. To construct an RGB image after light detection, an interpolation algorithm is used to determine the values of the two missing color components in each pixel based on the amounts of light of those missing colors detected in the neighboring pixels. In the second approach, incoming light is split by a prism into a red, a green, and a blue component, and each component is detected by a separate grayscale detector. The images acquired by the three grayscale detectors are then overlaid to produce an RGB image.

However, the fixed decomposition of light into its red, green, and blue components is not very well adapted to low-light fluorescence microscopy, where it is important to determine with great specificity the amount of fluorescence emitted by each of the fluorophores. We have seen in Section 10.7 how this is achieved with optical filters in a modern microscopy setup.

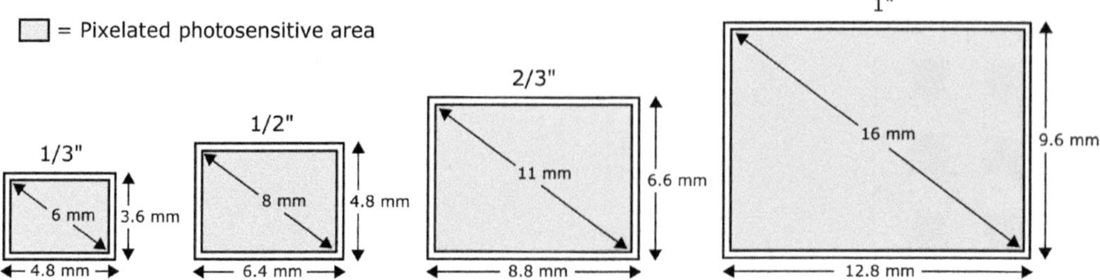

FIGURE 12.7
Examples of common detector sizes. The specification in inches is approximately 1.5 times the length of the diagonal of the detector.

12.6 Specifications of image detectors

We now briefly discuss a number of specifications of an image detector, including, for example, the detector size, the number and size of the pixels, the full well capacity of the pixels, the quantum efficiency, and the readout speed.

Detector size. The size of a detector is often given in inches and denotes approximately 1.5 times the length of the diagonal of the detector. This does not necessarily fully specify the detector's dimensions since the detector is typically not square. However, detector sizes are usually standardized so that the diagonal information in inches can be used to infer the dimensions of the rectangular area (Fig. 12.7).

Number and size of pixels. A typical detector for cellular microscopy has several hundred to over two thousand pixels per dimension. Common pixel sizes range from a relatively small 6.5 μm $\times$ 6.5 μm to a relatively large 24 μm $\times$ 24 μm. It is generally the case that the larger the number of pixels, the smaller the size of a pixel.

Full well capacity. The full well capacity is the maximum number of electrons that can be held in a pixel. Exceeding this number will lead to saturation of the given pixel, and may not only render the pixel useless for analysis, but might also lead to *blooming*, whereby the excess charge flows into neighboring pixels and corrupts the image information in those pixels.

Hardware binning. In hardware binning, the charges in several pixels are combined. Typically it would be an arrangement of 2×2 pixels or 4×4 pixels. This increases the effective pixel size, which also results in an increase in the effective full well capacity. Binning decreases the effective total number of pixels, thereby increasing the frame rate of the image acquisition. In a CCD detector, the fact that binning is an on-chip operation before readout is important. This means that the readout noise for a binned pixel is of the same level as the noise for one individual original pixel. This is not the case for a CMOS detector, where binning is performed after readout. In a CMOS detector, the variance of the readout noise for a binned pixel is equal to the sum of the variances of the readout noise for the component pixels.

Digitization level. The charge read out from each pixel needs to be digitized in an A/D to provide the final digital image. The digitization level of the digitization process is an important consideration. Consumer cameras often have pixels of bit depth of only 8, meaning that the value of a pixel is given by one of $2^8 = 256$ possible digital levels. In scientific cameras, 12-bit, 14-bit, and 16-bit digitization levels are often available. The fundamental question then relates to the relationship between the digital level and the actual signal level in units of electrons. The unit of an increment in the digital level in a pixel of the digital image is referred to as a *digital unit* (DU). For quantitation of the acquired images it is very important to know how many electrons correspond to one DU. This *electron-count-to-DU conversion factor* is often given in the quality control sheet that accompanies the purchased camera. We can, however, get a rough idea of this conversion factor by considering the full well capacity and bit depth of a pixel of the camera. If we assume that the full well capacity is 18000 electrons and the bit depth is 14 (such that there are $2^{14} = 16384$ possible digital levels for each pixel of a digital image taken by the camera), then about 18000/16384 ≈ 1.098 electrons correspond to one DU. This is only a very approximate calculation that does not take into account various offsets and pre-readout gains that might be part of the digitization process. Of course this calculation also assumes that the detector is highly linear. For discussions of how one can experimentally determine the electron-count-to-DU conversion factor for CCD/CMOS and EMCCD cameras, see Sections 12.7.1.3 and 12.7.2.2, respectively.

Quantum efficiency. Detection sensitivity is an important aspect of an image detector, especially in the low-light situations that fluorescence microscopy often presents. We know that the photoelectric effect is responsible for the conversion of the photons that impact a detector into electrons. However, we have so far not discussed the efficiency of this process in an actual imaging device. This conversion efficiency is described by the *quantum efficiency* of the detector, which is defined as the ratio of the number of electrons obtained in the detector to the number of photons that impact the detector, typically expressed as a percentage. Since the photoelectric effect implies that a detector can produce at most one electron for each photon that impacts the detector, the quantum efficiency is at most 100%. The quantum efficiency is also wavelength-dependent. Figure 12.8 shows the quantum efficiency plots for a number of typical detectors. Such plots are important for determining the detector that matches best the wavelength range for the experiments that are to be carried out. Unfortunately, some of the detectors with the highest quantum efficiency are also the most expensive. In the case of back-thinned or back-illuminated detectors (such as the Andor iXon detector; see Fig. 12.8), the higher cost is justified by a more complex manufacturing process that orientates the electronics associated with the pixels below the photosensitive surface of the detector. Compared to a front-illuminated orientation in which the electronics are situated above the photosensitive surface and act as a physical obstacle that prevents some incident light from reaching that surface, the back-illuminated design yields a significantly higher quantum efficiency by allowing incident light to reach the photosensitive surface unimpeded by the electronics.

Readout speed and frame rate. The readout speed that we have made mention of is more precisely called the pixel readout speed, and it refers to the rate at which pixels are read out by the detector. It is commonly given in units of millions of pixels per second, or megahertz (MHz). Typical CCD detectors have pixel readout speeds ranging from 1 MHz to 20 MHz, while typical CMOS detectors have substantially faster readout speeds of several hundred MHz. Together with the number of pixels comprising the image to be read out, the pixel readout speed is a primary determinant of the frame rate, i.e., the rate at which images can be acquired. To a first approximation, the frame rate is given by the inverse of the sum of two quantities — the image exposure time and the number of image pixels divided by the pixel readout speed. More precise calculations of the frame rate, however, require consideration of other factors that are specific to the given detector. The frame rate is typically given in units of frames per second. To give examples, the Hamamatsu ORCA-ER CCD camera in Fig. 12.8 has a pixel readout speed of 14.75 MHz and supports a frame rate of 8.3 frames per second for full-sized images of 1344×1024 pixels. The Andor Zyla 5.5 CMOS camera in the same figure has a significantly higher pixel readout speed of 560

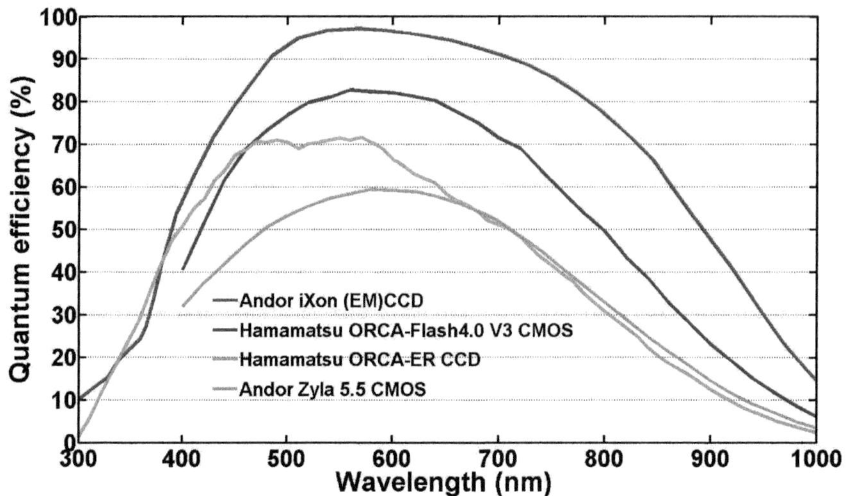

FIGURE 12.8
Quantum efficiency as a function of wavelength for four different detectors. [See Figure Credits.]

MHz, at which it can support a frame rate of 100 frames per second for full-sized images of 2560×2160 pixels.

Readout noise level. The readout noise level, usually specified as the standard deviation of the noise in units of electrons, typically increases with the readout speed at which a given detector is used. At the lowest readout speed, a readout noise level of 6 electrons is not uncommon for older and even some relatively new scientific CCD detectors. Some of the newest available CCD detectors, however, can operate at low readout noise levels of between 2 and 3 electrons. The newest commercially available scientific CMOS detectors have a readout noise level of around 1 electron. As each pixel of a CMOS detector is read out with its own electronic circuitry and therefore has its own readout noise level, the readout noise level reported for a CMOS detector is commonly the median of the noise levels of the individual pixels.

Dark current level. The typical scientific image detector is equipped with a cooling system that substantially reduces the dark current. When a detector is sufficiently cooled, the dark current level is reduced to significantly less than 1 electron per pixel per second. Many detectors have a dark current level that is on the order of 10^{-3} electrons per pixel per second.

Linearity of the response. Linearity of the response is commonly determined from a plot of the signal level versus the exposure time. By increasing the exposure time under illumination of stable and uniform intensity, the signal level is expected to increase linearly up to the full well capacity of a pixel. Linearity of the response is thus measured by fitting a straight line to the plot and quantifying the deviation of the data points from the best-fit line. A typical image detector exhibits excellent linearity with a deviation of no more than 1%.

Uniformity of the response. Uniformity of the response is determined by exposing the detector to light of stable and uniform intensity and calculating the deviation of the signal acquired in a pixel from the average acquired signal. A typical image detector is characterized by high uniformity represented by a deviation of less than 0.5%.

12.7 Measurements of detector specifications

A number of the specifications of a detector can be readily verified by a microscopist in a research laboratory. We will discuss here how one can assess the linearity of a detector, estimate its electron-count-to-DU conversion factor, and determine its readout noise level and dark current level. The bases for these measurements are the basic stochastic properties of the acquired signals, which allow for the exploitation of their nature as, for example, Poisson and Gaussian random variables. For all derivations we assume that we are dealing with a grayscale detector, although they can be easily adapted to the situation of a color detector depending on the specific implementation of the color imaging scheme.

12.7.1 Determination of CCD and CMOS detector specifications

12.7.1.1 Data model

In order to analyze the characteristics of a detector we need to have a good understanding of the nature of the data that is acquired. To this end we introduce the following notation. We assume that the data $I_k(\Delta T)$ acquired in the kth pixel of a CCD or CMOS detector, over an exposure time ΔT, is given by

$$I_k\left(\Delta T\right) = DS_k\left(\Delta T\right) + DB_k\left(\Delta T\right) + DD_k\left(\Delta T\right) + DW_k, \tag{12.1}$$

where $DS_k\left(\Delta T\right)$ is the digitized signal corresponding to photons emitted by the object of interest and detected in the kth pixel during the exposure of duration ΔT. Similarly, $DB_k\left(\Delta T\right)$ is the digitized signal in the kth pixel due to photons classified as background, such as scattered photons, and $DD_k\left(\Delta T\right)$ is the digitized signal in the same pixel due to the dark current that is generated in the detector. The final term DW_k stands for the digitized readout noise in the kth pixel. From the discussion in Section 12.4, we know that whereas DW_k is a sampled version of a Gaussian random variable, the photon signal components and the dark current component of the data are sampled versions of Poisson random variables. Importantly, the readout noise does not depend on the exposure time ΔT. For a CCD detector, the readout noise in every pixel has the same mean and the same variance, since all pixels share the same readout circuitry. For a CMOS detector, the dedicated readout circuitry for each pixel means that the readout noise in each pixel is characterized by its own mean and variance.

CCD and CMOS detectors typically also have an *offset*. This means that even in the absence of any signal, the mean value of the data in a pixel is not zero, but rather a typically small positive number. Since the offset is a property of the readout electronics, the offset is identical for all pixels in a CCD detector, but is different for each pixel of a CMOS detector. We do not include the offset explicitly in the model description, but rather absorb it as the mean $\mathrm{E}\left[DW_k\right]$ of the readout noise term DW_k.

Our standing assumption in this section is that the object of interest emits photons at the same rate over time. Therefore we immediately have that the mean of the digitized object signal in the kth pixel depends linearly on the exposure time ΔT, i.e.,

$$\mathrm{E}\left[DS_k\left(\Delta T\right)\right] = \Delta T \cdot \mathrm{E}\left[DS_k(1)\right],$$

where $DS_k(1)$ is the digitized object signal DS_k corresponding to unit exposure time, e.g., in most cases 1 second. The same relationship also applies to the corresponding Poisson-distributed photon signal $S_k = \frac{1}{c_k} \cdot DS_k$, where $c_k > 0$ is a constant representing the electron-count-to-DU conversion factor for the kth pixel, which converts between the digitized signal and the actual number of detected photoelectrons. (As in the case of the readout noise and the offset, the conversion factor c_k can be assumed to be the same for all pixels of a CCD detector, but has to be assumed to be different for each pixel of a CMOS detector due to the difference in the readout architectures of these two detector types.) We therefore have that

$$\mathrm{E}\left[S_k\left(\Delta T\right)\right] = \Delta T \cdot \mathrm{E}\left[S_k(1)\right].$$

In order to obtain similar relations for the variance of the digitized object signal, we need to exploit the key property of Poisson random variables, i.e., that their mean equals their variance. Using this property, we obtain

$$\begin{aligned}\mathrm{Var}\left(DS_k\left(\Delta T\right)\right) &= \mathrm{Var}\left(c_k \cdot S_k\left(\Delta T\right)\right) = c_k^2 \cdot \mathrm{Var}\left(S_k\left(\Delta T\right)\right) = c_k^2 \cdot \mathrm{E}\left[S_k\left(\Delta T\right)\right] \\ &= c_k^2 \cdot \Delta T \cdot \mathrm{E}\left[S_k(1)\right] = c_k^2 \cdot \Delta T \cdot \mathrm{Var}\left(S_k(1)\right) = \Delta T \cdot \mathrm{Var}\left(c_k \cdot S_k(1)\right) \\ &= \Delta T \cdot \mathrm{Var}\left(DS_k(1)\right).\end{aligned}$$

This shows that the variance of the digitized object signal $DS_k\left(\Delta T\right)$ is in fact linear in the exposure time ΔT.

12.7.1.2 Linearity of the response

For quantitative imaging, one of the most important properties of a detector is the linearity of its response. With this we mean that if the detector receives a certain multiple of input photons, then the measured digitized signal is increased by the same multiple. While it might be possible to also work with nonlinear detectors, the analysis of the data would become very involved.

To verify that a detector that is advertised as having a linear response is indeed linear, we can carry out the following analysis. We assume that we have a sample, including the object of interest and the background component, that provides a constant input to the detector. Under these circumstances, and given that the dark current also gives rise to electrons at a constant rate, we can develop a simple experiment to verify linearity by making acquisitions with varying exposure times. Under our assumptions, the mean $\mathrm{E}\left[I_k\left(\Delta T\right)\right]$ of the data acquired in the kth pixel over the exposure time ΔT is given by

$$\begin{aligned}\mathrm{E}\left[I_k\left(\Delta T\right)\right] &= \mathrm{E}\left[DS_k\left(\Delta T\right)\right] + \mathrm{E}\left[DB_k\left(\Delta T\right)\right] + \mathrm{E}\left[DD_k\left(\Delta T\right)\right] + \mathrm{E}\left[DW_k\right] \\ &= \Delta T \cdot \left(\mathrm{E}\left[DS_k(1)\right] + \mathrm{E}\left[DB_k(1)\right] + \mathrm{E}\left[DD_k(1)\right]\right) + \mathrm{E}\left[DW_k\right],\end{aligned} \quad (12.2)$$

where $DS_k(1)$, $DB_k(1)$, and $DD_k(1)$ denote the digitized signals with unit exposure time, typically 1 second. This expression clearly shows that the acquired data is affine in the exposure time ΔT and would be linear were it not for the constant offset $\mathrm{E}\left[DW_k\right]$ from the readout process. (Recall that an expression y is called *affine* in a variable x if $y = mx + b$, for some constants m and b. This expression is linear, if $b = 0$.) This offset does not impact the analysis very significantly, since it can be readily subtracted. In addition, in practice the offset is in most cases very small in comparison to the signals that are acquired and could even be ignored. Although strictly speaking the detector is not linear but affine, we nevertheless typically refer to the "linearity" of the detector.

Equation (12.2) can be used to verify the linearity, or more precisely, the affine property of the detector. We simply choose a number of exposure times such that the acquired signals will fall within the full range of the detector. For each exposure time ΔT, we take sufficiently many measurements so their mean $m\left(\Delta T\right)$ provides a good estimate of $\mathrm{E}\left[I_k\left(\Delta T\right)\right]$. We then plot the estimates $m\left(\Delta T\right)$ versus the exposure time ΔT. The resulting plot can then be investigated to determine whether or not the detector has the required performance regarding linearity/affinity. Absent nonuniform behavior of the detector, the plots should be analogous in all the pixels, assuming that each pixel was exposed in the same manner.

In case the resulting plots do not exhibit the anticipated linearity/affinity of the detector, it is important to verify that the light source was indeed constant over the full experiment.

12.7.1.3 Estimation of electron-count-to-DU conversion factor

The digital image produced by the camera does not give direct access to measurements of the number of electrons S_k, B_k, and D_k due to the object of interest, the background component, and the dark current, respectively, that arise in the kth pixel. Instead, the only available measurement is I_k, which is partially composed of the corresponding digitized counts DS_k, DB_k, and DD_k. As already seen in Section 12.7.1.1, the digitized object signal DS_k is related to the electron count S_k by the electron-count-to-DU conversion factor c_k, i.e., $DS_k(\Delta T) = c_k \cdot S_k(\Delta T)$. The same conversion factor also relates the digitized signals DB_k and DD_k to the electron counts B_k and D_k, respectively, so that $DB_k(\Delta T) = c_k \cdot B_k(\Delta T)$ and $DD_k(\Delta T) = c_k \cdot D_k(\Delta T)$.

For a careful quantitative analysis it is therefore important to know the electron-count-to-DU conversion factor c_k. Certainly the spec sheet for a detector will include the number, but it is important to be able to verify that number experimentally. The standard protocol for determining c_k exploits the Poisson nature of the underlying signals to relate the mean and the variance of the acquired data $I_k(\Delta T)$ in the kth pixel as

$$\begin{aligned}
\mathrm{Var}(I_k(\Delta T)) &= \mathrm{Var}(DS_k(\Delta T)) + \mathrm{Var}(DB_k(\Delta T)) + \mathrm{Var}(DD_k(\Delta T)) + \mathrm{Var}(DW_k) \\
&= \mathrm{Var}(c_k \cdot S_k(\Delta T)) + \mathrm{Var}(c_k \cdot B_k(\Delta T)) + \mathrm{Var}(c_k \cdot D_k(\Delta T)) + \mathrm{Var}(DW_k) \\
&= c_k^2 \cdot [\mathrm{Var}(S_k(\Delta T)) + \mathrm{Var}(B_k(\Delta T)) + \mathrm{Var}(D_k(\Delta T))] + \mathrm{Var}(DW_k) \\
&= c_k^2 \cdot (\mathrm{E}[S_k(\Delta T)] + \mathrm{E}[B_k(\Delta T)] + \mathrm{E}[D_k(\Delta T)]) + \mathrm{Var}(DW_k) \\
&= c_k \cdot (\mathrm{E}[c_k \cdot S_k(\Delta T)] + \mathrm{E}[c_k \cdot B_k(\Delta T)] + \mathrm{E}[c_k \cdot D_k(\Delta T)]) + \mathrm{Var}(DW_k) \\
&= c_k \cdot (\mathrm{E}[I_k(\Delta T)] - \mathrm{E}[DW_k]) + \mathrm{Var}(DW_k).
\end{aligned}$$

This last expression shows that if we plot the variance of the measured data versus its mean for different exposure times, we will obtain a straight line. Clearly, from the slope of that line the conversion factor c_k can be determined. Therefore, in practice we carry out an experiment for different exposure times, where for each exposure time we acquire multiple images. For each exposure time ΔT we then obtain a data point for the variance versus mean plot by calculating, from the multiple images, the mean and the variance of the digitized count $I_k(\Delta T)$ in the kth pixel. The conversion factor c_k is then determined from the slope of a line that best fits the obtained data points.

Note that in practice, variations of the procedure described here may need to be used in order to address issues that arise due to imperfect imaging conditions. For example, if the intensity of the light impinging on a pixel is not sufficiently stable over time, then the mean and variance of the multiple values of the pixel acquired with a given exposure time will likely not provide a good data point for the variance versus mean plot. In this case, if a CCD detector is used, then one can apply the procedure as described, but to the aggregate data from a multi-pixel region to minimize the variability of the signal level from image to image. This approach is suitable when a CCD detector is used, as all pixels of a CCD detector can justifiably be assumed to have the same electron-count-to-DU conversion factor. For another example, despite the high uniformity of the response for a CCD detector, it is often the case that small differences in responsivity exist between the pixels of a detector. These differences, although small, can produce differing estimates of the variance for the same signal level, which in turn can lead to significantly different estimates of the conversion factor. To minimize the effect of these responsivity differences, the described procedure can be modified so that the variance of the data is calculated from the difference of the pixel values between two successively acquired images. This approach uses the subtraction of one repeat image from another to effectively remove the responsivity differences between the pixels of a detector.

12.7.1.4 Estimation of readout noise mean and variance

Readout noise is a serious problem in low-light imaging. Therefore it is critical that the readout noise level is well understood. To measure the readout noise in terms of its mean and variance, we need to find an imaging configuration that allows us to access the readout noise component of the data. This can be achieved by using a shutter on the camera so that the components due to object and background photons drop out, i.e., $DS_k(\Delta T) = DB_k(\Delta T) = 0$. This still leaves the dark current component in the acquired data. The dark current component is typically quite small as discussed in Section 12.4. However, the dark current does depend on the exposure time. Therefore reducing the exposure time to the lowest possible value, typically in the range of tens of microseconds, will effectively remove the dark current from the acquired data. Under these experimental conditions, we are left with the acquired data being given by just the readout noise, i.e.,

$$I_k(\Delta T) = DW_k.$$

Taking repeat measurements of the kth pixel with the camera shutter closed and a minimal exposure time will therefore allow us to estimate the mean $\mathrm{E}[DW_k]$ and variance $\mathrm{Var}(DW_k)$ of the readout noise DW_k. In accordance with the data model discussion above, the mean determined in this manner is an estimate of the offset for the kth pixel. The calculated variance can be used to easily obtain the standard deviation in units of electrons, which is how the readout noise level is typically reported. Specifically, the square root of the variance is taken, and the resulting standard deviation is divided by the electron-count-to-DU conversion factor c_k.

Note that in the case of a CCD detector, instead of relying on a single pixel we could also calculate the mean and variance of multiple pixels and estimate the readout noise mean and variance as the averages of the means and variances for the different pixels. This is again due to the fact that a single readout circuitry is shared by all pixels of a CCD detector. For this same reason, we could in principle also compute the mean and variance of the pixel values of one exposure to obtain the desired estimate of the readout noise mean and variance.

Note also that even in the absence of light input, the calculation of the variance of the pixel data can be affected by systematic differences between the pixels of a CCD detector. Therefore, for the measurement of readout noise, one could also adopt the approach mentioned in Section 12.7.1.3 that calculates the variance using the difference of the pixel values between successively acquired images.

12.7.1.5 Estimation of mean of dark current

The dark current can be measured by keeping the camera shutter closed and by acquiring multiple images with a long exposure time. Averaging the resulting measurements and subtracting the offset will provide an estimate for the mean of the dark current in each pixel. Since the dark current is typically very low, very long exposure times might have to be used and a large number of images might have to be averaged for a reliable estimate.

12.7.2 Determination of EMCCD detector specifications

12.7.2.1 Data model

In an EMCCD detector, the signal components due to the object of interest, the background, and the dark current are amplified by an electron multiplication register before digitization.

We therefore assume that the data $I_k(\Delta T)$ acquired in the kth pixel of the detector, over an exposure time ΔT, is given by

$$I_k(\Delta T) = DS_{amp,k}(\Delta T) + DB_{amp,k}(\Delta T) + DD_{amp,k}(\Delta T) + DW_k,$$

where $DS_{amp,k}(\Delta T)$, $DB_{amp,k}(\Delta T)$, and $DD_{amp,k}(\Delta T)$ are the digitized signals corresponding to the electron counts resulting from the amplification of the electrons due to the object of interest, the background, and the dark current, respectively. As in the case of the CCD/CMOS data model (Eq. (12.1)), the term DW_k represents the digitized, or sampled, Gaussian-distributed readout noise in the kth pixel. The digitized signals $DS_{amp,k}(\Delta T)$, $DB_{amp,k}(\Delta T)$, and $DD_{amp,k}(\Delta T)$, however, are no longer sampled versions of Poisson random variables. Instead, they are sampled versions of random variables characterized by a more complex probability distribution that accounts for the amplification of the initially Poisson-distributed number of electrons in the pixel. The amplification by the detector's multiplication register is typically described by a branching process, and the precise probability distribution of the resulting electron counts $S_{amp,k}$, $B_{amp,k}$, and $D_{amp,k}$ depends on the assumed model of electron multiplication. As is the case for the CCD and CMOS detectors, the digitized signals are related to the actual electron counts by an electron-count-to-DU conversion factor $c_k > 0$, i.e., $DS_{amp,k}(\Delta T) = c_k \cdot S_{amp,k}(\Delta T)$, $DB_{amp,k}(\Delta T) = c_k \cdot B_{amp,k}(\Delta T)$, and $DD_{amp,k}(\Delta T) = c_k \cdot D_{amp,k}(\Delta T)$.

12.7.2.2 Estimation of electron-count-to-DU conversion factor

The electron-count-to-DU conversion factor c_k for the kth pixel of an EMCCD detector can be determined using an approach similar to the standard protocol described in Section 12.7.1.3 for a CCD or CMOS detector. Assuming that the signal amplification by the multiplication register is modeled as a geometrically multiplied branching process (Section 15.2.4), the mean and variance of the amplified object electron count are related by

$$\mathrm{Var}(S_{amp,k}(\Delta T)) = (2g-1)\cdot \mathrm{E}[S_{amp,k}(\Delta T)] \approx 2g\cdot \mathrm{E}[S_{amp,k}(\Delta T)],$$

where $g > 1$ is the EM gain. Note that the approximation holds provided that g is large, which is typically the case when imaging with an EMCCD detector. Using this relationship, which applies also to $B_{amp,k}(\Delta T)$ and $D_{amp,k}(\Delta T)$, the mean and variance of the acquired data $I_k(\Delta T)$ in the kth pixel can be related as

$$\begin{aligned}
&\mathrm{Var}(I_k(\Delta T))\\
&= \mathrm{Var}(DS_{amp,k}(\Delta T)) + \mathrm{Var}(DB_{amp,k}(\Delta T)) + \mathrm{Var}(DD_{amp,k}(\Delta T)) + \mathrm{Var}(DW_k)\\
&= \mathrm{Var}(c_k\cdot S_{amp,k}(\Delta T)) + \mathrm{Var}(c_k\cdot B_{amp,k}(\Delta T)) + \mathrm{Var}(c_k\cdot D_{amp,k}(\Delta T)) + \mathrm{Var}(DW_k)\\
&= c_k^2\cdot[\mathrm{Var}(S_{amp,k}(\Delta T)) + \mathrm{Var}(B_{amp,k}(\Delta T)) + \mathrm{Var}(D_{amp,k}(\Delta T))] + \mathrm{Var}(DW_k)\\
&\approx c_k^2\cdot 2g\cdot(\mathrm{E}[S_{amp,k}(\Delta T)] + \mathrm{E}[B_{amp,k}(\Delta T)] + \mathrm{E}[D_{amp,k}(\Delta T)]) + \mathrm{Var}(DW_k)\\
&= 2gc_k\cdot(\mathrm{E}[c_k\cdot S_{amp,k}(\Delta T)] + \mathrm{E}[c_k\cdot B_{amp,k}(\Delta T)] + \mathrm{E}[c_k\cdot D_{amp,k}(\Delta T)]) + \mathrm{Var}(DW_k)\\
&= 2gc_k\cdot(\mathrm{E}[I_k(\Delta T)] - \mathrm{E}[DW_k]) + \mathrm{Var}(DW_k).
\end{aligned}$$

As in the case of a CCD or CMOS detector, the last expression here shows that the conversion factor c_k can be determined from the slope of the line formed by plotting the variance versus the mean of the measured data. The slope in this case, however, includes an extra multiplicative factor of $2g$ that needs to be accounted for. In practice, to generate the plot, one would follow the same procedure as for the CCD or CMOS detector. Specifically, one would acquire sets of images for different values of the exposure time ΔT. Each point in the plot would then be obtained by calculating the mean and variance of the digitized counts in the kth pixel from the set of images corresponding to a particular exposure time. Note that the discussion in Section 12.7.1.3 regarding the use of the aggregate data in a multi-pixel region and the calculation of the variance of the data from the difference of the pixel values between consecutive frames is also applicable to the measurement procedure described here for the EMCCD detector.

12.7.2.3 Estimation of readout noise mean and variance

Estimation of the mean and variance of the readout noise of an EMCCD detector can be measured in essentially the same manner as described for the CCD detector in Section 12.7.1.4. It is important, however, that the images are acquired at the lowest EM gain setting possible. Otherwise, a significant amplification of even a single electron will result in a large pixel value that leads to measured readout noise statistics that are far from correct.

13

Geometrical Optics

In this chapter, we will introduce material on geometrical optics that is important for our treatment of the optics of microscopes. Geometrical optics does not make use of the wave nature of light, which will be explored in detail in Chapter 14. Instead, geometrical optics addresses situations in which it is sufficient to think of light as rays traveling from a light source, and whose direction is changed by passing through lenses and media with different optical properties. This approach is very well suited to studying questions related to magnification in a microscope system.

13.1 Reflection and refraction

The basic model of geometrical optics is that optical rays travel on straight lines in a homogeneous medium, for example in air, water, or homogeneous glass. However, the direction of a ray of light changes as it impacts another medium. The ray undergoes reflection and refraction as it encounters the second medium. It is these fundamental properties of light that we will discuss first.

13.1.1 Reflection

One of the most basic laws of geometrical optics relates to the *reflection* of light. For example, as the rays of light impact a planar mirror they are reflected. The *law of reflection* states that the impacting ray, the normal to the surface, and the reflected ray all lie in the same plane (Fig. 13.1). Moreover, the *angle of incidence* $\theta_{incidence}$, i.e., the angle between the incident ray and the normal to the plane, equals the *angle of reflection* $\theta_{reflection}$, i.e., the angle between the reflected ray and the normal:

$$\theta_{incidence} = \theta_{reflection}.$$

13.1.2 Refractive index

In order to discuss the behavior of light with respect to refraction, the notion of the *refractive index* of an optical medium is of importance. As light travels through space, its speed depends on the optical medium in which it travels. For example, in vacuum the speed of light is given by the well-known number $c = 299792.5$ km/s, whereas in water light travels at the lower speed of 224900.6 km/s. The refractive index n_m of an optical medium is defined as the ratio between the speed of light in vacuum and the speed of light in the medium, i.e.,

$$n_m = \frac{\text{speed of light in vacuum}}{\text{speed of light in the specific medium}} = \frac{c}{\text{speed of light in the specific medium}}. \quad (13.1)$$

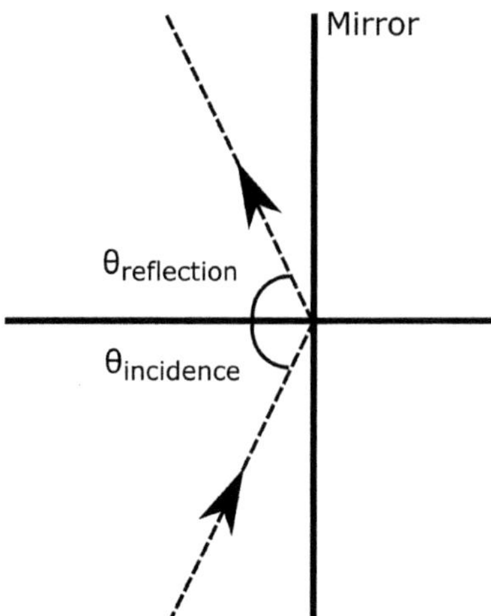

FIGURE 13.1
Reflection of a ray of light as it impacts a mirror.

Since light cannot travel faster than in vacuum, the refractive index is always greater than or equal to 1. Based on the above mentioned speed of light in water, the refractive index for water is 1.333. The refractive index for air is very close to 1 and typically assumed to be 1. The refractive index for glass varies depending on the type of glass but is often 1.520. Immersion oils for use in combination with an oil objective are available with different refractive indices that are matched to the refractive index of the cover glass. The refractive index of these oils is also typically temperature-dependent. It is therefore important to use immersion oil that is matched to the temperature at which the experiment is carried out to achieve the best possible image quality.

13.1.3 Snell's law

In a medium of constant refractive index, an optical ray follows a straight path. However, at the interface with a medium of a different refractive index, the ray will be reflected or refracted. *Snell's law* governs these phenomena and states that

$$n_1 \sin(\theta_1) = n_2 \sin(\theta_2), \tag{13.2}$$

where n_1 and n_2 are the refractive indices of the two media, θ_1 is the angle between the incident ray and the normal to the interface, and θ_2 is the angle between the refracted ray and the normal to the interface, as shown in Fig. 13.2.

It is instructive to consider a case in which a ray travels from a medium with lower refractive index n_1, such as air, to a medium with higher refractive index n_2, such as glass, i.e., when $n_1 < n_2$. In this case,

$$\sin(\theta_2) = \frac{n_1}{n_2} \sin(\theta_1) < \sin(\theta_1).$$

This shows immediately that the angle of refraction is smaller than the angle of incidence. In a situation where the refractive index of the first medium is higher than the refractive index of the second medium, i.e., when $n_1 > n_2$, the angle of refraction would instead be larger than the angle of incidence.

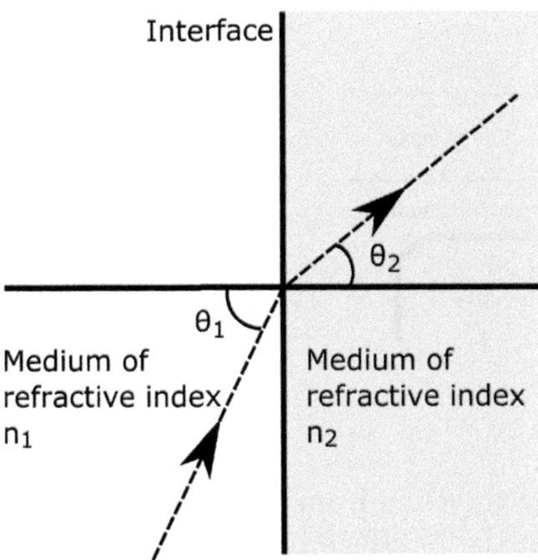

FIGURE 13.2
Refraction of a ray of light as it impacts an interface between two media with different refractive indices.

13.1.4 Total internal reflection

We now continue to consider the case where the ray is traveling from a medium of refractive index n_1 to a medium with refractive index n_2 that is lower than n_1, i.e., when $\frac{n_2}{n_1} < 1$. Assume that the angle of incidence θ_1 is such that $1 > \sin(\theta_1) > \frac{n_2}{n_1}$. Then, by Snell's law, the sine of the angle of refraction is given by $\sin(\theta_2) = \frac{n_1}{n_2}\sin(\theta_1) > \frac{n_1}{n_2} \cdot \frac{n_2}{n_1} = 1$, which is not possible. Hence refraction is not possible in this scenario, and the beam is fully reflected. This describes the phenomenon of *total internal reflection*, which occurs when the *critical angle* is exceeded. The critical angle θ_c is defined as

$$\theta_c = \arcsin\left(\frac{n_2}{n_1}\right), \tag{13.3}$$

where n_1 and n_2 are the refractive indices of the media.

As follows from the definition of the critical angle, the refractive index n_2 has to be less than the refractive index n_1, i.e., the ratio $\frac{n_2}{n_1}$ has to be less than 1, for total internal reflection to occur. For example, total internal reflection is possible if the light travels from glass to air, since the refractive index of glass is around 1.52, whereas the refractive index of air is 1. A ray is then totally internally reflected if it impacts the air interface at an angle larger than the critical angle of $\arcsin\left(\frac{n_2}{n_1}\right) = \arcsin\left(\frac{1}{1.52}\right) = \arcsin(0.658) = 41°$. However, no total internal reflection can be achieved when the ray of the light travels from air to glass, as in this case the ratio of the refractive indices is 1.52, which is well above the threshold of 1.

13.1.5 Extreme rays in microscopy optics

Total internal reflection fluorescence microscopy has been discussed in Section 11.4. This is an important imaging modality for studying events close to the cover glass, such as receptor dynamics on the plasma membrane of a cell. In this section, we will investigate the condition that must be satisfied by the objective in order to implement TIRFM.

Having the tools related to Snell's law available, we can now investigate the path of an *extreme ray* leaving a microscope objective. By an extreme ray we mean a ray that follows the envelope of the maximum cone of rays from the objective to an in-focus sample point.

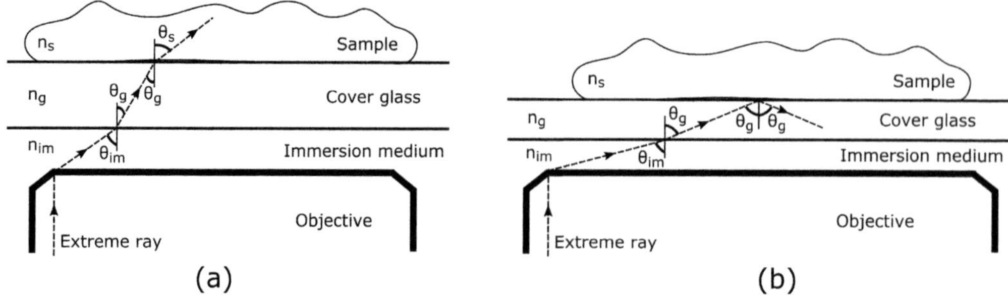

FIGURE 13.3
Light path of an extreme ray traveling from the objective through the immersion medium, the cover glass, and reaching the interface between the cover glass and the sample. In the case of conventional fluorescence microscopy (a), the ray is refracted at the interface and enters the sample. In the case of total internal reflection fluorescence microscopy (b), the ray is reflected at the interface and back into the cover glass. The symbols n_{im}, n_g, and n_s denote the refractive indices of the immersion medium, the cover glass, and the sample, respectively, and the symbols θ_{im}, θ_g, and θ_s denote the angles of incidence, refraction, and reflection at the interfaces between two media.

The maximum cone is determined by the numerical aperture of the objective as seen in Section 10.6.1 (see also Fig. 10.8). After exiting the objective, such a ray travels through the immersion medium, be it air, oil, or water, with refractive index n_{im}, and impacts the cover glass at an angle θ_{im} given by $\theta_{im} := \arcsin\left(\frac{n_a}{n_{im}}\right)$, where n_a is the numerical aperture of the objective. At the interface between the immersion medium and the cover glass, application of Snell's law yields

$$n_g \sin(\theta_g) = n_{im} \sin(\theta_{im}) = n_a, \tag{13.4}$$

where n_g is the refractive index of the cover glass and θ_g is the angle with respect to the normal at the interface with which the refracted ray travels in the cover glass. As this refracted ray subsequently impacts the interface between the cover glass and the sample, Snell's law again applies and yields

$$n_s \sin(\theta_s) = n_g \sin(\theta_g) = n_{im} \sin(\theta_{im}) = n_a,$$

where n_s is the refractive index of the sample, and θ_s is the angle with respect to the normal at the interface with which the refracted ray travels in the sample. The light path described is illustrated in Fig. 13.3(a).

Due to the importance of total internal reflection fluorescence microscopy, we are particularly interested in the condition under which the impacting ray undergoes total internal reflection at the interface between the cover glass and the sample (Fig. 13.3(b)). In order to obtain total internal reflection, the incident angle θ_g at this interface needs to exceed the critical angle θ_c. Using Eqs. (13.3) and (13.4), this condition can be expressed as

$$\sin(\theta_c) = \frac{n_s}{n_g} < \sin(\theta_g) = \frac{n_a}{n_g}.$$

This inequality is equivalent to

$$n_s < n_a,$$

which implies that total internal reflection is possible at the interface between the cover

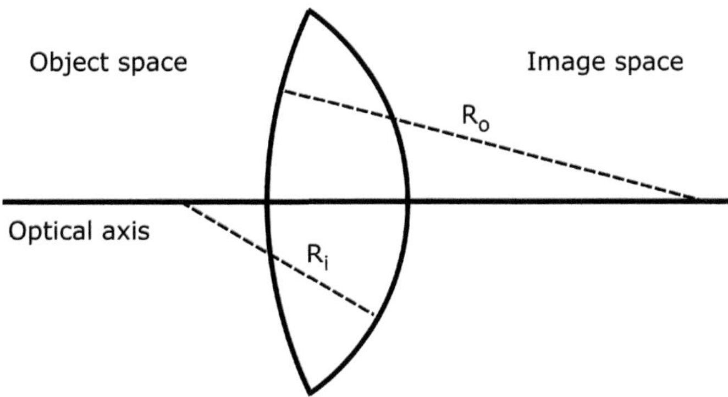

FIGURE 13.4
Spherical convex lens characterized by the two radii R_o and R_i, which define the curvatures of the two surfaces of the lens.

glass and the sample if the numerical aperture of the objective is larger than the refractive index of the sample.

As the refractive index for a live cell sample is around 1.33 (close to that of water), the numerical aperture of the objective has to be very high for the extreme ray to be able to achieve total internal reflection. As air objectives have a numerical aperture of at most 1, total internal reflection is not achievable with an air objective.

13.2 Lenses

At the heart of the optics of a microscope is its objective. A high-quality objective is typically made up of several optical elements. Many of the key properties of a microscope, however, can nevertheless be understood by considering an objective to be a thin lens. We will therefore first discuss the properties of a thin lens from the point of view of geometrical optics. Diffraction phenomena in the context of a thin lens will be the topic of Chapter 14.

The lenses that we will consider here are *spherical lenses* (Fig. 13.4), i.e., they are characterized by two surfaces that can be described as the surfaces of spheres with radii R_i and R_o. An important notion in the discussion of many optical systems is that of the *optical axis*, i.e., its main axis of symmetry. In the particular case of a spherical lens, it is the axis with respect to which the lens is radially symmetric and is given by the line that passes through the centers of the spheres that define the surfaces of the lens.

The term *thin lens* refers to a lens in which a ray leaves the lens at the same coordinates at which it has entered the lens, i.e., with respect to a coordinate system that is orthogonal to the optical axis, a ray is not significantly displaced by the lens. This will be the case for a lens that is physically "thin". However, this property can be satisfied for other optical systems, even if they consist of multiple optical elements or are not physically "thin". In fact, we will often use the insights we obtain for thin lenses in the context of microscopy optics, which includes both the objective and the tube lens.

Among lenses it is the *convex lenses* that are most important for our applications in microscopy, as such lenses are key elements in the focusing optics of a microscope. A convex lens is one whose two surfaces bend outwards from the body of the lens, while a *concave lens* is one whose two surfaces bend inwards toward the body of the lens.

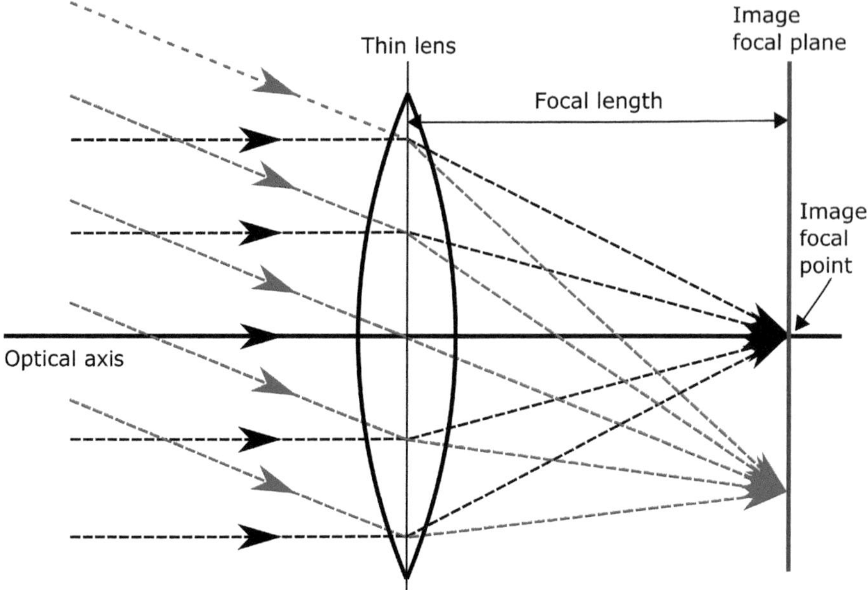

FIGURE 13.5
Illustration of the image focal point, image focal plane, and focal length of a thin lens using two sets of parallel rays.

13.2.1 Focal points and focal planes

We can now start to discuss the optical properties of lenses. Again we will consider convex lenses, and will assume they are thin lenses as this assumption simplifies the presentation and does not significantly limit the scope of the results. One of the most important characteristics of a lens are its focal points. In order to characterize them, we investigate the paths that parallel rays take as they enter the lens (Fig. 13.5). It can be shown that parallel rays have their paths changed by the lens in such a way that all the rays intersect in one point. Of all the sets of parallel rays, a particularly important one is the set of parallel rays that are themselves parallel to the optical axis. The point of intersection for this set of parallel rays is on the optical axis itself, since a ray that travels along the optical axis does not have its path changed by the lens. This point is called the *(image) focal point* of the lens. In fact, all the points of convergence for the various sets of parallel rays lie in one plane, the *(image) focal plane* of the lens, which is a plane that is orthogonal to the optical axis and passes through the (image) focal point of the lens (Fig. 13.5).

Here we speak of an image focal point and image focal plane. However, sending parallel sets of rays into the lens from the opposite side of the lens will lead to the identification of a further focal point and focal plane, which we refer to as the *object focal point* and the *object focal plane*. Since a lens does not have a natural orientation, distinguishing the two focal points and focal planes of a lens by being in the object space or image space is somewhat arbitrary. As many optical systems, in particular microscopy systems, have a natural orientation to their light paths, it does, however, provide a convenient way to identify which of the focal points or focal planes we are referring to. In fact, the object focal point and the object focal plane are sometimes referred to as the *front focal point* and the *front focal plane*, respectively. Analogously, the image focal point and the image focal plane are sometimes referred to as the *back focal point* and the *back focal plane*, respectively.

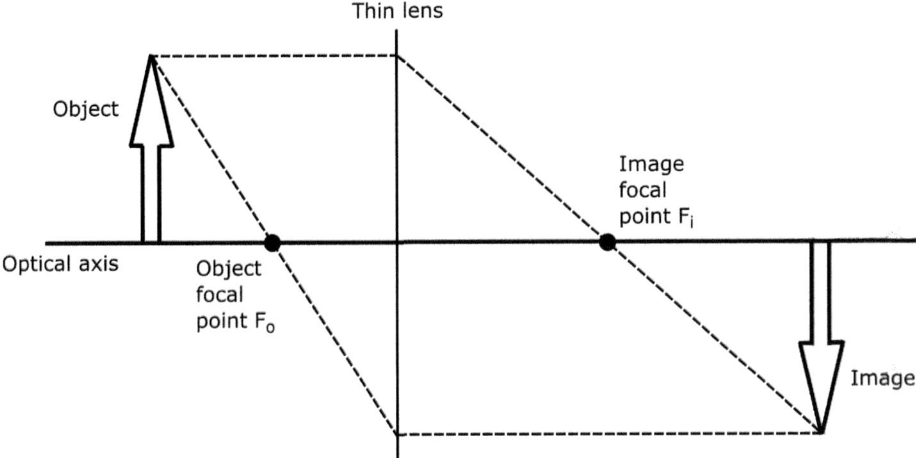

FIGURE 13.6
Image formation by a thin lens. The position of the image of an object is constructed using the basic principle that rays parallel to the optical axis of a thin lens travel through the conjugate focal point. Here we have one ray emanating from the tip of the object and traveling parallel to the optical axis to the lens. The lens changes the path of the ray to then travel through the image focal point. A second ray emanating from the tip of the object travels through the object focal point until it encounters the lens. There its path is redirected to travel parallel to the optical axis in the image space. The place at which these two rays intersect is the tip of the image of the object.

13.2.2 Image formation

The above characterization of the focal points of a thin convex lens as being the points of convergence of parallel rays allows us to geometrically construct the image of an object, provided that the object is located further from the lens than the object focal point. We start at a point on the object and consider two rays emanating from this point. The first ray is assumed to travel parallel to the optical axis until it encounters the lens. There the ray is refracted such that it continues its path through the image focal point. The second ray we consider is a ray that travels through the object focal point until it meets the lens. There its direction will be changed so that it continues its path parallel to the optical axis. The point where the two rays intersect is the image point corresponding to the point on the object where the rays started. In this fashion, the full image can be constructed using the elementary principle that rays parallel to the optical axis in the object space all intersect in the image focal point and, analogously, all rays that pass through the object focal point are parallel to the optical axis in the image space (Fig. 13.6).

Following this approach, we can immediately see that all points in a plane orthogonal to the optical axis will be imaged to a plane in the image space that is orthogonal to the optical axis. Such a pair of planes are called *conjugate planes*.

13.2.3 Lensmaker's formula and lens formula

In this section we will encounter some of the most important relations from geometrical optics that are fundamental for a quantitative understanding of the image formation process. The distances from the lens to its image and object focal points F_i and F_o, i.e., the *image and object focal lengths* f_i and f_o, are important distances for the design of optical systems

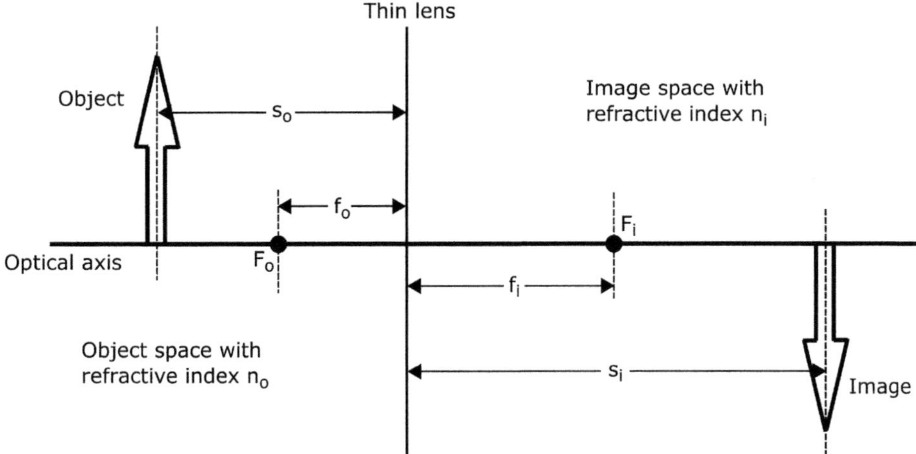

FIGURE 13.7
Illustration of distances involved in the lens formula. An object is located in the object space with refractive index n_o, a distance s_o from a convex thin lens with object (image) focal point F_o (F_i) and object (image) focal length f_o (f_i). The image of the object is located at distance s_i from the lens in the image space with refractive index n_i.

(Fig. 13.7). The lensmaker's formula gives expressions for the focal lengths of a lens in terms of the basic design parameters of the lens, i.e., the radii of the defining spheres and the refractive index of the lens, whereas the lens formula relates the focal distances of the lens to the positions of conjugate planes.

The lensmaker's formula enables calculation of the focal lengths of a lens with the equation

$$\frac{n_l - n_o}{R_o} + \frac{n_i - n_l}{R_i} = \frac{n_o}{f_o} = \frac{n_i}{f_i}, \tag{13.5}$$

where n_o is the refractive index of the medium on the object side of the lens, n_l is the refractive index of the lens, n_i is the refractive index of the medium on the image side of the lens, and R_o and R_i (Fig. 13.4) are the radii of the spherical surfaces of the lens on the object and image sides of the lens, respectively. It is important to address the *sign convention* for the radii R_o and R_i that are used in the lensmaker's formula. If, following the light from the object space to the image space, the lens surface is convex, then the associated radius is taken as a positive number. On the other hand, if the lens surface is concave, then the associated radius is taken as a negative number. For example, for convex lenses which we are concerned with, R_o is taken as positive, whereas R_i is taken as negative.

While the lensmaker's formula relates the radii of the spheres that define the lens to the focal lengths of the lens, the *lens formula* relates the distance s_o of the object to the lens and the distance s_i of the image to the lens (Fig. 13.7) to the focal lengths of the lens with the expression

$$\frac{n_o}{s_o} + \frac{n_i}{s_i} = \frac{n_o}{f_o} = \frac{n_i}{f_i}. \tag{13.6}$$

In the special case where $n_o = n_i$, such as when the medium is air on both sides of the lens, we obtain

$$\frac{1}{s_o} + \frac{1}{s_i} = \frac{1}{f_o} = \frac{1}{f_i} =: \frac{1}{f}.$$

This shows that, in the case of identical refractive indices, the object and the image focal lengths agree, and we can speak of a focal length f of the lens itself.

An immediate consequence of the lens formula is that if an object is positioned closer to a convex lens than the object focal point of the lens, no focused image will be obtained. This is a simple consequence of the fact that if $s_o < f_o$, then $\frac{n_o}{s_o} > \frac{n_o}{f_o}$, and therefore there is no positive s_i such that the lens formula is satisfied. If the object is located at any position further from the lens than the object focal point, then an image will be formed at a position beyond the image focal point of the lens. This is the case as $s_o > f_o$ implies that a distance s_i can be found so that the lens formula is satisfied. In this case, it follows from the lens formula that $\frac{n_i}{s_i} = \frac{n_i}{f_i} - \frac{n_o}{s_o} < \frac{n_i}{f_i}$, and therefore $s_i > f_i$, i.e., the image of the object is positioned beyond the image focal point.

13.3 Magnification

We can now discuss how imaging through a lens affects magnification.

13.3.1 Lateral magnification

The *lateral magnification*, or often simply the *magnification*, is defined as the ratio of the size of the image to the size of the object. More precisely, using the notation of Fig. 13.8, it is defined as

$$M := \frac{y_i}{y_o}. \tag{13.7}$$

Note that in this book, magnification is a positive quantity, as we take y_i to be positive despite its direction indicating an image that is inverted with respect to the object. In some books a convention may be used by which y_i is negative, so that a negative magnification explicitly conveys an inverted image.

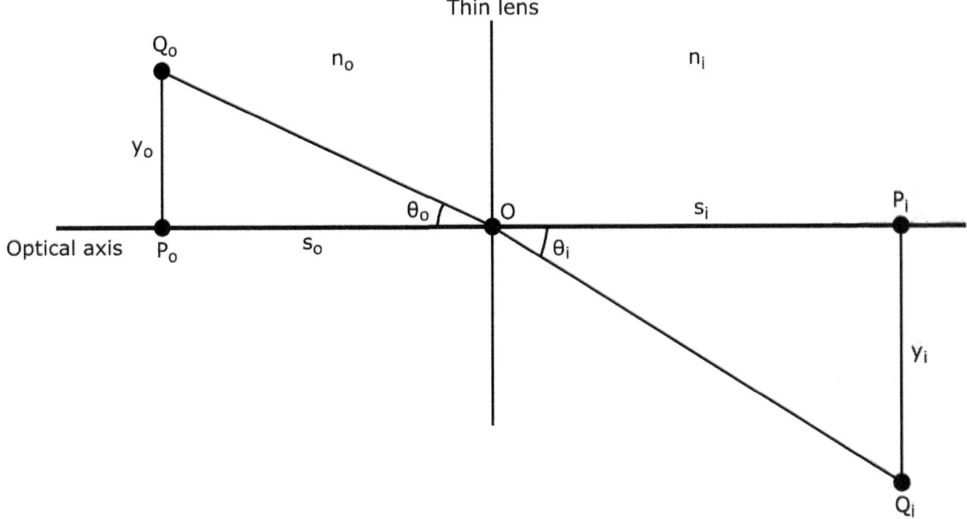

FIGURE 13.8
Description of image formation distances in the Gaussian formulation. The object point Q_o is imaged to the image point Q_i. The magnification is given by $M = \frac{y_i}{y_o} = \frac{n_o s_i}{n_i s_o}$ under the paraxial approximation (i.e., when the angles θ_o and θ_i are small).

The magnification can also be expressed in terms of the distance s_o of the object to the lens and the distance s_i of the image to the lens. To derive the expression, we begin by considering the refraction of light from an object of size y_o at the object-side spherical surface of a thin lens. Based on the illustration in Fig. 13.9, application of Snell's law at this spherical surface yields the equality

$$n_o \sin(\theta_o) = n_l \sin(\theta_l),$$

where n_o and n_l denote the refractive indices of the object space and the lens material, respectively, and θ_o and θ_i denote the angles of incidence and refraction, respectively. Under the *paraxial approximation*, which assumes the light rays from the object to lie close to, and therefore make small angles with, the optical axis, we have that the sine of each angle is well approximated by the tangent of the angle. Applying this approximation, the equality becomes

$$n_o \tan(\theta_o) = n_l \tan(\theta_l),$$

and is equivalent to

$$\frac{n_o y_o}{s_o} = \frac{n_l y_l}{s_l},$$

where s_o is the distance from the object to the spherical surface, and s_l is the distance between the spherical surface and the image of size y_l produced by the spherical surface. By Eq. (13.7), the magnification M_1 of the image of size y_l is then given by

$$M_1 = \frac{y_l}{y_o} = \frac{n_o s_l}{n_l s_o}.$$

To complete the derivation, we consider the fact that the image produced by the object-side spherical surface of the thin lens becomes the object that is imaged by the image-side spherical surface of the thin lens. Applying the same analysis at the image-side spherical surface, the lateral magnification M_2 of the image of size y_i in image space produced by this second spherical surface is given by

$$M_2 = \frac{y_i}{y_l} = \frac{n_l s_i}{n_i s_l},$$

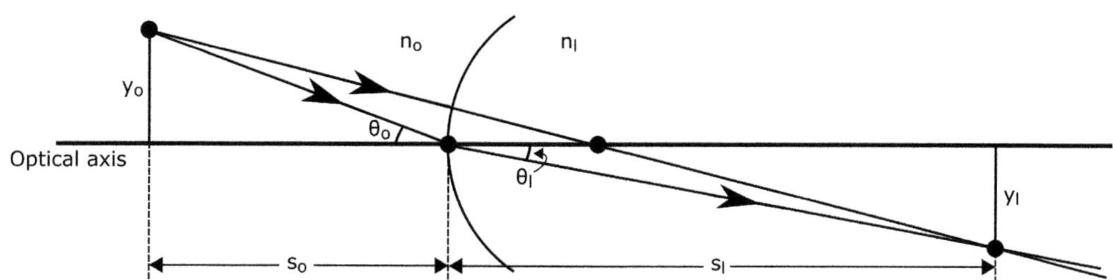

FIGURE 13.9
Refraction at the object-side spherical surface of a thin lens. The spherical surface is centered on the optical axis and produces an image of size y_l in lens material of refractive index n_l from an object of size y_o in object space of refractive index n_o. The object is located a distance s_o, and the image is located a distance s_l, from the point at which the spherical surface intersects the optical axis. The angles θ_o and θ_l denote the angles of incidence and refraction, respectively, of the ray that begins at the tip of the object and impacts the spherical surface at its intersection with the optical axis. A second ray is shown that passes through the center of the spherical surface. This ray is not refracted at the spherical surface as it is perpendicular to the tangent at the point it impacts the surface.

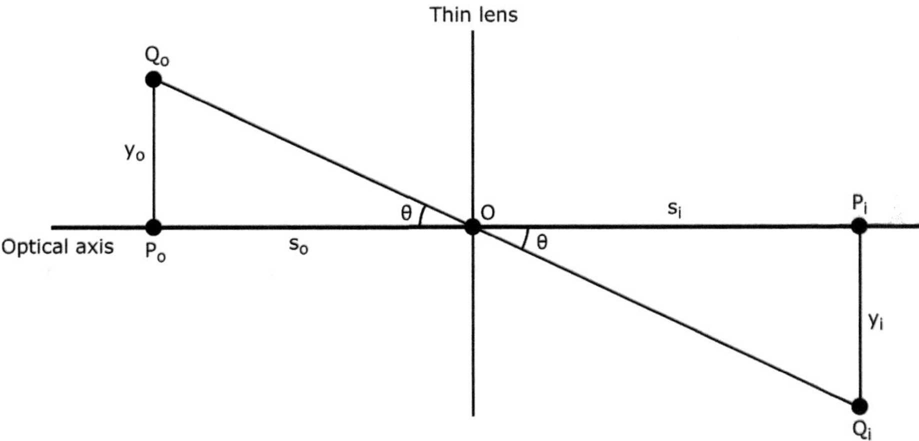

FIGURE 13.10
Gaussian formulation of image formation when the object space and image space refractive indices are equal. The triangles Q_oP_oO and OP_iQ_i are similar, and the magnification is given by $M = \frac{y_i}{y_o} = \frac{s_i}{s_o}$.

where n_i is the refractive index of the image space medium, s_i is the distance of the image from the second spherical surface, and s_l is the distance between the second spherical surface and the image of size y_l produced by the first spherical surface. Note that the distance s_l in the expression for M_1 is taken to be equal to the distance s_l in the expression for M_2, as the distance between the two spherical surfaces of a thin lens is considered to be negligible.

Solving M_1 for y_o and M_2 for y_i, and again using Eq. (13.7), we obtain the lateral magnification M of the image produced by the thin lens comprising the two spherical surfaces as

$$M = \frac{y_i}{y_o} = \frac{\frac{n_l s_i y_l}{n_i s_l}}{\frac{n_l s_o y_l}{n_o s_l}} = \frac{n_o s_i}{n_i s_o}. \tag{13.8}$$

In the special case when the object space and image space media have the same refractive index, Eq. (13.8) becomes

$$M := \frac{y_i}{y_o} = \frac{s_i}{s_o}, \tag{13.9}$$

and the triangles Q_oP_oO and OP_iQ_i in Fig. 13.8 become similar, as shown in Fig. 13.10.

Equation (13.8) is in terms of the *Gaussian formulation* of geometrical optics, which is based on measuring distances from the lens. Another formulation, the *Newtonian formulation*, is also often very useful. In this approach, distances are primarily defined in reference to the focal points of the lens. In Fig. 13.11, triangles are indicated with reference to the focal points. In particular, we see that the triangles $Q_oF_oP_o$ and F_oOU are similar, which implies that

$$\frac{y_o}{\Delta_o} = \frac{y_i}{f_o},$$

where Δ_o denotes the distance between the object and the object focal point. Analogously from the similarity of the triangles VOF_i and $F_iP_iQ_i$, we have that

$$\frac{y_i}{\Delta_i} = \frac{y_o}{f_i},$$

where Δ_i denotes the distance between the image and the image focal point. The first of

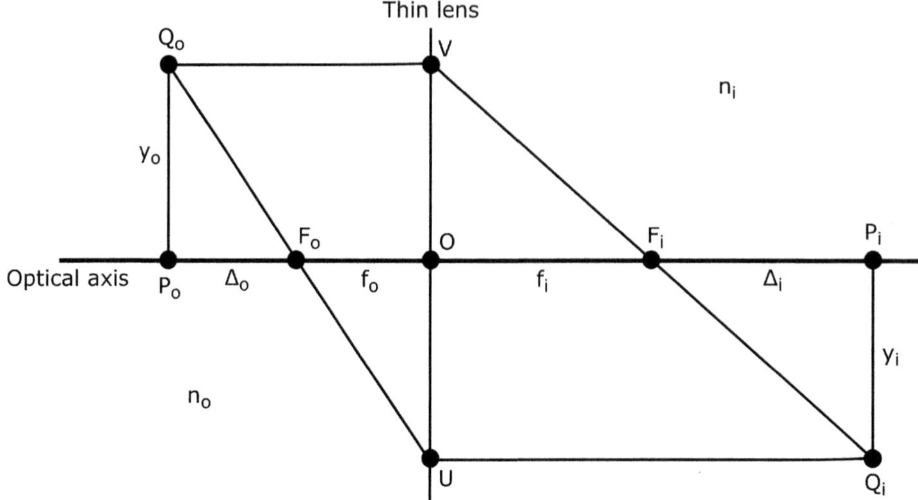

FIGURE 13.11
Description of image formation distances in the Newtonian formulation. The magnification is given by $M = \frac{\Delta_i}{f_i} = \frac{f_o}{\Delta_o}$.

these equations yields

$$M = \frac{y_i}{y_o} = \frac{f_o}{\Delta_o},$$

whereas the second equation implies that

$$M = \frac{y_i}{y_o} = \frac{\Delta_i}{f_i}. \tag{13.10}$$

These two equations provide expressions for the magnification in terms of the Newtonian formulation. Combining them yields the useful identity

$$f_o f_i = \Delta_o \Delta_i.$$

Since the focal lengths of a lens are fixed, this identity immediately implies that changing the position of the object with respect to the object focal plane will produce a reciprocal change of the distance of the image to the image focal plane. We see from this identity that if the object is moved towards the object focal plane, its image moves towards infinity. In fact, when the object is located at the object focal point, there is no image at a finite distance to the lens and we say that the object is imaged to infinity.

13.3.2 Axial magnification

We now address the problem of *axial magnification*, i.e., magnification in the third dimension, or along the optical axis. The way this is interpreted is to think of two objects that are positioned along the optical axis, separated by a distance δ_o (Fig. 13.12). The question then arises as to what the distance δ_i is that separates the images of these two objects. To analyze this question we consider the two pairs of conjugate planes corresponding to the two objects and their images.

We assume that the positions of the two objects, in Gaussian coordinates, are given by s_{o1} and s_{o2}, with $s_{o2} > s_{o1}$, and that the corresponding image planes have Gaussian

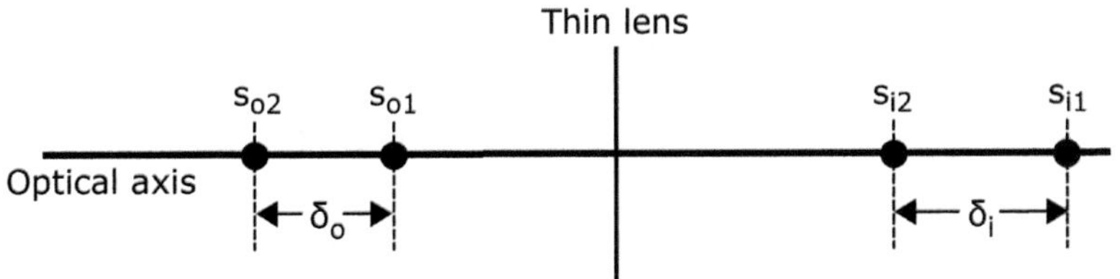

FIGURE 13.12
Notation for the derivation of the axial magnification. Two objects located at positions s_{o1} and s_{o2}, separated by a distance of δ_o, have images at positions s_{i1} and s_{i2}, respectively, separated by a distance of δ_i.

coordinates s_{i1} and s_{i2} which, by the lens formula, are such that $s_{i1} > s_{i2}$. The distances between the pairs of object and image positions are then given by

$$\delta_o := s_{o2} - s_{o1},$$
$$\delta_i := s_{i1} - s_{i2}.$$

We start by considering the lens formula and, in particular, the equation

$$\frac{n_i}{s_{i2}} = \frac{n_i}{f_i} - \frac{n_o}{s_{o2}} = \frac{n_i s_{o2} - n_o f_i}{f_i s_{o2}},$$

which allows us to write

$$s_{i1} - \delta_i = s_{i2} = \frac{n_i f_i s_{o2}}{n_i s_{o2} - n_o f_i},$$

and hence

$$\delta_i = \frac{n_i f_i s_{o2}}{n_o f_i - n_i s_{o2}} + s_{i1}. \tag{13.11}$$

In order to obtain our desired connection between δ_o and δ_i, we need to find expressions for f_i, s_{o1}, and s_{i1} in terms of M, Δ_{i1} (the distance between the image of object 1 and the image focal point), and the refractive indices n_o and n_i. We have from Eq. (13.10) that

$$f_i = \frac{\Delta_{i1}}{M}$$

and

$$s_{i1} = f_i + \Delta_{i1} = \frac{\Delta_{i1}}{M} + \Delta_{i1} = \Delta_{i1}\frac{1}{M}(1 + M).$$

We will use the lens formula again to find the expression for s_{o1} that we need, i.e.,

$$\frac{n_o}{s_{o1}} = \frac{n_i}{f_i} - \frac{n_i}{s_{i1}} = \frac{n_i M}{\Delta_{i1}} - \frac{n_i}{\Delta_{i1}}\frac{M}{1+M} = \frac{n_i M}{\Delta_{i1}}\left(1 - \frac{1}{1+M}\right) = \frac{n_i M^2}{\Delta_{i1}(1+M)}$$

and therefore

$$s_{o1} = \frac{n_o}{n_i}\frac{\Delta_{i1}(1+M)}{M^2}.$$

Now using the expressions for f_i and s_{i1}, we obtain from Eq. (13.11)

$$\begin{aligned}\delta_i &= \frac{n_i \frac{\Delta_{i1}}{M} s_{o2}}{n_o \frac{\Delta_{i1}}{M} - n_i s_{o2}} + \Delta_{i1}\frac{1}{M}(1+M) = \frac{\Delta_{i1}}{M}\left(\frac{M n_i s_{o2}}{n_o \Delta_{i1} - M n_i s_{o2}} + (1+M)\right) \\ &= \Delta_{i1}\left(\frac{n_o\Delta_{i1} + M n_o \Delta_{i1} - M^2 n_i s_{o2}}{M n_o \Delta_{i1} - M^2 n_i s_{o2}}\right).\end{aligned}$$

If we then use the expression for s_{o1} in the definition of δ_o, we obtain

$$s_{o2} = s_{o1} + \delta_o = \frac{n_o}{n_i}\frac{\Delta_{i1}(1+M)}{M^2} + \delta_o = \frac{n_o}{n_i M^2}\left(\Delta_{i1}(1+M) + \frac{n_i}{n_o}M^2\delta_o\right).$$

Substituting this expression into the expression for δ_i then yields

$$\delta_i = \Delta_{i1}\left(\frac{n_o\Delta_{i1} + Mn_o\Delta_{i1} - n_o\left(\Delta_{i1}(1+M) + \frac{n_i}{n_o}M^2\delta_o\right)}{Mn_o\Delta_{i1} - n_o\left(\Delta_{i1}(1+M) + \frac{n_i}{n_o}M^2\delta_o\right)}\right) = \frac{\Delta_{i1}n_i M^2\delta_o}{n_o\Delta_{i1} + n_i M^2\delta_o}, \tag{13.12}$$

which provides the desired relationship between δ_o and δ_i. This expression is more easily interpreted if we make an approximation that is typically justified in microscopy applications (Section 13.4). If we assume that the distance δ_o between the objects is very small in comparison to $\frac{n_o\Delta_{i1}}{n_i M^2}$, then

$$\delta_i = \frac{\Delta_{i1}n_i M^2\delta_o}{n_o\Delta_{i1} + n_i M^2\delta_o} = \frac{\Delta_{i1}\delta_o}{\frac{n_o\Delta_{i1}}{n_i M^2} + \delta_o} \approx \frac{\Delta_{i1}\delta_o}{\frac{n_o\Delta_{i1}}{n_i M^2}} = \frac{n_i}{n_o}M^2\delta_o. \tag{13.13}$$

An important immediate consequence of this expression is the dependence of the axial magnification on the refractive indices and on the square of the lateral magnification.

13.3.3 Dependence of lateral magnification on axial position

We have just seen how the distance between the axial positions of two objects is related to the distance between their images. We now consider how the lateral magnification is changed if we move an object from one axial position to another. We continue with the notation of the prior section and assume that the object's location is moved from position 1 defined by the distance s_{o1} from the lens to position 2 at a distance s_{o2} from the lens. We know from Eq. (13.10) that the lateral magnification M_1 for object position 1 is given as the ratio of the image position in Newtonian coordinates to the focal length in image space, i.e.,

$$M_1 := \frac{\Delta_{i1}}{f_i}.$$

For the magnification M_2 for object position 2, we therefore have

$$M_2 := \frac{\Delta_{i1} - \delta_i}{f_i} = \frac{\Delta_{i1} - \delta_i}{\Delta_{i1}}M_1. \tag{13.14}$$

This expression, not surprisingly, indicates that the lateral magnification is indeed dependent on the axial position of the object.

13.4 Applications to microscopy

The results on geometrical optics that we have just presented allow us to make a number of inferences that are of relevance to microscopy. As we have seen in Section 10.1.1 and as illustrated in Fig. 10.1, a modern microscope is infinity-corrected. We will now consider the objective and the tube lens as one optical microscopy system and apply the lens formula to this combined system.

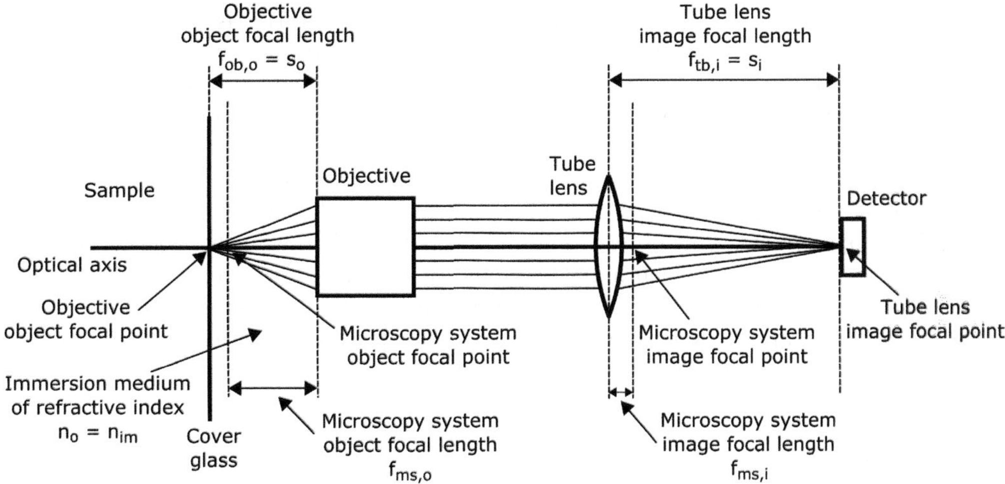

FIGURE 13.13
Infinity-corrected microscopy system. An objective with object focal length $f_{ob,o}$ and a tube lens with image focal length $f_{tb,i}$ positioned in infinity-corrected configuration. This is achieved by positioning the object that is to be imaged at the object focal point of the objective. As a result the image of the object is projected to infinity by the objective. The detector is positioned at the image focal point of the tube lens. Therefore the tube lens focuses any object, such as the image of the sample, from infinity onto the detector. If the objective and the tube lens are considered as a unified system, i.e., a microscopy system, the object and image focal points of the system, with object focal length $f_{ms,o}$ and image focal length $f_{ms,i}$, can be determined.

If we apply the basic result for the magnification of a thin lens (Eq. (13.8)) to our case, then we have for the magnification M_{ms} of the microscopy system, consisting of the objective and the tube lens,

$$M_{ms} := \frac{n_o s_i}{n_i s_o} = \frac{n_o f_{tb,i}}{n_i f_{ob,o}},$$

where we have used that the distance s_o to the object in a microscope with infinity-corrected light path is given by the object focal length $f_{ob,o}$ of the objective, and the distance s_i to the image, i.e., the distance to the detector, is given by the image focal length $f_{tb,i}$ of the tube lens (Fig. 13.13). In this section, we will also assume, for simplicity, that the object space has uniform refractive index given by $n_o = n_{im}$, the refractive index of the immersion medium.

These identities immediately give us an interesting result, i.e.,

$$f_{ob,o} = s_o = \frac{n_o s_i}{n_i M_{ms}} = \frac{n_o f_{tb,i}}{n_i M_{ms}}.$$

This expression allows us to determine the object focal length of commercial objectives for which typically only the magnification M_{ms} of the associated microscopy system and the image focal length $f_{tb,i}$ of the tube lens are easily available.

For example, assuming the typical scenario where the image space medium is air and therefore has refractive index $n_i = 1$, if the image focal length of the tube lens is $f_{tb,i} = 200$ mm (the tube lens focal lengths employed by commercial manufacturers range from 165 mm to 200 mm (Section 10.1.1)) and the system magnification, provided by an objective designed for use with immersion oil of refractive index $n_o = 1.515$, is $M_{ms} = 100$, then the object focal length of the objective, or the distance of the objective to the object, is $f_{ob,o} = 3.03$ mm. However, if the system magnification is provided by an air objective and is only $M_{ms} = 10$, then the object focal length is $f_{ob,o} = 20$ mm, i.e., about 6.6 times larger than that for the 100× magnification objective.

The tube lens of a microscope is typically fixed. This therefore shows that the distance between the objective and the object depends significantly on the overall magnification M_{ms} of the microscope system. This observation also helps to explain why the working distance (Note 13.1) of an objective, which indicates, roughly speaking, how deep an objective can image into a sample, depends significantly on the magnification of the microscopy system.

We can now also apply the lens formula to determine the focal lengths of the microscope system. For this purpose, we consider

$$\frac{n_o}{f_{ob,o}} + \frac{1}{f_{tb,i}} = \frac{n_o}{f_{ms,o}} = \frac{1}{f_{ms,i}},$$

where $f_{ms,o}$ is the object focal length of the microscope system and $f_{ms,i}$ is its image focal length. We have again assumed here that the medium in the image space is air (i.e., $n_i = 1$). For the 100× oil objective with object focal length $f_{ob,o} = 3.03$ mm and the tube lens with image focal length $f_{tb,i} = 200$ mm, we obtain

$$\frac{1.515}{f_{ms,o}} = \frac{1}{f_{ms,i}} = \frac{1.515}{3.03 \text{ mm}} + \frac{1}{200 \text{ mm}} = 0.505 \text{ mm}^{-1}.$$

Therefore

$$f_{ms,i} \approx 1.980 \text{ mm},$$
$$f_{ms,o} = 3 \text{ mm}.$$

For the 10× air objective with object focal length $f_{ob,o} = 20$ mm and the same tube lens with image focal length $f_{tb,i} = 200$ mm, we obtain

$$\frac{1}{f_{ms,o}} = \frac{1}{f_{ms,i}} = \frac{1}{20 \text{ mm}} + \frac{1}{200 \text{ mm}} = 0.055 \text{ mm}^{-1}.$$

Therefore,

$$f_{ms,i} = f_{ms,o} \approx 18.182 \text{ mm},$$

and we have focal lengths that are substantially longer than those for the 100× oil objective.

Having established the object and image focal lengths $f_{ms,o}$ and $f_{ms,i}$ of the combined microscopy system, it is now also instructive to consider the distances $\Delta_{ms,o}$ and $\Delta_{ms,i}$ between the object and image focal points and the positions of the object and the image/detector, respectively. Note that by the above discussion we have that the distance to the object equals the object focal length of the objective, i.e., $s_o = f_{ob,o}$, and therefore

$$\Delta_{ms,o} = s_o - f_{ms,o} = f_{ob,o} - f_{ms,o}.$$

Similarly,

$$\Delta_{ms,i} = s_i - f_{ms,i} = f_{tb,i} - f_{ms,i}.$$

For the two examples considered above, we therefore obtain

$$\Delta_{ms,o} = f_{ob,o} - f_{ms,o} = 3.03 \text{ mm} - 3 \text{ mm} = 0.03 \text{ mm},$$
$$\Delta_{ms,i} = f_{tb,i} - f_{ms,i} = 200 \text{ mm} - 1.980 \text{ mm} = 198.02 \text{ mm}$$

for the first microscope system consisting of a 100× oil objective and a tube lens with an image focal length of 200 mm, and

$$\Delta_{ms,o} = f_{ob,o} - f_{ms,o} = 20 \text{ mm} - 18.182 \text{ mm} = 1.818 \text{ mm},$$
$$\Delta_{ms,i} = f_{tb,i} - f_{ms,i} = 200 \text{ mm} - 18.182 \text{ mm} = 181.818 \text{ mm}$$

for the second microscope system consisting of a 10× air objective and a tube lens with an image focal length of 200 mm.

We can now also address to what extent the approximation of Eq. (13.13) was valid in Section 13.3.2, where we analyzed axial magnification. The condition was that the difference δ_o between object positions is very small compared to $\frac{n_o \Delta_{i1}}{n_i M^2}$, where $\Delta_{i1} = \Delta_{ms,i}$ is the distance between the image focal point of the microscope system and the position of the image itself. For our examples, we have that

$$\frac{n_o \Delta_{i1}}{n_i M^2} = \frac{1.515 \times 198.02 \text{ mm}}{1 \times 100^2} \approx 0.030 \text{ mm} = 30\ \mu\text{m}$$

for the 100× oil objective, and

$$\frac{n_o \Delta_{i1}}{n_i M^2} = \frac{1 \times 181.818 \text{ mm}}{1 \times 10^2} \approx 1.818 \text{ mm} = 1818\ \mu\text{m}$$

for the 10× air objective. These results show that the approximation is valid for a few micrometers of position change of the object along the optical axis when the 100× oil objective is used, and that it is valid for significantly larger changes in the object axial position when the 10× air objective is used. It is therefore important to carefully evaluate the applicability of the approximation by considering the specifics of a given microscopy experiment. If in doubt, however, the non-approximated expression that relates changes in object axial position to changes in image position (Eq. (13.12)) can be used.

14

Diffraction

In the prior chapter, we have introduced methods in geometrical optics and have seen how they can be employed to investigate various aspects of image formation in an optical microscope. While these methods, based on analyzing pathways of rays of light, are powerful tools in many regards, they cannot explain the diffraction patterns that are characteristic of image formation in high-resolution microscopy systems. To understand such phenomena, we need to rely on the wave nature of light. Here we will not delve in detail into a full theoretical description of electromagnetic waves. Rather, we will start from a basic description of wave forms, proceed to discuss diffraction, derive expressions for point spread functions, and present a number of additional consequences.

14.1 Wave description of light

The electromagnetic waves that govern the propagation of light are solutions to Maxwell's equations. Considering light from this point of view provides a high level of generality. However, for what we will need to achieve, it will be sufficient to consider a more elementary approach that is based on the Huygens-Fresnel principle, which we will introduce later in this chapter. This principle will be the major tool for us to analyze the propagation of waves through space.

We will, however, first introduce two special forms of waves, plane waves and spherical waves, and illustrate a number of important concepts with these examples. We will see that plane waves play a role analogous to that of parallel rays in geometrical optics. Spherical waves, on the other hand, are waves analogous to converging and diverging rays.

14.1.1 Plane waves

A *plane wave* ψ is given by the following expression for the electromagnetic field at position r and time t,

$$\psi(r,t) = Ae^{i(\langle\overline{k},r\rangle - \omega t + \varphi)} = Ae^{i\langle\overline{k},r\rangle}e^{-i(\omega t - \varphi)}, \quad r := (r_x, r_y, r_z) \in \mathbb{R}^3, \quad t \geq t_0.$$

Here the vector $\overline{k} := (k_x, k_y, k_z)$ is called the 3D *wave vector*, and its Euclidean norm

$$k := \|\overline{k}\| = \sqrt{k_x^2 + k_y^2 + k_z^2}$$

is called the *wave number*. The value of the wave depends on the inner or scalar product $\langle\cdot,\cdot\rangle$ of the position vector r and the wave vector $\overline{k}$, i.e., $\langle\overline{k}, r\rangle = r_x k_x + r_y k_y + r_z k_z$. The scalar A is called the *amplitude*, ω the *angular frequency*, and φ the *phase shift*, of the wave. We have defined the wave as a complex valued function. However, we do need to remember that this is primarily done as it provides significant advantages for our computations. The actual objects of interest are the real parts of the waves that we investigate.

A plane wave can be decomposed into a *spatial part*, which depends on the position vector r, and a *temporal part*, which depends on time t. For example, we can take $Ae^{i\langle \bar{k},r\rangle}$ as the spatial part and $e^{-i(\omega t-\varphi)}$ as the *temporal part*. The spatial and temporal parts are clearly not unique, as the amplitude A and the phase shift term $e^{i\varphi}$ could each be associated with either the spatial or the temporal part. Intimately associated with the temporal part of the wave is the *period* of the wave, which is the smallest nonzero time T such that $\psi(r, t+T) = \psi(r, t)$ for any time t and position r. Examining the exponent of the temporal part we immediately obtain that the following relationship is satisfied for the angular frequency of the wave and its period, i.e.,

$$\omega T = 2\pi.$$

The above expression for the plane wave defines for each point r in space and each point t in time a complex number that indicates the state of the wave at that point and time. As the wave function is a complex function we can always represent it in polar coordinates, i.e., $\psi(r,t) = \alpha e^{i\beta}$, with amplitude $\alpha \geq 0$ and phase $\beta \in \mathbb{R}$. In what follows, we will see that computing the phase for specific circumstances is crucial to analyzing the propagation of waves. It is, however, important to note that the energy of a wave depends on its amplitude. Standard detectors, such as those routinely used in microscopy (Chapter 12), can only detect the square of the modulus of the amplitude (Section 14.2).

14.1.1.1 Planes of identical phase

Clearly a plane wave has equal amplitude at any point in space and time. Its phase, on the other hand, changes with space and time. Note, however, that for each time point $t \geq t_0$, the temporal portion $-(\omega t - \varphi)$ of the phase is independent of the position r and therefore constant throughout space.

We would like to determine the *wavefronts* of a plane wave, i.e., those points in space that have identical phase. The spatial portion $\langle \bar{k}, r\rangle$ of the phase at the point r_0 depends on the scalar product $\langle \bar{k}, r_0\rangle = k_x r_{0x} + k_y r_{0y} + k_z r_{0z}$. Therefore the phase of the plane wave is identical to that at r_0 for all points r in space such that the scalar product $\langle \bar{k}, r\rangle$ equals the scalar product $\langle \bar{k}, r_0\rangle$ corresponding to the point r_0. But elementary linear algebra tells us that this is the case for all positions of the form

$$r = r_0 + \zeta_1 r_{p1} + \zeta_2 r_{p2}, \quad \zeta_1, \zeta_2 \in \mathbb{R},$$

where r_{p1} and r_{p2} are independent vectors that are orthogonal to the wave vector $\bar{k}$, i.e., $\langle \bar{k}, r_{p1}\rangle = \langle \bar{k}, r_{p2}\rangle = 0$. This means that all points in the plane that passes through the point r_0 and is orthogonal to $\bar{k}$ have identical phase.

In fact, adding multiples of 2π to the exponent $\langle \bar{k}, r_0\rangle$ will also produce the same results, i.e., $e^{i\langle \bar{k},r_0\rangle} = e^{i\langle \bar{k},r_0\rangle + i2\pi m} = e^{i\langle \bar{k},r_0\rangle + \frac{i2\pi m}{k^2}\langle \bar{k},\bar{k}\rangle}$, for $m = \ldots, -1, 0, 1, \ldots$. All points r in space that satisfy this equation can be obtained by translating the above plane in the direction of the wave vector $\bar{k}$ at distances that are multiples of 2π, i.e., for $m = \ldots, -1, 0, 1, \ldots$,

$$r = r_0 + \frac{2\pi m}{k^2}\bar{k} + \zeta_1 r_{p1} + \zeta_2 r_{p2}, \quad \zeta_1, \zeta_2 \in \mathbb{R}.$$

This gives us a description of points in space that have the same phase as the point r_0. We see that these points are given by parallel planes that are offset from one another by a distance of 2π (Fig. 14.1), thereby also explaining why this wave is called a plane wave.

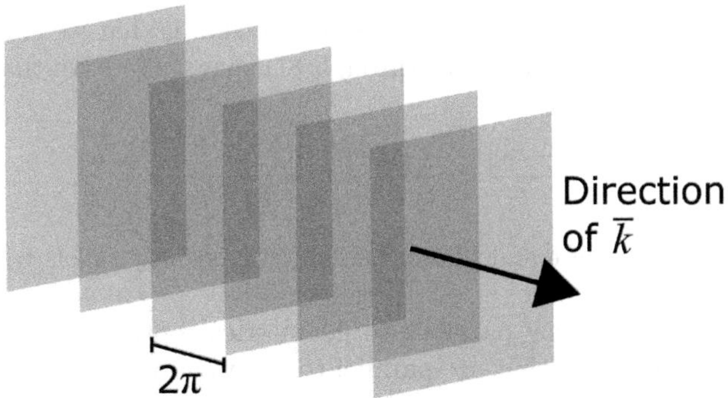

FIGURE 14.1
Plane wave. A plane wave is characterized by wavefronts consisting of parallel planes that are orthogonal to the wave vector $\bar{k}$ and offset from one another by a distance of 2π. The wave propagates in the direction of $\bar{k}$.

14.1.1.2 Speed of wave propagation

If we now consider the propagation of the wave as a function of time, it is instructive to investigate how the above characterized planes of equal phase travel through space. Consider again the point r_0. At time $t_1 \geq t_0$, the phase at this point is given by

$$\langle \bar{k}, r_0 \rangle - \omega t_1 + \varphi.$$

Here we have assumed that A is real, but if it is not, we can always absorb it in the phase shift term φ. At a later time $t_2 > t_1$ and a point in space $r_1 := r_0 + \Delta r$, the phase is accordingly

$$\langle \bar{k}, (r_0 + \Delta r) \rangle - \omega t_2 + \varphi.$$

The difference in phase is therefore given by

$$\langle \bar{k}, (r_0 + \Delta r) \rangle - \omega t_2 + \varphi - \left(\langle \bar{k}, r_0 \rangle - \omega t_1 + \varphi \right) = \langle \bar{k}, \Delta r \rangle - \omega (t_2 - t_1).$$

We would like to find the points in space for which this difference is 0 at time t_2. For this to be the case, Δr needs to satisfy

$$\langle \bar{k}, \Delta r \rangle = \omega (t_2 - t_1).$$

This equality is achieved if we set Δr to be a multiple of the wave vector $\bar{k}$, i.e., if

$$\Delta r := \frac{\omega (t_2 - t_1)}{k^2} \bar{k}.$$

Therefore all points on the plane

$$r_0 + \frac{\omega (t_2 - t_1)}{k^2} \bar{k} + \zeta_1 r_{p1} + \zeta_2 r_{p2}, \quad \zeta_1, \zeta_2 \in \mathbb{R},$$

where r_{p1} and r_{p2} are again independent vectors that are orthogonal to $\bar{k}$, have the same phase as the point r_1 at time t_2. We can therefore consider the wave to move in the direction of the wave vector (Fig. 14.1). In fact, if we restrict ourselves to the line in space spanned by the wave vector and passing through the arbitrarily chosen point r_0, we can consider the wave to be one-dimensional (1D).

An additional important consequence of this expression is that we can infer the speed v with which a plane of equal phase moves through 3D space. We have that

$$v = \frac{\|r_1 - r_0\|}{t_2 - t_1} = \frac{\|\Delta r\|}{t_2 - t_1} = \frac{\left\|\frac{\omega(t_2-t_1)}{k^2}\overline{k}\right\|}{t_2 - t_1} = \frac{\omega}{k},$$

which shows that the plane of equal phase, i.e., the wavefront, travels with speed $\frac{\omega}{k}$ in the direction of the wave vector.

This is a good place to draw parallels to geometrical optics. If we define normals to the wavefronts, we obtain lines that are parallel to the wave vector. If we assign directions to these parallel lines to match the direction of propagation of the wave, we can interpret these lines as rays that are parallel and point in the direction of the light propagation. This illustrates how we can associate rays with a propagating wave. In the present case, we have shown that the rays associated with a plane wave are parallel. This provides a justification as to why in the theory of the wave propagation of light, plane waves are considered to play a role analogous to that played by parallel rays in geometrical optics.

14.1.1.3 Wave number and wavelength

We can also relate the wave number to the wavelength λ associated with the wave. By definition of the wavelength, for a fixed time $t \geq t_0$, the wave at point r_0 is identical to the wave at a point translated by the wavelength λ along the direction of the wave, i.e., the point $r_0 + \frac{\lambda}{k}\overline{k}$. We therefore have the equality

$$\psi(r_0, t) = Ae^{i(\langle \overline{k}, r_0\rangle - \omega t + \varphi)} = Ae^{i(\langle \overline{k}, (r_0 + \frac{\lambda}{k}\overline{k})\rangle - \omega t + \varphi)} = \psi\left(r_0 + \frac{\lambda}{k}\overline{k}, t\right).$$

But this implies that the phases at the two points differ by 2π, i.e.,

$$2\pi = \left\langle \overline{k}, \left(r_0 + \frac{\lambda}{k}\overline{k}\right)\right\rangle - \omega t + \varphi - (\langle \overline{k}, r_0\rangle - \omega t + \varphi) = \frac{\lambda}{k}\langle \overline{k}, \overline{k}\rangle = \frac{\lambda}{k}k^2 = \lambda k.$$

Therefore, between the wave number k and the wavelength λ, we have the important relationship

$$k = \frac{2\pi}{\lambda}.$$

14.1.1.4 Propagation in different media

We now want to examine what happens if the wave travels through vacuum or through a medium with refractive index n_1 for the distance of precisely one wavelength λ. For the wave to travel this distance takes exactly the time T of one period for the wave. We know that the angular frequency of the wave ω, and therefore also the period T, are independent of the medium in which the wave travels. If we calculate the ratio between the wavelength λ_1 of the wave in the medium with refractive index n_1 and the wavelength λ_{vac} in vacuum, we obtain

$$\frac{\lambda_1}{\lambda_{vac}} = \frac{v_1 T}{v_{vac} T} = \frac{v_1}{v_{vac}} = \frac{1}{n_1}, \quad \text{and therefore} \quad n_1\lambda_1 = \lambda_{vac},$$

where v_1 is the speed of a plane of identical phase in medium 1 and $v_{vac} = c$ is its speed in vacuum. Here we have used the definition of the refractive index as the ratio between the speed of light in vacuum and the speed of light in the medium (Eq. (13.1)).

This identity has important consequences. It shows that while the angular frequency of the wave does not depend on the medium in which it travels, the wavelength does depend on the refractive index of the medium. In fact, the higher the refractive index, the smaller the wavelength, i.e., the smaller the distance that the wave travels in the medium during the same time interval. As the wave number depends on the wavelength, we also have that the wave number depends on the refractive index of the medium in which the wave travels. When the wave travels in a medium with refractive index n_1, we have the wave number

$$k_1 = \frac{2\pi}{\lambda_1} = \frac{2\pi n_1}{\lambda_{vac}} = n_1 k_{vac},$$

where k_{vac} is the wave number when the wave travels in vacuum.

14.1.1.5 Optical path length

Following on from the previous discussion, if we consider the distance d_1 traveled by a wave over the time duration Δt, we obtain

$$d_1 = v_1 \Delta t = \frac{\omega}{k_1}\Delta t = \frac{\omega}{n_1 k_{vac}}\Delta t = \frac{1}{n_1} v_{vac}\Delta t = \frac{1}{n_1} d_{vac},$$

where d_{vac} is the distance that the wave would travel in vacuum during the same amount of time. The distance traveled in the medium is therefore shorter by a factor of $\frac{1}{n_1}$ in comparison to the distance traveled in vacuum over the same time duration.

Conversely, if d_1 is the distance traveled in the medium with refractive index n_1, then $n_1 d_1$ immediately provides the equivalent distance traveled in vacuum. In fact, the weighted distance $n_1 d_1$ is known as the *optical path length* (OPL), i.e., OPL $= n_1 d_1$. One important aspect of the optical path length is that dividing it by the speed of light $v_{vac} = c$ in vacuum gives the amount of time $\Delta t = \text{OPL}/c$ that the wave takes to travel the physical distance associated with the optical path length, i.e., in this case the distance d_1.

14.1.2 Spherical waves

Having discussed plane waves in some detail we can now come to investigate spherical waves, which are of significance in analyzing the behavior of either point sources of light, i.e., sources that emanate light uniformly, or focused light, such as light focused by a lens. In fact, spherical waves play an analogous role in the wave treatment of light as converging or diverging rays play in geometrical optics.

While for a plane wave the amplitude A was assumed to be independent of position and time, for a spherical wave the amplitude is obtained by replacing A with $\frac{A}{\|r-r_0\|}$, which depends reciprocally on the distance $\|r - r_0\|$ from the source $r_0 \in \mathbb{R}^3$ of the wave. In addition, the phase does not depend on the inner product of the position in space and the wave vector as in the case of the plane wave, but instead depends on the distance $\|r - r_0\|$. A spherical wave ψ with source r_0 is given by

$$\psi(r,t) = \frac{A}{\|r-r_0\|} e^{i(k\|r-r_0\| - \omega t + \varphi)} = \frac{A}{\|r-r_0\|} e^{ik\|r-r_0\|} e^{-i(\omega t - \varphi)}, \quad r \in \mathbb{R}^3, \quad t \geq t_0,$$

where A is a constant, k is the wave number, ω is the angular frequency, and φ is the phase shift. As in the case of a plane wave, the wave number is given by $k = 2\pi/\lambda$, where λ is the wavelength of the wave.

Like the description of the plane wave, the description of the spherical wave naturally decomposes into a spatial part and a temporal part. It can be verified easily that each sphere with center r_0 forms a wavefront for the spherical wave (Fig. 14.2).

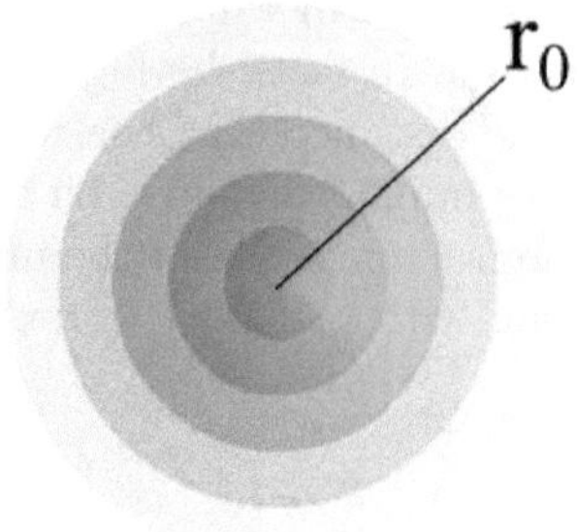

FIGURE 14.2
Spherical wave. A spherical wave emanating from a point source located at r_0 in 3D space is characterized by wavefronts consisting of spheres centered at r_0.

14.1.2.1 Converging and diverging spherical waves

Let us now consider the phase that defines a wavefront, i.e.,

$$k\|r - r_0\| - \omega t + \varphi.$$

We want to understand how the wavefront travels in space as time increases. A wavefront is defined by this quantity being constant. Therefore as the term ωt increases with time, the term $k\|r - r_0\|$ also needs to increase to ensure that the phase remains constant. But this means that the distance of the wavefront from the source of the wave increases with increasing time, and hence that the wavefront is moving outwards from the source. The wave is therefore *diverging*. Changing the sign of $k\|r - r_0\|$ to a minus sign, i.e., changing the phase to $-k\|r - r_0\| - \omega t + \varphi$, implies a significantly different behavior of the wavefronts of the wave. Here, for increasing time t the term $k\|r - r_0\|$ needs to decrease to ensure that the phase remains constant. This implies that the wavefront converges to the source of the wave. The wave is thus *converging*, modeling light being focused to the point r_0.

In order to draw a connection with geometrical optics, as in the case of a plane wave we introduce rays, defining them as orthogonal to the wavefronts. For diverging spherical waves this means that the rays originate at the source of the wave and travel radially outwards. For converging spherical waves the rays remain in the same location but change direction, indicating that they converge to the source point of the wave.

14.1.3 Spatial part of a wave

We have seen in the above discussion of the plane and spherical waves that they decompose into a spatial part and a temporal part. For our later developments, we will assume that this decomposition exists for all waves that we encounter, i.e., that for a wave $\psi(r, t)$, $r \in \mathbb{R}^3$, $t \geq t_0$, we have that

$$\psi(r, t) = U(r)e^{-i\omega t}$$

for some time-independent function $U(r)$ that represents the spatial part of the wave, where $\omega \in \mathbb{R}$ is the angular frequency of the wave.

An important aspect of wave-based analyses of optical systems is that phenomena such as diffraction can be explained by the interactions or *interference* of waves. This amounts to studying the sum of waves. Now assume we are dealing with a number of *monochromatic* waves ψ_l, $l = 1, \ldots, L$, i.e., waves whose temporal parts all have the same angular frequency

ω. Accordingly, these waves have the representations

$$\psi_l(r,t) = U_l(r)e^{-i\omega t}, \quad r \in \mathbb{R}^3, \quad t \geq t_0, \quad l = 1, \ldots, L,$$

with spatial parts $U_l(r)$, $r \in \mathbb{R}^3$, $l = 1, \ldots, L$. If we want to study the interference of these waves we need to consider their sum

$$\sum_{l=1}^{L} \psi_l(r,t) = \sum_{l=1}^{L} U_l(r)e^{-i\omega t} = e^{-i\omega t}\sum_{l=1}^{L} U_l(r), \quad r \in \mathbb{R}^3, \quad t \geq t_0.$$

This expression shows that the sum also decomposes into a spatial part and a temporal part. Moreover, it shows that the spatial part of the sum of the monochromatic waves is simply the sum of the spatial parts of the individual waves. As a result, we can restrict our further developments to just the spatial parts of the waves. The temporal part can then simply be multiplied with the result of any calculations involving the spatial parts.

14.2 What does a camera detect?

The topic of this chapter is the exploration of the wave nature of light to gain further insights into the image formation process of a microscope. As we have seen with the plane wave and the spherical wave, we model the propagation of light as the propagation of a wave through space and time. Ultimately, we will want to detect the image of the object through a detector, such as a CCD camera. When discussing detectors we employed the particle model of light to understand how light that impacts the detector is recorded (Section 12.1). We now need to try to understand how a wave impacting the detector leads to an image, and reconcile it with the particle description of image detection. Image detectors such as a CCD camera cannot detect the phase of a wave. In fact, it is the square of the modulus of the wave that impacts a pixel of the detector that governs the "amount" of light that is detected. More precisely, assume that $\psi_{C,t}(x,y)$ is the wave at coordinate (x,y) on the detector $C \subseteq \mathbb{R}^2$ at time $t \geq t_0$. If C_p is a pixel of the detector, then

$$\int_{C_p} |\psi_{n,t}(x,y)|^2 \, dxdy,$$

i.e., the integral of the square of the modulus of the normalized wave function $\psi_{n,t}$ over the pixel, defines the probability that a photon that impacts the detector at time t will be registered in the pixel. Here the normalized wave function $\psi_{n,t}$ is given by

$$\psi_{n,t}(x,y) = \frac{\psi_{C,t}(x,y)}{\|\psi_{C,t}\|}, \quad (x,y) \in C,$$

with

$$\|\psi_{C,t}\| = \left(\int_{(x,y)\in C} |\psi_{C,t}(x,y)|^2 \, dxdy\right)^{\frac{1}{2}}.$$

Note that with this normalization, $\|\psi_{n,t}\|^2$ defines a probability density over the detector C at time $t \geq t_0$. This setting will form the basis for the formal discussion of the spatial aspect of the image detection process in Section 15.1.2. If we ignore pixelation and the stochastics of the photon emission and detection processes, we can consider $|\psi_{n,t}(x,y)|^2$ as the image acquired at time t and at position (x,y) in the detector plane.

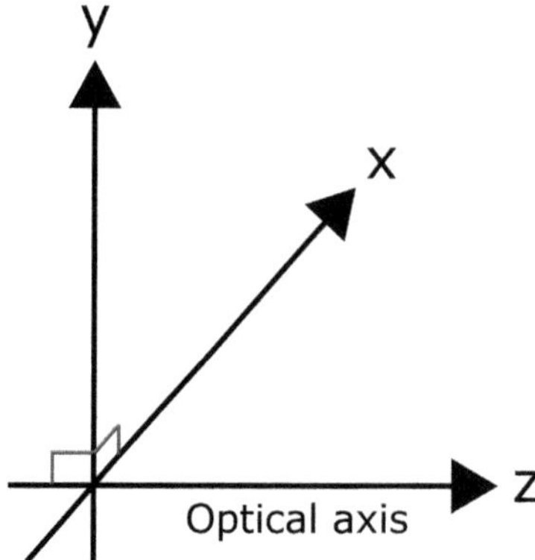

FIGURE 14.3
Coordinate system. The xy-plane is orthogonal to the optical (z-)axis.

Since the image is given by the square of the modulus of the wave function, it does not depend on the phase of the wave. For this reason, we will often ignore constant phase terms. However, we need to be aware that we cannot indiscriminately neglect any phase terms. As we will see, when dealing with the interference of waves, it is exactly the phases of the interfering waves that govern the interference pattern. We should also add that we will often ignore constants of the wave whose image we are interested in. This is due to the fact that the image model that we will ultimately employ in Chapter 15 for the quantitative analysis of image data will be scaled accordingly so that the image intensity will be given by the number of photons that are collected from the sample.

14.3 Effect of a thin lens on waves

The thin lens is our prime example of an optical system, and we will use it as a simplistic but informative model for a microscope. It is therefore important for us to understand how a thin lens affects a light wave. The main objective of this chapter is to derive an expression for the image of an object, when imaged with a thin lens. One of the crucial steps in the derivation will be the analysis of the behavior of a wave as it travels through the lens itself, i.e., from immediately in front of the lens to immediately behind the lens. The development of appropriate tools for the study of this question will be the content of this section.

In order to develop an appropriate approach for the influence of a thin lens on the wave that passes through it, we need to remember the characterization of a thin lens. In our discussion on geometrical optics, we defined a thin lens as a lens that does not displace a ray (Section 13.2). More precisely, throughout the discussion we assumed a coordinate system that is orthogonal to the optical axis (Fig. 14.3). In this coordinate system, a ray that enters the thin lens at certain coordinates will leave the lens at the same coordinates. The equivalent characterization for a thin lens in terms of its actions on waves is that the wave is not displaced by the lens, but that only its phase is changed. Therefore, if U_1 is the field at the plane directly before the lens and U_2 is the field at the plane directly behind the lens (Fig. 14.4), for a thin lens there exists a real function ϕ such that

$$U_2(x, y) = e^{i\phi(x,y)} P(x, y) U_1(x, y), \quad (x, y) \in \mathbb{R}^2,$$

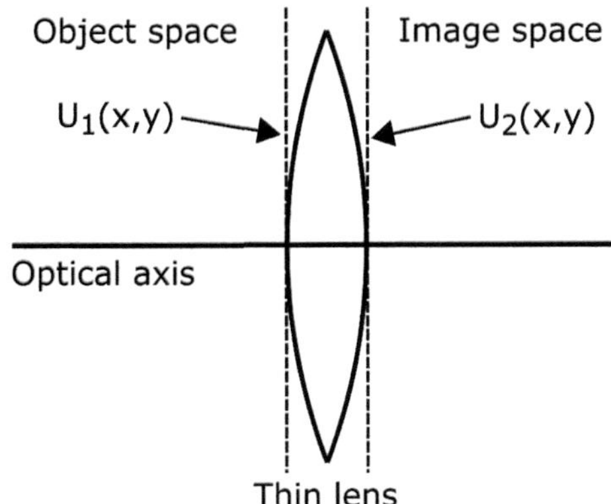

FIGURE 14.4
Transmittance of a thin lens. The transmittance is given by the ratio of the field $U_2(x, y)$ at the plane immediately behind the thin lens to the field $U_1(x, y)$ at the plane immediately before the thin lens.

where $P(x, y)$, $(x, y) \in \mathbb{R}^2$, is the *pupil function* for the lens, i.e., $P(x, y) = 1$ if (x, y) is a point within the aperture of the lens (Fig. 14.5) and $P(x, y) = 0$ if (x, y) is a point outside the aperture. The pupil function thus indicates that light can only pass through the lens if it is within the lens' aperture. (An aperture is an opening through which light passes, and in the case of a lens, it is simply the physical extent of the lens itself.) More compactly written, $P(x, y) = I_A(x, y)$, where $A \subseteq \mathbb{R}^2$ is the set that defines the aperture corresponding to the lens, and I_A is the *indicator function*, i.e., $I_A(x, y) = 1$ for $(x, y) \in A$ and $I_A(x, y) = 0$ for $(x, y) \notin A$.

We can now define the *transmittance* $\mathcal{T}$ of the lens as

$$\mathcal{T}(x, y) := P(x, y)e^{i\phi(x,y)}, \quad (x, y) \in \mathbb{R}^2.$$

The definition of the transmittance immediately implies that it is the ratio between the wave in the plane immediately behind the lens and the wave in the plane immediately before the lens (Fig. 14.4), i.e.,

$$\mathcal{T}(x, y) = \frac{U_2(x, y)}{U_1(x, y)}, \quad (x, y) \in \mathbb{R}^2.$$

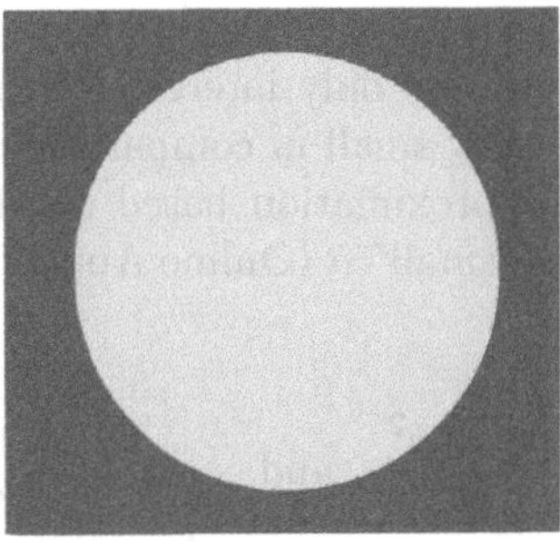

FIGURE 14.5
Circular aperture. Light can only pass through the lens if it is within the circular area corresponding to the physical extent of the lens.

The phase change in a wave that passes through a thin lens can be analyzed by considering the optical path in detail between a point before the lens and a point behind the lens. At a point (x, y), the path from the plane before, to the plane behind, the lens is made up of three parts, i.e., the path of length $D_o(x, y)$ in the object space, between the plane before the lens and the lens, the path of length $D_l(x, y)$ within the lens, and the path of length $D_i(x, y)$ in the image space, between the lens and the plane behind the lens (Fig. 14.6). The medium of each of these three paths may have a different refractive index, i.e., we have refractive index n_o for the medium in the object space, refractive index n_l for the glass of the lens, and refractive index n_i for the medium of the image space. Therefore the optical path length at the point (x, y) in the aperture of the lens, i.e., the phase $\phi(x, y)$ of the transmittance $\mathcal{T}(x, y)$, is given by

$$\phi(x,y) = k_{vac}\left(n_o D_o(x,y) + n_l D_l(x,y) + n_i D_i(x,y)\right), \quad (x,y) \in A,$$

where k_{vac} is the wave number in vacuum.

We now need to calculate expressions for D_o, D_l, and D_i based on the geometry of the lens. Elementary calculations (Fig. 14.7), exploiting that the surfaces of the lens are spheres of radii R_o and R_i, show that for $(x, y) \in A$, we have

$$D_o(x,y) = R_o - \sqrt{R_o^2 - x^2 - y^2} = R_o\left(1 - \sqrt{1 - \frac{x^2+y^2}{R_o^2}}\right),$$

$$D_i(x,y) = -R_i - \sqrt{R_i^2 - x^2 - y^2} = -R_i\left(1 - \sqrt{1 - \frac{x^2+y^2}{R_i^2}}\right),$$

$$\begin{aligned}
D_l(x,y) &= D_{l,o} - D_o(x,y) + D_{l,i} - D_i(x,y) \\
&= D_{l,o} - R_o\left(1 - \sqrt{1 - \frac{x^2+y^2}{R_o^2}}\right) + D_{l,i} - \left(-R_i\left(1 - \sqrt{1 - \frac{x^2+y^2}{R_i^2}}\right)\right) \\
&= D_{l,o} + D_{l,i} - R_o\left(1 - \sqrt{1 - \frac{x^2+y^2}{R_o^2}}\right) + R_i\left(1 - \sqrt{1 - \frac{x^2+y^2}{R_i^2}}\right).
\end{aligned}$$

Here $D_{l,o}$ is the thickness of the lens at the optical axis on the part that is attributed to the object space, and $D_{l,i}$ is the thickness of the lens at the optical axis on the part that is attributed to the image space (Fig. 14.6). Note that we are dealing with a convex lens and therefore by the sign convention described in Section 13.2.3, the radius R_o is taken to be positive and the radius R_i is taken to be negative.

These expressions, while accurate, are difficult to handle in later parts of our development. We therefore assume that we are only interested in points close to the optical axis, or more precisely that $x^2 + y^2$ is very small in comparison to R_o^2 and R_i^2. This assumption allows us to make the paraxial approximation based on the Taylor series approximation $\sqrt{1+a} \approx 1 + \frac{a}{2}$, which is valid for "small" a (Online Appendix Section B.2). For $(x, y) \in A$, this approximation yields

$$\sqrt{1 - \frac{x^2+y^2}{R_o^2}} \approx 1 - \frac{x^2+y^2}{2R_o^2} \quad \text{and} \quad \sqrt{1 - \frac{x^2+y^2}{R_i^2}} \approx 1 - \frac{x^2+y^2}{2R_i^2},$$

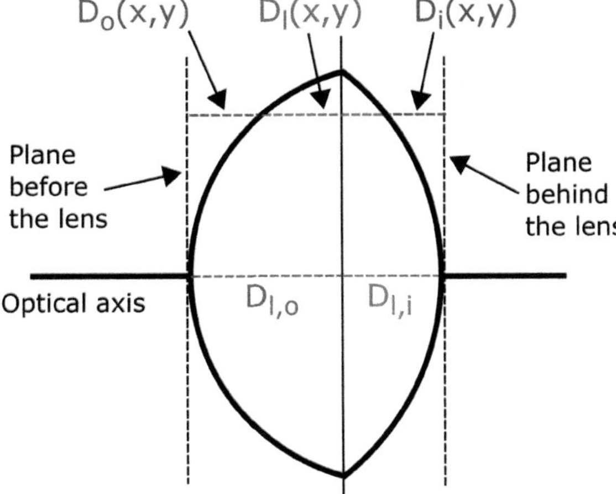

FIGURE 14.6
Path from the plane before, to the plane behind, a lens. The path comprises three segments of lengths $D_o(x, y)$, $D_l(x, y)$, and $D_i(x, y)$. At the optical axis, the thickness of the object space portion of the lens is denoted by $D_{l,o}$, and the thickness of the image space portion of the lens is denoted by $D_{l,i}$.

and therefore

$$
\begin{aligned}
D_o(x, y) &\approx \frac{x^2 + y^2}{2R_o}, \\
D_i(x, y) &\approx -\frac{x^2 + y^2}{2R_i}, \\
D_l(x, y) &\approx D_{l,o} + D_{l,i} - \frac{x^2 + y^2}{2R_o} + \frac{x^2 + y^2}{2R_i} = D_{l,o} + D_{l,i} - \frac{x^2 + y^2}{2}\left(\frac{1}{R_o} - \frac{1}{R_i}\right).
\end{aligned}
$$

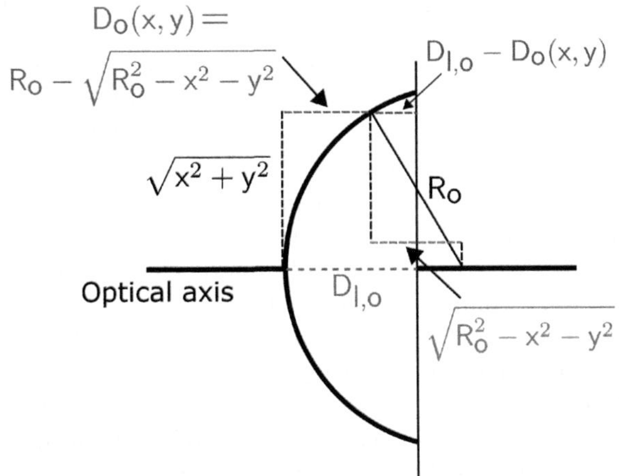

FIGURE 14.7
Object-side geometry of a lens in relation to its transmittance. An analogous geometry for the image side of the lens can be obtained based on the radius R_i of the lens' image-side spherical surface.

We thus obtain, for the phase term ϕ, the expression

$$\begin{aligned}
\phi(x,y) &= k_{vac}\left(n_o D_o(x,y) + n_l D_l(x,y) + n_i D_i(x,y)\right)\\
&= k_{vac}\left(n_o\frac{x^2+y^2}{2R_o} - n_i\frac{x^2+y^2}{2R_i} + n_l\left(D_{l,o} + D_{l,i} - \frac{x^2+y^2}{2}\left(\frac{1}{R_o} - \frac{1}{R_i}\right)\right)\right)\\
&= k_{vac}\left(n_l D_{l,o} + n_l D_{l,i} + \frac{x^2+y^2}{2}\left(\frac{n_o}{R_o} - \frac{n_i}{R_i} - \frac{n_l}{R_o} + \frac{n_l}{R_i}\right)\right)\\
&= k_{vac}\left(n_l\left(D_{l,o} + D_{l,i}\right) - \frac{x^2+y^2}{2}\left(\frac{n_l - n_o}{R_o} + \frac{n_i - n_l}{R_i}\right)\right)\\
&= k_{vac}\left(n_l\left(D_{l,o} + D_{l,i}\right) - \frac{x^2+y^2}{2}\frac{n_i}{f_i}\right), \quad (x,y) \in A,
\end{aligned}$$

where in the last step we have used the lensmaker's formula (Eq. (13.5)), and f_i is the image focal length of the lens.

Summarizing, we have obtained an analytical expression for the transmittance of a convex lens as

$$\mathcal{T}(x,y) = P(x,y)e^{i\phi(x,y)} = P(x,y)e^{ik_{vac}\left(n_l(D_{l,o}+D_{l,i}) - \frac{x^2+y^2}{2}\frac{n_i}{f_i}\right)}, \quad (x,y) \in \mathbb{R}^2.$$

In applications of this expression, we will follow our standard practice and omit the constant phase term $k_{vac}n_l\left(D_{l,o} + D_{l,i}\right)$. We will therefore use

$$\mathcal{T}(x,y) = P(x,y)e^{-ik_{vac}\frac{x^2+y^2}{2}\frac{n_i}{f_i}}, \quad (x,y) \in \mathbb{R}^2. \tag{14.1}$$

It is interesting to note that the phase term is quadratic in the coordinates and depends inverse proportionally on the image focal length f_i. Therefore, within the aperture, the phase of this expression increases with increasing distance from the center of the lens.

14.4 Huygens-Fresnel principle and Fresnel integral

Our objective in this chapter is to investigate image formation in a microscope, which we model as a thin lens. Among many other topics, this will allow us to derive the Airy profile introduced in Section 4.1 as the image of a point source. As discussed earlier, the propagation of waves is governed by Maxwell's equations, and one approach would be to set up the specific equations corresponding to our particular image formation problem. This is, however, a notoriously difficult problem, and analytical expressions for the solutions cannot generally be expected. We now live in a time of unparalleled computational resources, and numerical solutions to very complex differential equations are possible. While this is a very important approach to simulate image formation, we will follow another path. One of the most successful approaches to the study of image formation relies on the *Huygens-Fresnel principle*, which is based on the idea that a given wave can be thought of as giving rise to spherical waves at each of its points. The behavior of the wave at other places is then the superposition of these spherical waves. It is this principle that we will use to analyze image formation in our setting. We do, however, need to remain very much aware that there are a number of approximations inherent to this approach. One of the most important approximations is the paraxial approximation, by which we restrict our analysis to points that are close to the optical axis (Note 14.3). What we gain with the Huygens-Fresnel approach is a derivation of the Airy profile and a 3D point spread function in terms of optically relevant parameters such as the wavelength and magnification. In contrast to numerical solutions, such analytical expressions can give us insight into the dependence of the image on the specific optically relevant parameters.

14.4.1 Huygens-Fresnel principle

We can state the Huygens-Fresnel principle as follows.

- Each point of a wavefront defines the origin of a *wavelet*, i.e., of a spherical wave, whose characteristics, i.e., phase and amplitude, are given by the corresponding properties of the wave at this particular point.
- The wave at different points in space is given by the superposition of these spherical waves.

The Huygens-Fresnel principle is easily translated into the following mathematical description. If $U(r_1)$ denotes an electromagnetic wave at the point r_1, then following the Huygens-Fresnel principle, the wavelet that originates at r_1 gives rise to a spherical wave of the form

$$\frac{e^{ik\|r-r_1\|}}{\|r-r_1\|}U(r_1), \quad r \in \mathbb{R}^3.$$

We are particularly interested in understanding the behavior of waves as they pass through an aperture such as a microscope objective, modeled here as a thin lens. Let A stand for an aperture in a 2D plane in 3D space. If we assume that outside this aperture no light can pass from one half-space to the other, and if $U(r_1)$ describes the wave in the aperture, i.e., $r_1 \in A$, then at a point r_2 away from the aperture, the wave $U(r_2)$ will be the superposition of all wavelets originating from the aperture, i.e.,

$$U(r_2) = \int_{r_1 \in A} \frac{e^{ik\|r_2-r_1\|}}{\|r_2-r_1\|}U(r_1)dr_1. \tag{14.2}$$

We should note that we ignore here any constant multiple of the integral that might need to be used, for example, to ensure preservation of energy. In our developments, we are not concerned about such scalings, as our main interest is in obtaining expressions for the "shapes" of the functions. In fact, in Section 15.1.2 we will introduce specific scalings for *image functions* that will allow us to analyze imaging experiments in a quantitative setting.

The integral expression of Eq. (14.2) is nothing else other than a formalization of the "summing" of all of the values, at the point $r_2 = (x_2, y_2, z_2)$, of the spherical waves that originate from all points $r_1 = (x_1, y_1, z_1)$ comprising the aperture A. (See Section 14.1.3 for a discussion of the summing of waves and for why we can restrict our analysis here to the spatial parts of the spherical waves.) If we assume a coordinate system (Fig. 14.8) such that the aperture lies in a plane positioned at the coordinate z_1, then the integral can be written as

$$U(r_2) = \int_{(x_1,y_1,z_1)\in A} \frac{e^{ik(\|(x_2,y_2,z_2)-(x_1,y_1,z_1)\|)}}{\|(x_2,y_2,z_2)-(x_1,y_1,z_1)\|}U(x_1,y_1,z_1)dx_1dy_1.$$

As z_1 and z_2 are constant, we have that $z := z_2 - z_1$ is constant, and we can write

$$U(r_2) = \int_{(x_1,y_1,z_1)\in A} \frac{e^{ik\sqrt{(x_2-x_1)^2+(y_2-y_1)^2+z^2}}}{\sqrt{(x_2-x_1)^2+(y_2-y_1)^2+z^2}}U(x_1,y_1,z_1)dx_1dy_1.$$

This integral expression can be made computationally much more tractable by introducing

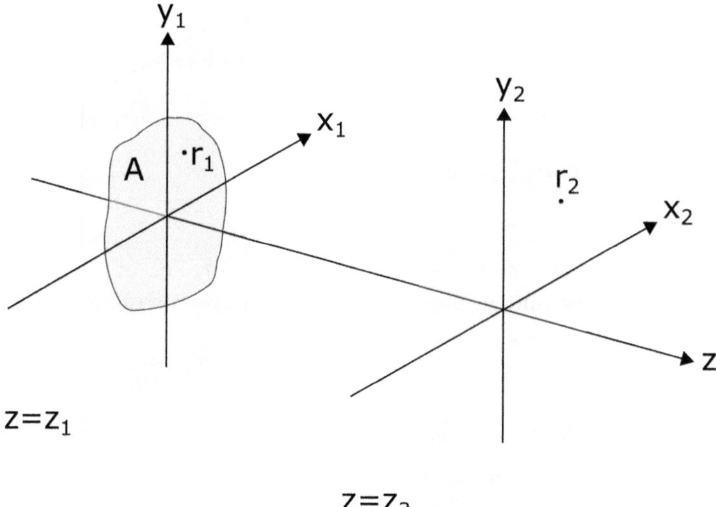

FIGURE 14.8
Coordinate system for the Huygens-Fresnel principle. The wave at a point $r_2 = (x_2, y_2, z_2)$ is the superposition of all spherical waves that originate from all points $r_1 = (x_1, y_1, z_1)$ comprising the aperture A.

two modifications based on the paraxial approximation that allow us to separate out the dependence on the distance parameter z. We assume here that the distance between the aperture and the point we are interested in is large. Therefore

$$z^2 \gg (x_2 - x_1)^2 + (y_2 - y_1)^2,$$

and hence we can approximate the denominator by

$$\sqrt{(x_2 - x_1)^2 + (y_2 - y_1)^2 + z^2} \approx z.$$

For the same reason, we can also obtain an approximation of the same expression in the term $e^{-ik\sqrt{(x_2-x_1)^2+(y_2-y_1)^2+z^2}}$. Here we use the reformulation

$$(x_2 - x_1)^2 + (y_2 - y_1)^2 + z^2 = z^2 \left(1 + \frac{1}{z^2}\left((x_2 - x_1)^2 + (y_2 - y_1)^2\right)\right).$$

Taking the square root and using the Taylor series approximation $\sqrt{1+a} \approx 1 + \frac{a}{2}$ (Online Appendix Section B.2), we obtain the so-called *Fresnel approximation*

$$\sqrt{(x_2 - x_1)^2 + (y_2 - y_1)^2 + z^2} \approx z \left(1 + \frac{1}{2z^2}\left((x_2 - x_1)^2 + (y_2 - y_1)^2\right)\right).$$

Therefore

$$U(r_2) = \frac{1}{z} e^{ikz} \int_{(x_1,y_1,z_1)\in A} e^{ik\frac{1}{2z}\left((x_2-x_1)^2+(y_2-y_1)^2\right)} U(x_1, y_1, z_1) dx_1 dy_1, \quad r_2 \in \mathbb{R}^3, \quad (14.3)$$

which is known as the *Fresnel diffraction integral*.

The importance of the Fresnel diffraction integral lies in the fact that it provides a tool to calculate the phase of a wave based on knowledge of the phase at a different spatial location. For example, we will be confronted with the situation where we know the wave at the object we want to image. Using the Fresnel diffraction integral, we can now calculate the wave right before the lens that is used for the imaging. Use of the Fresnel diffraction integral and the result on the transmittance of a lens to calculate the wave corresponding to the image of an object will be the subject of the next section.

14.5 Imaging through a thin lens

To discuss imaging through a thin lens, we will consider the propagation of light emanating from an object in three parts. The first part consists of the propagation of light from the object to the plane immediately in front of the lens. The second part is the transmission of the light through the lens, which was analyzed in Section 14.3. The third and final part addresses the propagation of the light from the plane immediately behind the lens to the detector. We will then assemble these segments of light propagation to obtain a description of the light wave on the detector as determined by the object.

To carry out the first of these steps, we use the Fresnel diffraction integral (Eq. (14.3)) to obtain the wave $U_l^o(x_l, y_l)$, $(x_l, y_l) \in \mathbb{R}^2$, in the plane immediately in front of the lens, which is a distance s_o away from the object (see Fig. 14.9 for an illustration of the notation). Denoting the field that characterizes the object at its location by $o(x_o, y_o)$, $(x_o, y_o) \in \mathbb{R}^2$, we have

$$U_l^o(x_l, y_l) = \frac{1}{s_o} e^{in_o k_{vac} s_o} \int_{-\infty}^{\infty} \int_{-\infty}^{\infty} o(x_o, y_o) e^{\frac{in_o k_{vac}}{2s_o}\left((x_l - x_o)^2 + (y_l - y_o)^2\right)} dx_o dy_o, \quad (x_l, y_l) \in \mathbb{R}^2,$$

where k_{vac} is the wave number in vacuum and n_o is the refractive index of the object space. Note that notationally we have replaced the 3D field U in Eq. (14.3) with the 2D fields U_l^o and o located at their respective fixed axial positions. In addition, the use of the infinite integrals means that the aperture A is now specified by the definition of the object field o.

In Section 14.3, we characterized the effect of a thin lens on a wave that passes through it. Specifically, the effect was characterized in terms of the lens' transmittance, with pupil function P and a phase term ϕ that depends on the image focal length f_i of the lens and the refractive index n_i of the image space (Eq. (14.1)). Applying that result, the field U_l^i in the plane immediately behind the thin lens is given by

$$\begin{aligned} U_l^i(x_l, y_l) &= P(x_l, y_l) e^{i\phi(x_l, y_l)} U_l^o(x_l, y_l) \\ &= \frac{1}{s_o} e^{in_o k_{vac} s_o} e^{\frac{-in_i k_{vac}}{f_i}\left(\frac{x_l^2 + y_l^2}{2}\right)} P(x_l, y_l) \\ &\quad \int_{-\infty}^{\infty} \int_{-\infty}^{\infty} o(x_o, y_o) e^{\frac{in_o k_{vac}}{2s_o}\left((x_l - x_o)^2 + (y_l - y_o)^2\right)} dx_o dy_o, \quad (x_l, y_l) \in \mathbb{R}^2. \end{aligned}$$

We now need just one more application of the Fresnel diffraction integral to obtain the field U_i on the detector plane in terms of the field U_l^i behind the lens. The detector plane is a distance s_i from the lens. Therefore

$$U_i(x_i, y_i) = \frac{1}{s_i} e^{in_i k_{vac} s_i} \int_{-\infty}^{\infty} \int_{-\infty}^{\infty} U_l^i(x_l, y_l) e^{\frac{in_i k_{vac}}{2s_i}\left((x_i - x_l)^2 + (y_i - y_l)^2\right)} dx_l dy_l, \quad (x_i, y_i) \in \mathbb{R}^2.$$

This expression now gives us the field at the detector in terms of the field just behind the lens, but not in terms of the object. Ultimately, we want to determine how the field at the

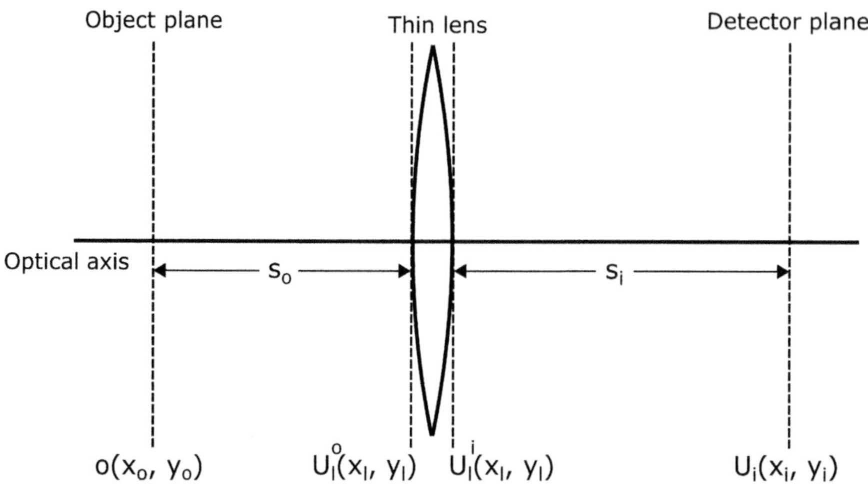

FIGURE 14.9
Notation for the propagation of light in thin lens imaging. The field at the object plane located a distance s_o from the thin lens is represented by the function $o(x_o, y_o)$, $(x_o, y_o) \in \mathbb{R}^2$, and the field at the detector plane located a distance of s_i from the thin lens is given by the function $U_i(x_i, y_i)$, $(x_i, y_i) \in \mathbb{R}^2$. The fields at the planes immediately in front of the lens and immediately behind the lens are denoted by $U_l^o(x_l, y_l)$ and $U_l^i(x_l, y_l)$, $(x_l, y_l) \in \mathbb{R}^2$, respectively.

detector is given through the field that characterizes the object. In order to do this, we need to insert the full expression for U_l^i into the expression for U_i. This, at first, gives the lengthy expression

$$U_i(x_i, y_i) = \frac{1}{s_o}\frac{1}{s_i} e^{in_i k_{vac} s_i} e^{in_o k_{vac} s_o} \int_{-\infty}^{\infty}\int_{-\infty}^{\infty}\int_{-\infty}^{\infty}\int_{-\infty}^{\infty} P(x_l, y_l) o(x_o, y_o) e^{\frac{-in_i k_{vac}}{f_i}\left(\frac{x_l^2+y_l^2}{2}\right)}$$
$$e^{\frac{in_o k_{vac}}{2s_o}\left((x_l-x_o)^2+(y_l-y_o)^2\right)} e^{\frac{in_i k_{vac}}{2s_i}\left((x_i-x_l)^2+(y_i-y_l)^2\right)} dx_o dy_o dx_l dy_l, \ (x_i, y_i) \in \mathbb{R}^2.$$

We now proceed to simplify the integrand, starting with

$$\begin{aligned}
&o(x_o, y_o) P(x_l, y_l) e^{-\frac{in_i k_{vac}}{f_i}\left(\frac{x_l^2+y_l^2}{2}\right)} e^{\frac{in_o k_{vac}}{2s_o}\left((x_l-x_o)^2+(y_l-y_o)^2\right)} e^{\frac{in_i k_{vac}}{2s_i}\left((x_i-x_l)^2+(y_i-y_l)^2\right)} \\
&= o(x_o, y_o) P(x_l, y_l) e^{-\frac{in_i k_{vac}}{2f_i}(x_l^2+y_l^2)} e^{\frac{in_o k_{vac}}{2s_o}(x_l^2+y_l^2)} e^{\frac{in_o k_{vac}}{2s_o}(x_o^2+y_o^2)} e^{-\frac{in_o k_{vac}}{s_o}(x_l x_o + y_l y_o)} \\
&\qquad e^{\frac{in_i k_{vac}}{2s_i}(x_i^2+y_i^2)} e^{\frac{in_i k_{vac}}{2s_i}(x_l^2+y_l^2)} e^{-\frac{in_i k_{vac}}{s_i}(x_l x_i + y_l y_i)} \\
&= o(x_o, y_o) P(x_l, y_l) e^{\frac{in_i k_{vac}}{2s_i}(x_i^2+y_i^2)} e^{-\frac{ik_{vac}}{2}\left(\frac{n_i}{f_i} - \frac{n_o}{s_o} - \frac{n_i}{s_i}\right)(x_l^2+y_l^2)} \\
&\qquad e^{-ik_{vac}\left(\frac{n_o}{s_o}(x_l x_o + y_l y_o) + \frac{n_i}{s_i}(x_l x_i + y_l y_i)\right)} e^{\frac{in_o k_{vac}}{2s_o}(x_o^2+y_o^2)}.
\end{aligned}$$

The term $\Delta f := \frac{n_i}{f_i} - \frac{n_o}{s_o} - \frac{n_i}{s_i}$, which we refer to as the *defocus*, is of course well known from the lens formula (Eq. (13.6)). If $\Delta f = 0$, the image and the object are in focus with respect to one another. Since we do not want to assume, in general, that we are in the in-focus situation, we introduce the concept of the *effective pupil function*

$$P_{eff}(x_l, y_l) := P(x_l, y_l) e^{-\frac{ik_{vac}\Delta f}{2}(x_l^2+y_l^2)}, \quad (x_l, y_l) \in \mathbb{R}^2.$$

We clearly see from this definition that, in the in-focus situation, the effective pupil function reduces to the pupil function. With this notation the integrand can be written as

$$o(x_o, y_o)P_{eff}(x_l, y_l)e^{\frac{in_ik_{vac}}{2s_i}(x_i^2+y_i^2)}e^{-ik_{vac}\left(\frac{n_o}{s_o}(x_lx_o+y_ly_o)+\frac{n_i}{s_i}(x_lx_i+y_ly_i)\right)}e^{\frac{in_ok_{vac}}{2s_o}(x_o^2+y_o^2)}.$$

Note that the magnification is given by $M := \frac{n_os_i}{n_is_o}$ (Eq. (13.8)). Then for $(x_i, y_i) \in \mathbb{R}^2$,

$$\begin{aligned}U_i(x_i, y_i) &= \frac{n_o}{Ms_o^2n_i}e^{ik_{vac}s_o\left(n_o+\frac{n_i^2M}{n_o}\right)}e^{\frac{in_ok_{vac}}{2Ms_o}(x_i^2+y_i^2)}\int_{-\infty}^{\infty}\int_{-\infty}^{\infty}\int_{-\infty}^{\infty}\int_{-\infty}^{\infty}\\ &\quad o(x_o, y_o)P_{eff}(x_l, y_l)e^{\frac{in_ok_{vac}}{2s_o}(x_o^2+y_o^2)}e^{\frac{-in_ok_{vac}}{s_o}\left(x_l\left(x_o+\frac{x_i}{M}\right)+y_l\left(y_o+\frac{y_i}{M}\right)\right)}dx_ody_odx_ldy_l\\ &= \frac{n_o}{Ms_o^2n_i}e^{ik_{vac}s_o\left(n_o+\frac{n_i^2M}{n_o}\right)}e^{\frac{in_ok_{vac}}{2Ms_o}(x_i^2+y_i^2)}\int_{-\infty}^{\infty}\int_{-\infty}^{\infty}o(x_o, y_o)e^{\frac{in_ok_{vac}}{2s_o}(x_o^2+y_o^2)}\\ &\quad\left(\int_{-\infty}^{\infty}\int_{-\infty}^{\infty}P_{eff}(x_l, y_l)e^{-\frac{in_ok_{vac}}{s_o}\left(x_l\left(x_o+\frac{x_i}{M}\right)+y_l\left(y_o+\frac{y_i}{M}\right)\right)}dx_ldy_l\right)dx_ody_o. \qquad (14.4)\end{aligned}$$

We now have the desired relationship between the wave at the object and the wave at the detector. Inspecting the expression more closely we see that the term in the inner brackets does not depend on the object, and we can obtain further insights into this expression by analyzing this term in some detail. This will be done in the next subsection.

14.5.1 Amplitude point spread function

We now examine the bracketed term in the integrand of Eq. (14.4). We refer to it as the *amplitude point spread function*, and denote it by

$$h(x, y) := \int_{-\infty}^{\infty}\int_{-\infty}^{\infty}P_{eff}(x_l, y_l)e^{-\frac{in_ok_{vac}}{s_o}(x_lx+y_ly)}dx_ldy_l, \quad (x, y) \in \mathbb{R}^2. \qquad (14.5)$$

We introduce it at this point with the purpose of using notation to simplify the integral expression of Eq. (14.4). As we will see very soon, there are important interpretations associated with this function.

With the notation for the point spread function, we now rewrite Eq. (14.4) as

$$\begin{aligned}U_i(x_i, y_i) &= \frac{n_o}{Ms_o^2n_i}e^{ik_{vac}s_o\left(n_o+\frac{n_i^2M}{n_o}\right)}e^{\frac{in_ok_{vac}}{2Ms_o}(x_i^2+y_i^2)}\int_{-\infty}^{\infty}\int_{-\infty}^{\infty}o(x_o, y_o)\\ &\quad e^{\frac{in_ok_{vac}}{2s_o}(x_o^2+y_o^2)}h\left(x_o+\frac{x_i}{M}, y_o+\frac{y_i}{M}\right)dx_ody_o, \quad (x_i, y_i) \in \mathbb{R}^2. \qquad (14.6)\end{aligned}$$

The point spread function is often thought of as an impulse response in analogy to the system-theoretic notion of the impulse response of a linear system. By investigating the image of an "impulse"-like object, we can easily show that this notion is justified. In analogy with system theory we consider an "input", i.e., an object, that is very small, but has very large power. The mathematical idealization of this idea is to set the object to be a product of 1D delta functions (Online Appendix Section B.1), i.e.,

$$o(x_o, y_o) := \delta_0(x_o)\delta_0(y_o), \quad (x_o, y_o) \in \mathbb{R}^2.$$

In practice, the object could, for example, be a very small bead with high fluorescence intensity. With this point object, the field at the detector is given by

$$
\begin{aligned}
U_i(x_i, y_i) &= \frac{n_o}{M s_o^2 n_i} e^{i k_{vac} s_o \left(n_o + \frac{n_i^2 M}{n_o}\right)} e^{\frac{i n_o k_{vac}}{2 M s_o}(x_i^2 + y_i^2)} \\
&\quad \int_{-\infty}^{\infty} \int_{-\infty}^{\infty} \delta_0(x_o)\delta_0(y_o) e^{\frac{i n_o k_{vac}}{2 s_o}(x_o^2 + y_o^2)} h\left(x_o + \frac{x_i}{M}, y_o + \frac{y_i}{M}\right) dx_o dy_o \\
&= \frac{n_o}{M s_o^2 n_i} e^{i k_{vac} s_o \left(n_o + \frac{n_i^2 M}{n_o}\right)} e^{\frac{i n_o k_{vac}}{2 M s_o}(x_i^2 + y_i^2)} e^{\frac{i n_o k_{vac}}{2 s_o}(0^2 + 0^2)} h\left(0 + \frac{x_i}{M}, 0 + \frac{y_i}{M}\right) \\
&= \frac{n_o}{M s_o^2 n_i} e^{i k_{vac} s_o \left(n_o + \frac{n_i^2 M}{n_o}\right)} e^{\frac{i n_o k_{vac}}{2 M s_o}(x_i^2 + y_i^2)} h\left(\frac{x_i}{M}, \frac{y_i}{M}\right), \quad (x_i, y_i) \in \mathbb{R}^2.
\end{aligned}
$$

This shows that the absolute value of the image of the point object is a scaled version of the point spread function h, thereby justifying the notion of "point spread function".

14.5.2 Convolution description

To obtain the final representation of the input-output relationship between the object and its image, we need to carry out one more approximation. In the integral of Eq. (14.6), we have the expression $e^{\frac{i n_o k_{vac}}{2 s_o}(x_o^2 + y_o^2)}$ as a factor in the integrand. In the absence of this factor, the integral is a convolution integral between the function o that describes the object and the point spread function h. Note that the convolution integral is written in a slightly unconventional fashion, but a simple reformulation will show the more conventional difference of the arguments (Exercise 14.4).

Different arguments can be made that will allow us to remove this quadratic phase term from the integrand. The first argument is predicated on the extent of the object being very small in comparison to the distance s_o of the object to the lens. This is typically the case in high-resolution subcellular microscopy where the object is often no larger than very few micrometers. As the expression is only evaluated over the size of the object, the argument is that the exponent will be negligibly small when the object is small, and hence the entire quadratic phase term can be ignored.

The second argument is based on the premise that the point spread function h will decay rapidly to 0 away from $(0, 0)$. This is justified by the understanding that h is the image of a point source, and as such it should appear point-like if the image is considered to be good. This means that $h(x, y) \approx 0$ if (x, y) is a point away from $(0, 0)$, and $h(x, y) \neq 0$ if (x, y) is close to $(0, 0)$. In the integral of Eq. (14.6), the argument of h is $\left(x_o + \frac{x_i}{M}, y_o + \frac{y_i}{M}\right)$, and its being close to $(0, 0)$ implies that $x_o \approx -\frac{x_i}{M}$ and $y_o \approx -\frac{y_i}{M}$. Therefore, for all (x_o, y_o) and (x_i, y_i) that satisfy this condition, we have that

$$
e^{\frac{i n_o k_{vac}}{2 s_o}(x_o^2 + y_o^2)} h\left(x_o + \frac{x_i}{M}, y_o + \frac{y_i}{M}\right) \approx e^{\frac{i n_o k_{vac}}{2 s_o}\left(\left(\frac{x_i}{M}\right)^2 + \left(\frac{y_i}{M}\right)^2\right)} h\left(x_o + \frac{x_i}{M}, y_o + \frac{y_i}{M}\right).
$$

Note that this identity also applies when the argument of h is away from $(0, 0)$, i.e., when $x_o \not\approx -\frac{x_i}{M}$ or $y_o \not\approx -\frac{y_i}{M}$. In this case, both sides of the identity are 0, as h is assumed to have a value of 0 or a value that's very nearly 0 when its argument is away from $(0, 0)$. We can therefore assume that this identity holds for all (x_o, y_o) and (x_i, y_i). It should, however, be noted that we have to be careful with this assumption. For example, if we have a significant defocus Δf, then h may not be point-like and the identity may not hold.

The second argument is then that the quadratic phase term becomes effectively independent of the integration variables when the point spread function is point-like, and therefore can be taken out of the double integral. Using the above identity, Eq. (14.6) becomes

$$\begin{aligned}
U_i(x_i,y_i) &= \frac{n_o}{M s_o^2 n_i} e^{ik_{vac}s_o\left(n_o+\frac{n_i^2 M}{n_o}\right)} e^{\frac{i n_o k_{vac}}{2Ms_o}(x_i^2+y_i^2)} \int_{-\infty}^{\infty}\int_{-\infty}^{\infty} o(x_o,y_o) \\
&\quad e^{\frac{i n_o k_{vac}}{2s_o}\left(\left(\frac{x_i}{M}\right)^2+\left(\frac{y_i}{M}\right)^2\right)} h\left(x_o+\frac{x_i}{M}, y_o+\frac{y_i}{M}\right) dx_o dy_o \\
&= \frac{n_o}{M s_o^2 n_i} e^{ik_{vac}s_o\left(n_o+\frac{n_i^2 M}{n_o}\right)} e^{\frac{i n_o k_{vac}}{2Ms_o}(x_i^2+y_i^2)} e^{\frac{i n_o k_{vac}}{2s_o}\left(\left(\frac{x_i}{M}\right)^2+\left(\frac{y_i}{M}\right)^2\right)} \\
&\quad \int_{-\infty}^{\infty}\int_{-\infty}^{\infty} o(x_o,y_o) h\left(x_o+\frac{x_i}{M}, y_o+\frac{y_i}{M}\right) dx_o dy_o \\
&= \frac{n_o}{M s_o^2 n_i} e^{ik_{vac}s_o\left(n_o+\frac{n_i^2 M}{n_o}\right)} e^{\frac{i n_o k_{vac}}{2Ms_o}\left(1+\frac{1}{M}\right)(x_i^2+y_i^2)} \\
&\quad \int_{-\infty}^{\infty}\int_{-\infty}^{\infty} o(x_o,y_o) h\left(x_o+\frac{x_i}{M}, y_o+\frac{y_i}{M}\right) dx_o dy_o, \quad (x_i,y_i)\in\mathbb{R}^2. \qquad (14.7)
\end{aligned}$$

This shows that the field obtained at the detector from an object in front of a thin lens can be computed by a convolution of the object function with the point spread function. The factors in front of the convolution integral are phase factors or constant scalars and are therefore not significant for the image generation.

14.5.3 Relationship to geometrical optics

We have discussed above that the amplitude point spread function, being the image of a point source, should have the appearance of a somewhat broadened point object. In the current consideration we will therefore make the assumption that the point spread function itself is given by a delta function, i.e., $h(x,y) \approx \delta_0(x)\delta_0(y)$, $(x,y) \in \mathbb{R}^2$.

Then from Eq. (14.7) the image of the object o is given by

$$\begin{aligned}
U_i(x_i,y_i) &= \frac{n_o}{M s_o^2 n_i} e^{ik_{vac}s_o\left(n_o+\frac{n_i^2 M}{n_o}\right)} e^{\frac{i n_o k_{vac}}{2Ms_o}\left(1+\frac{1}{M}\right)(x_i^2+y_i^2)} \\
&\quad \int_{-\infty}^{\infty}\int_{-\infty}^{\infty} o(x_o,y_o)\delta_0\left(x_o+\frac{x_i}{M}\right)\delta_0\left(y_o+\frac{y_i}{M}\right) dx_o dy_o \\
&= \frac{n_o}{M s_o^2 n_i} e^{ik_{vac}s_o\left(n_o+\frac{n_i^2 M}{n_o}\right)} e^{\frac{i n_o k_{vac}}{2Ms_o}\left(1+\frac{1}{M}\right)(x_i^2+y_i^2)} o\left(-\frac{x_i}{M}, -\frac{y_i}{M}\right), \quad (x_i,y_i)\in\mathbb{R}^2.
\end{aligned}$$

This shows that, up to the usual phase factors and the scalar multiple, the field in the detector plane is the inverted and magnified field of the object. This is, of course, consistent with the results that are obtained using geometrical optics for imaging with a thin lens, which show that the image is an inverted and magnified version of the object (Fig. 13.6).

14.5.4 Point spread function and Fourier transformation

Having established that the point spread function is the scaled image of a point source (Section 14.5.1), we now analyze the amplitude point spread function itself (Eq. (14.5)) in more detail and find that

$$\begin{aligned} h(x,y) &:= \int_{-\infty}^{\infty}\int_{-\infty}^{\infty} P_{eff}(x_l,y_l)e^{-\frac{in_ok_{vac}}{s_o}(x_lx+y_ly)}dx_ldy_l \\ &= \int_{-\infty}^{\infty}\int_{-\infty}^{\infty} P_{eff}(x_l,y_l)e^{-2\pi i\left(x_l\left(\frac{n_ok_{vac}}{2\pi s_o}x\right)+y_l\left(\frac{n_ok_{vac}}{2\pi s_o}y\right)\right)}dx_ldy_l \\ &= (\mathcal{F}(P_{eff}))\left(\frac{n_ok_{vac}}{2\pi s_o}x, \frac{n_ok_{vac}}{2\pi s_o}y\right), \quad (x,y)\in\mathbb{R}^2. \end{aligned} \tag{14.8}$$

The last identity shows the important result that the point spread function is in fact given by the 2D Fourier transform of the effective pupil function P_{eff}, since the 2D Fourier transform of a function $f(x_1,y_1)$, $(x_1,y_1)\in\mathbb{R}^2$, is defined as

$$(\mathcal{F}(f))(x_2,y_2) = \int_{-\infty}^{\infty}\int_{-\infty}^{\infty} f(x_1,y_1)e^{-2\pi i(x_1x_2+y_1y_2)}dx_1dy_1, \quad (x_2,y_2)\in\mathbb{R}^2. \tag{14.9}$$

This result is one of many in the field of *Fourier optics*, which exploits the strong relationship between Fourier analysis and optics.

If the pupil function is circularly symmetric, as is the case for a standard lens, the point spread function can be expressed in terms of a single integral through the use of the polar coordinate system. We next show that this is a result that follows from the definition of the Fourier transform itself. Assume that the function f is circularly symmetric, such that for some single-variable function f^s, we have that for all $(x_1,y_1)\in\mathbb{R}^2$,

$$f(x_1,y_1) = f^s\left(\sqrt{x_1^2+y_1^2}\right) = f^s(r), \quad r=\sqrt{x_1^2+y_1^2}.$$

Then, by expressing both $(x_1,y_1)\in\mathbb{R}^2$ and $(x_2,y_2)\in\mathbb{R}^2$ in Eq. (14.9) in polar coordinates, such that

$$\begin{aligned} (x_1,y_1) &=: (r\cos(\phi), r\sin(\phi)), \quad r\geq 0, \quad \phi\in[0,2\pi), \\ (x_2,y_2) &=: (\rho\cos(\psi), \rho\sin(\psi)), \quad \rho\geq 0, \quad \psi\in[0,2\pi), \end{aligned}$$

and by applying the change of variables theorem (Online Appendix Section B.3), we obtain the 2D Fourier transform

$$\begin{aligned} (\mathcal{F}(f))(x_2,y_2) &= \int_0^{\infty}\int_0^{2\pi} rf(r\cos(\phi),r\sin(\phi))e^{-2\pi i(r\cos(\phi)x_2+r\sin(\phi)y_2)}d\phi dr \\ &= \int_0^{\infty}\int_0^{2\pi} rf^s(r)e^{-2\pi i(r\rho\cos(\phi)\cos(\psi)+r\rho\sin(\phi)\sin(\psi))}d\phi dr \\ &= \int_0^{\infty}\int_0^{2\pi} rf^s(r)e^{-2\pi ir\rho\cos(\phi-\psi)}d\phi dr = \int_0^{\infty} rf^s(r)\int_0^{2\pi} e^{-2\pi ir\rho\cos(\phi)}d\phi dr \\ &= \int_0^{\infty} rf^s(r)\int_0^{2\pi} e^{2\pi ir\rho\cos(\phi)}d\phi dr = 2\pi\int_0^{\infty} rf^s(r)J_0(2\pi r\rho)dr, \end{aligned} \tag{14.10}$$

where $\rho = \sqrt{x_2^2+y_2^2}$. In the last step we made use of the zeroth order Bessel function identity

$$J_0(x) = \frac{1}{2\pi}\int_0^{2\pi} e^{ix\cos(\phi)}d\phi, \quad x\in\mathbb{R}.$$

The above derivation implies that the 2D Fourier transform of a circularly symmetric function is itself circularly symmetric.

We now assume the lens to have a circular aperture of radius a, such that the effective pupil function is given by

$$P_{eff}(x_l, y_l) = P(x_l, y_l)e^{-\frac{ik_{vac}\Delta f}{2}(x_l^2+y_l^2)}, \quad (x_l, y_l) \in \mathbb{R}^2,$$

where

$$P(x_l, y_l) := \begin{cases} 1, & \sqrt{x_l^2 + y_l^2} \le a \\ 0, & \sqrt{x_l^2 + y_l^2} > a \end{cases}.$$

Letting $r = \sqrt{x_l^2 + y_l^2}$, we can express P_{eff} in terms of the single-variable function $P^s_{eff}(r)$, $r \in \mathbb{R}$, given by

$$P^s_{eff}(r) = P^s(r)e^{-\frac{ik_{vac}\Delta f}{2}r^2}, \quad \text{where} \quad P^s(r) := \begin{cases} 1, & r \le a \\ 0, & r > a \end{cases}.$$

Using P^s_{eff} in the application of the result of Eq. (14.10) to Eq. (14.8), we obtain, for a lens with a circular aperture of radius a, the point spread function

$$\begin{aligned} h(x, y) &= (\mathcal{F}(P_{eff}))\left(\frac{n_o k_{vac}}{2\pi s_o}x, \frac{n_o k_{vac}}{2\pi s_o}y\right) \\ &= 2\pi \int_0^\infty r P^s_{eff}(r) J_0\left(2\pi r \frac{n_o k_{vac}}{2\pi s_o}\sqrt{x^2 + y^2}\right) dr \\ &= 2\pi \int_0^\infty r P^s(r) e^{-\frac{ik_{vac}\Delta f}{2}r^2} J_0\left(r\frac{n_o k_{vac}}{s_o}\sqrt{x^2 + y^2}\right) dr \\ &= 2\pi \int_0^a r e^{-\frac{ik_{vac}\Delta f}{2}r^2} J_0\left(r\frac{n_o k_{vac}}{s_o}\sqrt{x^2 + y^2}\right) dr, \quad (x, y) \in \mathbb{R}^2. \end{aligned} \tag{14.11}$$

This gives an expression for the point spread function as a 1D integral over a zeroth order Bessel function of the first kind. It is important to remember that in this expression we have made no assumption that the imaging system is in focus, as the integrand also depends on the defocus Δf.

14.5.4.1 In-focus point spread function

If we specialize Eq. (14.11) to the case where the image is in focus, the expression for the point spread function simplifies as $\Delta f = 0$. We have for the *in-focus amplitude point spread function*

$$h_{infocus}(x, y) = 2\pi \int_0^a r J_0\left(r\frac{n_o k_{vac}}{s_o}\sqrt{x^2 + y^2}\right) dr, \quad (x, y) \in \mathbb{R}^2.$$

Evaluating this integral by making use of the Bessel function identity

$$\int_0^c x J_0(x) dx = c J_1(c), \quad c > 0,$$

we obtain

$$h_{infocus}(x, y) = \frac{2\pi a s_o}{n_o k_{vac}\sqrt{x^2 + y^2}} J_1\left(\frac{a n_o k_{vac}}{s_o}\sqrt{x^2 + y^2}\right), \quad (x, y) \in \mathbb{R}^2.$$

This expression is not that useful to us since we typically do not know the distance s_o from the lens to the object that is being imaged. Neither is the radius a of the aperture typically easily available to us. However, we can relate the quotient of these numbers to the numerical aperture of the objective with

$$\frac{an_o k_{vac}}{s_o} = \frac{2\pi n_o}{\lambda_{vac}} \frac{a}{s_o} = \frac{2\pi n_o}{\lambda_{vac}} \tan(\theta_o) = \frac{2\pi n_o}{\lambda_{vac}} \frac{\sin(\theta_o)}{\cos(\theta_o)} = \frac{2\pi n_a}{\lambda_{vac}} \frac{1}{\cos(\theta_o)},$$

where $n_a = n_o \sin(\theta_o)$ is the numerical aperture of the objective, with θ_o the half angle of the cone defined by the aperture of the objective as shown in Fig. 14.10 (see also Section 10.6.1). Therefore the in-focus point spread function is given by

$$h_{infocus}(x, y) = \frac{2\pi a s_o}{n_o k_{vac}\sqrt{x^2 + y^2}} J_1\left(\frac{2\pi n_a}{\lambda_{vac}} \frac{1}{\cos(\theta_o)} \sqrt{x^2 + y^2}\right), \quad (x, y) \in \mathbb{R}^2. \quad (14.12)$$

In Section 14.2, we pointed out that the image that is registered by a detector is given by the modulus squared of the wave that impacts the detector. Therefore, the registered image of the in-focus amplitude point spread function on the detector is given by (a possible scalar multiple of) the *intensity point spread function*

$$\begin{aligned} psf_{infocus}(x, y) &= (h_{infocus}(x, y))^2 \\ &= \left(\frac{2\pi a s_o}{n_o k_{vac}\sqrt{x^2 + y^2}}\right)^2 J_1^2\left(\frac{2\pi n_a}{\lambda_{vac}} \frac{1}{\cos(\theta_o)} \sqrt{x^2 + y^2}\right), \quad (x, y) \in \mathbb{R}^2. \end{aligned} \quad (14.13)$$

We introduced the Airy profile in Section 4.1. The only significant difference between the in-focus intensity point spread function derived here and the Airy profile of Eq. (4.1) is the $\frac{1}{\cos(\theta_o)}$ factor in the argument of the Bessel function. However, for small values of θ_o, i.e., for small values of the numerical aperture, this factor can be ignored and the two functions can be considered to be identical.

This result brings us to where we wanted to arrive at, i.e., the derivation of the Airy

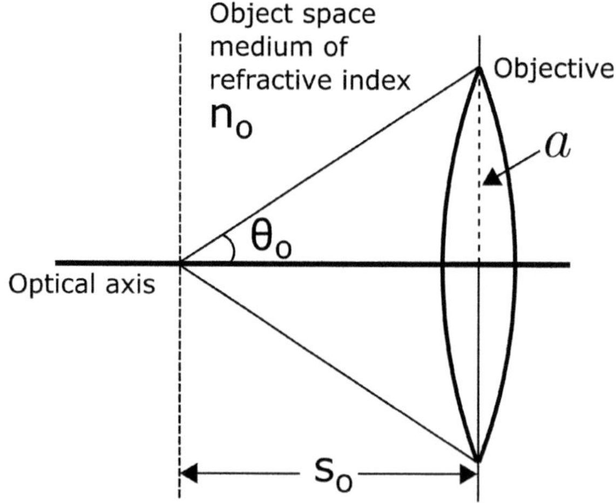

FIGURE 14.10
The numerical aperture n_a of an objective is given by the product of the refractive index n_o of the object space medium and the sine of the half angle θ_o of the maximum cone of light defined by the aperture of the objective. The radius a of the objective aperture and the distance s_o between an in-focus object and the objective are also shown.

profile, from first principles. We have now seen how a point in space that is emitting light can give rise to a highly complex image such as an Airy profile (Fig. 14.11). What we had seen in our discussion of the plane wave and the spherical wave are examples of waves that propagate in space without leading to such diffraction patterns. It is when waves pass through apertures that we see diffraction patterns.

14.5.5 Imaging with defocus and the 3D point spread function

The effective pupil function as it is used in the point spread function of Eq. (14.11) contains the phase term $e^{-\frac{ik_{vac}\Delta f}{2}r^2}$, where Δf depends on the distance from being in focus as defined by the lens formula, i.e., $\Delta f = \frac{n_i}{f_i} - \frac{n_o}{s_o} - \frac{n_i}{s_i}$. If the system is in focus, then $\Delta f = 0$ and the phase term disappears in the expression for the point spread function (Section 14.5.4.1).

We will obtain very useful expressions if we express the point spread function in terms of distances from the default in-focus position of the optical system. Assume that the planes given by the distances s_{o0} and s_{i0} are in focus, i.e.,

$$\frac{n_o}{s_{o0}} + \frac{n_i}{s_{i0}} = \frac{n_i}{f_i}.$$

With this notation we now define the actual positions of the object and the image in relation to the positions s_{o0} and s_{i0}. We let z_o be the distance of the object to the default position s_{o0} and z_i be the distance of the image to the default position s_{i0}, such that

$$s_o = s_{o0} - z_o,$$
$$s_i = s_{i0} + z_i.$$

Using these identities in the lens formula yields

$$\Delta f = \frac{n_i}{f_i} - \frac{n_o}{s_o} - \frac{n_i}{s_i} = \frac{n_i}{f_i} - \frac{n_o}{s_{o0} - z_o} - \frac{n_i}{s_{i0} + z_i} = \frac{n_i}{f_i} - \frac{n_o}{s_{o0}}\left(\frac{1}{1 - \frac{z_o}{s_{o0}}}\right) - \frac{n_i}{s_{i0}}\left(\frac{1}{1 + \frac{z_i}{s_{i0}}}\right)$$
$$\approx \frac{n_i}{f_i} - \frac{n_o}{s_{o0}}\left(1 + \frac{z_o}{s_{o0}}\right) - \frac{n_i}{s_{i0}}\left(1 - \frac{z_i}{s_{i0}}\right),$$

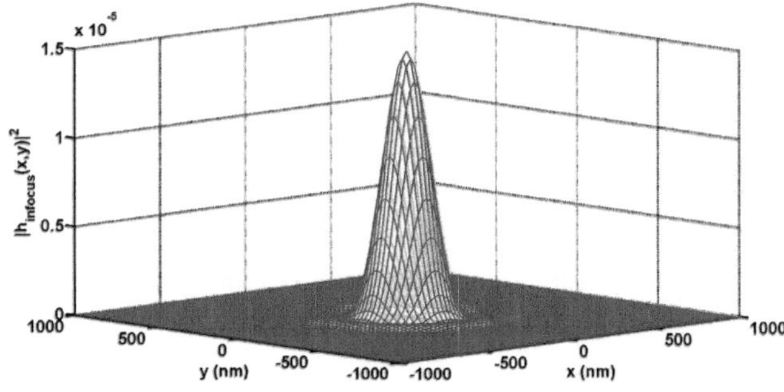

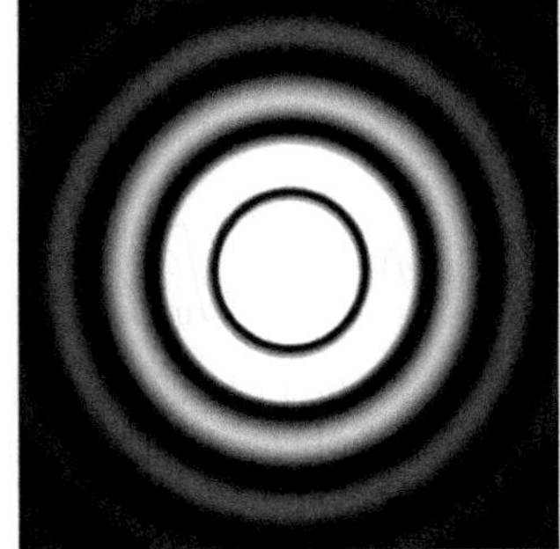

FIGURE 14.11
Airy profile. Mesh and intensity representations of the in-focus intensity point spread function of Eq. (14.13), computed with wavelength $\lambda_{vac} = 573$ nm and numerical aperture $n_a = 1.25$. The scalar constant $2\pi a s_o/n_o k_{vac}$ in Eq. (14.13) is ignored, and the function is instead normalized so that it integrates to 1 over $\mathbb{R}^2$. The factor $1/\cos(\theta_o)$ in the argument of the Bessel function is also ignored. In the intensity representation, a nonlinear scaling is used to emphasize the rings of the Airy pattern.

where we have used that $1 + \frac{z_o}{s_{o0}}$ are the first two terms of the infinite geometric series whose sum is $\frac{1}{1-\frac{z_o}{s_{o0}}}$. An analogous approximation applies to the last part of the expression. In these approximations, we have assumed that $|z_o| < s_{o0}$ and $|z_i| < s_{i0}$. Continuing the calculation, and using that $\frac{n_i}{f_i} - \frac{n_o}{s_{o0}} - \frac{n_i}{s_{i0}} = 0$, we obtain

$$\Delta f \approx \frac{n_i}{f_i} - \frac{n_o}{s_{o0}} - \frac{n_i}{s_{i0}} - \frac{n_o z_o}{s_{o0}^2} + \frac{n_i z_i}{s_{i0}^2} = \frac{1}{s_{o0}^2}\left(\frac{z_i n_o^2}{M_0^2 n_i} - n_o z_o\right), \tag{14.14}$$

where $M_0 := \frac{n_o s_{i0}}{n_i s_{o0}}$ is the magnification associated with the default in-focus position of the system.

We can now use this expression for the defocus Δf in the expression for the amplitude point spread function of Eq. (14.11), obtaining

$$\begin{aligned} h(x,y) &= 2\pi \int_0^a r e^{-\frac{ik_{vac}\Delta f}{2} r^2} J_0\left(r\frac{n_o k_{vac}}{s_o}\sqrt{x^2+y^2}\right) dr \\ &= 2\pi a^2 \int_0^1 r e^{-\frac{ik_{vac}\Delta f}{2} a^2 r^2} J_0\left(r\frac{a n_o k_{vac}}{s_o}\sqrt{x^2+y^2}\right) dr \\ &= 2\pi a^2 \int_0^1 r e^{-\frac{ik_{vac}a^2}{2s_{o0}^2}\left(\frac{z_i n_o^2}{M_0^2 n_i} - n_o z_o\right) r^2} J_0\left(r\frac{a n_o k_{vac}}{s_o}\sqrt{x^2+y^2}\right) dr \\ &= 2\pi a^2 \int_0^1 r e^{-\frac{ik_{vac}a^2 n_o^2}{2s_{o0}^2}\left(\frac{z_i}{M_0^2 n_i} - \frac{z_o}{n_o}\right) r^2} J_0\left(r\frac{a n_o k_{vac}}{s_o}\sqrt{x^2+y^2}\right) dr, \quad (x,y) \in \mathbb{R}^2. \end{aligned}$$

As in Section 14.5.4.1, we can use $\frac{a}{s_{o0}} = \tan(\theta_o)$, where θ_o is the half angle of the cone defined by the aperture of the objective (Fig. 14.10). For small θ_o, i.e., in the situation where the paraxial approximation is valid, $\tan(\theta_o) \approx \sin(\theta_o)$, and we can use the approximation

$$\frac{n_o a}{s_{o0}} = n_o \tan(\theta_o) \approx n_o \sin(\theta_o) = n_a,$$

where n_a is the numerical aperture of the objective. For small defocus distances and in the paraxial approximation, we can also assume that $s_o \approx s_{o0}$, and therefore we have

$$\frac{n_o a}{s_o} \approx \frac{n_o a}{s_{o0}} = n_o \tan(\theta_o) \approx n_o \sin(\theta_o) = n_a.$$

Hence we arrive at the amplitude point spread function

$$h(x,y) = 2\pi a^2 \int_0^1 r e^{-\frac{ik_{vac}n_a^2}{2}\left(\frac{z_i}{M_0^2 n_i} - \frac{z_o}{n_o}\right) r^2} J_0\left(r k_{vac} n_a \sqrt{x^2+y^2}\right) dr, \quad (x,y) \in \mathbb{R}^2.$$

This result shows how the point spread function is dependent on the defocus distance z_o of the object and the defocus distance z_i of the detector. If we introduce the defocus parameter

$$\Delta z := \frac{z_i}{M_0^2 n_i} - \frac{z_o}{n_o}$$

and write

$$h_{\Delta z}(x,y) := 2\pi a^2 \int_0^1 r e^{-\frac{ik_{vac}n_a^2}{2} r^2 \Delta z} J_0\left(r k_{vac} n_a \sqrt{x^2+y^2}\right) dr, \quad (x,y) \in \mathbb{R}^2, \tag{14.15}$$

we see that the point spread function has an important symmetry property in Δz. (Note that we refer to both Δf and Δz as the defocus, and we will specify which quantity is referred to when it is important to do so.) Replacing Δz with $-\Delta z$ yields the complex conjugate of the amplitude point spread function, i.e.,

$$h_{-\Delta z}(x,y) = \overline{h_{\Delta z}(x,y)}, \quad (x,y) \in \mathbb{R}^2.$$

This means that the images of these two point spread functions, i.e., their corresponding intensity point spread functions, are identical, as the image is given by the square of the modulus of the wave function. For example, if the detector is at the focus position, i.e., if $z_i = 0$, then the result implies that defocusing the point source by some distance above the focal plane produces an image that is identical to the image obtained by defocusing the point source below the focal plane by the same distance. We should note, however, that this is not always the case with experimental data. In practical situations, we can see different results due to aberrations, such as refractive index changes in the object space (Note 14.5), that are not accounted for by this point spread function. Note that if we set $\Delta z = 0$, the in-focus point spread function of Eq. (14.12) is obtained by applying the Bessel function identity presented in Section 14.5.4.1. (The factor $\frac{1}{\cos(\theta_o)}$ will not show up as the paraxial approximation has already been applied to the argument of the Bessel function here.)

Figure 14.12 shows the intensity point spread function at different defocus levels. We can see that the diffraction pattern changes with changing defocus level. In general, the diffraction rings become larger as the magnitude of the defocus increases, but even more pronounced is the decrease in amplitude of the central lobe. In fact, at the two largest defocus levels shown, we see that the central lobe has significantly decreased and appears to completely disappear. This will be studied in more detail next.

14.5.5.1 3D point spread function evaluated on the optical axis

We have just seen that the 3D point spread function, if analyzed in its dependence on the defocus, demonstrates interesting behavior along the optical axis by the disappearance of its central lobe (Fig. 14.12). In fact, the central lobe repeatedly disappears and reappears with increasing defocus, and we will analytically investigate this behavior by evaluating the point spread function of Eq. (14.15) at its central point, i.e., at $(x,y) = (0,0)$, and considering the dependence of its value there on the defocus Δz. At $(x,y) = (0,0)$, we have

$$\begin{aligned} h_{\Delta z}(0,0) &= 2\pi a^2 \int_0^1 r e^{-\frac{ik_{vac}n_a^2}{2} r^2 \Delta z} J_0\left(r k_{vac} n_a \sqrt{0}\right) dr = 2\pi a^2 \int_0^1 r e^{-\frac{ik_{vac}n_a^2}{2} r^2 \Delta z} J_0(0) dr \\ &= 2\pi a^2 \int_0^1 r e^{-\frac{ik_{vac}n_a^2}{2} r^2 \Delta z} dr = \pi a^2 \int_0^1 e^{-\frac{ik_{vac}n_a^2}{2} u \Delta z} du \\ &= \frac{\pi a^2}{-ik_{vac}n_a^2 \frac{\Delta z}{2}} \left(e^{-\frac{ik_{vac}n_a^2 \Delta z}{2} u} \right) \Bigg|_0^1 = -\frac{2\pi a^2}{ik_{vac}n_a^2 \Delta z} \left(e^{-\frac{ik_{vac}n_a^2 \Delta z}{2}} - 1 \right). \end{aligned}$$

The intensity point spread function at $(x,y) = (0,0)$ is then given by

$$\begin{aligned} psf_{\Delta z}(0,0) = |h_{\Delta z}(0,0)|^2 &= \left| -\frac{2\pi a^2}{ik_{vac}n_a^2 \Delta z} \left(e^{-\frac{ik_{vac}n_a^2 \Delta z}{2}} - 1 \right) \right|^2 \\ &= \frac{8\pi^2 a^4}{k_{vac}^2 n_a^4 (\Delta z)^2} \left(1 - \cos\left(k_{vac} n_a^2 \frac{\Delta z}{2} \right) \right). \end{aligned}$$

Using the trigonometric identity $\sin^2(\theta) = (1 - \cos(2\theta))/2$, we can rewrite the expression as

$$psf_{\Delta z}(0,0) = \frac{16\pi^2 a^4}{k_{vac}^2 n_a^4 (\Delta z)^2} \sin^2\left(k_{vac} n_a^2 \frac{\Delta z}{4} \right) = \pi^2 a^4 \left[\frac{4}{k_{vac} n_a^2 \Delta z} \sin\left(\frac{k_{vac}}{4} n_a^2 \Delta z \right) \right]^2. \tag{14.16}$$

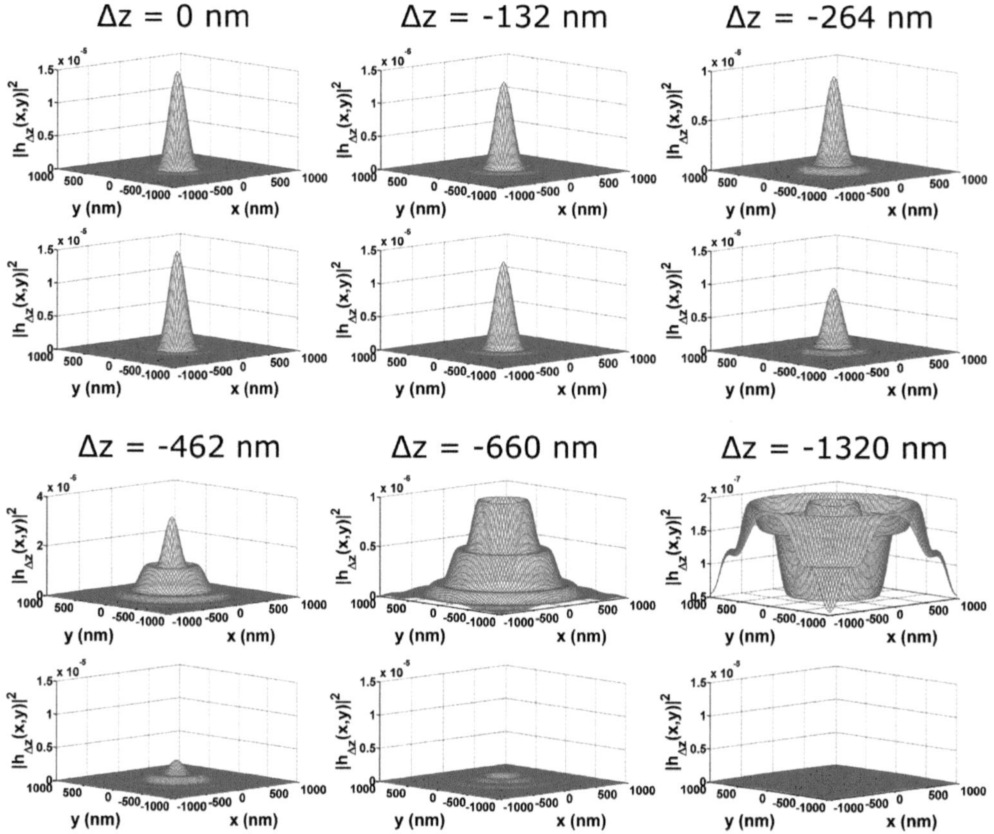

FIGURE 14.12
3D intensity point spread function at different defocus levels. Mesh representations are shown for the square of the modulus of the point spread function of Eq. (14.15) at different defocuses Δz, computed with wavelength $\lambda_{vac} = 573$ nm, numerical aperture $n_a = 1.25$, and object space refractive index $n_o = 1.515$. The detector is assumed to be positioned at the focus position, so that $z_i = 0$ and $\Delta z = -z_o/n_o$. The Δz values shown therefore correspond to point source positions $z_o = 0$ nm (in focus), 200 nm, 400 nm, 700 nm, 1000 nm, and 2000 nm. The scalar constant $2\pi a^2$ in Eq. (14.15) is ignored, and the square of the modulus of the function is instead normalized so that it integrates to 1 over $\mathbb{R}^2$. For each defocus Δz, the top mesh representation is displayed with a z-axis that best shows the shape of the point spread function, while the bottom mesh representation uses the z-axis of $\Delta z = 0$ nm to show how the intensity level decreases with increasing defocus.

This shows that the intensity of the point spread function along the optical axis displays oscillatory behavior, with zeros occurring at defocus values of Δz such that

$$\frac{k_{vac}}{4} n_a^2 \Delta z = m\pi, \quad m = \pm 1, \pm 2, \ldots.$$

This analysis thus indicates that the central lobe of the intensity point spread function will continually wax and wane with increasing defocus, and will completely disappear at defocus values

$$\Delta z = \frac{4\pi}{k_{vac} n_a^2} m, \quad m = \pm 1, \pm 2, \ldots.$$

An interesting difference between the spacings of the zeros here and the spacings of the zeros of the Airy profile (Section 4.2) is that these zeros are regularly spaced.

14.5.5.2 Depth of field and depth of focus

In our discussion of geometrical optical methods, we encountered the concept of conjugate focal planes. This means that if an object, such as a point source, is in an object focal plane and the detector is positioned in the corresponding conjugate image focal plane, the object will be in focus in the acquired image. An important question in this context is how sensitive these positions are, meaning how far the object can be moved without moving out of focus in the detector, or conversely, how far the detector can be moved and still be able to obtain an in-focus image of the object.

To analyze this problem, we need a criterion for how large a defocus can be and still produce an image of acceptable quality. A test object that is suitable for such an analysis is a point source. Let the largest allowable defocus be the Δz value that produces an image intensity value $psf_{\Delta z}(0,0)$ (Eq. (14.16)) that is about 80% of the intensity value corresponding to the in-focus scenario when $\Delta z = 0$. The 20% loss in intensity is, of course, an arbitrary cutoff, but it is typically considered to be an appropriate value that indicates a tolerable loss in intensity, and that the point source is still sufficiently in focus.

We first note from Eq. (14.16) that the image intensity $psf_{\Delta z}(0,0)$ equals $\pi^2 a^4$ when $\Delta z = 0$. Now if we set Δz such that

$$\frac{k_{vac}}{4} n_a^2 \Delta z = \pm \frac{\pi}{4},$$

we obtain

$$\begin{aligned} psf_{\Delta z}(0,0) &= \pi^2 a^4 \left(\frac{\sin\left(\frac{k_{vac}}{4} n_a^2 \Delta z\right)}{\frac{k_{vac}}{4} n_a^2 \Delta z} \right)^2 = \pi^2 a^4 \left(\frac{\sin\left(\pm\frac{\pi}{4}\right)}{\left(\pm\frac{\pi}{4}\right)} \right)^2 \\ &\approx 0.81 \cdot \pi^2 a^4 = 0.81 \cdot psf_0(0,0). \end{aligned}$$

We therefore consider a defocus level of at most

$$\Delta z = \pm \frac{\pi}{k_{vac} n_a^2}$$

to be acceptable without unduly decreasing the quality of the focused image.

The definition of the defocus Δz as

$$\Delta z = \frac{z_i}{M_0^2 n_i} - \frac{z_o}{n_o}$$

suggests that both changes in object distance and detector distance could contribute to the maximum allowable defocus. We consider two special cases: one in which we assume the detector to be in focus, the other in which we assume the object to be in focus. In the first case, the detector is in focus and hence $z_i = 0$. Therefore,

$$\Delta z = \pm \frac{\pi}{k_{vac} n_a^2} = -\frac{z_o}{n_o},$$

and hence

$$z_o = \pm \frac{\pi n_o}{k_{vac} n_a^2},$$

which we refer to as the *depth of field*, the maximum axial distance by which the object can be moved without appreciably changing the image quality (Fig. 14.13).

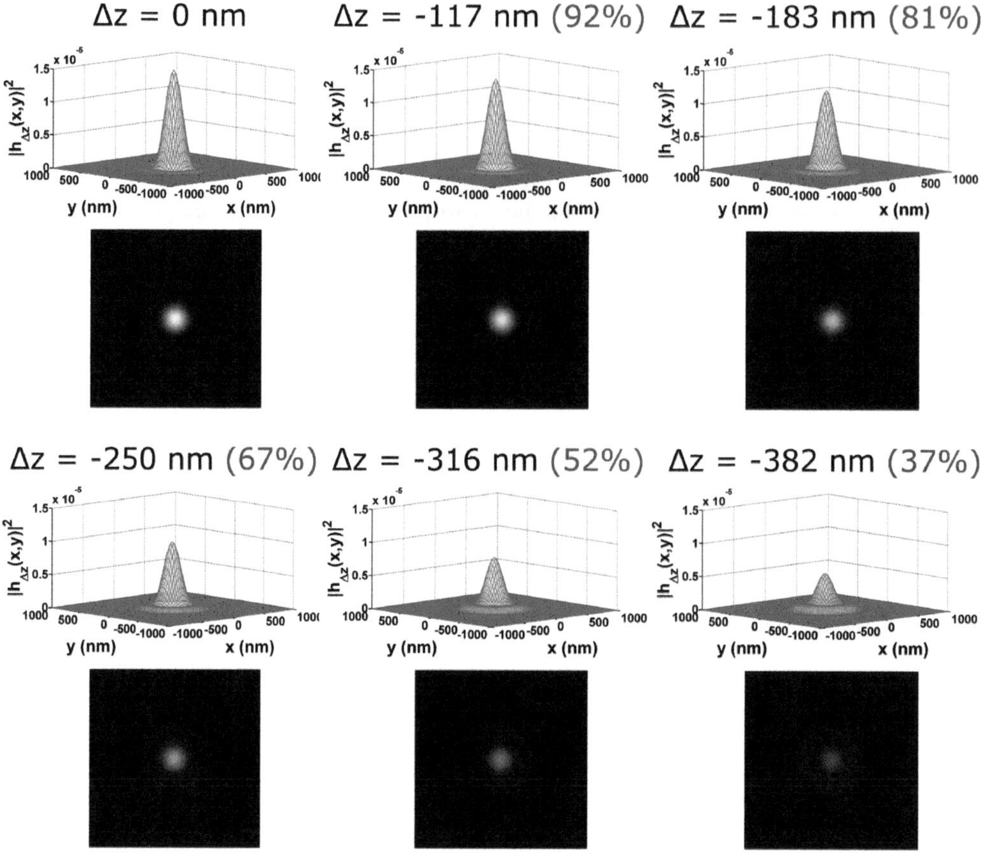

FIGURE 14.13
Intensity-based depth of field heuristic. Mesh and intensity representations are shown for the square of the modulus of the point spread function of Eq. (14.15) at different defocuses Δz, computed with wavelength $\lambda_{vac} = 573$ nm, numerical aperture $n_a = 1.25$, and object space refractive index $n_o = 1.515$. The detector is assumed to be positioned at the focus position, so that $z_i = 0$ and $\Delta z = -z_o/n_o$. The Δz values shown therefore correspond to point source positions $z_o = 0$ nm (in focus), 178 nm, 278 nm, 378 nm, 478 nm, and 578 nm. The scalar constant $2\pi a^2$ in Eq. (14.15) is ignored, and the square of the modulus of the function is instead normalized so that it integrates to 1 over $\mathbb{R}^2$. The percentage shown in blue for each $\Delta z \neq 0$ nm gives the ratio of the intensity $|h_{\Delta z}(0,0)|^2$ at the center of the point spread function at the particular defocus Δz to the intensity $|h_0(0,0)|^2$ at the center of the in-focus point spread function. In this example, the defocus of $\Delta z = -183$ nm, which corresponds to point source position $z_o = 278$ nm, approximately represents the 80% cutoff above which the point source is considered to be sufficiently in focus in the acquired image.

In the second case, we assume that the object is in focus, i.e., $z_o = 0$, and therefore

$$\Delta z = \pm \frac{\pi}{k_{vac} n_a^2} = \frac{z_i}{M_0^2 n_i}.$$

This means that the *depth of focus*, i.e., the maximum distance that the detector can be moved without unduly changing the quality of the image, is given by

$$z_i = \pm \frac{\pi M_0^2 n_i}{k_{vac} n_a^2}.$$

14.5.5.3 Heuristic 3D resolution criterion

In Section 4.2 we introduced Rayleigh's resolution criterion. The basis for that resolution criterion is to declare two point sources in a focal plane as resolvable if the maximum of the Airy profile of the second point source is farther away than the first zero of the Airy profile of the first point source. We can apply an analogous approach to deriving a heuristic resolution criterion for two point sources in three dimensions. We can declare two point sources in 3D object space to be *3D-resolvable* if, along the optical axis, the second point source is positioned such that the maximum of its point spread function is farther than the first zero of the point spread function of the first point source. Using the characterization of the zeros of the 3D point spread function above, this translates to a defocus of

$$|\Delta z| \geq \frac{4\pi}{k_{vac} n_a^2}. \tag{14.17}$$

To analyze the 3D resolution problem, we assume that the first point source is located on the optical axis in the object focal plane and that its image is in focus in the detector plane. The second point source is assumed to be located on the optical axis at a distance z_o from the first point source. As the detector is in the in-focus position, we have $z_i = 0$, and therefore $\Delta z = -z_o/n_o$. The 3D resolution criterion, which we will refer to as the *classical 3D resolution limit*, then suggests that the two point sources are distinguishable if

$$|z_o| \geq \frac{4\pi n_o}{k_{vac} n_a^2} = \frac{2\lambda_{vac} n_o}{n_a^2}. \tag{14.18}$$

It is instructive to compare Rayleigh's resolution criterion with the 3D criterion we have just derived. For example, for $\lambda_{vac} = 520$ nm and an objective with $n_a = 1.4$, Rayleigh's criterion gives a value of $0.61\lambda_{vac}/n_a \approx 227$ nm. In addition to λ_{vac} and n_a, the 3D criterion depends on the refractive index n_o of the immersion medium. If we assume that $n_o = 1.515$ for oil, then the 3D criterion gives a value of $\sim$804 nm, which is more than three and a half times higher than Rayleigh's criterion. This is consistent with our knowledge that the imaging properties of an optical microscope are inferior when we are dealing with out-of-focus situations.

14.6 Convolution for intensity profiles

The impact that the optical system has on the object as it is registered by the detector is of major importance for our understanding of any imaging methodology, and certainly microscopy as we are interested in it here. We have already considered this situation in Section 14.5, and in Section 14.5.2 we arrived at a convolution-type relationship between the object and the amplitude point spread function at the detector plane. This result was based on a careful analysis of the diffraction properties of the microscope imaging system when modeled as a thin lens, and importantly took into account the phase properties of the object.

As we know, image detectors do not measure phase information. We can therefore ask the question what the relationship is between the intensity distribution of the object and the resulting intensity of the wave at the detector, which will be registered as the image of the object by the detector. If we are interested in modeling the imaging properties of a fluorescence microscope, we need to remember that fluorescent objects do not comprise coherent emitters. This means that there is no phase relationship between fluorescent emitters at different parts of the object. For fluorescent samples, we therefore do not expect any benefits from an analysis that takes phase properties of the emitters into account. It is therefore sufficient to concentrate on the intensity profile of the sample.

In principle the intensity of the wave at the detector, i.e., the image of the object registered by the detector, can be obtained by taking the square of the modulus of the convolution expression of Eq. (14.7). However, here we are looking to obtain a convolution expression for the image of the object involving the intensity point spread function and the intensity object function.

To derive the relationship between object and image intensities, we are going to use a different approach to the one taken in Section 14.5. Here we are going to assume that the image at the detector is the superposition of the images for the points on the object. In Section 14.5.5, we derived the 3D amplitude point spread function $h_{\Delta z}(x,y)$, $(x,y) \in \mathbb{R}^2$, for an arbitrary point in object space given a lens with a circular aperture (Eq. (14.15)). Further, we denoted the square of its modulus (i.e., the intensity point spread function), and therefore the image of a point, by $psf_{\Delta z}(x,y)$, $(x,y) \in \mathbb{R}^2$. Under the assumption that the emission of light from the object is incoherent, we can obtain the image of the object by summing, or more precisely, integrating, $psf_{\Delta z}$ over the object.

Using $psf(x,y,\Delta z) = psf_{\Delta z}(x,y)$, $(x,y) \in \mathbb{R}^2$, for notational convenience, we have that the intensity image $I_{(x_o,y_o,z_o)}$ of a point source of intensity $O(x_o,y_o,z_o)$ located at (x_o,y_o,z_o) in object space is given by

$$\begin{aligned} I_{(x_o,y_o,z_o)}(x,y,z) &= psf\left(\frac{x}{M}+x_o, \frac{y}{M}+y_o, \Delta z\right) O(x_o,y_o,z_o) \\ &= psf\left(\frac{x}{M}+x_o, \frac{y}{M}+y_o, \frac{z}{M_0^2 n_i}-\frac{z_o}{n_o}\right) O(x_o,y_o,z_o). \end{aligned}$$

This expression states that the intensity image of a point is simply a magnified version of the intensity point spread function that is shifted according to the position of the object and weighted by the intensity of the point. Note also that both the object O and the image I are defined as 3D functions, as opposed to in Section 14.5 where the expressions relate 2D slices of the object and the image at their respective axial positions.

We can now proceed to state the intensity image $I(x,y,z)$, $(x,y,z) \in \mathbb{R}^3$, of an intensity object $O(x,y,z)$, $(x,y,z) \in \mathbb{R}^3$. The function O that stands for the intensity object is a positive piecewise continuous function in $\mathbb{R}^3$. The intensity image I at a point $(x,y) \in \mathbb{R}^2$ in the detector plane, located at $z_i = z$, is given by integrating $I_{(x_o,y_o,z_o)}$ over the object space, i.e.,

$$\begin{aligned} I(x,y,z) &= \int_{-\infty}^{\infty}\int_{-\infty}^{\infty}\int_{-\infty}^{\infty} I_{(x_o,y_o,z_o)}(x,y,z)dx_o dy_o dz_o \\ &= \int_{-\infty}^{\infty}\int_{-\infty}^{\infty}\int_{-\infty}^{\infty} psf\left(\frac{x}{M}+x_o, \frac{y}{M}+y_o, \frac{z}{M_0^2 n_i}-\frac{z_o}{n_o}\right) O(x_o,y_o,z_o)dx_o dy_o dz_o. \end{aligned}$$

Note that an important implicit assumption is made here that the optical system is *translation-invariant*, or *shift-invariant*, meaning that the intensity point spread function does not depend on the particular position in the object plane. In most practical situations, this condition is satisfied when the object is in reasonable proximity to the optical axis.

If this translational invariance is not satisfied, a spatially dependent point spread function would need to be used.

The convolution expression that we have obtained for the intensity image, given an intensity object and an intensity point spread function, is not in the form that is conventionally understood as a convolution in the mathematical literature, as the independent variables are added rather than subtracted. However, as mentioned for Eq. (14.7) which is of the same form, the classical convolution description can be easily obtained with a reformulation. To be consistent with the mathematical conventions, we will be using the reformulation in Chapter 21 when we discuss deconvolution, the process of recovering the object from the acquired image.

Notes

Chapter 10

1. Prior to the introduction of phase-contrast microscopy, cells had to be stained in order to visualize subcellular structures using a transmitted light microscope. A significant drawback of this visualization approach was that the staining process killed the cells. The advent of phase-contrast microscopy therefore represented an important breakthrough as it removed the staining requirement and allowed the observation of living cells at a high level of structural detail using transmitted light. For his invention of phase-contrast microscopy, Dutch physicist Frits Zernike was awarded the Nobel Prize in Physics in 1953. For details relating to the principle and use of phase-contrast microscopy as well as different implementations of the technique, see, e.g., [70, 80, 105].
2. A detailed discussion of the principle and use of DIC microscopy can be found in, e.g., [70, 80].

Chapter 11

1. The pulse-chase experiment presented in Section 11.1.3 for the verification of the intracellular fate of mutated IgG is described in detail in [74].
2. The work by Daniel Axelrod was instrumental in introducing total internal reflection techniques in fluorescence microscopy. In Section 11.4, we could only give a brief overview of the topic, and the reader is referred to, for example, reviews by Axelrod et al. [9, 8], which also discuss implementations of TIRFM that utilize a prism. It should be mentioned that the polarization properties of the evanescent field generated in a TIRFM setup, which play an important role in polarization-sensitive experiments, are particularly complex [19].
3. For more details on the dual-color TIRFM experiment described in Section 11.4.2 for the study of FcRn-mediated exocytosis of IgG, see [73].
4. For a study that uses GFP-tagged FcRn and mRFP1-tagged FcRn to investigate the presence of FcRn on the membrane and in the lumen of lysosomes, see [39].
5. For additional information relating to pH determination using ratiometric imaging, particularly in the context of the measurement of lysosomal pH, see [21].
6. For details concerning the development and properties of ratiometric pHluorin and ecliptic pHluorin, see [68].
7. For more information on the TIRFM imaging of the post-exocytosis trajectories of individual IgG and FcRn molecules as presented in Section 11.6.2, see [73].
8. The 2014 Nobel Prize in Chemistry was awarded for advances in high-resolution microscopy. Eric Betzig and W. E. Moerner contributed to various aspects of localization-based super-resolution microscopy [13, 50, 99, 47], with W. E. Moerner also having been the first to image a single molecule in 1989. This classical experiment was carried out in a low-temperature environment using tools very

different from those employed in current experiments [69]. It formed the foundation for subsequent major developments in single molecule microscopy. Stefan Hell developed stimulated emission depletion (STED) microscopy, an imaging technique that is not based on single molecule techniques. STED microscopy is a point scanning technique that addresses the crowding problem by introducing an implementation of the excitation profile of the microscope that generates a significantly smaller illumination spot than a conventional point scanning device [49, 126].

9. For a compilation and discussion of organic dyes and fluorescent proteins used in localization-based super-resolution microscopy, see [116] and [77, 103], respectively.

10. Detailed descriptions and discussions relating to the implementation of localization-based super-resolution microscopy can be found in, e.g., [116, 13, 50].

11. For information relating to the photochemistry of photoactivatable and photoswitchable fluorescent proteins, see, e.g., [78, 5, 119, 64, 109].

12. Additional details regarding the photophysics of the stochastic excitation of organic fluorophores can be found in [115, 116, 114].

13. Multifocal plane microscopy has been a major focus of research for the research group of the authors. The technique was first described and demonstrated by the group in [83]. In that work, a MUM setup was implemented that allowed the simultaneous imaging of two distinct focal planes within the sample. This setup was achieved by splitting the microscope's emission light path with a beam splitter and adding a second detector that was positioned at a customizable distance from the tube lens. Using this setup, the principle of MUM was demonstrated with the imaging of bead samples and a live cell sample. The example with the live cell illustrated the visualization and analysis of the movement of a tubule of interest from a sorting endosome to the plasma membrane, an investigation which would have been very difficult, if not impossible, to carry out with a standard microscope due to the large axial distance involved. In this example, one focal plane was positioned to coincide with the plasma membrane while the other was positioned to image the interior of the cell. Using a dually labeled sample, the dynamics at the plasma membrane were imaged in TIRFM mode in one color, and the dynamics in the cell interior were imaged in widefield mode in another color.

    The flexibility and utility of MUM were demonstrated further in [82] with the tracking and characterization of intracellular events on the recycling pathway. In that extensive study, two different MUM configurations were used to image the trafficking pathways of interest in live cells. Using one configuration, up to four distinct focal planes covering a larger depth than the live cell configuration of [83] were simultaneously imaged to record the trafficking pathway of FcRn. Using the other configuration, two distinct focal planes were simultaneously imaged, but unlike in [83], each focal plane was imaged by two detectors, each capturing the fluorescence associated with a different protein of interest. The second configuration importantly enabled the investigation of the trafficking pathway of two closely associated proteins — the receptor FcRn and its ligand IgG. As was the case with the live cell configuration of [83], both configurations used in [82] simultaneously imaged the plasma membrane in TIRFM mode and the cell interior in widefield mode.

    From a quantitative standpoint, information theory was used in [86] to show that

MUM is an imaging modality capable of overcoming the depth discrimination problem. Specifically, it was demonstrated that by combining the information contained in images simultaneously acquired of multiple focal planes, the axial location of an object, such as a single molecule, can be determined with a high level of certainty even when the object is in-focus or near-focus with respect to a focal plane. Putting theory into practice, the MUM localization algorithm (MUMLA) was introduced in [88] as an important tool for the quantitative analysis of MUM image data. MUMLA is a parameter estimation algorithm that incorporates data from simultaneously acquired images of distinct focal planes to produce a highly accurate estimate of the 3D location of an object of interest at a given point in time. By globally fitting a model to images of distinct focal planes, MUMLA enables the determination of the axial coordinate of an object that is in or near focus with respect to a focal plane, a task that is difficult to carry out with reasonable accuracy based on a single image acquired using a conventional microscope. Using MUMLA, it was demonstrated in [88] that the trajectory of an antibody molecule imaged with a MUM setup could be determined with high quantitative accuracy in three dimensions, and that the result allowed for a detailed characterization of the molecule's movement and associated events such as the molecule's internalization by a cell. MUMLA was again demonstrated in [87], where the trafficking of transferrin molecules in a thick cellular specimen is investigated using a four-plane MUM setup that supports the imaging of trafficking pathways over a depth of up to 10 μm.

A further application of the principle of MUM resulted in the development of *dual objective multifocal plane microscopy*, or dMUM [89]. Like MUM, dMUM supports the simultaneous imaging of distinct focal planes. dMUM, however, utilizes two opposing objectives to significantly increase the number of photons detected from the sample. Using both information theory and MUMLA, it was shown in [89] that by virtue of collecting more signal, a dMUM setup allows the axial location of an object to be estimated with an even greater level of certainty than a comparable MUM setup.

In [111], an approach based on information theory was introduced for designing a MUM setup in terms of the number of focal planes to use and the spacings between the planes. Given experimental parameters such as magnification, numerical aperture, and detector type and pixel size, this approach allows the determination of an appropriate set of focal planes based on how well the 3D position of a point source can be estimated.

14. The multifocal plane microscopy principle as discussed in Section 11.7 can be implemented in a number of different ways. We have described the projection of images of distinct focal planes onto different detectors positioned at distinct distances from the microscope tube lens [83]. The identical optical approach can, of course, also be realized by projecting the different split emission light paths onto the same detector, by suitably arranging the focusing optics [112, 57]. A different optical approach for the splitting of the emission light path can be achieved with a diffractive optical element as proposed by Blanchard and Greenaway [15] (see also [2]).
15. For more on the four-plane MUM experiment described in Section 11.7.2 for the investigation of transferrin trafficking in epithelial cells, see [87].
16. The two-plane MUM experiment presented in Section 11.7.3, which uses simultaneous total internal reflection and widefield illumination to image FcRn-mediated

IgG exocytosis and the events preceding exocytosis, is described and discussed in more detail in [82].

Chapter 12

1. See [53] for the description and characterization of a prototype EMCCD detector.
2. For more information on the various noise sources associated with a CCD detector and the determination of the readout noise statistics and the electron-count-to-DU conversion factor for a CCD detector, see, e.g., [56, 72].

Chapter 13

1. The working distance of an objective is the distance between the objective's front lens element and the closest surface of the cover glass when the sample is in focus. In our discussion, we have assumed that it is given by the object focal length of the objective, which we model as a thin lens. In practice, however, as the front lens element of the objective is typically plano-convex (i.e., a lens with one flat surface and one surface that bends outwards from the lens body), the principal plane from which the object focal length is measured is farther from the sample than the surface of the lens. Therefore, in actuality the working distance of an objective is typically a little shorter than its object focal length.

Chapter 14

1. For more details regarding the basics of the wave nature of light, the reader is referred to the book *Optics* by E. Hecht [46].
2. A good introduction to diffraction is given by M. Gu in [44], which inspired the development presented in Chapter 14. The books by M. Born and E. Wolf [16] and J. W. Goodman [41] have also formed the basis for some of the content of Chapter 14. All three sources also contain important additional material on diffraction theory and image formation relevant to optical microscopy.
3. An implication of the use of the paraxial approximation in the application of the Huygens-Fresnel principle is that the results obtained cannot be assumed to be applicable to high numerical aperture objectives, which can collect light over a large range of angles. While we do use high numerical aperture values in illustrations involving the Airy profile and the 3D point spread function derived using this approach, it is important to note that more advanced models of image formation are available that do not use the paraxial approximation (see, e.g., [44]) and therefore can be expected to be more appropriate for analyses involving high numerical aperture objectives.
4. The Huygens-Fresnel integral of Eq. (14.2) is often derived (e.g., [44]) from the Rayleigh-Sommerfeld diffraction integral, which leads to an analogous expression but with a multiplicative cosine term in the integrand that allows the integral to be used in situations that are not paraxial. In paraxial situations this cosine term is dropped as it is close to 1. Since our derivation is confined to the paraxial situation, we directly introduced Eq. (14.2).
5. A 3D point spread function that models the propagation of light through an object space comprising media of different refractive indices is described in [40].
6. For more on the classical 3D resolution limit, see, for example, [16, 79, 55].

Exercises

Chapter 10

1. Assume that a light source, such as a laser, supplies 100000 photons per second in a specific wavelength band. What optical density does an optical filter need to have to reduce the photon flux rate in this wavelength band to 20 photons per second?
2. Consider the emission light path of a fluorescence microscope. Assume 500 photons fall in a wavelength band for which the dichroic beam splitter has a transmission of 90%, the emission filter has a transmission of 95%, and the camera has a quantum efficiency of 60%. How many photons will be detected in the camera?
3. For a confocal microscope, determine the pixel integration time given a certain frame rate and a certain number of pixels per frame.
4. When designing a multi-color experiment, it is important to consider the extent of bleed-through, or crosstalk, by which we mean the amount of signal that is detected from a fluorophore through an incorrect channel. Consider, for example, the dual-color setup of Fig. 10.21, which enables the simultaneous imaging of GFP and Alexa Fluor 546 with two detectors, one for each fluorophore. Given the specific dichroic beam splitters and emission filters indicated in parentheses, fluorescence in the wavelength range of 573 nm to 613 nm is passed through to the detector for Alexa Fluor 546, and fluorescence in the wavelength range of 503 nm to 526 nm is passed through to the detector for GFP. Using graphical means, i.e., by estimating the area under the emission spectrum of GFP, evaluate the percentage of GFP signal that is detected through the Alexa Fluor 546 channel, and compare it with the percentage that is detected through the GFP channel.

Chapter 11

1. Assume that in a ratiometric imaging experiment with fluorescein as described in Section 11.5, fluorescence counts of $F_{490} = 1080$ and $F_{440} = 216$ are detected with excitation wavelengths 490 nm and 440 nm, respectively, resulting in a ratio of $R_{490/440} = F_{490}/F_{440} = 5$. Now assume that these counts are not corrected for background fluorescence, and further assume that the same background level is observed at the two excitation wavelengths. Denoting the background fluorescence count by F_b, calculate the background-corrected ratio $R_{490/440,C} = (F_{490} - F_b)/(F_{440} - F_b)$ for different values of F_b and assess the error in pH measurement based on the curve in Fig. 11.12(b).

Chapter 12

1. Calculate the storage requirement for the following images:
   (a) Grayscale, 8-bit, 1000×1024 pixels.
   (b) Grayscale, 16-bit, 256×256 pixels.
   (c) Grayscale, 16-bit, 4000×4000 pixels.

(d) RGB, 8-bit, 128×256 pixels.
(e) RGB, 16-bit, 1024×1024 pixels.
(f) RGB, double precision, 1024×1024 pixels.

2. A live cell imaging experiment acquires 10 images per second for 10 minutes. The detector is a grayscale CCD detector with 1000×1024 pixels. If the grayscale data is stored in 16-bit format, calculate the minimum storage requirement in megabytes (MB) for the acquired data.
3. Assuming a CCD detector, discuss the advantages of binning as an on-chip operation, in comparison to software binning, i.e., the addition of pixels after the image has been read out into the computer's memory.
4. Simulate the readout noise associated with a pixel in an image detector assuming an offset of 100 electrons and a readout noise standard deviation of 6 electrons, by generating a Gaussian random variable with mean 100 and variance 36. How many data samples do you need in your simulation so that the estimate of the mean is within 10%, 5%, and 1% of the actual mean of 100 electrons? Similarly, how many data samples do you need in your simulation so that the estimate of the variance is within 10%, 5%, and 1% of the actual variance of 36 electrons squared?
5. Consider a CCD detector whose characteristics we want to investigate. Assume that for the kth pixel the rate of photoelectron acquisition is 500 photoelectrons per second from the sample, 30 photoelectrons per second from the background, and 0.1 electrons per second from the dark current. Further assume that the readout noise is characterized by a mean (offset) of 100 DU and a standard deviation of 8 DU. The electron-count-to-DU conversion factor is given by $c = 0.3$ DU/electron.

   (a) Under the above conditions, for the kth pixel simulate 50 repeat experiments for each of the exposure times $\Delta T = 100$ ms, 500 ms, 1 s, 2 s, 5 s, and 10 s. Calculate the mean of the data for each exposure time and plot the result as a function of the exposure time.
   (b) Calculate the variance of the data for each exposure time and plot the result as a function of the exposure time. Compare this plot with the plot from (a) and comment on the similarities and differences.
   (c) Use the results from (a) and (b) to obtain an estimate of the electron-count-to-DU conversion factor.

6. Simulate data, similar to that generated in Exercise 12.5, for the purpose of estimating the properties of the readout noise and the dark current.
7. Calculate/estimate the electron-count-to-DU conversion factors for the following detectors:

   (a) Bit depth of 12, full well capacity of 20000 electrons.
   (b) Bit depth of 14, full well capacity of 157000 electrons.
   (c) Bit depth of 16, full well capacity of 200000 electrons.

8. For the following CCD detectors, determine the size in object space of the full detector and an individual pixel. Assume system magnifications of 5×, 40×, 63×, 100×, and 160×.

   (a) Detector size of 512×512 pixels, pixel size of 16 μm × 16 μm.
   (b) Detector size of 1344×1024 pixels, pixel size of 6.45 μm × 6.45 μm.

(c) Detector size of 1024×1024 pixels, pixel size of 13 μm × 13 μm.

9. Using the CCD detectors and system magnifications in Exercise 12.8, assume that a z-stack of 100 images is acquired with steps of 50 nm. Calculate in image space the height of the full z-stack and the height of a single 50-nm step. (See Section 13.3.2 for a treatment of magnification in the z dimension, and state any assumptions that you make in your calculations.)

10. Assume a cell is "flat" and round with a diameter of 3 μm. How large does a region of interest (i.e., a rectangular array of pixels) need to be in order to cover the cell? Do this calculation for the CCD detectors and magnifications in Exercise 12.8. State any assumptions that you make.

11. Calculate the probability of one electron of dark current being detected in a pixel of a typical CCD camera during an exposure of 50 ms. For the rate at which dark current electrons are generated, look in the specifications of a typical CCD camera.

Chapter 13

1. Find the largest incident angle at the interface between the cover glass and the sample for an air objective with numerical aperture $n_a = 0.9$, if the refractive index of the cover glass is $n_g = 1.52$. How is this angle changed if the air objective has a numerical aperture of $n_a = 0.4$? How is this angle changed if the air objective is replaced by an oil objective with numerical aperture $n_a = 1.45$?

2. Let f_o and f_i be the object and image space focal lengths for a thin lens. Let s_o and s_i be the object and image distances, respectively, associated with a pair of conjugate planes. Show that necessarily $s_o \geq f_o$ and $s_i \geq f_i$.

3. Let f_o and f_i be the object and image space focal lengths for a thin lens. Assume that s_{o1} and s_{o2} are the distances of two objects from the lens, such that object 1 is closer to the lens than object 2, i.e., $s_{o1} < s_{o2}$. If s_{i1} and s_{i2} are the distances of the images to the lens, determine whether $s_{i1} < s_{i2}$, i.e., determine whether the image for object 1 is closer to the lens than the image for object 2. Use the lens formula to support your claim.

4. Consider an object of size y_o. Using analytical or numerical methods, analyze the magnification of the corresponding image in terms of the object location s_o.

5. Assume that an oil objective has numerical aperture $n_a = 1.4$. Assuming that its object focal length is 2 mm. Use the definition of the numerical aperture to determine the diameter of the aperture associated with this objective. Now assume that the tube lens in the same microscopy system has an image focal length of 200 mm. Determine the numerical aperture of the tube lens assuming that it shares the same aperture with the objective.

6. Consider an infinity-corrected microscopy system consisting of an objective and a tube lens with an image focal length of 200 mm. Assuming that the medium associated with the image side of the tube lens is air and the magnification of the microscopy system is 100×, determine how the distance s_o to an in-focus object changes if the immersion medium of the objective is changed from air, to water, to oil. How will these distances change if the magnification of the microscopy system is changed to 40×?

7. Given an object point (x, y, z), calculate the Gaussian image point. Then assume that the object point is shifted to $(x, y, z+\delta_z)$ and determine the resulting shifted image point.

Chapter 14

1. Let $\psi(r,t) = Ce^{i(\langle \overline{k}, r\rangle - \omega t + \varphi_0)}$, $r \in \mathbb{R}^3, t \geq t_0$, be a plane wave with wave vector $\overline{k}$, angular frequency ω, phase shift φ_0, and possibly complex constant C. Writing C in polar form as $C = Ae^{i\theta}$ with A real and positive, show that $\psi(r,t)$ can also be interpreted as a plane wave with wave vector $\overline{k}$, angular frequency ω, and phase shift $\varphi_0 + \theta$.

2. Let ψ be a spherical wave with source r_0. Show that the wavefronts of the wave are given by spheres with center r_0.

3. Using a computational package, simulate the superposition (Section 14.1.3) of
   (a) two monochromatic spherical waves with different sources, and
   (b) two monochromatic plane waves with different wave vectors.

4. Consider the expression

$$U_i(x_i, y_i) = \int_{-\infty}^{\infty}\int_{-\infty}^{\infty} o(x_o, y_o)\, h\left(x_o + \frac{x_i}{M}, y_o + \frac{y_i}{M}\right) dx_o, dy_o, \quad (x_i, y_i) \in \mathbb{R}^2.$$

   Perform the substitution $(x_o', y_o') = (-x_o, -y_o)$ and show that with $o'(x_o', y_o') = o(-x_o', -y_o')$,

$$U_i(x_i, y_i) = \int_{-\infty}^{\infty}\int_{-\infty}^{\infty} o'(x_o', y_o')\, h\left(\frac{x_i}{M} - x_o', \frac{y_i}{M} - y_o'\right) dx_o', dy_o', \quad (x_i, y_i) \in \mathbb{R}^2.$$

5. Determine the Fresnel diffraction integral for a point object, located in an aperture $A \subseteq \mathbb{R}^2$ at $r_0 = (r_x, r_y) \in A$, given by $U(x_1, y_1, z_1) = \delta_{r_x}(x_1)\,\delta_{r_y}(y_1)$, $(x_1, y_1) \in \mathbb{R}^2$.

6. Assume that aperture A axially located at z_1 is a full plane, i.e., $A = \mathbb{R}^2$, meaning that in fact there is no spatial restriction to the passage of light. Assume that a wave U is given by the constant $U(x_1, y_1) = U_0$ for all $(x_1, y_1) \in \mathbb{R}^2$. Determine the Fresnel diffraction integral if the initial wave in A is given by U. Compare, using a computational software package, the resulting expression to a spherical wave originating at $r_0 = (0, 0, z_1)$.

7. Assume objectives with numerical apertures $n_a = 1.4$, $n_a = 0.8$, and $n_a = 0.4$. Assume the immersion medium has a refractive index of $n_o = 1.515$ for the case of $n_a = 1.4$ and $n_o = 1$ for the cases of $n_a = 0.8$ and $n_a = 0.4$. Plot the main 1D cross section of the intensity point spread function of Eq. (14.13) with and without the factor $\frac{1}{\cos(\theta_o)}$ in the argument of the Bessel function.

Part IV

Data Analysis

Overview

In the prior parts, we discussed the scientific background to microscopy experiments and introduced several important types of experiments. In this last part, we will address the fundamental problem of the analysis of the image data that is obtained from microscopy experiments. Such an analysis can take many different forms, ranging from a simple presentation of the acquired data to a sophisticated algorithmic investigation to obtain detailed and precise quantitative measurements. In the spirit of this textbook, we are not concerned with the more elementary processing of image data, but are instead concerned with a presentation of a careful analytical framework within which advanced quantitative image analysis approaches can be carried out. In particular, we will present the necessary probabilistic and statistical background for the treatment of advanced parameter estimation problems, such as the estimation of the location of an imaged object. We will close this part with introductions to two additional topics in microscopy image analysis. The first is deconvolution, a commonly employed computational approach for removing the blurring of images of a 3D sample due to out-of-focus haze. The second is spatial statistics, a field of study providing tools that can be applied for the investigation of the spatial distribution of, for example, molecules of interest in a cell.

15

From Photons to Image: Data Models

We provide in this chapter a rigorous introduction to the different statistical data models that will form the basis for the treatment of the estimation problems that will be analyzed later. In the earlier parts of the book, we have learned that light naturally leads to a stochastic description of the acquired data in terms of Poisson random variables. The readout process and other noise sources inherent in the various types of detectors further complicate this stochastic description. In order to gain a precise understanding of how these various stochastic phenomena influence the quality of the quantitative analysis of the acquired data, we first need to carefully describe these phenomena in rigorous mathematical language. More specifically, the task of this chapter is to present and illustrate models that provide a stochastic description of the acquired image data in terms of the photons that are emitted by the imaged object and captured by the detector, and in terms of the noise that is potentially introduced.

15.1 Accounting for each photon: fundamental data model

An illustration of image acquisition by a microscope is given in Fig. 15.1. Quite simply, the schematic shows that light from an object is collected and focused by the microscope optics (i.e., the objective and tube lens) onto a detector. A data model provides a description of the object's "image", or more specifically the object's emitted photons, as captured by the detector, and also depends on the characteristics of the detector.

We begin by introducing the *fundamental data model*, which assumes that an idealized detector is used which is unpixelated and records, with arbitrarily high precision, the locations at which photons from the imaged object are detected. In addition, this idealized detector records, also with arbitrarily high precision, the time points at which the photons are detected. The fundamental data model further assumes that the idealized detector, which can be infinite in size, does not introduce noise such as detector readout noise to the captured image.

In practice, an actual image detector is pixelated (Section 12.3) and can only record the location of a detected photon with a precision that is limited by the dimensions of a pixel. In other words, with an actual pixelated detector, the precision is lower because the extent to which a photon's location is known is reduced to the pixel in which it is detected. A practical detector also does not record the precise times at which the photons are detected. Instead, each photon is only known to be detected at some point during the time interval over which the image was acquired. Furthermore, as we have seen in Chapter 12, a practical detector deteriorates the quality of the captured image by introducing noise such as that due to its readout process.

By assuming an idealized detector, the fundamental data model thus represents a useful basis for comparison with data models entailing practical detectors. In particular, as we will see in Chapters 18, 19, and 20 in relation to the extraction of quantities of interest from

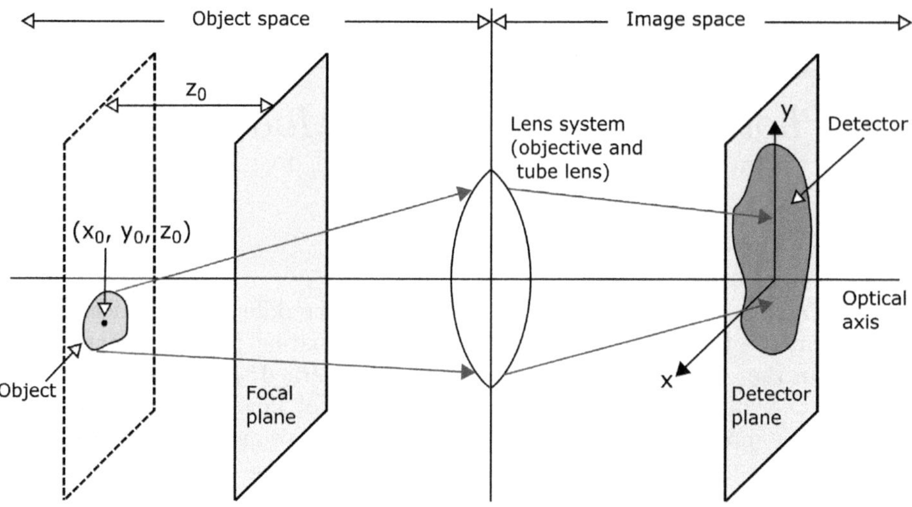

FIGURE 15.1
Image acquisition by a microscope. Light emitted by an object is collected and focused by the microscope lens system onto a detector. The imaged object resides in 3D space, and is located at (x_0, y_0, z_0), where z_0 specifies the axial displacement of the object from the focal plane. The detector is located in the detector plane.

image data, performance measures obtained under the fundamental data model provide useful benchmarks for the assessment of analogous measures calculated for images acquired with practical detectors. It is for its usefulness as a benchmark scenario that we refer to this data model as "fundamental".

Under the fundamental data model, an "image", i.e., the data, formed from N_0 photons comprises the time points $\{\tau_1, \ldots, \tau_{N_0}\}$ and the locations $\{(x_1, y_1), \ldots, (x_{N_0}, y_{N_0})\}$ at which the N_0 photons are detected (Fig. 15.2). The mathematical description of the fundamental data model thus deals with both the temporal and the spatial aspects of photon detection.

In time and space, the detection of photons is inherently a stochastic phenomenon. Photons are emitted by the imaged object at random points in time, and are therefore detected at random points in time. Similarly, the location on the detector at which a photon is detected is random in nature. A mathematical representation of the process of photon detection therefore requires a stochastic description. For the temporal component of the description, the time points at which photons are detected by the detector are described by a *Poisson process*. This follows directly from the typical assumption that photon emission occurs according to a Poisson process. For the spatial component of the description, the location on the detector at which a photon is detected is specified probabilistically with a 2D probability density function. The temporal and spatial components of a photon detection process are assumed to be independent of each other, and are described in more detail below (Note 15.1).

15.1.1 Temporal component of photon detection — Poisson process

A Poisson process (Note 15.2) is a counting process, containing a sequence of random variables that count the number of events that occur over continuous time. It is typically denoted by $\{\mathcal{N}(\tau); \tau \geq t_0\}$, where $\mathcal{N}(\tau)$ is the number of events that have occurred by time τ since the start time t_0, and is Poisson distributed. At $\tau = t_0$, the number of occurrences

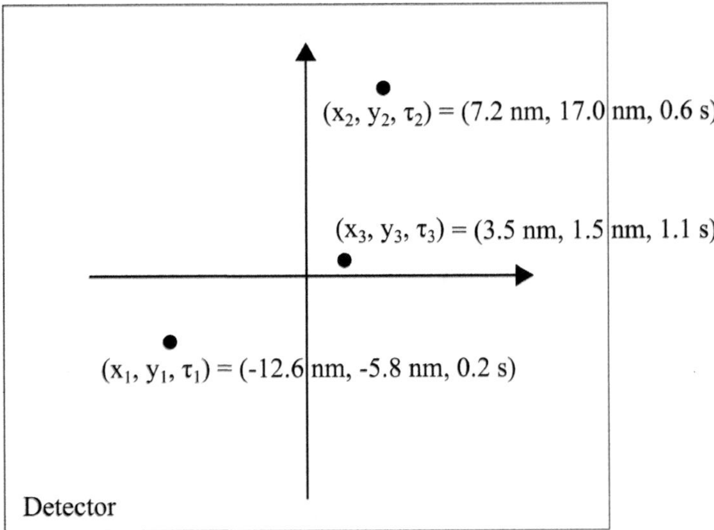

FIGURE 15.2
Data acquired by an idealized detector under the fundamental data model. In this data model, an "image" comprises the positional coordinates and the time points at which the detected photons impact the detector. The example shown depicts the locations and time points at which three photons are detected.

is $\mathcal{N}(t_0) = 0$. In a Poisson process, no two events occur simultaneously. In our microscopy application, an event is the detection of a photon at the detector, and the count is therefore the total number of photons that have been detected at the detector up to a particular point in time.

By definition, a Poisson process satisfies the property of independent increments, meaning that the numbers of photons collected during two disjoint time intervals are stochastically independent of each other. For example, if $t_1 < t_2 < t_3 < t_4$, then the number of photons $\mathcal{N}(t_2) - \mathcal{N}(t_1)$ collected during the time interval $[t_1, t_2]$ and the number of photons $\mathcal{N}(t_4) - \mathcal{N}(t_3)$ collected during the time interval $[t_3, t_4]$ are stochastically independent.

An *inhomogeneous Poisson process* $\{\mathcal{N}(\tau); \tau \geq t_0\}$ is characterized by the rate function $\Lambda(\tau)$, $\tau \geq t_0$, also known as its *intensity function*, such that the number of photons $\mathcal{N}(t) - \mathcal{N}(t_1)$ collected during the time interval $[t_1, t]$, $t > t_1 \geq t_0$, follows the Poisson probability distribution

$$P(\mathcal{N}(t) - \mathcal{N}(t_1) = k) = \frac{e^{-\lambda_{t_1,t}} \lambda_{t_1,t}^k}{k!}, \quad k = 0, 1, \ldots, \tag{15.1}$$

where $\lambda_{t_1,t}$ is the integral of the rate function over the time interval, given by

$$\lambda_{t_1,t} = \int_{t_1}^{t} \Lambda(\tau) d\tau. \tag{15.2}$$

The rate function $\Lambda(\tau)$, $\tau \geq t_0$, allows the modeling of the *photon detection rate*, which can in general vary over time. In fluorescence microscopy, photobleaching is a phenomenon that occurs when the photon emission from the object of interest decreases over time, leading to a corresponding decrease in the number of photons that are detected by the detector (Section 8.4). Imaging done at the beginning of a microscopy experiment may therefore yield higher intensity images of the object than imaging done at a later time. In such

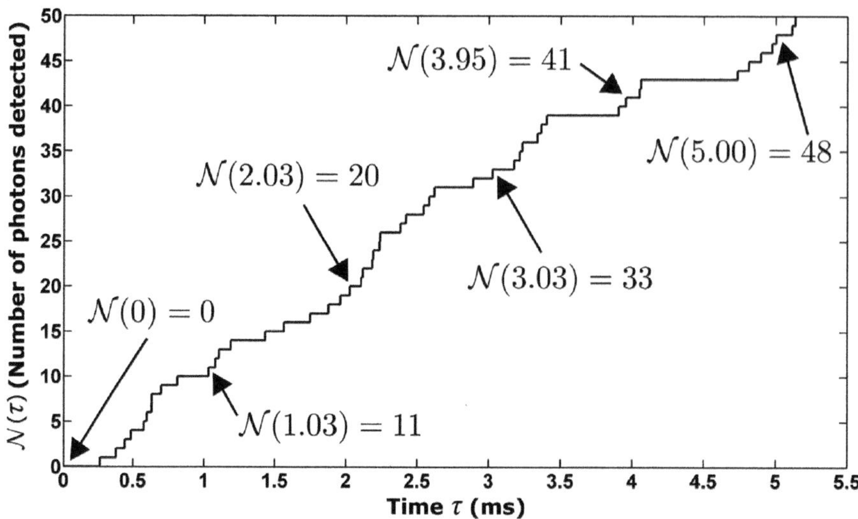

FIGURE 15.3
Photon detection according to a homogeneous Poisson process $\{\mathcal{N}(\tau); \tau \geq 0\}$. In this example, the photon detection rate of the process is $\Lambda_0 = 10000$ photons/s, which means that on average 10 photons are detected every millisecond. In the realization shown, the arrows indicate detections that occur at time points nearest an integer millisecond.

situations, the rate function can be defined to be a decreasing function, such as one that describes an exponential decay.

In the case where the object of interest is relatively photostable, such that its photons are emitted, and therefore detected, at the same intensity level over the duration of the image acquisition, the rate function $\Lambda(\tau)$ can be defined to be a constant, i.e., $\Lambda(\tau) = \Lambda_0$, $\tau \geq t_0$. Given the constant rate Λ_0, the integral of Eq. (15.2) evaluates to $\lambda_{t_1,t} = \Lambda_0(t - t_1)$, and accordingly, Eq. (15.1) becomes

$$P\left(\mathcal{N}(t) - \mathcal{N}(t_1) = k\right) = e^{-\Lambda_0(t-t_1)} \frac{(\Lambda_0(t - t_1))^k}{k!}, \quad k = 0, 1, \ldots. \tag{15.3}$$

In this special case, the inhomogeneous Poisson process reduces to a *homogeneous Poisson process*. A homogeneous Poisson process is characterized by a constant rate Λ_0 and, as a result, satisfies the property of stationary increments. This property refers to the fact that the probability distribution of the number of photons collected during a given time interval depends on the length of that time interval, and not the interval itself. This can be seen in the Poisson parameter $\Lambda_0(t - t_1)$ in Eq. (15.3) being a function of the length $t - t_1$ of the time interval, and not of the time interval $[t_1, t]$. This is in contrast to the Poisson parameter for an inhomogeneous Poisson process, which Eq. (15.2) shows to be an integral that depends on the time interval $[t_1, t]$. Figure 15.3 shows a realization of a homogeneous Poisson process that is characterized by a photon detection rate of $\Lambda_0 = 10000$ photons/s, or 10 photons/ms. The plot shows the accumulation of photons detected over an interval of approximately 5 ms, and it can be seen that about 10 photons are indeed detected every millisecond.

15.1.1.1 Mean number of detected photons

One important way to characterize an image is by the mean, or expected, number of photons $N_{photons}$ that form the image, assuming the acquisition of the image can be repeated multiple times under the same conditions. To determine the expression for $N_{photons}$, we start by looking at the mean number of photons that are detected during the acquisition time interval.

Consider the acquisition time interval $[t_0, t]$. From the above presentation, we know that during this interval photons are detected according to the Poisson process $\{\mathcal{N}(\tau); \tau \geq t_0\}$, and that the number of photons detected during the interval is modeled by the Poisson random variable $\mathcal{N}(t)$. The mean number of photons detected during $[t_0, t]$ is therefore the expectation of $\mathcal{N}(t)$. We see from Eqs. (15.1) and (15.2) that this expectation is given by $\mathrm{E}[\mathcal{N}(t)] = \int_{t_0}^{t} \Lambda(\tau)d\tau$, where Λ is the rate function of the Poisson process and the photon detection rate in our context.

The expectation $\mathrm{E}[\mathcal{N}(t)]$ is the mean number of photons detected *during the acquisition time interval*, but not the mean number of photons $N_{photons}$ that *form an image* acquired during the interval. The distinction between $\mathrm{E}[\mathcal{N}(t)]$ and $N_{photons}$ is subtle. We should, however, point out that this difference is, as we will see, only important in the extreme low photon count regime. For most imaging experiments that are practically significant, the distinction is not relevant. The expectation $\mathrm{E}[\mathcal{N}(t)]$ is simply the mean of the Poisson random variable $\mathcal{N}(t)$, and therefore takes into account the scenario of $\mathcal{N}(t) = 0$, in which no photons are detected during the acquisition time interval and hence no image is formed. The quantity $N_{photons}$, on the other hand, is subject to the condition that an image as defined by our data model is formed from $N_0 > 0$ photons, as no photon detection locations and time points can be recorded unless at least one photon is detected during the acquisition time interval. In determining its expression, we must therefore take care to exclude the $\mathcal{N}(t) = 0$ scenario (Note 15.3).

In accordance with our description, we determine $N_{photons}$ as the mean of the Poisson random variable $\mathcal{N}(t)$, provided that $\mathcal{N}(t) > 0$. It is effectively the mean of what is often referred to as a *zero-truncated Poisson distribution*. This is the distribution we obtain from the Poisson distribution by taking into account only the probabilities with which positive counts occur. The zero-truncated Poisson distribution is described by the probability mass function

$$P(X = k \mid X > 0) = \frac{e^{-\lambda}\lambda^k}{k!\left(1 - e^{-\lambda}\right)}, \quad k = 1, 2, \ldots, \tag{15.4}$$

where X is a Poisson random variable with mean λ, and its mean is given by

$$\mathrm{E}[X \mid X > 0] = \frac{\lambda}{1 - e^{-\lambda}}. \tag{15.5}$$

The mass function and the mean of the zero-truncated Poisson distribution are easily obtained by applying the definitions of conditional probability and expectation. Their derivations are left as an exercise for the reader (Exercise 15.1).

From Eq. (15.4) we see that the zero-truncated Poisson probability mass function is simply the probability mass function of the Poisson random variable X divided by the factor $1 - e^{-\lambda}$. Similarly, Eq. (15.5) shows that the mean of the zero-truncated Poisson random variable is just the mean of X divided by the same factor. Since $1 - e^{-\lambda}$ quickly approaches one with increasing values of λ, we can expect the zero-truncated Poisson random variable to be similar to the Poisson random variable X when λ is not very small. We will explore this further after rewriting Eqs. (15.4) and (15.5) using the notation for our current problem.

By letting $X = \mathcal{N}(t)$ and $\lambda = \mathrm{E}[\mathcal{N}(t)]$, and by replacing k with N_0, we have, in the context of photon detection, the zero-truncated Poisson probability mass function

$$P(\mathcal{N}(t) = N_0 \mid \mathcal{N}(t) > 0) = \frac{e^{-\mathrm{E}[\mathcal{N}(t)]}\left(\mathrm{E}[\mathcal{N}(t)]\right)^{N_0}}{N_0!\left(1 - e^{-\mathrm{E}[\mathcal{N}(t)]}\right)} = \frac{e^{-\int_{t_0}^{t}\Lambda(\tau)d\tau}\left(\int_{t_0}^{t}\Lambda(\tau)d\tau\right)^{N_0}}{N_0!\left(1 - e^{-\int_{t_0}^{t}\Lambda(\tau)d\tau}\right)}, \tag{15.6}$$

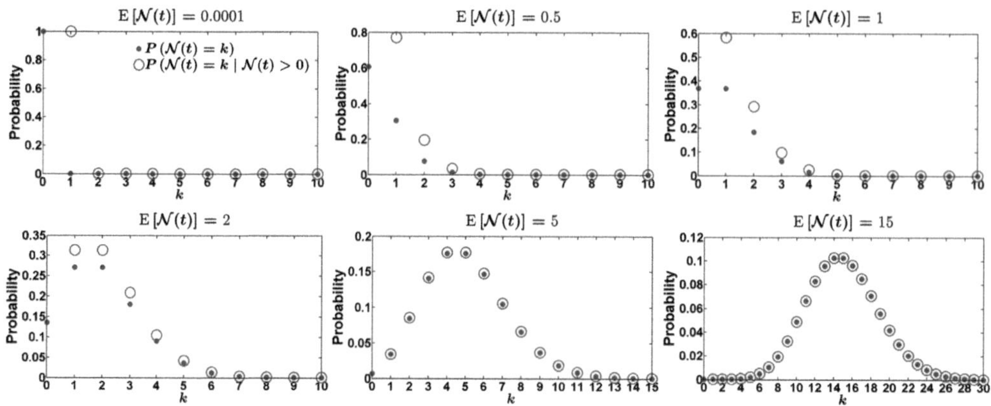

FIGURE 15.4
Comparing the Poisson and zero-truncated Poisson probability distributions. In each plot, the probability mass function for the Poisson random variable $\mathcal{N}(t)$ is shown together with the corresponding zero-truncated Poisson probability mass function. The two mass functions can be seen to become increasingly similar with increasing values of the Poisson parameter $\mathrm{E}[\mathcal{N}(t)]$.

$N_0 = 1, 2, \ldots$, with mean $N_{photons}$ given by

$$N_{photons} := \mathrm{E}[\mathcal{N}(t) \mid \mathcal{N}(t) > 0] = \frac{\mathrm{E}[\mathcal{N}(t)]}{1 - e^{-\mathrm{E}[\mathcal{N}(t)]}} = \frac{\int_{t_0}^{t} \Lambda(\tau)d\tau}{1 - e^{-\int_{t_0}^{t} \Lambda(\tau)d\tau}}. \tag{15.7}$$

Demonstrating that we can expect the zero-truncated Poisson random variable to be similar to the Poisson random variable $\mathcal{N}(t)$ when the mean number of photons $\mathrm{E}[\mathcal{N}(t)]$ detected during the acquisition interval $[t_0, t]$ is not very small, Fig. 15.4 shows that, at a value of $\mathrm{E}[\mathcal{N}(t)]$ as low as 5 photons, the zero-truncated Poisson probability mass function already takes on values very close to those taken on by the probability mass function for $\mathcal{N}(t)$. Not unexpectedly, Fig. 15.5 shows that from $\mathrm{E}[\mathcal{N}(t)] = 5$ and onwards, the mean $N_{photons}$ of the zero-truncated Poisson distribution is practically equal to $\mathrm{E}[\mathcal{N}(t)]$. Therefore, for example, for regimes beyond very small numbers of expected photons, we can assume that $N_{photons} = \mathrm{E}[\mathcal{N}(t) \mid \mathcal{N}(t) > 0] \approx \mathrm{E}[\mathcal{N}(t)] = \Lambda_0(t - t_0)$ for the case of a constant photon detection rate Λ_0. As in most practical situations we deal with mean photon counts well above a few photons, it becomes clear that the conditioning on $\mathcal{N}(t) > 0$, while important to guarantee that results are theoretically correctly derived, does not have practical significance in most experimentally relevant cases.

15.1.2 Spatial component of photon detection — spatial density function

Having addressed how to stochastically model the temporal aspect of a photon detection process by describing the photon arrival times at the detector as a temporal Poisson process, we now come to consider the spatial aspect of the photon detection process. In the same way that the arrival time of a photon is random, so is its arrival location on a detector. The spatial part of an acquired fundamental data model "image" is therefore composed of the random locations at which photons impact the detector. Under the fundamental data model, we think of a detector C to be represented by a subset of a 2D plane, i.e., $C \subseteq \mathbb{R}^2$. Accordingly, the distribution of the arrival location of a photon is described by a 2D probability density

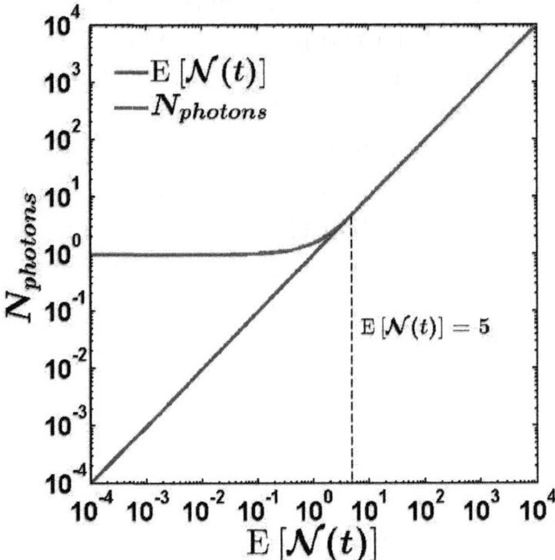

FIGURE 15.5
Expectations of the Poisson and zero-truncated Poisson probability distributions. The mean $\mathrm{E}[\mathcal{N}(t)]$ of the Poisson random variable $\mathcal{N}(t)$ is shown together with the mean $N_{photons} := \mathrm{E}[\mathcal{N}(t) \mid \mathcal{N}(t) > 0]$ of the corresponding zero-truncated Poisson random variable for values of $\mathrm{E}[\mathcal{N}(t)]$ ranging from 10^{-4} to 10^4.

function. Specifically, at time $\tau \geq t_0$, the probability density function for the location (x, y) at which a photon impacts the detector C is given by $f_\tau(x, y)$, $(x, y) \in C$, a 2D function that is also referred to as a *photon distribution profile*. As it is a probability density defined over the detector C, the photon distribution profile f_τ at time $\tau \geq t_0$ is a function such that

1. $f_\tau(x, y) \geq 0, \quad (x, y) \in C, \quad$ and
2. $\int_C f_\tau(x, y) dx dy = 1$.

The distribution f_τ specifies the probability with which a photon will impact an arbitrary subset A of the detector C at time $\tau \geq t_0$, i.e.,

$$P(U_\tau \in A) = \int_A f_\tau(x, y) dx dy,$$

where U_τ is the random variable that denotes the impact location of the photon on the detector C at time $\tau \geq t_0$ (Note 15.4). In our modeling of a photon detection process, it is assumed that the impact locations of different photons are stochastically mutually independent.

Note that in Section 14.2 we stated that the intensity image at time t on the detector is given by $|\psi_{n,t}(x, y)|^2$, $(x, y) \in C$, where $\psi_{n,t}$ is the normalized wave function at time t. If we now set $f_\tau(x, y) = |\psi_{n,\tau}(x, y)|^2$, $(x, y) \in C$, we have interpreted the image formation process in the current framework of a fundamental data model.

15.1.2.1 Translational invariance and image function

In Section 14.6 we discussed translation-invariant image formation. This means that if an object is shifted by a certain amount in the object plane, its image in the detector plane is simply a shifted version of the original image. If the imaging system is translation-invariant, we can somewhat simplify the description of the photon distribution profile by introducing the notion of an *image function*. With the image function we can model that the image of an object does not change if the object position is shifted orthogonal to the optical axis. Consider a detector having an infinitely large detection area consisting of the entire detector plane, so that $C = \mathbb{R}^2$. In this special, but important, case where every photon emitted by the imaged object reaches the detector, an image function, by definition, describes the image of the object, at unit magnification, when the object is located on the optical axis of the microscope. An image function q further satisfies the following properties:

1. $q(x, y) \geq 0, \quad (x, y) \in \mathbb{R}^2$, and
2. $\int_{\mathbb{R}^2} q(x, y) dx dy = 1$.

The photon distribution profile at time $\tau \geq t_0$ is then given in terms of the image function by

$$f_\tau(x, y) = \frac{1}{M^2} q\left(\frac{x}{M} - x_{0,\tau}, \frac{y}{M} - y_{0,\tau}\right), \quad (x, y) \in \mathbb{R}^2, \tag{15.8}$$

where $M > 0$ is the system magnification and the coordinates $(x_{0,\tau}, y_{0,\tau})$ denote the location of the object at time τ. The photon distribution profile is thus simply the image function scaled by the magnification and shifted by the object's positional coordinates. Note that if the imaged object is stationary, such that its positional coordinates are given by (x_0, y_0), then the photon distribution profile is no longer time-dependent and τ can be dropped from Eq. (15.8). We should also mention that, in practice, a moving object is often assumed to be stationary during an image acquisition time interval $[t_0, t]$ that is considered to be sufficiently short. In a tracking application, for example, the trajectory of an object of interest is often constructed by estimating its location from each image of a time sequence under the assumption that it is stationary in each image.

We next present the image functions corresponding to three different image profiles, among them are the Airy profile and the 3D point spread function which we have encountered in previous chapters.

Airy image function. The Airy profile gives the image of an in-focus point source according to the standard diffraction model (Section 14.5.4.1). We define the *Airy image function* as

$$q(x, y) = \frac{J_1^2(\alpha\sqrt{x^2 + y^2})}{\pi(x^2 + y^2)}, \quad (x, y) \in \mathbb{R}^2, \tag{15.9}$$

with J_1 the first order Bessel function of the first kind, and $\alpha = \frac{2\pi n_a}{\lambda}$, where n_a is the numerical aperture of the objective and λ is the wavelength of the photons detected from the point source. A quick comparison with Eq. (4.1) in Section 4.1 shows that the Airy image function is just the Airy intensity point spread function with the magnification M set to 1. In Fig. 15.6(a), an Airy image function computed with $n_a = 1.4$ and $\lambda = 520$ nm is shown. Note the rings of the Airy function, which are due to the zeros of the Bessel function J_1 (see Section 4.2 for a discussion).

2D Gaussian image function. In cellular microscopy, 2D Gaussian functions are often used to model the images of different types of objects, such as organelles and vesicles. In addition, a 2D Gaussian function is sometimes used as a simple approximation of the Airy profile in the modeling of the image of an in-focus molecule. We therefore also consider the 2D Gaussian function as a point source image model that we will analyze. For $\sigma_g > 0$, the function $q : \mathbb{R}^2 \to \mathbb{R}$ given by

$$q(x, y) = \frac{1}{2\pi\sigma_g^2} e^{-\frac{1}{2\sigma_g^2}(x^2+y^2)}, \quad (x, y) \in \mathbb{R}^2, \tag{15.10}$$

is defined as the *2D Gaussian image function*. A 2D Gaussian image function with $\sigma_g =$

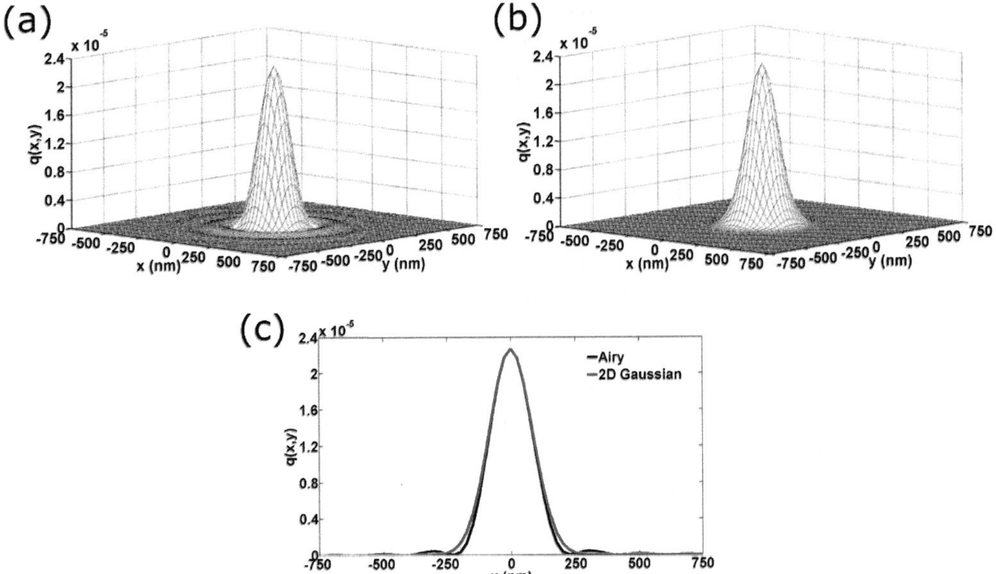

FIGURE 15.6
Airy and 2D Gaussian image functions. (a) Airy image function $q(x, y)$ for a point source emitting photons of wavelength $\lambda = 520$ nm which are collected with an objective lens of numerical aperture $n_a = 1.4$. (b) 2D Gaussian image function $q(x, y)$ with a width of $\sigma_g = 83.61$ nm. The particular width is chosen to yield an image that matches the Airy image of (a) in terms of height. In both (a) and (b), the image function is defined over the detector plane $\mathbb{R}^2$, but the plot shows its values over the 1500 nm $\times$ 1500 nm region over which it is centered. (c) Comparison of the center cross sections ($y = 0$ nm) of the image functions of (a) and (b).

83.61 nm is shown in Fig. 15.6(b). A Gaussian profile is not as steep as an Airy profile. The Gaussian image function of Fig. 15.6(b), for example, has a height that matches that of the Airy image function of Fig. 15.6(a), and can be seen, upon close inspection, to be a little broader and therefore not as steep as the Airy image function. By comparing the center cross sections of the two image functions, Fig. 15.6(c) provides a clearer visualization of this difference. Nevertheless, the use of Gaussian profiles as approximations of Airy profiles is often justified by their relative computational simplicity.

Image function for an out-of-focus point source. When a point source is not located in the focal plane of the microscope system, a 3D point spread function is used to describe its image. A 3D point spread function is dependent on the point source's axial (i.e., z) location, which specifies to what extent the point source is out of focus. An example of such a point spread function was derived in Section 14.5.5, and here we will present its corresponding image function. We first note that the intensity point spread function corresponding to our 3D point spread function is given by the square of the modulus of Eq. (14.15). Then, assuming the detector is located in the design image plane, such that $z_i = 0$, this intensity point spread function yields the image function $q_{z_0} : \mathbb{R}^2 \to \mathbb{R}$, given by

$$q_{z_0}(x, y) = \frac{1}{C_{z_0}} \left| \int_0^1 J_0\left(\frac{2\pi n_a}{\lambda}\sqrt{x^2 + y^2}r\right) e^{\frac{i\pi n_a^2 z_0}{\lambda n_o} r^2} r dr \right|^2, \quad (x, y) \in \mathbb{R}^2, \quad z_0 \in \mathbb{R}, \quad (15.11)$$

where J_0 is the zeroth order Bessel function of the first kind, n_a and λ are as defined for the Airy image function, n_o is the refractive index of the immersion medium for the objective, z_0 is the axial coordinate of the point source in object space, and C_{z_0} is a normalization term given by

$$C_{z_0} = \int_{\mathbb{R}^2} \left| \int_0^1 J_0\left(\frac{2\pi n_a}{\lambda}\sqrt{x^2+y^2}r\right) e^{\frac{i\pi n_a^2 z_0}{\lambda n_o}r^2} rdr \right|^2 dxdy,$$

which ensures that the image function integrates to 1 over $\mathbb{R}^2$. Note that z_0 has been added as a subscript to the symbol for the function to emphasize the function's dependence on the axial position of the point source. Note also that notationally we have replaced z_o and k_{vac} in Eq. (14.15) with z_0 and $2\pi/\lambda$, respectively. Further, we have replaced the multiplicative constant in Eq. (14.15) with the inverse of the normalization term C_{z_0}.

While we have motivated the image function q_{z_0} as modeling the situation where the detector is in focus and the object is located at the position z_0, the identical image function could arise if we were to assume that the point source is in focus (i.e., located at $z_0 = 0$) while the detector position z_i is changed.

Examples of the image function of Eq. (15.11) are shown in Fig. 15.7. The examples have been computed with $n_a = 1.4$, $\lambda = 520$ nm, and $n_o = 1.515$ for three different values of z_0. Note in Fig. 15.7(a) the similarity of this image function to the Airy image function when the point source is only $z_0 = 200$ nm away from the focal plane, i.e., when it is close to being in focus. As illustrated by Figs. 15.7(b) and 15.7(c), however, the image can change significantly when the point source is located further away from the focal plane. A comparison of the center cross sections of the three images is shown in Fig. 15.7(d).

15.1.3 Background component

Even though the fundamental data model assumes a detector that does not introduce noise to the acquired image, it does allow for noise sources that do not originate from the detector and are often collectively referred to as a background component. Examples of such noise sources are sample autofluorescence and scattered photons. The detection of photons from these noise sources is independent of the detection of photons from the object of interest, but can be described in the same manner. The temporal component can be modeled as a Poisson process with photon detection rate $\Lambda^b(\tau)$, $\tau \geq t_0$, while the spatial component can be described with a photon distribution profile $f_\tau^b(x,y)$, $(x,y) \in C$, $\tau \geq t_0$. Formally, the two independent photon detection processes can be considered as one combined process whose photon detection rate and distribution profile are simple functions of the photon detection rates and distribution profiles of the two component processes. Using $\Lambda(\tau)$, $\tau \geq t_0$, and $f_\tau(x,y)$, $(x,y) \in C$, $\tau \geq t_0$, to denote the rate and distribution profile that describe the detection of photons from the object of interest, the photon detection rate Λ^c of the combined process is given by

$$\Lambda^c(\tau) = \Lambda(\tau) + \Lambda^b(\tau), \quad \tau \geq t_0, \tag{15.12}$$

and the photon distribution profile f_τ^c of the combined process is given by (Note 15.5)

$$f_\tau^c(x,y) = \frac{\Lambda(\tau)}{\Lambda^c(\tau)} f_\tau(x,y) + \frac{\Lambda^b(\tau)}{\Lambda^c(\tau)} f_\tau^b(x,y), \quad (x,y) \in C, \quad \tau \geq t_0. \tag{15.13}$$

We leave it to the reader to show that Eq. (15.13) is in fact a photon distribution profile (Exercise 15.8).

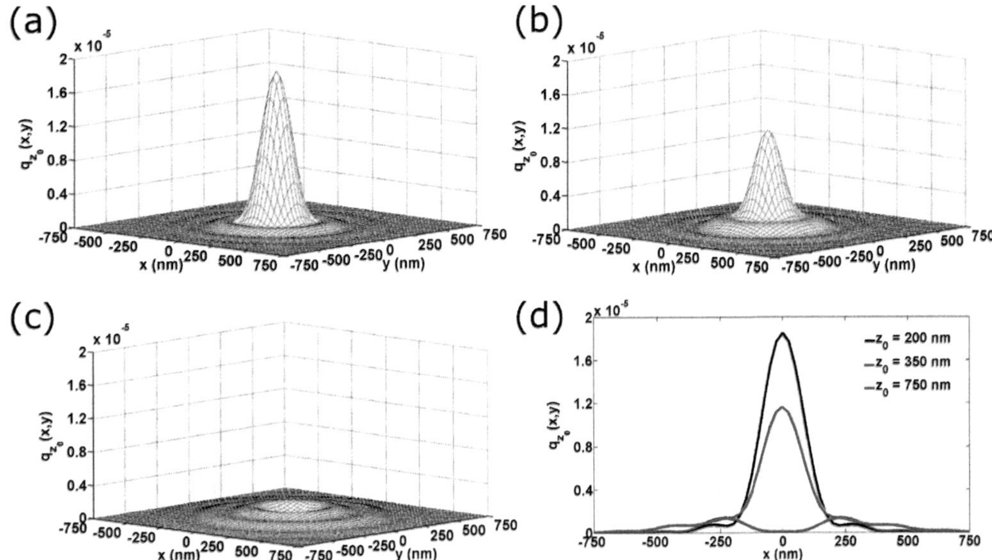

FIGURE 15.7
Image function for an out-of-focus point source. The image function q_{z_0} of Eq. (15.11) for a stationary point source axially positioned at (a) $z_0 = 200$ nm, (b) $z_0 = 350$ nm, and (c) $z_0 = 750$ nm. It is assumed that the point source emits photons of wavelength $\lambda = 520$ nm, and that an objective is used which has a numerical aperture of $n_a = 1.4$. It is also assumed that the objective immersion medium has a refractive index of $n_o = 1.515$. The image function is defined over the detector plane $\mathbb{R}^2$, but the plot shows its values over the 1500 nm $\times$ 1500 nm region over which it is centered. (d) Comparison of the center cross sections ($y = 0$ nm) of the images of (a), (b), and (c).

15.1.4 Examples

Assuming the imaged object to be a stationary point source that emits photons at a constant rate, two examples of data acquired under the fundamental data model, assuming an infinitely large detector $C = \mathbb{R}^2$, are given in Fig. 15.8. The two examples differ by the photon distribution profile, but in both cases the locations of photon detection are represented by dots, and color coding is used to denote the time sub-interval during which a photon is detected, in lieu of the difficult task of graphically specifying each photon's unique detection time point (which is technically required, in addition to each photon's detection location, for depicting an "image" under the fundamental data model). Note that in Fig. 15.8(a), the characteristic diffraction rings of an Airy profile can be seen, as the photon distribution profile is specified with the Airy image function. In Fig. 15.8(b) where the photon distribution profile is specified with a 2D Gaussian image function, no such rings are present.

In these two examples, we have not included a background component. Inclusion of a background component, however, would simply entail the analogous but independent simulation of photon detection using the background photon detection rate and the background photon distribution profile, and the inclusion of the resulting simulated background photons as additional color-coded dots.

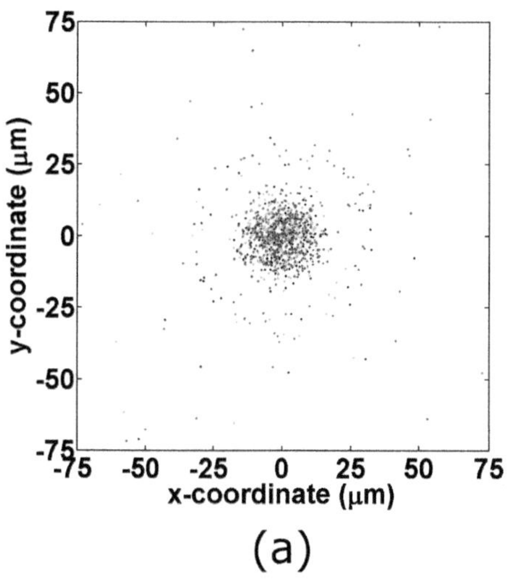

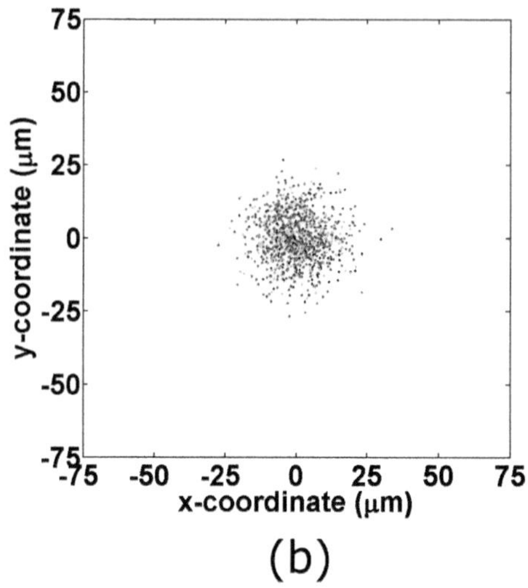

FIGURE 15.8
Illustration of a fundamental data model "image" of a stationary and in-focus point source located at $(x_0, y_0) = (0 \text{ nm}, 0 \text{ nm})$ in object space, simulated using the photon distribution profile of Eq. (15.8) with a magnification of $M = 100$ and with (a) the Airy image function of Fig. 15.6(a), and (b) the 2D Gaussian image function of Fig. 15.6(b). In both (a) and (b), each dot represents the location at which a photon is detected, and the data is assumed to be acquired over the exposure interval $[0 \text{ s}, 0.1 \text{ s}]$, during which photons from the point source are detected at a constant rate of $\Lambda_0 = 10000$ photons/s. While the expected number of detected photons is $N_{photons} = \Lambda_0(0.1 - 0)/\left(1 - e^{-\Lambda_0(0.1-0)}\right) = 1000$, the particular simulated image of (a) contains 1060 photons, 61 of which are detected outside of the 150 μm $\times$ 150 μm region shown, and the particular simulated image of (b) contains 986 photons, all of which are detected inside the region shown. To provide an idea of the time points of detection, photons detected during the exposure sub-intervals $[0 \text{ s}, 0.02 \text{ s}]$, $(0.02 \text{ s}, 0.04 \text{ s}]$, $(0.04 \text{ s}, 0.06 \text{ s}]$, $(0.06 \text{ s}, 0.08 \text{ s}]$, and $(0.08 \text{ s}, 0.1 \text{ s}]$ are shown in black, magenta, green, cyan, and red, respectively.

15.2 Practical data models

A data model that is used to describe the image data obtained from an actual microscopy experiment is referred to as a *practical data model*. With practical data models we seek to describe data that is acquired using the standard image detectors (CCD, CMOS, EMCCD) that were introduced in Chapter 12 (Note 15.6). Unlike the ideal imaging conditions assumed for the fundamental data model, in practical situations we have to allow for detector pixelation, which decreases the precision with which the location of each detected photon is known, the detector's inability to track time points of photon arrival, which results in a poorer precision with which the detection time of each photon is known, and noise introduced by the detector, which leads to the acquired image being a corrupted version of the photon signal detected from the object. In practice, we of course also do not have the option of an infinitely large detection area, which implies that data is lost whenever a photon from the imaged object impacts the detector plane at a point outside the finite-sized detector.

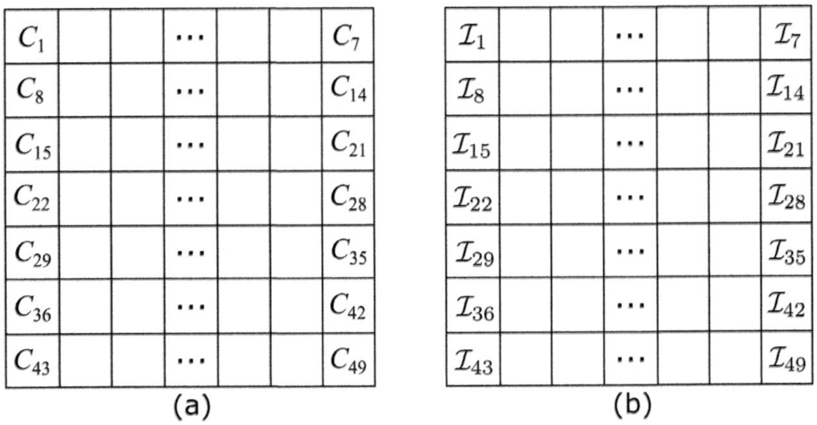

C_1			$\cdots$			C_7
C_8			$\cdots$			C_{14}
C_{15}			$\cdots$			C_{21}
C_{22}			$\cdots$			C_{28}
C_{29}			$\cdots$			C_{35}
C_{36}			$\cdots$			C_{42}
C_{43}			$\cdots$			C_{49}

(a)

$\mathcal{I}_1$			$\cdots$			$\mathcal{I}_7$
$\mathcal{I}_8$			$\cdots$			$\mathcal{I}_{14}$
$\mathcal{I}_{15}$			$\cdots$			$\mathcal{I}_{21}$
$\mathcal{I}_{22}$			$\cdots$			$\mathcal{I}_{28}$
$\mathcal{I}_{29}$			$\cdots$			$\mathcal{I}_{35}$
$\mathcal{I}_{36}$			$\cdots$			$\mathcal{I}_{42}$
$\mathcal{I}_{43}$			$\cdots$			$\mathcal{I}_{49}$

(b)

FIGURE 15.9
Modeling of practical data. (a) The detector C is a collection of N_p non-overlapping pixels $\{C_1, \ldots, C_{N_p}\}$. In the example shown, $N_p = 49$ and the pixels C_k, $k = 1, \ldots, 49$, are regions corresponding to squares of identical size arranged in a 2D array, as is commonly the case for an image detector. In practice, a detector has many more pixels, and a 49-pixel region as depicted typically represents a subset of the pixels, or a region of interest, that is used in an analysis. Such a 7×7-pixel region of interest might, for example, contain the image of a fluorescent molecule of interest. (b) Continuing with the example of (a), this illustration shows that the data acquired in pixel C_k of a detector is modeled by a random variable $\mathcal{I}_k$, $k = 1, \ldots, N_p$.

The detector C under a practical data model is therefore a collection of N_p non-overlapping pixels $\{C_1, \ldots, C_{N_p}\}$, such that $\bigcup_{k=1}^{N_p} C_k = C$ (Fig. 15.9(a)). Accordingly, the general form of a practical data model comprises data from the pixels of an image, described by a collection of random variables $\{\mathcal{I}_1, \ldots, \mathcal{I}_{N_p}\}$. Here $\mathcal{I}_k$, $k = 1, \ldots, N_p$, is a random variable that describes the data acquired in the kth pixel (Fig. 15.9(b)). Specific models corresponding to different detector types differ in terms of the probability distributions for $\mathcal{I}_k$, $k = 1, \ldots, N_p$, depending on, for example, the type of noise the detector introduces to the acquired image. Every practical data model, however, has at its basis the temporal and spatial components of photon arrival at the detector plane described in Sections 15.1.1 and 15.1.2, as the physical nature of photon emission and arrival remains the same regardless of the specific detector used. The primary difference between the use of an ideal detector and a practical detector is that while an ideal detector will produce an "image" consisting of the precise arrival times and locations of the detected photons, a practical detector will produce a pixelated image where each pixel is represented by a single value representing the number of photons (assuming the absence of detector noise) that were detected over the acquisition time interval and within the boundaries of the pixel. The loss of precision in the practical case, both in time and space, is thus apparent from the lack of precise time and location information for each detected photon.

15.2.1 Poisson data model

We begin with a specific practical data model, which we refer to as the Poisson data model, that forms the basis for the more complex models that follow. The Poisson data model describes an image produced by a pixelated detector that introduces no readout noise, thereby preserving the signals detected in its pixels as they are. Although a detector without readout noise is hypothetical in nature, the Poisson data model may be appropriate in cases where the detected signals are large enough that the detector's readout noise can be considered a negligible component of the acquired data. Additionally, this data model is useful for analytical purposes, as it can be viewed as a baseline model against which other practical data models, which account for readout and potentially other types of detector noise, can be compared (Section 18.6.2).

In the Poisson data model, the data in each image pixel is the sum of two independent components — the photons detected from the imaged object and the photons due to the background component (Section 15.1.3). Therefore, under the Poisson data model, the data in the kth pixel of an N_p-pixel image $\{\mathcal{I}_1, \ldots, \mathcal{I}_{N_p}\}$ is given by

$$\mathcal{I}_k = S_k + B_k, \tag{15.14}$$

where S_k and B_k are random variables representing the number of photons detected from the object and the background component, respectively.

In Section 15.1.3, we saw that the detection of photons from the imaged object and the detection of photons from the background component are described by independent Poisson processes. Consequently, in the kth image pixel, the number of object photons S_k and the number of background photons B_k, collected over a given acquisition time interval, are independent Poisson random variables. The mean of the object photon count S_k is determined by integrating the photon detection rate $\Lambda(\tau)$, $\tau \geq t_0$, and the photon distribution profile $f_\tau(x, y)$, $(x, y) \in C$, $\tau \geq t_0$, both temporally over the acquisition interval and spatially over the region on the detector corresponding to the kth pixel. Specifically, given that the image is acquired over the time interval $[t_0, t]$, the mean μ_k of the number of object photons detected in the kth pixel, i.e., the mean of the Poisson random variable S_k, is given by

$$\mu_k = \int_{t_0}^{t} \Lambda(\tau) \int_{C_k} f_\tau(x, y)\,dx\,dy\,d\tau, \tag{15.15}$$

where C_k is again the region in the detector plane occupied by the kth pixel (Note 15.7). In the case where the object is stationary or can be assumed to be stationary during the acquisition of its image, the photon distribution profile no longer depends on time and can simply be denoted as f. Eq. (15.15) then becomes

$$\mu_k = \int_{t_0}^{t} \Lambda(\tau)\,d\tau \cdot \int_{C_k} f(x, y)\,dx\,dy, \tag{15.16}$$

a product of an integral over time and an integral over space. This expression further reduces to

$$\mu_k = \Lambda_0 (t - t_0) \cdot \int_{C_k} f(x, y)\,dx\,dy \tag{15.17}$$

if the photons emitted by the object are detected at a constant rate Λ_0 over the acquisition time interval $[t_0, t]$.

By calculating μ_k for $k = 1, \ldots, N_p$, we obtain the N_p-pixel image that we would acquire, on average, in the scenario where no background component is present. An example of such a *mean image* is shown in Fig. 15.10(a) for a stationary point source whose emitted photons are detected at a constant rate and spatially distributed according to the Airy image function of Fig. 15.6(a). Each pixel of this mean image was computed using Eq. (15.17), with f given by the time-independent version of the photon distribution profile of Eq. (15.8) (Note 15.8). We can see that, as per Eq. (15.17), this image is effectively a binned representation of a magnified and shifted version of the Airy image function whose height is scaled by the expected number of detected photons. Note that the unit for the value of each pixel of the

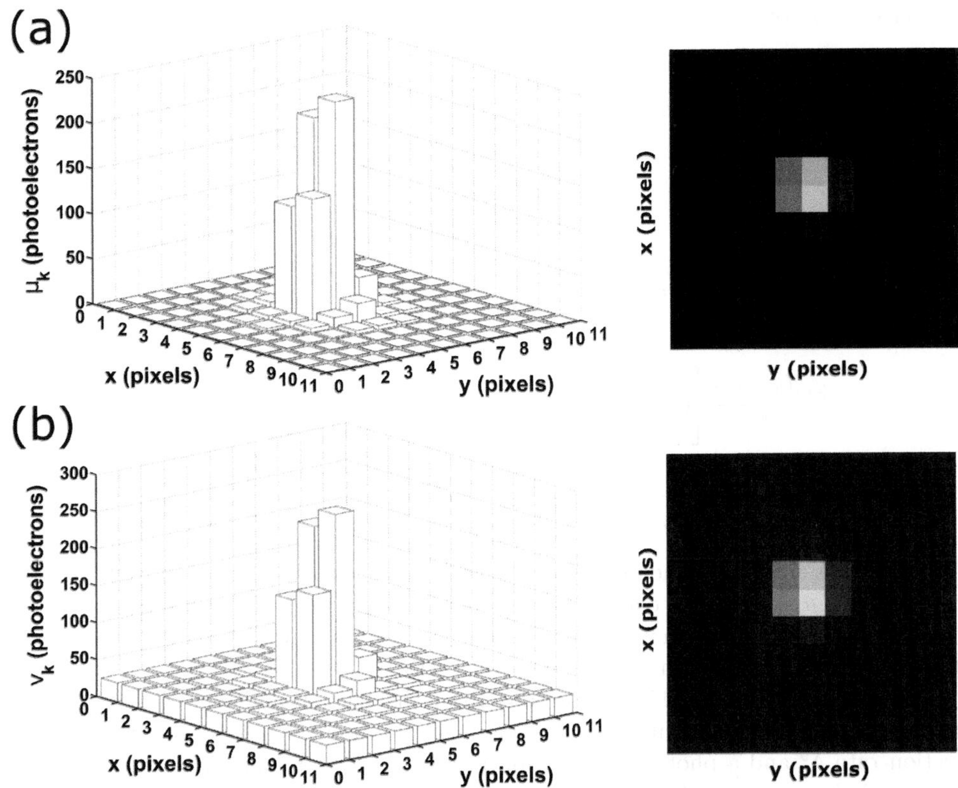

FIGURE 15.10
Bar graph and intensity representations of mean pixelated images of a stationary and in-focus point source located at $(x_0, y_0) = (656.5 \text{ nm}, 682.5 \text{ nm})$ in object space, calculated using the time-independent version of the photon distribution profile of Eq. (15.8) with a magnification of $M = 100$ and the Airy image function of Fig. 15.6(a). In both (a) and (b), the image consists of an 11×11-pixel region with a pixel size of 13 μm $\times$ 13 μm. The point source's object space location corresponds to (5.05 pixels, 5.25 pixels) in image space, assuming that $(0, 0)$ in both spaces coincides with the upper left corner of the 11×11-pixel region. The image is acquired over an exposure interval of length $t - t_0 = 0.1$ s, during which photons from the point source are detected at a constant rate of $\Lambda_0 = 10000$ photons/s. While the expected number of detected photons is $\Lambda_0(t - t_0) = 1000$ over the detector plane, the expected number of detected photons within the confines of the 11×11-pixel image is 952.3. The mean number μ_k of object photons is the same in (a) and (b) for each pixel k, but in (a) there is no background component, whereas in (b) the mean background photon count is $\beta_k = 25$ for each pixel k.

image is photoelectrons because each detected photon is converted to an electron as per the photoelectric effect (Section 12.1).

Computation of μ_k given the Airy image function requires the numerical integration of the photon distribution profile over the pixel C_k (Note 15.9). If the image profile is given by the 2D Gaussian image function, however, it is possible to express μ_k analytically in terms of the error function $\text{erf}(u) = \frac{2}{\sqrt{\pi}} \int_0^u e^{-u^2} du$. Consider, for example, a detector with rectangular or square pixels. If the region C_k occupied by the kth pixel is delimited by the interval $[x_{k,1}, x_{k,2}]$ in the x direction and the interval $[y_{k,1}, y_{k,2}]$ in the y direction, then μ_k

of Eq. (15.17) becomes

$$\begin{aligned}
\mu_k &= \Lambda_0(t-t_0)\cdot\int_{C_k} f(x,y)dxdy \\
&= \frac{\Lambda_0(t-t_0)}{M^2}\int_{C_k} q\left(\frac{x}{M}-x_0,\frac{y}{M}-y_0\right)dydy \\
&= \frac{\Lambda_0(t-t_0)}{2\pi M^2\sigma_g^2}\int_{C_k} e^{-\frac{1}{2\sigma_g^2}\left(\left(\frac{x}{M}-x_0\right)^2+\left(\frac{y}{M}-y_0\right)^2\right)}dxdy \\
&= \frac{\Lambda_0(t-t_0)}{2\pi M^2\sigma_g^2}\int_{x_{k,1}}^{x_{k,2}} e^{-\frac{1}{2\sigma_g^2}\left(\frac{x}{M}-x_0\right)^2}dx\cdot\int_{y_{k,1}}^{y_{k,2}} e^{-\frac{1}{2\sigma_g^2}\left(\frac{y}{M}-y_0\right)^2}dy \\
&= \frac{\Lambda_0(t-t_0)}{4}\left[\operatorname{erf}\left(\frac{1}{\sqrt{2}\sigma_g}\left(\frac{x_{k,2}}{M}-x_0\right)\right)-\operatorname{erf}\left(\frac{1}{\sqrt{2}\sigma_g}\left(\frac{x_{k,1}}{M}-x_0\right)\right)\right] \\
&\quad\times\left[\operatorname{erf}\left(\frac{1}{\sqrt{2}\sigma_g}\left(\frac{y_{k,2}}{M}-y_0\right)\right)-\operatorname{erf}\left(\frac{1}{\sqrt{2}\sigma_g}\left(\frac{y_{k,1}}{M}-y_0\right)\right)\right].
\end{aligned}$$

This expression eliminates the need for numerical integration over C_k, and allows a faster computation of μ_k. The faster computational speed is of particular importance for parameter estimation, in which the mean image may need to be computed iteratively for different values of parameters of interest such as x_0 and y_0 (Note 15.10).

As discussed in Section 15.1.3, photons from the background component can be accounted for in the same manner as photons from the imaged object, but with a photon detection rate Λ^b and a photon distribution profile f_τ^b that are specific to the detection of background photons. Therefore, the mean β_k of the number of background photons detected in the kth pixel of an image, i.e., the mean of the Poisson random variable B_k, is given by expressions that are analogous to the above expressions for μ_k. Note that we generally speak of photons in the context of background because autofluorescence and other sources of spurious photons are typically the primary constituents of background noise. In the case of a practical data model, however, the background component can more broadly include sources of spurious electrons such as the detector's dark current (Section 12.4), since such electrons are usually also assumed to be generated according to a Poisson process.

As S_k and B_k are independent and Poisson-distributed with mean μ_k and β_k, respectively, their sum $\mathcal{I}_k$, i.e., the total number of photons detected in the kth pixel of an image, is Poisson-distributed the mean ν_k given by (Online Appendix Exercise A.1)

$$\nu_k = \mu_k + \beta_k. \tag{15.18}$$

Figure 15.10(b) provides an example of a mean image that includes photons from a background component. The image shown was obtained by adding a constant level of $\beta_k = 25$ background photons to each pixel k of the image in Fig. 15.10(a). The constant level of background can be obtained by, for example, defining a time-independent photon distribution profile f^b to be a uniform probability density function over the image, and by choosing the appropriate photon detection rate Λ^b for the given exposure time interval.

Under the Poisson data model, the data $\mathcal{I}_k$ in the kth pixel of an image is therefore distributed according to the Poisson probability mass function

$$p_k(z) = \frac{\nu_k^z e^{-\nu_k}}{z!}, \quad z = 0, 1, \ldots. \tag{15.19}$$

The probability distribution for the data in the pixels of an image is required for many of the analyses in the chapters that follow. An example is the estimation of parameters of

Data model	Probability distribution $p_k(z)$	Domain
Poisson	$\frac{\nu_k^z e^{-\nu_k}}{z!}$	$z = 0, 1, \ldots$
CCD/CMOS	$\frac{e^{-\nu_k}}{\sqrt{2\pi}\sigma_k} \sum_{l=0}^{\infty} \frac{\nu_k^l}{l!} e^{-\frac{1}{2}\frac{(z-l-\eta_k)^2}{\sigma_k^2}}$	$z \in \mathbb{R}$
Deterministic	$\frac{1}{\sqrt{2\pi\xi_k^2}} e^{-\frac{(z-D_k-\zeta_k)^2}{2\xi_k^2}}$	$z \in \mathbb{R}$
CCD/CMOS Gaussian approximation	$\frac{1}{\sqrt{2\pi(\nu_k+\sigma_k^2)}} e^{-\frac{(z-\nu_k-\eta_k)^2}{2(\nu_k+\sigma_k^2)}}$	$z \in \mathbb{R}$
EMCCD	$\frac{e^{-\nu_k}}{\sqrt{2\pi}\sigma_k} \left[e^{-\left(\frac{z-\eta_k}{\sqrt{2}\sigma_k}\right)^2} + \sum_{l=1}^{\infty} e^{-\left(\frac{z-l-\eta_k}{\sqrt{2}\sigma_k}\right)^2} \sum_{j=0}^{l-1} \frac{\binom{l-1}{j}\left(1-\frac{1}{g}\right)^{l-j-1}}{(j+1)!\left(\frac{g}{\nu_k}\right)^{j+1}} \right]$	$z = 0, 1, \ldots$
EMCCD high gain approximation	$\frac{e^{-\nu_k}}{\sqrt{2\pi}\sigma_k} \left[e^{-\left(\frac{z-\eta_k}{\sqrt{2}\sigma_k}\right)^2} + \int_0^{\infty} e^{-\left(\frac{z-u-\eta_k}{\sqrt{2}\sigma_k}\right)^2 - \frac{u}{g}} \frac{\sqrt{\frac{\nu_k u}{g}} I_1\left(2\sqrt{\frac{\nu_k u}{g}}\right)}{u} du \right]$	$z \in \mathbb{R}$
EMCCD Gaussian approximation	$\frac{1}{\sqrt{2\pi\left(\left(2g^2-g\right)\nu_k+\sigma_k^2\right)}} e^{-\frac{(z-g\nu_k-\eta_k)^2}{2\left(\left(2g^2-g\right)\nu_k+\sigma_k^2\right)}}$	$z \in \mathbb{R}$

TABLE 15.1
Probability distribution function for the kth pixel of an image under various practical data models.

interest from image data using the maximum likelihood estimator (Section 16.1). For ease of referencing, the distribution of Eq. (15.19) is reproduced in Table 15.1 along with the probability distributions for the other practical data models discussed below.

To simulate the data in the kth pixel of an image under the Poisson data model, one simply has to compute ν_k and provide it as input to a Poisson random number generator to obtain a realization of the random variable $\mathcal{I}_k$. An example of an image of a stationary and in-focus point source simulated in this way is shown in Fig. 15.11(a). In this example, the Poisson random number in the kth pixel was generated using ν_k from the corresponding pixel of the mean image of Fig. 15.10(b). Note the randomness of the pixel values of this simulated image and contrast it with the mean image of Fig. 15.10(b), whose pixel values were obtained by computing the deterministic mathematical function ν_k for $k = 1, \ldots, N_p$.

In Section 15.1.1.1, we discussed, in the context of the fundamental data model, the situation in which no photon emitted by the object of interest is detected. We showed there that this situation is only of concern when very low photon counts are expected during the acquisition time interval. As such, it is of no relevance under practical experimental conditions. In our discussion of the practical data models, we will therefore assume that we have a sufficient number of expected photons impacting the detector, such that the probability of detecting no photons emitted by the object of interest during an acquisition time interval is negligible (Note 15.11). Accordingly, under the practical data models, we have that the mean photon count detected over the acquisition time interval $[t_0, t]$ (over the detector plane) is given by $N_{photons} = \int_{t_0}^{t} \Lambda(\tau)d\tau$, or in the case of a constant photon detection rate Λ_0, $N_{photons} = \Lambda_0(t - t_0)$.

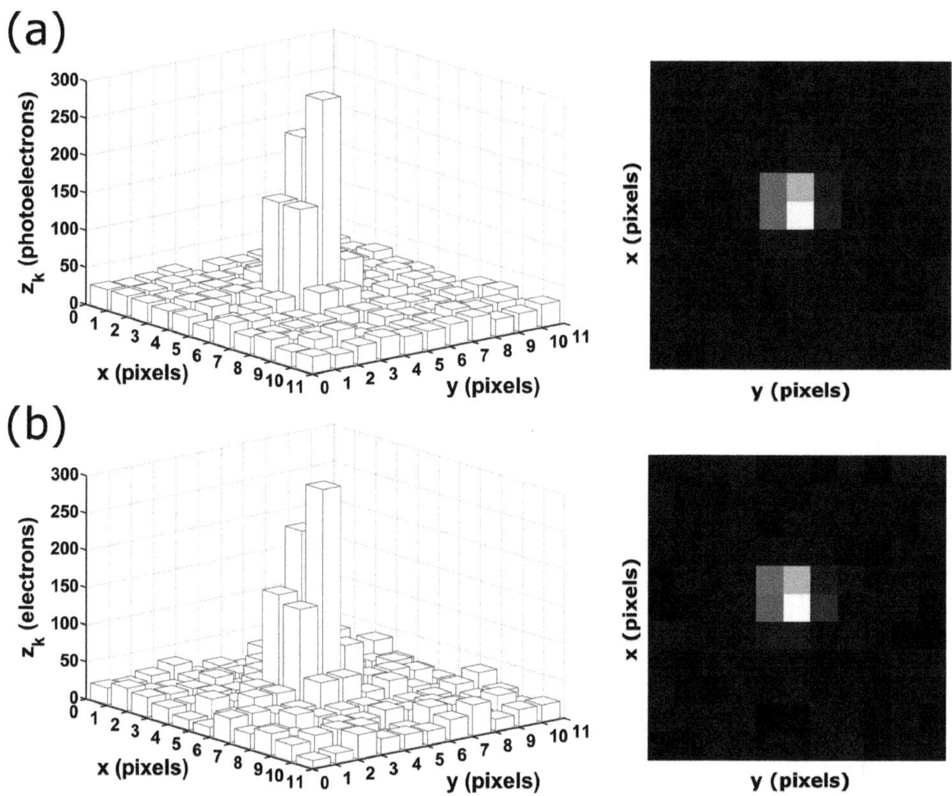

FIGURE 15.11
Bar graph and intensity representations of images of a stationary and in-focus point source simulated under (a) the Poisson data model and (b) the CCD/CMOS data model. In (a), the data in each pixel is a realization of a Poisson random variable with mean given by the value of the corresponding pixel in the mean image of Fig. 15.10(b). In (b), the data in each pixel consists of the sum of a Poisson component and a Gaussian component. The Poisson component is given by the value of the corresponding pixel in the image of (a), and for each pixel k, the Gaussian component is a realization of a Gaussian random variable generated by assuming readout noise with a mean of $\eta_k = 0$ electrons and a variance of $\sigma_k^2 = 36$ electrons squared.

15.2.2 CCD/CMOS data model

The CCD/CMOS data model provides a description of an image produced by a CCD or CMOS detector. CCD and CMOS detectors are the most prevalent detectors currently used in imaging experiments, and they are discussed in some detail in Sections 12.3.1 and 12.3.2. As in the Poisson data model, the data in each CCD or CMOS image pixel includes photons detected from the object of interest and photons/electrons from the background component. Additionally and importantly, the data in each CCD or CMOS image pixel contains the detector's measurement noise, which is added when the object and background signals are read out.

Accordingly, by adding a readout noise component to Eq. (15.14), we model, for $k = 1, \ldots, N_p$, the data in the kth pixel of an N_p-pixel CCD or CMOS image as the random variable

$$\mathcal{I}_k = S_k + B_k + W_k, \tag{15.20}$$

where S_k and B_k are again the Poisson random variables with means μ_k and β_k, representing the signals due to the object and the background component, and W_k is a Gaussian random variable with mean η_k and variance σ_k^2, representing the detector readout noise. For a given CCD or CMOS detector, the mean and variance of W_k can be determined as described in Section 12.7.1.4. Note that as discussed in Chapter 12, due to fundamental design differences in the readout circuitry for CCD and CMOS detectors, the readout noise in every pixel of a CCD detector can often be assumed to be described by the same probability distribution (i.e., for an N_p-pixel image, $\eta_1 = \eta_2 = \cdots = \eta_{N_p}$ and $\sigma_1^2 = \sigma_2^2 = \cdots = \sigma_{N_p}^2$), whereas the readout noise in each pixel of a CMOS detector has to be modeled with great care as the CMOS pixels do not share the same readout electronics. Different mean and variance values may therefore have to be assumed and determined for the pixels of a CMOS detector.

Equation (15.20) shows that to simulate the data in the kth pixel of a CCD or CMOS image, one has to generate two random numbers and add them. The first random number is a realization of the Poisson random variable $S_k + B_k$, which can be obtained as described in Section 15.2.1, and the second random number is a realization of the Gaussian random variable W_k, which can be obtained using a Gaussian random number generator. An example of a CCD image of a stationary and in-focus point source simulated in this manner is shown in Fig. 15.11(b). In this example, the Poisson random number in the kth pixel was simply obtained from the corresponding pixel of the Poisson data model image of Fig. 15.11(a), and the Gaussian random number was generated using a mean of $\eta_k = 0$ electrons and a variance of $\sigma_k^2 = 36$ electrons squared. We should note here that in the technical literature for detectors the readout noise variance is typically characterized through the standard deviation, i.e., the square root of the variance.

Based on the data model of Eq. (15.20), the data $\mathcal{I}_k$ in the kth pixel of a CCD or CMOS image, being the sum of a Poisson random variable (with mean $\nu_k = \mu_k + \beta_k$) and a Gaussian random variable (with mean η_k and variance σ_k^2) that are independent of each other, has a probability density function p_k that is given by the convolution of the probability mass function of the Poisson random variable and the probability density function of the Gaussian random variable, i.e.,

$$p_k(z) = \frac{e^{-\nu_k}}{\sqrt{2\pi}\sigma_k} \sum_{l=0}^{\infty} \frac{\nu_k^l}{l!} e^{-\frac{1}{2}\frac{(z-l-\eta_k)^2}{\sigma_k^2}}, \quad z \in \mathbb{R}. \tag{15.21}$$

15.2.3 Deterministic data model

Due to the ease with which they can be used, data models based on Gaussian distributions are widely used in many areas of application. For this reason, it is also useful to consider under which circumstances a Gaussian stochastic data model can be appropriate for the analysis of image data. This leads us immediately to the consideration of the deterministic data model.

The deterministic data model considers the data in an image pixel to be the sum of a deterministic signal and a Gaussian noise component. For $k = 1, \ldots, N_p$, the random variable $\mathcal{I}_k$ representing the data in the kth pixel of an N_p-pixel image is given by

$$\mathcal{I}_k = D_k + U_k, \tag{15.22}$$

where D_k is the deterministic signal detected in the pixel, and U_k is a Gaussian random variable representing some noise component. If we denote the mean and variance of U_k by

ζ_k and ξ_k^2, respectively, then the probability distribution of the data $\mathcal{I}_k$ is given by the Gaussian probability density function

$$p_k(z) = \frac{1}{\sqrt{2\pi\xi_k^2}} e^{-\frac{(z-D_k-\zeta_k)^2}{2\xi_k^2}}, \quad z \in \mathbb{R}. \tag{15.23}$$

Both D_k and U_k can be more specifically defined as needed. In the simplest case, the noise component U_k does not depend on the deterministic signal D_k. This is the model that is often used in the least squares formulation of the deconvolution problem, as we will see in Section 21.3. In what follows, however, we will see an important example in which the variance of the noise component U_k is specified in terms of the signal D_k.

15.2.3.1 Gaussian approximation for the CCD/CMOS data model

When high object and/or background signal levels are detected using a CCD or CMOS detector, the acquired image may be described using a deterministic data model. To arrive at appropriate specifications for the deterministic signal D_k and the Gaussian noise component U_k in the kth image pixel, we start with the fact that a Poisson random variable with a large mean can be approximated well by a Gaussian random variable with mean and variance both equal to that large mean. This is illustrated in Fig. 15.12, which compares Poisson distributions with different mean values with their corresponding Gaussian distributions. For the Poisson signal $S_k + B_k$ detected in the kth image pixel, we make the approximation

$$S_k + B_k \approx \nu_k + V_k,$$

where ν_k (Eq. (15.18)) is the mean of $S_k + B_k$, and V_k is a Gaussian random variable with mean 0 and variance ν_k. This approximation amounts to using a Gaussian random variable that has the same mean and variance as the Poisson signal, i.e.,

$$\begin{aligned} \mathrm{E}[\nu_k + V_k] &= \mathrm{E}[\nu_k] + \mathrm{E}[V_k] = \nu_k + 0 = \nu_k = \mathrm{E}[S_k + B_k], \\ \mathrm{Var}(\nu_k + V_k) &= \mathrm{Var}(\nu_k) + \mathrm{Var}(V_k) = 0 + \nu_k = \nu_k = \mathrm{Var}(S_k + B_k). \end{aligned}$$

Next, by taking $D_k = \nu_k$ as the deterministic signal, and by combining V_k with the readout noise W_k (which is Gaussian with mean η_k and variance σ_k^2) to yield a Gaussian random variable U_k with mean $\zeta_k = \eta_k$ and variance $\xi_k^2 = \nu_k + \sigma_k^2$, we arrive at the deterministic data model $\mathcal{I}_k = D_k + U_k$.

Given that $D_k = \nu_k$, $\zeta_k = \eta_k$, and $\xi_k^2 = \nu_k + \sigma_k^2$, the probability distribution for $\mathcal{I}_k$ is the Gaussian probability density function

$$p_k(z) = \frac{1}{\sqrt{2\pi(\nu_k + \sigma_k^2)}} e^{-\frac{(z-\nu_k-\eta_k)^2}{2(\nu_k+\sigma_k^2)}}, \quad z \in \mathbb{R}. \tag{15.24}$$

Note that in this data model, the variance $\nu_k + \sigma_k^2$ of the Gaussian noise component U_k depends on the particular image, since the ν_k portion is signal-dependent, and therefore can in general vary from image to image over the duration of a microscopy experiment. This is in contrast to the variance σ_k^2 of the Gaussian readout noise component W_k in the CCD/CMOS data model, which accounts solely for the variance of the detector's readout noise, and can therefore generally be assumed to remain constant from image to image.

15.2.4 EMCCD data model

The EMCCD data model describes an image produced by an EMCCD detector, the basic principles of which are discussed in Section 12.3.3. Similar to the CCD/CMOS data model, it

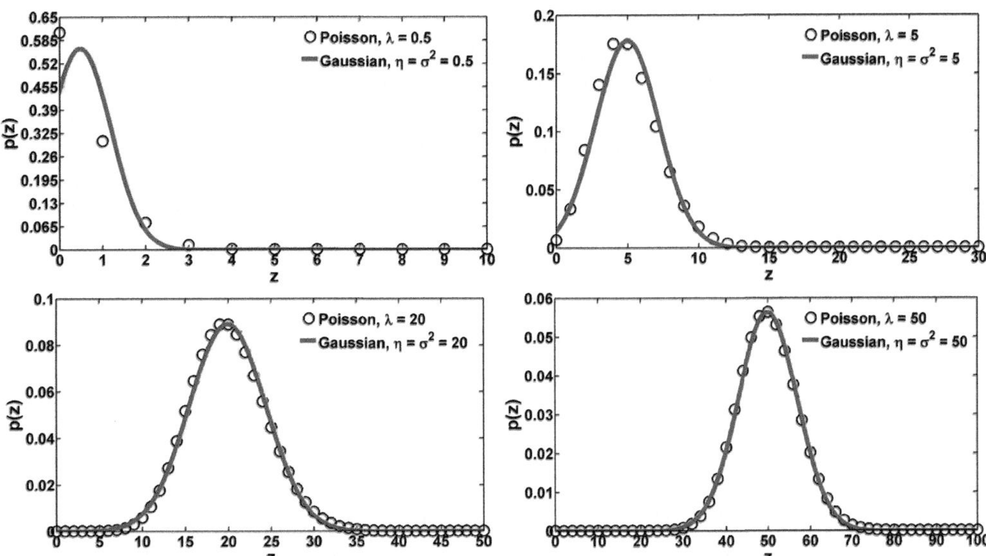

FIGURE 15.12
Comparing the Poisson and Gaussian probability distributions. In each plot, a Poisson distribution with mean λ is shown together with a Gaussian distribution with mean η and variance σ^2 both equal to λ. For larger values of λ, the Poisson distribution can be seen to be very well approximated by the Gaussian distribution. For smaller values of λ, significant differences can be seen between the two distributions.

models the data in each image pixel as the sum of three mutually independent quantities that arise from the imaged object, the background component, and the detector readout noise. In place of the object and background signals, however, are the stochastically amplified versions of these signals. Specifically, for $k = 1, \ldots, N_p$, the data in the kth pixel of an N_p-pixel image is modeled by the random variable

$$\mathcal{I}_k = S_{amp,k} + B_{amp,k} + W_k, \tag{15.25}$$

where $S_{amp,k}$ and $B_{amp,k}$ are random variables representing the results of the amplification of the signals modeled by the Poisson random variables S_k and B_k from Eq. (15.20), and W_k is, as in the case of the CCD/CMOS data model, a Gaussian random variable with mean η_k and variance σ_k^2 that models the detector's readout noise in the pixel. Note that since an EMCCD detector is architecturally a CCD-based detector, the amplified signals in all pixels are read out by the same electronic circuitry. The Gaussian random variables W_k, $k = 1, \ldots, N_p$, can therefore typically be modeled in the same fashion as for a CCD detector.

EMCCD signal amplification as a branching process. To obtain the probability distributions for the amplified signals $S_{amp,k}$ and $B_{amp,k}$, $k = 1, \ldots, N_p$, we make use of branching processes from probability theory. Classically, branching processes are used in biology to model reproduction in studies that are concerned with how a population grows or declines over generations. However, they have found application in many areas of investigation, including the modeling of nuclear chain reactions, DNA amplification by the PCR (Section 7.2.2), and signal amplification or, more precisely, electron multiplication, in devices such as an EMCCD detector. In Chapter G of the Online Appendix, we give a detailed derivation of our models of electron multiplication using the theory of branching processes. Here, we will simply present the results.

When using a branching process to model signal amplification by an EMCCD detector, the most concise results are obtained if we model the multiplication of electrons in each stage of the detector's multiplication register with a geometric probability distribution. Given that multiplication of the electrons accumulated in an EMCCD pixel is modeled in this way, the amplified object signal $S_{amp,k}$ in the kth image pixel has the probability mass function (Online Appendix Section G.1)

$$p_{\mu_k,g}(x) = \begin{cases} e^{-\mu_k}, & x = 0, \\ e^{-\mu_k} \sum_{j=0}^{x-1} \frac{\binom{x-1}{j}}{(j+1)!} \left(1 - \frac{1}{g}\right)^{x-1-j} \left(\frac{\mu_k}{g}\right)^{j+1}, & x = 1, 2, \ldots, \end{cases} \tag{15.26}$$

where $g > 1$ is the EM gain used to capture the image, defined as the mean number of electrons that result from the amplification of a single initial electron, and μ_k is the mean of the Poisson-distributed object photon count S_k in the pixel prior to amplification, and is as defined in Eq. (15.15). Since the electrons due to the background signal are multiplied the same way along with the electrons due to the object of interest, the amplified background signal $B_{amp,k}$ in the kth pixel has a probability mass function of the same form, but with μ_k replaced by the mean β_k of the Poisson-distributed background signal B_k in the pixel prior to amplification.

We will in fact consider the sum $S_{amp,k} + B_{amp,k}$ as a single amplified signal, having originated from the initial Poisson signal $S_k + B_k$, which has mean ν_k as given in Eq. (15.18). The probability mass function for $S_{amp,k} + B_{amp,k}$ is then given by Eq. (15.26) with μ_k replaced by ν_k. This can be verified by convolving the probability mass functions for $S_{amp,k}$ and $B_{amp,k}$ (i.e., by convolving Eq. (15.26) as given and Eq. (15.26) with μ_k replaced by β_k).

An example of a simulated EMCCD image of a stationary and in-focus point source is shown in Fig. 15.13. The value of the kth pixel in this image was generated by first obtaining a realization of the amplified signal $S_{amp,k} + B_{amp,k}$ using Eq. (15.26), with μ_k replaced by ν_k from the image of Fig. 15.10(b), and with an EM gain of $g = 500$. This realization was then added to a Gaussian random number generated using a mean of $\eta_k = 0$ electrons and a variance of $\sigma_k^2 = 576$ electrons squared. Compared to the variance used to simulate the CCD image of Fig. 15.11, the larger variance used here reflects the typically higher readout noise level of an EMCCD detector. Note the substantially higher electron counts in the EMCCD image compared to those in the CCD image.

By the independence of the random variables $S_{amp,k} + B_{amp,k}$ and W_k, the probability distribution of the data $\mathcal{I}_k$ in the kth pixel is simply the convolution of the probability mass function of $S_{amp,k} + B_{amp,k}$ and the Gaussian probability density function with mean η_k and variance σ_k^2. This convolution can be written as

$$p_k(z) = \frac{e^{-\nu_k}}{\sqrt{2\pi}\sigma_k} \left[e^{-\left(\frac{z-\eta_k}{\sqrt{2}\sigma_k}\right)^2} + \sum_{l=1}^{\infty} e^{-\left(\frac{z-l-\eta_k}{\sqrt{2}\sigma_k}\right)^2} \sum_{j=0}^{l-1} \frac{\binom{l-1}{j}\left(1-\frac{1}{g}\right)^{l-j-1}}{(j+1)!\left(\frac{g}{\nu_k}\right)^{j+1}} \right], \quad z \in \mathbb{R}. \tag{15.27}$$

15.2.4.1 High gain approximation for the EMCCD data model

When the EM gain g is large, which is typically the case when imaging with an EMCCD detector, the geometric model of multiplication is well approximated by an exponential model of multiplication. Using the exponential approximation, the probability mass function of Eq. (15.26) may be replaced by the probability density function (Online Appendix

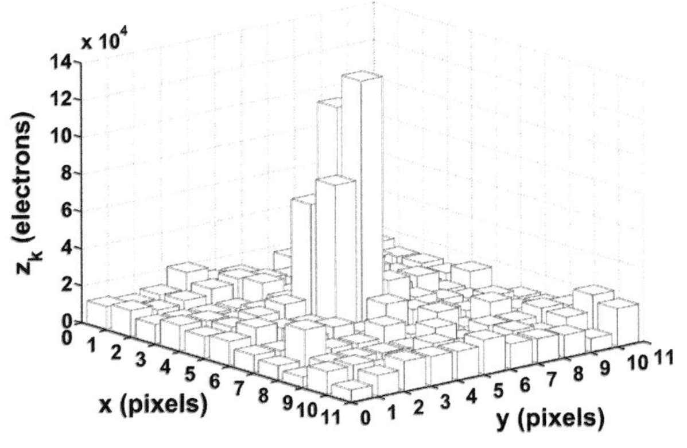

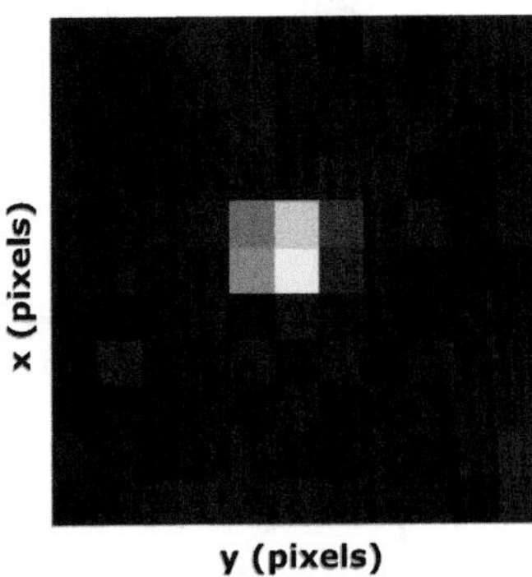

FIGURE 15.13
Bar graph and intensity representations of a stationary and in-focus point source simulated under the EMCCD data model. The amplified signal component of the data in each pixel k is a realization of a random variable distributed according to the probability mass function of Eq. (15.26), with μ_k replaced by ν_k from the corresponding pixel in Fig. 15.10(b), and with an EM gain of $g = 500$. The Gaussian component of the data in each pixel k is generated by assuming readout noise with a mean of $\eta_k = 0$ electrons and a variance of $\sigma_k^2 = 576$ electrons squared.

Section G.2)

$$p_{\mu_k,g}(x) = \begin{cases} e^{-\mu_k}, & x = 0, \\ \frac{e^{-\left(\frac{x}{g}+\mu_k\right)}\sqrt{\mu_k x/g}\, I_1\left(2\sqrt{\mu_k x/g}\right)}{x}, & x > 0, \end{cases} \tag{15.28}$$

which can be computationally more efficient than the probability mass function. Note that I_1 is the first order modified Bessel function of the first kind (Online Appendix Eq. (G.15)). Since this density function is appropriate only when g is large, as demonstrated by the comparison of the two distributions in Fig. 15.14, it is referred to as the *high gain approximation*.

Under the high gain approximation, the probability mass function of $S_{amp,k} + B_{amp,k}$ can be replaced with Eq. (15.28), with μ_k replaced by ν_k. Convolution with the Gaussian probability distribution with mean η_k and variance σ_k^2 will then yield the probability density function for $\mathcal{I}_k$ as

$$p_k(z) = \frac{e^{-\nu_k}}{\sqrt{2\pi}\sigma_k}\left[e^{-\left(\frac{z-\eta_k}{\sqrt{2}\sigma_k}\right)^2} + \int_0^\infty e^{-\left(\frac{z-u-\eta_k}{\sqrt{2}\sigma_k}\right)^2 - \frac{u}{g}} \frac{\sqrt{\frac{\nu_k u}{g}}\, I_1\left(2\sqrt{\frac{\nu_k u}{g}}\right)}{u} du\right], \quad z \in \mathbb{R}. \tag{15.29}$$

15.2.4.2 Gaussian approximation for the EMCCD data model

As in the case of a CCD/CMOS image, an EMCCD image may also be described with a deterministic data model when the detected object and/or background signal levels (i.e., the signal levels prior to amplification) are high. More specifically, we will use the fact that when the mean of the initial Poisson signal is large, the distribution of Eq. (15.27) is approximated well by a Gaussian distribution with mean and variance given by the mean and variance of the distribution of Eq. (15.27). For a comparison of these two distributions at different levels of the initial Poisson signal, see Fig. 15.15.

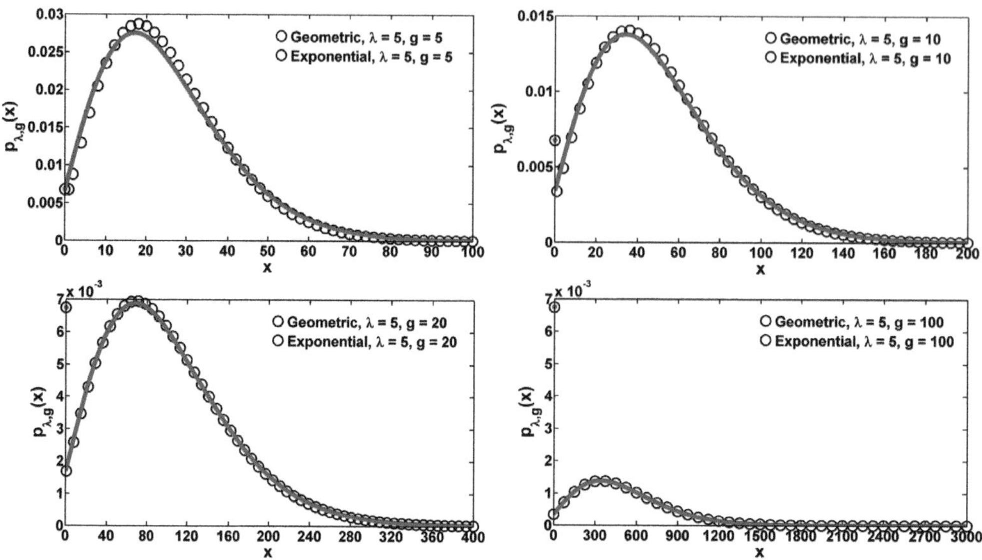

FIGURE 15.14
Comparing the geometric multiplication-based and exponential multiplication-based EMCCD probability distributions. For the same mean initial Poisson signal of $\lambda = 5$, the plots show the distributions of Eq. (15.26) ("Geometric") and Eq. (15.28) ("Exponential") for different values of the EM gain g. With increasing values of g, the exponential multiplication-based distribution can be seen to better approximate the geometric multiplication-based distribution.

Proceeding in analogous fashion as for the approximation for the CCD/CMOS data model in Section 15.2.3.1, we first approximate the amplified signal $S_{amp,k} + B_{amp,k}$ in the kth image pixel as

$$S_{amp,k} + B_{amp,k} \approx g\nu_k + V_k,$$

where $g\nu_k$ is the mean of $S_{amp,k} + B_{amp,k}$, with $g > 1$ the EM gain and ν_k (Eq. (15.18)) the mean of the Poisson signal $S_k + B_k$ prior to amplification, and V_k is a Gaussian random variable with mean 0 and variance $\left(2g^2 - g\right)\nu_k$. This variance is that of the amplified signal $S_{amp,k} + B_{amp,k}$, assuming that the amplification is modeled as a branching process that multiplies electrons according to a geometric probability distribution in each stage of the EMCCD detector's multiplication register. (Note that the variance corresponding to other models of amplification can also be used.) To show that this approximation is the equivalent of a Gaussian random variable having the same mean and variance as the amplified signal, we have

$$\begin{aligned}
\mathrm{E}\left[g\nu_k + V_k\right] &= \mathrm{E}\left[g\nu_k\right] + \mathrm{E}\left[V_k\right] = g\nu_k + 0 = g\nu_k = \mathrm{E}\left[S_{amp,k} + B_{amp,k}\right], \\
\mathrm{Var}\left(g\nu_k + V_k\right) &= \mathrm{Var}\left(g\nu_k\right) + \mathrm{Var}\left(V_k\right) = 0 + \left(2g^2 - g\right)\nu_k = \left(2g^2 - g\right)\nu_k \\
&= \mathrm{Var}\left(S_{amp,k} + B_{amp,k}\right).
\end{aligned}$$

We next take $D_k = g\nu_k$ to be the deterministic signal, and combine V_k with the Gaussian readout noise W_k to yield a Gaussian random variable U_k with mean $\zeta_k = \eta_k$ and variance $\xi_k^2 = \left(2g^2 - g\right)\nu_k + \sigma_k^2$, where η_k and σ_k^2 are the mean and variance of W_k. We thus arrive at the deterministic data model $\mathcal{I}_k = D_k + U_k$, characterized by the Gaussian probability

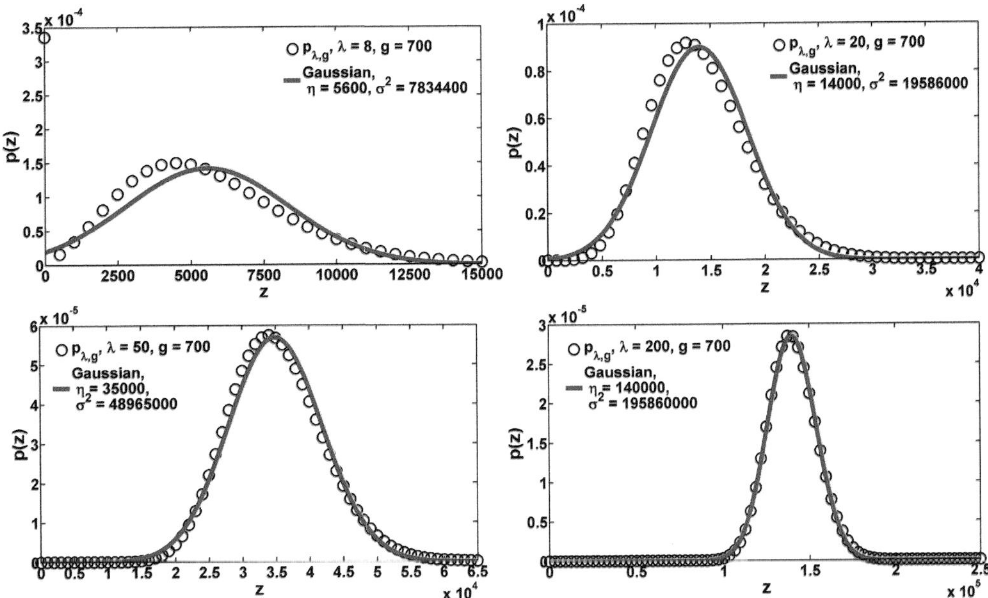

FIGURE 15.15
Comparing the geometric multiplication-based EMCCD probability distribution with the Gaussian probability distribution. For the same EM gain g but different values for the mean λ of the initial Poisson signal, the plots show the distribution of Eq. (15.26), denoted here by $p_{\lambda,g}$, overlaid with a Gaussian distribution with mean $\eta = g\lambda$ and variance $\sigma^2 = (2g^2 - g)\lambda$. The expressions for η and σ^2 are the mean and variance of $p_{\lambda,g}$. With increasing initial signal level λ, the Gaussian distribution can be seen to become a better approximation for $p_{\lambda,g}$.

density function

$$p_k(z) = \frac{1}{\sqrt{2\pi\left(\left(2g^2 - g\right)\nu_k + \sigma_k^2\right)}} e^{-\frac{(z - g\nu_k - \eta_k)^2}{2\left(\left(2g^2 - g\right)\nu_k + \sigma_k^2\right)}}, \quad z \in \mathbb{R}. \tag{15.30}$$

16

Parameter Estimation

In the previous chapter, we introduced a rigorous framework to stochastically describe the data that is acquired in imaging experiments. We therefore have the necessary background to carefully analyze questions of quantitation. As the data we obtain is stochastic in nature, we are confronted with statistical parameter estimation problems. It is the discussion of approaches to the parameter estimation problem, such as maximum likelihood estimation, that will be the main focus of this chapter.

16.1 Maximum likelihood estimation

An important aspect of quantitative fluorescence microscopy is parameter estimation, whereby quantities of interest are estimated from the acquired image data. For example, quantities that need to be estimated and considered during the tracking of an object at a given point in time might be the location of the object, the orientation of the object, the size of the object, and the rate at which photons were detected from the object.

A general method to estimate parameters is *maximum likelihood estimation* (Note 16.1). The principle of maximum likelihood estimation is to find the value of the parameter for which the observed outcome is most likely to occur. Suppose we have K measurements $z_1, z_2, \ldots, z_K$, with joint probability density function $p_{\theta,joint}(z_1, ..., z_K)$, where $\theta \in \Theta$ is the vector of parameters, i.e., quantities of interest, to be estimated from the measurements, and Θ is the parameter space. By parameter space we mean the set of all candidate parameter values amongst which we are going to find a value that leads to the best model of the data, i.e., the maximum likelihood estimate. Depending on whether the parameter is a scalar or in fact a vector, the parameter space is a subset of the set of real numbers $\mathbb{R}$ or a subset of $\mathbb{R}^n$, with n the number of components of the parameter (vector). When viewed as a function of θ instead of the measurements, the joint density function is referred to as the *likelihood function.* Taking its logarithm, we obtain the *log-likelihood function*

$$\mathcal{L}(\theta \mid z_1, ..., z_K) = \ln p_{\theta,joint}(z_1, ..., z_K). \tag{16.1}$$

The *maximum likelihood estimate* of θ, denoted by $\hat{\theta}_{MLE}$, is the value of θ that maximizes the log-likelihood function, and can be written as

$$\hat{\theta}_{MLE} = \underset{\theta \in \Theta}{\arg\max}\, \mathcal{L}(\theta \mid z_1, \ldots, z_K), \tag{16.2}$$

where the notation $\underset{\theta \in \Theta}{\arg\max}$ stands for the value of the parameter θ from the parameter space Θ that maximizes the expression.

A question that is often asked is why the logarithm of the likelihood function, rather than the likelihood function itself, is maximized. The answer is that both yield the same result as the logarithm is a monotonically increasing function. However, as we will see later, using the logarithm can simplify some calculations, especially if they involve Gaussian probability densities as the logarithm "removes" their exponential function.

When the K measurements are stochastically independent, the joint density function is given by the product of the measurements' respective probability density functions $p_{\theta,k}(z_k)$, $k = 1, \ldots, K$. In this case, the log-likelihood function can be written as

$$\mathcal{L}(\theta \mid z_1, ..., z_K) = \ln p_{\theta,joint}(z_1, ..., z_K) = \ln \prod_{k=1}^{K} p_{\theta,k}(z_k) = \sum_{k=1}^{K} \ln p_{\theta,k}(z_k), \quad (16.3)$$

where the logarithm of the product of density functions is expressed as the sum of the logarithms of the individual density functions. As the measurement in a pixel of an image detector can be assumed to be stochastically independent of the measurements in the other pixels, Eq. (16.3) gives the general form of the log-likelihood functions for practical image data models (Section 16.2.2; Note 16.2). The stochastic independence, and hence Eq. (16.3), also apply to measurements in the same pixel that are obtained at different times (see the examples of Sections 16.1.1 and 16.1.2).

Note that in cases where the measurements take values from a finite or countable set, the joint probability mass function is used in place of the joint probability density function. If in addition the measurements are stochastically independent, then the product of the individual probability density functions accordingly becomes a product of the individual probability mass functions.

We will next calculate the maximum likelihood estimate in two important examples.

16.1.1 Example 1: mean of a Poisson random variable

Suppose we have acquired N_{im} images with equal exposure to a constant level of light, and suppose the images are not corrupted by any extraneous noise. Then the data in each image pixel is a Poisson-distributed signal, and for a given pixel, the mean of the detected signal can be estimated from the N_{im} measurements in that pixel.

We therefore have, for a given image, the Poisson data model (Section 15.2.1)

$$\mathcal{I}_{\theta,k} = S_{\theta,k}, \quad k = 1, \ldots, N_p, \quad (16.4)$$

where $S_{\theta,k}$ is a Poisson random variable representing the number of photons detected in the kth pixel from the imaged object, and N_p is the number of pixels comprising the image. Note that this is just Eq. (15.14) with the random variable representing the background component removed, as we assume here the absence of noise of any kind. Note also the added subscript θ, which denotes the dependence of the Poisson signal on the parameters of interest.

We would like to estimate the mean $\theta = \mu_k$ of the signal detected in the kth pixel, i.e., the mean of the random variable $S_{\theta,k}$, from the N_{im} measurements in that pixel. We assume these measurements to be independent and denote them by $z_1, \ldots, z_{N_{im}}$. In order to obtain the maximum likelihood estimate $\hat{\theta}_{MLE}$, we need to maximize, with respect to θ, the log-likelihood function

$$\begin{aligned}\mathcal{L}(\theta \mid z_1, \ldots, z_{N_{im}}) &= \sum_{i=1}^{N_{im}} \ln p_\theta(z_i) = \sum_{i=1}^{N_{im}} \ln \frac{e^{-\theta}\theta^{z_i}}{z_i!} = \sum_{i=1}^{N_{im}} \left(\ln e^{-\theta} + \ln \theta^{z_i} - \ln z_i!\right) \\ &= -\theta N_{im} + \left(\sum_{i=1}^{N_{im}} z_i\right) \cdot \ln\theta - \sum_{i=1}^{N_{im}} \ln z_i!. \quad (16.5)\end{aligned}$$

This expression has the general form given by Eq. (16.3), with the distribution p_θ of each measurement given by the Poisson probability mass function with mean θ. To maximize this log-likelihood function, we differentiate it with respect to θ, and set the derivative equal to 0:

$$0 = \frac{\partial}{\partial\theta}\mathcal{L}(\theta \mid z_1, \ldots, z_{N_{im}}) = \frac{\partial}{\partial\theta}\left(-\theta N_{im} + \left(\sum_{i=1}^{N_{im}} z_i\right) \cdot \ln\theta - \sum_{i=1}^{N_{im}} \ln z_i!\right)$$

$$= -N_{im} + \frac{1}{\theta}\sum_{i=1}^{N_{im}} z_i - 0.$$

Solving for θ, we obtain

$$\hat{\theta}_{MLE} := \frac{1}{N_{im}}\sum_{i=1}^{N_{im}} z_i \tag{16.6}$$

as the maximum likelihood estimator of the mean μ_k of the Poisson signal $S_{\theta,k}$. The maximum likelihood estimator in this case is therefore the average of the measured values. An example of this estimator is given in Fig. 16.1(a).

16.1.2 Example 2: mean of a Gaussian random variable

We now revisit the estimation of the mean of the readout noise in a CCD or CMOS camera pixel, which we discussed in Section 12.7.1.4. Suppose we have acquired N_{im} images with a CCD or CMOS camera, with the camera shutter closed and the shortest possible exposure time. The closing of the shutter prevents light from impacting the detector, hence eliminating the detection of photons. The use of a minimal exposure time reduces the influence of the camera's dark current. A given image pixel can therefore be assumed to contain just the Gaussian-distributed camera readout noise. The mean of this noise, also referred

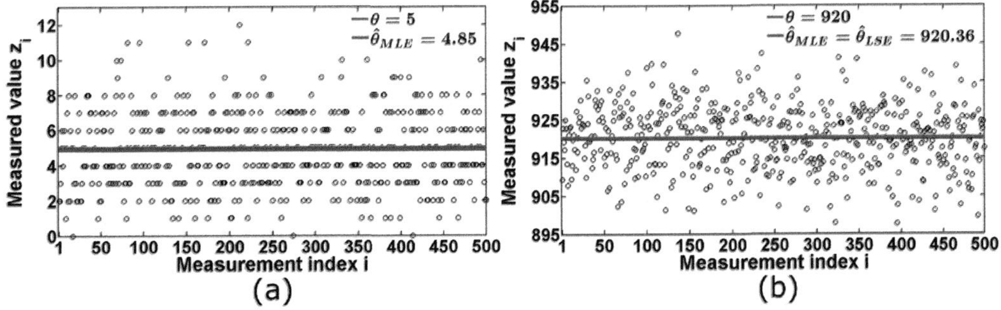

FIGURE 16.1
Examples of maximum likelihood estimation of the mean of Poisson and Gaussian random variables. In (a), it is assumed that the mean of the Poisson-distributed number of photons detected in an image pixel is $\theta = \mu_k = 5$ photons, and that there are $N_{im} = 500$ independent measurements of the photon count in that pixel. In (b), it is assumed that the camera offset, i.e., the mean of the Gaussian-distributed number of readout noise electrons in an image pixel, is $\theta = \eta_k = 920$ electrons, and that there are $N_{im} = 500$ independent measurements of the readout noise in that pixel. The variance of the number of readout electrons is assumed to be $\sigma_k^2 = 64$ electrons squared. In both examples, the mean of the random variable is shown along with its maximum likelihood estimate given the 500 measurements.

to as the offset for the given pixel (Section 12.7.1.1), can then be estimated from the N_{im} measurements of the pixel.

A given image can thus be described by

$$\mathcal{I}_{\theta,k} = W_{\theta,k}, \quad k = 1, \ldots, N_p, \tag{16.7}$$

where $W_{\theta,k}$ is a Gaussian random variable representing the number of readout noise electrons in the kth pixel, and N_p is the number of pixels comprising the image. This data model is the result of removing the object photon signal and the background noise component from the CCD/CMOS model of Section 15.2.2, achieved by the closing of the camera shutter and the minimization of the exposure time. The subscript θ indicates the dependence of the readout noise on the parameters to be estimated.

Let the N_{im} measurements in the kth pixel be denoted by $z_1, \ldots, z_{N_{im}}$. By the stochastic independence of the measurements, the mean $\theta = \eta_k$ of the readout noise in the kth pixel, i.e., the mean of the Gaussian random variable $W_{\theta,k}$, can be estimated by maximizing, with respect to θ, the log-likelihood function

$$\begin{aligned}
\mathcal{L}(\theta \mid z_1, ..., z_{N_{im}}) &= \sum_{i=1}^{N_{im}} \ln p_\theta(z_i) = \sum_{i=1}^{N_{im}} \ln \left(\frac{1}{\sqrt{2\pi\sigma_k^2}} e^{-\frac{1}{2\sigma_k^2}(\theta - z_i)^2} \right) \\
&= \sum_{i=1}^{N_{im}} \left[\ln 1 - \ln \sqrt{2\pi\sigma_k^2} + \ln e^{-\frac{1}{2\sigma_k^2}(\theta - z_i)^2} \right] \\
&= -N_{im} \cdot \ln \sqrt{2\pi\sigma_k^2} + \sum_{i=1}^{N_{im}} \left[-\frac{1}{2\sigma_k^2}(\theta - z_i)^2 \right].
\end{aligned} \tag{16.8}$$

As in the case of Eq. (16.5), this function has the general form of Eq. (16.3), but here the distribution p_θ of each measurement is given by the Gaussian probability density function with mean θ and variance σ_k^2.

The maximum likelihood estimator $\hat{\theta}_{MLE}$ is then given by

$$\begin{aligned}
\hat{\theta}_{MLE} &= \arg\max_{\theta \in \Theta} \mathcal{L}(\theta \mid z_1, ..., z_{N_{im}}) \\
&= \arg\max_{\theta \in \Theta} \left(-N_{im} \cdot \ln \sqrt{2\pi\sigma_k^2} + \sum_{i=1}^{N_{im}} \left[-\frac{1}{2\sigma_k^2}(\theta - z_i)^2 \right] \right) \\
&= \arg\max_{\theta \in \Theta} \left(-\sum_{i=1}^{N_{im}} \frac{1}{2\sigma_k^2}(\theta - z_i)^2 \right) = \arg\min_{\theta \in \Theta} \left(\frac{1}{2\sigma_k^2} \sum_{i=1}^{N_{im}} (\theta - z_i)^2 \right) \\
&= \arg\min_{\theta \in \Theta} \sum_{i=1}^{N_{im}} (\theta - z_i)^2 \\
&= \hat{\theta}_{LSE},
\end{aligned}$$

which shows that in this case, maximum likelihood estimation is equivalent to the minimization of the sum of the squares of the residuals $\theta - z_i$, $i = 1, \ldots, N_{im}$, and the maximum likelihood estimator is therefore the least squares estimator $\hat{\theta}_{LSE}$. To solve the least squares problem, we differentiate the sum of the squares of the residuals with respect to θ, and set the result equal to zero:

$$0 = \frac{\partial}{\partial\theta} \sum_{i=1}^{N_{im}} (\theta - z_i)^2 = \sum_{i=1}^{N_{im}} 2\,(\theta - z_i) = 2N_{im}\theta - 2\sum_{i=1}^{N_{im}} z_i.$$

Solving for θ, we obtain

$$\hat{\theta}_{MLE} = \hat{\theta}_{LSE} = \frac{1}{N_{im}} \sum_{i=1}^{N_{im}} z_i. \tag{16.9}$$

This shows that, as in the example of Section 16.1.1, the maximum likelihood estimator here is the average of the measured values. An example of this estimator is shown in Fig. 16.1(b).

16.2 Log-likelihood functions for the image data models

We have seen from the above descriptions and examples that a key step in maximum likelihood estimation is finding an expression for the log-likelihood function for the specific parameter estimation problem. We will therefore discuss here the log-likelihood functions for the estimation of parameters from image data that are described by the fundamental and practical data models presented in Chapter 15.

16.2.1 Log-likelihood function for the fundamental data model

Although the experimental data that we obtain leads to an estimation problem involving a practical data model, we nevertheless also want to briefly discuss parameter estimation in the context of the fundamental data model. As we will see in later chapters, the analysis of estimation problems involving the fundamental data model can produce very useful insights.

We will now sketch the derivation of the log-likelihood function for parameter estimation given data described by the fundamental data model. From an acquired fundamental data image, we might, for example, want to estimate the location $\theta = (x_0, y_0)$ of an object such as an organelle or a single molecule. To arrive at the log-likelihood function for such a problem, we begin with the mathematical description of an acquired image as presented in Section 15.1.

Let $\{r_1, \ldots, r_{N_0}\}$, where $r_k = (x_k, y_k) \in C$, $k = 1, \ldots, N_0$, be the observed locations of photon impact on the detector C during the exposure time interval $[t_0, t]$, and let $\{\tau_1, \ldots, \tau_{N_0}\}$ be the time points at which these impacts are detected. If $U_1, \ldots, U_{N_0}$ are the random variables describing the photon detection locations on the detector and $T_1, \ldots, T_{N_0}$ are the random variables describing the photon detection time points, the log-likelihood function $\mathcal{L}$ corresponding to the acquired "image" is given by

$$\begin{aligned}
&\mathcal{L}\left(\theta \mid (r_1, \tau_1), \ldots, (r_{N_0}, \tau_{N_0})\right) \\
&\qquad = \ln p_\theta\left(U_1 = r_1, \ldots, U_{N_0} = r_{N_0}, T_1 = \tau_1, \ldots, T_{N_0} = \tau_{N_0}, \mathcal{N}(t) = N_0 \mid \mathcal{N}(t) > 0\right),
\end{aligned}$$

where p_θ is the probability density that depends on the true parameter θ, and $\mathcal{N}(\tau)$, $\tau \geq t_0$, is the Poisson process that describes the detection of photons over time. Note that p_θ is a density conditioned on the number of photons $\mathcal{N}(t)$ detected over the exposure time interval $[t_0, t]$ being greater than zero. This importantly excludes the scenario of $\mathcal{N}(t) = 0$, in which no photons are detected during $[t_0, t]$ and therefore no image is actually acquired. Writing this expression in terms of a joint probability distribution, we have

$$\begin{aligned}
&\mathcal{L}\left(\theta \mid (r_1, \tau_1), \ldots, (r_{N_0}, \tau_{N_0})\right) \\
&\quad = \ln\left[p_\theta\left(U_1 = r_1, \ldots, U_{N_0} = r_{N_0}, T_1 = \tau_1, \ldots, T_{N_0} = \tau_{N_0}, \mathcal{N}(t) = N_0, \mathcal{N}(t) > 0\right)\right. \\
&\qquad\qquad \left. / P\left(\mathcal{N}(t) > 0\right)\right] \\
&\quad = \ln\left[p_\theta\left(U_1 = r_1, \ldots, U_{N_0} = r_{N_0}, T_1 = \tau_1, \ldots, T_{N_0} = \tau_{N_0}, \mathcal{N}(t) = N_0\right) / P\left(\mathcal{N}(t) > 0\right)\right],
\end{aligned}$$

where we have used that $\mathcal{N}(t) = N_0$ implies $\mathcal{N}(t) > 0$.

Rewriting the log-likelihood function in terms of conditional probability densities, we obtain

$$\begin{aligned}
&\mathcal{L}(\theta \mid (r_1, \tau_1), \ldots, (r_{N_0}, \tau_{N_0})) \\
&\quad = \ln \left[p_\theta \left(U_1 = r_1, \ldots, U_{N_0} = r_{N_0} \mid T_1 = \tau_1, \ldots, T_{N_0} = \tau_{N_0}, \mathcal{N}(t) = N_0 \right) \right. \\
&\qquad \left. \times p_\theta \left(T_1 = \tau_1, \ldots, T_{N_0} = \tau_{N_0} \mid \mathcal{N}(t) = N_0 \right) P\left(\mathcal{N}(t) = N_0\right) / P\left(\mathcal{N}(t) > 0\right) \right] \\
&\quad = \ln \left[p_\theta \left(U_1 = r_1 \mid T_1 = \tau_1, \ldots, T_{N_0} = \tau_{N_0} \right) p_\theta \left(U_2 = r_2 \mid T_1 = \tau_1, \ldots, T_{N_0} = \tau_{N_0} \right) \right. \\
&\qquad \cdots p_\theta \left(U_{N_0} = r_{N_0} \mid T_1 = \tau_1, \ldots, T_{N_0} = \tau_{N_0} \right) \\
&\qquad \left. \times p_\theta \left(T_1 = \tau_1, \ldots, T_{N_0} = \tau_{N_0} \mid \mathcal{N}(t) = N_0 \right) P\left(\mathcal{N}(t) = N_0\right) / P\left(\mathcal{N}(t) > 0\right) \right] \\
&\quad = \ln \left[p_\theta \left(U_1 = r_1 \mid T_1 = \tau_1 \right) p_\theta \left(U_2 = r_2 \mid T_2 = \tau_2 \right) \cdots p_\theta \left(U_{N_0} = r_{N_0} \mid T_{N_0} = \tau_{N_0} \right) \right. \\
&\qquad \left. \times p_\theta \left(T_1 = \tau_1, \ldots, T_{N_0} = \tau_{N_0} \mid \mathcal{N}(t) = N_0 \right) P\left(\mathcal{N}(t) = N_0\right) / P\left(\mathcal{N}(t) > 0\right) \right].
\end{aligned}$$

Note here that we have used the mutual independence of the photon locations $U_1, \ldots, U_{N_0}$ to write their joint density as the product of their individual densities. Now, using the definition of the photon distribution profiles $\{f_{\theta,\tau_k}\}_{k=1,\ldots,N_0}$ (Section 15.1.2) and of the intensity function Λ_θ of the Poisson process (i.e., the photon detection rate (Section 15.1.1)), we have

$$\begin{aligned}
&\mathcal{L}\left(\theta \mid (r_1, \tau_1), \ldots, (r_{N_0}, \tau_{N_0})\right) \\
&\quad = \ln \left[\left(\prod_{k=1}^{N_0} f_{\theta,\tau_k}(r_k) \right) \right. \\
&\qquad \left. \times p_\theta \left(T_1 = \tau_1, \ldots, T_{N_0} = \tau_{N_0} \mid \mathcal{N}(t) = N_0 \right) P\left(\mathcal{N}(t) = N_0\right) / P\left(\mathcal{N}(t) > 0\right) \right] \\
&\quad = \ln \left[\left(\prod_{k=1}^{N_0} f_{\theta,\tau_k}(r_k) \right) \right. \\
&\qquad \left. \times \frac{N_0! \prod_{k=1}^{N_0} \Lambda_\theta(\tau_k)}{\left(\int_{t_0}^{t} \Lambda_\theta(\tau) d\tau \right)^{N_0}} \cdot \frac{e^{-\int_{t_0}^{t} \Lambda_\theta(\tau) d\tau} \left(\int_{t_0}^{t} \Lambda_\theta(\tau) d\tau \right)^{N_0}}{N_0!} \cdot \frac{1}{1 - e^{-\int_{t_0}^{t} \Lambda_\theta(\tau) d\tau}} \right] \\
&\quad = \ln \left[\left(\prod_{k=1}^{N_0} f_{\theta,\tau_k}(r_k) \right) \left(\prod_{k=1}^{N_0} \Lambda_\theta(\tau_k) \right) \cdot \frac{e^{-\int_{t_0}^{t} \Lambda_\theta(\tau) d\tau}}{1 - e^{-\int_{t_0}^{t} \Lambda_\theta(\tau) d\tau}} \right] \\
&\quad = \sum_{k=1}^{N_0} \ln f_{\theta,\tau_k}(r_k) + \sum_{k=1}^{N_0} \ln \Lambda_\theta(\tau_k) - \int_{t_0}^{t} \Lambda_\theta(\tau) d\tau - \ln \left(1 - e^{-\int_{t_0}^{t} \Lambda_\theta(\tau) d\tau} \right). \qquad (16.10)
\end{aligned}$$

Here we have used that $P\left(\mathcal{N}(t) > 0\right) = 1 - P\left(\mathcal{N}(t) = 0\right)$, and the established result that the probability density of the arrival times of a Poisson process, conditioned on the number of arrivals, is given by (Note 16.3)

$$p_\theta \left(T_1 = \tau_1, \ldots, T_{N_0} = \tau_{N_0} \mid \mathcal{N}(t) = N_0 \right) = \frac{N_0! \prod_{k=1}^{N_0} \Lambda_\theta(\tau_k)}{\left(\int_{t_0}^{t} \Lambda_\theta(\tau) d\tau \right)^{N_0}}.$$

Equation (16.10) provides a general form of the log-likelihood function for a parameter estimation problem if the acquired data can be modeled by the fundamental data model. As pointed out in Section 15.1, the fundamental data model is most frequently used for theoretical and baseline considerations. As we will see, analyzing estimation problems for the fundamental data model will provide important insights. Note that unlike in Chapter 15, the functions f_{θ,τ_k} and Λ_θ used here have the subscript θ to denote them as functions of the parameters to be estimated.

16.2.1.1 Example 3: Localization of an object with a 2D Gaussian image profile

We consider here an "image" of a stationary point source that is acquired under the fundamental data scenario in which the detector $C = \mathbb{R}^2$ is infinitely large. An image of N_0 photons will therefore comprise the time points $\tau_1, \ldots, \tau_{N_0}$ and the mutually independent coordinates $r_k = (x_k, y_k) \in \mathbb{R}^2$, $k = 1, \ldots, N_0$, at which the photons are detected during some acquisition time interval $[t_0, t]$. The imaging system is translation-invariant. We will show that if the image of the point source is described by a 2D Gaussian profile, then the maximum likelihood estimate $\hat{\theta}_{MLE}$ of the point source's positional coordinates $\theta = (x_0, y_0) \in \mathbb{R}^2$ is the magnification-corrected *center of mass* of the spatial coordinates of the N_0 detected photons.

To derive the maximum likelihood estimate $\hat{\theta}_{MLE}$, we first need to determine the log-likelihood function for the data in question. As our image is described by the fundamental data model, we can start with the general expression of Eq. (16.10). Noting that in the estimation problem considered here the photon detection rate does not depend on the parameter θ, the terms involving Λ_θ are constants with respect to the maximization of θ, and we can immediately drop them from Eq. (16.10). As our point source is immobile, the photon distribution profile is independent of time, and we can write $f_{\theta,\tau_k} = f_\theta$ for $k = 1, \ldots, N_0$. Moreover, as the image of the point source is a 2D Gaussian profile acquired with a translation-invariant imaging system, we have, for $k = 1, \ldots, N_0$,

$$f_\theta(r_k) = \frac{1}{M^2} q\left(\frac{x_k}{M} - x_0, \frac{y_k}{M} - y_0\right) = \frac{1}{M^2}\frac{1}{2\pi\sigma_g^2} e^{-\frac{1}{2\sigma_g^2}\left(\left(\frac{x_k}{M}-x_0\right)^2+\left(\frac{y_k}{M}-y_0\right)^2\right)}. \tag{16.11}$$

This expression is obtained as the stationary object version of Eq. (15.8), with the image function q given by the 2D Gaussian function of Eq. (15.10). Our log-likelihood function is therefore given by

$$\begin{aligned}\mathcal{L}(\theta \mid (r_1, \tau_1), \ldots, (r_{N_0}, \tau_{N_0})) &= \sum_{k=1}^{N_0} \ln f_{\theta,\tau_k}(r_k) = \sum_{k=1}^{N_0} \ln f_\theta(r_k) \\ &= \sum_{k=1}^{N_0} \ln\left[\frac{1}{M^2}\frac{1}{2\pi\sigma_g^2} e^{-\frac{1}{2\sigma_g^2}\left(\left(\frac{x_k}{M}-x_0\right)^2+\left(\frac{y_k}{M}-y_0\right)^2\right)}\right]. \end{aligned}\tag{16.12}$$

Note that due to the time-independent photon distribution profile, the time points of photon detection $\tau_1, \ldots, \tau_{N_0}$ do not appear in this expression and are of no relevance to this estimation problem.

By definition, the maximum likelihood estimate $\hat{\theta}_{MLE}$ is the value of θ that maximizes the log-likelihood function of Eq. (16.12). We therefore start with the expression

$$\begin{aligned}\hat{\theta}_{MLE} &= \arg\max_{\theta\in\mathbb{R}^2} \mathcal{L}(\theta \mid r_1, \ldots, r_{N_0}) \\ &= \arg\max_{\theta\in\mathbb{R}^2} \sum_{k=1}^{N_0} \ln\left(\frac{1}{M^2}\frac{1}{2\pi\sigma_g^2} e^{-\frac{1}{2\sigma_g^2}\left(\left(\frac{x_k}{M}-x_0\right)^2+\left(\frac{y_k}{M}-y_0\right)^2\right)}\right) \\ &= \arg\max_{\theta\in\mathbb{R}^2} \sum_{k=1}^{N_0} \left[\ln\left(\frac{1}{M^2}\frac{1}{2\pi\sigma_g^2}\right) - \frac{1}{2\sigma_g^2}\left(\left(\frac{x_k}{M} - x_0\right)^2 + \left(\frac{y_k}{M} - y_0\right)^2\right)\right] \\ &= \arg\max_{\theta\in\mathbb{R}^2} \sum_{k=1}^{N_0} \left[-\frac{1}{2\sigma_g^2}\left(\left(\frac{x_k}{M} - x_0\right)^2 + \left(\frac{y_k}{M} - y_0\right)^2\right)\right]\end{aligned}$$

$$= \arg\min_{\theta \in \mathbb{R}^2} \sum_{k=1}^{N_0} \left[\left(\frac{x_k}{M} - x_0 \right)^2 + \left(\frac{y_k}{M} - y_0 \right)^2 \right],$$

which gives $\hat{\theta}_{MLE}$ as the value of θ that minimizes a sum. To find this value of θ, we differentiate the sum with respect to $\theta = (x_0, y_0)$ and set the result equal to zero, arriving at

$$\begin{aligned}
[0 \quad 0] &= \frac{\partial}{\partial \theta} \sum_{k=1}^{N_0} \left[\left(\frac{x_k}{M} - x_0 \right)^2 + \left(\frac{y_k}{M} - y_0 \right)^2 \right] \\
&= \left[\sum_{k=1}^{N_0} \frac{\partial}{\partial x_0} \left(\left(\tfrac{x_k}{M} - x_0 \right)^2 + \left(\tfrac{y_k}{M} - y_0 \right)^2 \right) \quad \sum_{k=1}^{N_0} \frac{\partial}{\partial y_0} \left(\left(\tfrac{x_k}{M} - x_0 \right)^2 + \left(\tfrac{y_k}{M} - y_0 \right)^2 \right) \right] \\
&= \left[\sum_{k=1}^{N_0} \left(-2 \left(\tfrac{x_k}{M} - x_0 \right) \right) \quad \sum_{k=1}^{N_0} \left(-2 \left(\tfrac{y_k}{M} - y_0 \right) \right) \right] \\
&= \left[-2 \left(\sum_{k=1}^{N_0} \tfrac{x_k}{M} - N_0 x_0 \right) \quad -2 \left(\sum_{k=1}^{N_0} \tfrac{y_k}{M} - N_0 y_0 \right) \right].
\end{aligned}$$

Finally, we solve the equation to obtain the maximum likelihood estimate as

$$\hat{\theta}_{MLE} = [\hat{x}_0 \quad \hat{y}_0] = \left[\frac{1}{MN_0} \sum_{k=1}^{N_0} x_k \quad \frac{1}{MN_0} \sum_{k=1}^{N_0} y_k \right], \tag{16.13}$$

which is just the magnification-corrected center of mass of the spatial coordinates of the detected photons.

We should be careful to note that the derivation of the general log-likelihood function of Eq. (16.10) takes into account the fact that a fundamental data image is the result of photons arriving at the detector according to a Poisson process $\mathcal{N}(\tau)$, $\tau \geq t_0$. Specifically, it incorporates the probability that $\mathcal{N}(t)$, the random number of photons detected over the acquisition time interval $[t_0, t]$, is equal to N_0, given that $\mathcal{N}(t) > 0$. The inclusion of this probability addresses the fact that if one were to repeat the image acquisition under the same conditions, the value of N_0 would potentially be different. When considering properties such as the expectation of an estimator like $\hat{\theta}_{MLE}$, we should therefore take care to recognize N_0 as a realization of the random variable $\mathcal{N}(t)$, given that $\mathcal{N}(t) > 0$.

An example of the estimator of Eq. (16.13) is given in Fig. 16.2, which shows the results obtained for 500 repeat images of a stationary point source located at $(x_0, y_0) = (250 \text{ nm}, -750 \text{ nm})$ in the object space, simulated using a magnification of $M = 63$. Each image is obtained by assuming a Poisson process by which $\mathrm{E}[\mathcal{N}(t)] = 50$ photons are, on average, detected during the exposure time interval $[t_0, t]$. Spatially, the locations of the detected photons are randomly distributed according to a 2D Gaussian image function with a width of $\sigma_g = 95.00$ nm. Part (a) of the figure shows one of the 500 repeat images and the estimate $(\hat{x}_0, \hat{y}_0)$ obtained for it using Eq. (16.13). Part (b) shows a plot of the estimates $(\hat{x}_0, \hat{y}_0)$ obtained for all 500 images.

16.2.2 Log-likelihood functions for the practical data models

We have seen in Section 16.1 that in the case of a practical data model, the log-likelihood function for the values $z_1, \ldots, z_{N_p}$ acquired in the pixels of an N_p-pixel image is given by

$$\mathcal{L}(\theta \mid z_1, \ldots, z_{N_p}) = \ln p_\theta(z_1, \ldots, z_{N_p}) = \ln \prod_{k=1}^{N_p} p_{\theta,k}(z_k) = \sum_{k=1}^{N_p} \ln p_{\theta,k}(z_k), \tag{16.14}$$

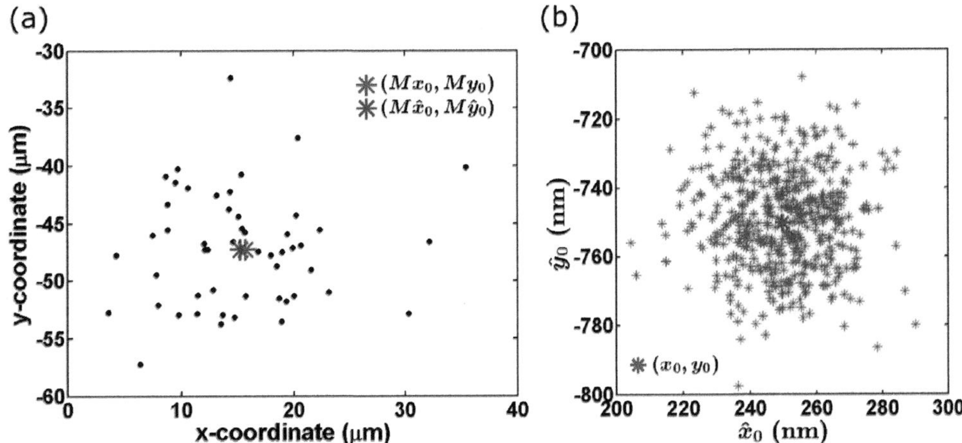

FIGURE 16.2
Example of the maximum likelihood estimator of the location (x_0, y_0) of a stationary point source with a 2D Gaussian image profile. (a) One of 500 repeat "images" of the point source under the fundamental data model, each simulated with a magnification of $M = 63$ and a 2D Gaussian image function with width $\sigma_g = 95.00$ nm. An "image" consists of the (x, y) locations (black dots) of photons detected from the point source, randomly distributed according to the spatial density based on the image function. The simulation assumes that, on average, $\mathrm{E}[\mathcal{N}(t)] = 50$ photons are detected during the exposure time interval $[t_0, t]$. A fundamental data image technically also includes the time points at which the photons are detected, but the time information is omitted here. For the image shown, the number of detected photons is $N_0 = 51$. The object space location of the point source is $(x_0, y_0) = (250\text{ nm}, -750\text{ nm})$, and for the image shown the maximum likelihood estimate obtained using Eq. (16.13) is $(\hat{x}_0, \hat{y}_0) = (242.66\text{ nm}, -750.01\text{ nm})$. After mapping to the image space and a unit conversion to microns, these locations are indicated in the image with stars. (b) The estimates $(\hat{x}_0, \hat{y}_0)$ obtained for all 500 "images" of the point source. For reference, the true location (x_0, y_0) is also shown.

where $p_{\theta,k}$ is the probability density or mass function for the data in the kth pixel of the image. From this expression, it is easily seen that the log-likelihood functions for different practical data models will differ via the log-likelihood function $\ln p_{\theta,k}$ for the kth pixel, $k = 1, \ldots, N_p$. In Table 16.1, the log-likelihood function $\ln p_{\theta,k}$ is shown that corresponds to the probability distribution p_k for each practical data model presented in Section 15.2. In all cases, the typical situation is assumed in which information about the quantities we wish to estimate is contained only within the signal detected from the imaged sample, and not within noise components such as the detector's readout noise. Mathematically, this means that the parameter θ parameterizes only the signal component of the data in each pixel k. For the simple form of the deterministic data model, in which the Gaussian noise component in a pixel does not depend on the signal detected in that pixel, we therefore denote the deterministic signal D_k in the kth pixel as $D_{\theta,k}$. For all the other practical data models, including the Gaussian approximations for the CCD/CMOS and EMCCD data models which represent more complex forms of the deterministic data model, we likewise denote the mean ν_k of the Poisson component (i.e., the signal component) of the data in the kth pixel as $\nu_{\theta,k}$. Note that the log-likelihood function for the simple form of the deterministic data model has in fact the form of a least squares expression. We will use

Data model	Log-likelihood function $\ln p_{\theta,k}(z_k)$
Poisson	$z_k \ln(\nu_{\theta,k}) - \nu_{\theta,k} - \ln(z_k!)$
CCD/CMOS	$-\nu_{\theta,k} - \ln\left(\sqrt{2\pi}\sigma_k\right) + \ln\left(\sum_{l=0}^{\infty} \frac{\nu_{\theta,k}^{l}}{l!} e^{-\frac{1}{2}\frac{(z_k-l-\eta_k)^2}{\sigma_k^2}}\right)$
Deterministic	$-\ln\left(\sqrt{2\pi\xi_k^2}\right) - \frac{(z_k - D_{\theta,k} - \zeta_k)^2}{2\xi_k^2}$
CCD/CMOS Gaussian approximation	$-\ln\left(\sqrt{2\pi(\nu_{\theta,k} + \sigma_k^2)}\right) - \frac{(z_k - \nu_{\theta,k} - \eta_k)^2}{2(\nu_{\theta,k} + \sigma_k^2)}$
EMCCD	$-\nu_{\theta,k} - \ln\left(\sqrt{2\pi}\sigma_k\right) + \ln\left(e^{-\left(\frac{z_k-\eta_k}{\sqrt{2}\sigma_k}\right)^2} + \sum_{l=1}^{\infty} e^{-\left(\frac{z_k-l-\eta_k}{\sqrt{2}\sigma_k}\right)^2} \sum_{j=0}^{l-1} \frac{\binom{l-1}{j}\left(1-\frac{1}{g}\right)^{l-j-1}}{(j+1)!\left(\frac{g}{\nu_{\theta,k}}\right)^{j+1}}\right)$
EMCCD high gain approximation	$-\nu_{\theta,k} - \ln\left(\sqrt{2\pi}\sigma_k\right) + \ln\left(e^{-\left(\frac{z_k-\eta_k}{\sqrt{2}\sigma_k}\right)^2} + \int_0^{\infty} e^{-\left(\frac{z_k-u-\eta_k}{\sqrt{2}\sigma_k}\right)^2 - \frac{u}{g}} \frac{\sqrt{\frac{\nu_{\theta,k} u}{g}} I_1\left(2\sqrt{\frac{\nu_{\theta,k} u}{g}}\right)}{u} du\right)$
EMCCD Gaussian approximation	$-\ln\left(\sqrt{2\pi\left(\left(2g^2 - g\right)\nu_{\theta,k} + \sigma_k^2\right)}\right) - \frac{(z_k - g\nu_{\theta,k} - \eta_k)^2}{2\left(\left(2g^2 - g\right)\nu_{\theta,k} + \sigma_k^2\right)}$

TABLE 16.1
Log-likelihood function for the kth pixel of an image under various practical data models. For each log-likelihood function, the corresponding probability distribution for the data in the kth image pixel is given in Table 15.1.

this observation in Section 16.4 to investigate the relationship between maximum likelihood estimation and least squares estimation.

16.3 Obtaining the maximum likelihood estimate

Now that we have established the log-likelihood functions for the different image data models we had previously introduced, the maximum likelihood estimate $\hat{\theta}_{MLE}$ of the parameter θ can be found, given the acquired image data, by maximizing the appropriate log-likelihood function with respect to θ (Note 16.4). In the two introductory examples (Sections 16.1.1 and 16.1.2) where θ is the mean of a Poisson or Gaussian random variable, we have seen that we can in fact solve the maximization problem analytically by treating it as a problem in calculus. Likewise, in the example of Section 16.2.1.1, we have shown that for a specific scenario under the fundamental data model, we can analytically determine the maximum likelihood estimate of the location of a point source. In all three cases, we obtained the derivative of the log-likelihood function with respect to θ, set this derivative equal to zero, and managed to solve the resulting equation analytically. Although not shown in the examples, one should confirm that in each case the solution indeed maximizes the log-likelihood function. This can be done by, for example, verifying the standard second order criteria for a maximum.

Unfortunately, typically we cannot assume that a maximum likelihood estimate can be analytically derived. In those cases numerical optimization routines need to be employed, and they are readily available in standard numerical analysis software packages.

16.4 Maximum likelihood estimation and least squares estimation

In Table 16.1, the log-likelihood function for the kth image pixel is shown for the simple form of the deterministic data model in which the Gaussian noise component of the data is independent of the signal component of the data (Section 15.2.3). We will now analyze the maximum likelihood estimator for this problem, as in this particular case we can correlate the maximum likelihood estimation problem to a different optimization problem, namely the least squares optimization problem. In fact, in Section 16.1.2 we already considered a special case of this problem.

Let us assume that we have data $z_1, \ldots, z_{N_p}$ from N_p pixels, and that the data obeys the deterministic data model such that the data z_k in the kth pixel is a realization of the random variable $D_{\theta,k} + U_k$, where $D_{\theta,k}$ is a deterministic quantity that depends on the parameter θ, and U_k is a Gaussian random variable with mean ζ_k and variance ξ_k^2. Then the log-likelihood function is given by

$$\mathcal{L}(\theta \mid z_1, \ldots, z_{N_p}) = \sum_{k=1}^{N_p} \ln p_{\theta,k}(z_k) = -\sum_{k=1}^{N_p} \ln\left(\sqrt{2\pi\xi_k^2}\right) - \sum_{k=1}^{N_p} \frac{(z_k - D_{\theta,k} - \zeta_k)^2}{2\xi_k^2},$$

where we have substituted into Eq. (16.14) the expression shown for the deterministic data model in Table 16.1.

To obtain the maximum likelihood estimate, we need to maximize the above expression or, equivalently, minimize its negative

$$\sum_{k=1}^{N_p} \ln\left(\sqrt{2\pi\xi_k^2}\right) + \sum_{k=1}^{N_p} \frac{(z_k - D_{\theta,k} - \zeta_k)^2}{2\xi_k^2}$$

over all possible values of θ. As its first term does not depend on θ, minimization of this expression is achieved by minimizing

$$\sum_{k=1}^{N_p} \frac{(z_k - D_{\theta,k} - \zeta_k)^2}{2\xi_k^2}.$$

But this expression is the sum of the squares of the residuals between the observed data and the model. Therefore, we have shown that for this simple form of the deterministic data model, the maximum likelihood estimator is given by the parameter that minimizes this least squares criterion, i.e., it is given by the least squares estimator. In this particular case, we have a weighted least squares problem, where the weights are the inverses of the known variances of the Gaussian noise component in the data. If these variances are all identical (i.e., $\xi_1^2 = \xi_2^2 = \cdots = \xi_{N_p}^2$), we can instead minimize the function

$$\sum_{k=1}^{N_p} (z_k - D_{\theta,k} - \zeta_k)^2,$$

which is the classical unweighted least squares criterion. It is immediately seen that this provides a generalization of the result in Section 16.1.2, where we saw an example in which the maximum likelihood estimator is a least squares estimator. Specifically, if the data in all N_p pixels of the image are identically distributed, i.e., if for $k = 1, \ldots, N_p$, $D_{\theta,k} + \zeta_k = D_{\theta,0} + \zeta_0$ and $\xi_k^2 = \xi_0^2$, where D_{θ_0}, ζ_0, and ξ_0 are constants, then we obtain the problem of Section 16.1.2 by taking $\theta = D_{\theta,0} + \zeta_0$, the mean of a Gaussian random variable, as the parameter to be estimated.

16.5 Unbiased estimator

We have now introduced maximum likelihood estimation to obtain estimates of quantities that we would like to obtain from the acquired data. The question then immediately arises whether these estimates are of good quality. To answer this question we first need to establish criteria for the quality of the estimates. As the first criterion for the quality of an estimation approach, we will consider the *bias* of the estimator, meaning the amount by which, on average, the estimates deviate from the true parameter. The bias of an estimator $\hat{\theta}$ of the parameter θ is thus the difference $\mathrm{E}\left[\hat{\theta}\right] - \theta$, where $\mathrm{E}\left[\hat{\theta}\right]$ is the expected value of the estimator. An estimator $\hat{\theta}$ whose bias equals zero for all values of θ is called *unbiased*.

In this section, we will discuss a number of specific examples for which the question of bias can be analyzed in detail. Such analyses, however, are unfortunately not possible in general. For example, when we consider in subsequent chapters the estimation of the location of an object from image data described by the practical data models, we will make empirical assessments of bias through simulations.

16.5.1 Example 1: sample mean

First, we will show that the *sample mean* is an unbiased estimator of the population mean. Suppose we have a sample of K realizations of a random variable with mean m, denoted by $z_1, \ldots, z_K$. The sample mean $\hat{m}$ is then given by

$$\hat{m} = \frac{1}{K}\sum_{i=1}^{K} z_i, \tag{16.15}$$

and its expectation is

$$\mathrm{E}\left[\hat{m}\right] = \mathrm{E}\left[\frac{1}{K}\sum_{i=1}^{K} z_i\right] = \frac{1}{K}\sum_{i=1}^{K}\mathrm{E}[z_i] = \frac{1}{K}\cdot K\cdot m = m,$$

which is the population mean.

The maximum likelihood estimators of Sections 16.1.1 and 16.1.2 are sample means of realizations of a Poisson and a Gaussian random variable, respectively, and are therefore unbiased estimators of the mean of those random variables (Fig. 16.3).

16.5.2 Example 2: sample variance

We now turn our attention to the estimation of the variance of a random variable. By definition, the variance of a random variable X is given by

$$\mathrm{Var}(X) = \mathrm{E}\left[(X - \mathrm{E}[X])^2\right]. \tag{16.16}$$

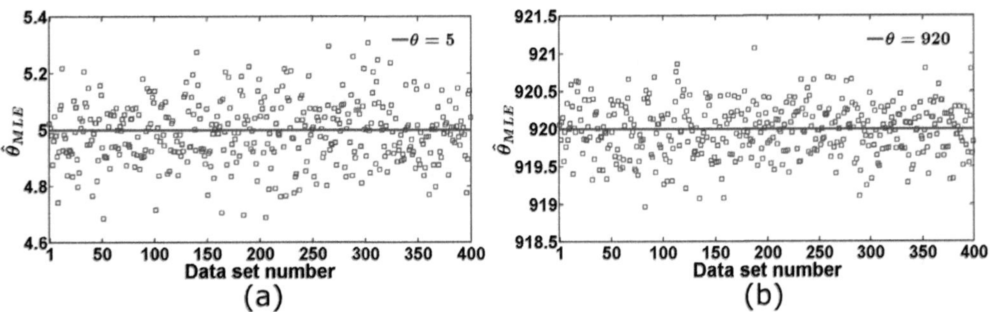

FIGURE 16.3
The sample mean is an unbiased estimator of the population mean. In both (a) and (b), the unbiasedness of a sample mean can be seen through the even spread of estimates about the mean θ of a random variable. Shown in (a) are estimates of the mean of a Poisson random variable, obtained by applying the maximum likelihood estimator $\hat{\theta}_{MLE}$ of Eq. (16.6) to 400 data sets. Each data set consists of $N_{im} = 500$ realizations of a Poisson random variable with mean $\theta = \mu_k = 5$. See Fig. 16.1(a) for an example data set and its associated estimate. Shown in (b) are estimates of the mean of a Gaussian random variable, obtained by applying the maximum likelihood estimator $\hat{\theta}_{MLE}$ of Eq. (16.9) to 400 data sets. Each data set comprises $N_{im} = 500$ realizations of a Gaussian random variable with mean $\theta = \eta_k = 920$ and variance $\sigma_k^2 = 64$. See Fig. 16.1(b) for an example data set and its associated estimate.

Suppose we have a sample of K realizations, denoted by $z_1, \ldots, z_K$, of the random variable X which has mean m and variance σ^2. Following the definition of the variance, an estimate $\widehat{\sigma^2}$ of the variance could be introduced as

$$\widehat{\sigma^2} = \frac{1}{K}\sum_{i=1}^{K}\left(z_i - \frac{1}{K}\sum_{l=1}^{K} z_l\right)^2 = \frac{1}{K}\sum_{i=1}^{K}(z_i - \hat{m})^2,$$

where $\hat{m} = \frac{1}{K}\sum_{l=1}^{K} z_l$ is the sample mean. We refer to this estimate $\widehat{\sigma^2}$ as the *sample variance*. Expanding the square, we obtain

$$\begin{aligned}\widehat{\sigma^2} &= \frac{1}{K}\sum_{i=1}^{K}(z_i - \hat{m})^2 = \frac{1}{K}\sum_{i=1}^{K}\left(z_i^2 - 2z_i\hat{m} + \hat{m}^2\right) = \frac{1}{K}\sum_{i=1}^{K} z_i^2 - 2\hat{m}\frac{1}{K}\sum_{i=1}^{K} z_i + \frac{1}{K}\sum_{i=1}^{K}\hat{m}^2 \\ &= \frac{1}{K}\sum_{i=1}^{K} z_i^2 - 2\hat{m}^2 + \hat{m}^2 = \frac{1}{K}\sum_{i=1}^{K} z_i^2 - \hat{m}^2.\end{aligned}$$

Taking the expectation then yields

$$\begin{aligned}\mathrm{E}\left[\widehat{\sigma^2}\right] &= \mathrm{E}\left[\frac{1}{K}\sum_{i=1}^{K} z_i^2 - \hat{m}^2\right] = \frac{1}{K}\sum_{i=1}^{K}\mathrm{E}\left[z_i^2\right] - \mathrm{E}\left[\hat{m}^2\right] = \frac{1}{K}K\mathrm{E}\left[z_i^2\right] - \mathrm{E}\left[\hat{m}^2\right] \\ &= \mathrm{E}\left[z_i^2\right] - \mathrm{E}\left[\hat{m}^2\right] = \mathrm{Var}(z_i) + (\mathrm{E}[z_i])^2 - \left(\mathrm{Var}(\hat{m}) + (\mathrm{E}[\hat{m}])^2\right) \\ &= \sigma^2 + m^2 - \left(\mathrm{Var}\left(\frac{1}{K}\sum_{i=1}^{K} z_i\right) + m^2\right) = \sigma^2 + m^2 - \left(\frac{1}{K^2}\sum_{i=1}^{K}\mathrm{Var}(z_i) + m^2\right) \\ &= \sigma^2 + m^2 - \left(\frac{K}{K^2}\sigma^2 + m^2\right) = \sigma^2 + \frac{\sigma^2}{K} = \left(1 - \frac{1}{K}\right)\sigma^2,\end{aligned}$$

where we have used the identity $\mathrm{Var}(X) = \mathrm{E}\left[X^2\right] - (\mathrm{E}[X])^2$, and the fact that the sample mean $\hat{m}$ is unbiased as shown in Section 16.5.1. Since $\mathrm{E}\left[\hat{\sigma^2}\right] \neq \sigma^2$, we see that the sample variance $\hat{\sigma^2}$ is biased. Note, however, that when the size of the sample increases, the amount of bias decreases, since

$$\mathrm{E}\left[\hat{\sigma^2}\right] \to \sigma^2 \quad \text{as} \quad K \to \infty.$$

Therefore, for a large sample, the bias is probably not significant from a practical point of view. We can obtain an unbiased estimator, however, by introducing a constant c such that

$$\mathrm{E}\left[c\hat{\sigma^2}\right] = c\mathrm{E}\left[\hat{\sigma^2}\right] = c\left(1 - \frac{1}{K}\right)\sigma^2 = \sigma^2.$$

Choosing $c = 1/(1 - 1/K)$, the unbiased sample variance $\hat{\sigma_u^2}$ is then given by

$$\hat{\sigma_u^2} = c\hat{\sigma^2} = \frac{1}{1 - \frac{1}{K}} \cdot \frac{1}{K}\sum_{i=1}^{K}(z_i - \hat{m})^2 = \frac{1}{K-1}\sum_{i=1}^{K}(z_i - \hat{m})^2.$$

16.5.3 Example 3: center of mass as an object location estimator under the fundamental data model

We now consider an estimator whose bias can be determined analytically under the fundamental data model. This estimator is the object location estimator that returns as the object's location the magnification-corrected center of mass of the locations of the photons detected from the object. In Section 16.2.1.1, we saw that under the fundamental data model the maximum likelihood estimate of the location of an object with a 2D Gaussian image profile is in fact such an estimator (Eq. (16.13)). Obviously, the center of mass could also be used as an estimate of the location of an object with any other image profile.

Suppose an "image" of a stationary object is captured by a detector $C = \mathbb{R}^2$ over some time interval $[t_0, t]$ under the fundamental data scenario. Let N_0 be the number of photons comprising this image, and let the spatial coordinates of the N_0 photons be denoted by $r_k = (x_k, y_k) \in \mathbb{R}^2$, $k = 1, \ldots, N_0$. Let $\theta = (x_0, y_0) \in \mathbb{R}^2$ be the location of the object, and let $\hat{\theta}_{CM}$ be the estimate of θ given by the magnification-corrected center of mass of the spatial coordinates of the N_0 photons, i.e.,

$$\hat{\theta}_{CM} = \left[\frac{1}{MN_0}\sum_{k=1}^{N_0} x_k \quad \frac{1}{MN_0}\sum_{k=1}^{N_0} y_k\right]. \tag{16.17}$$

Let $U = [X\ Y]$ be the random vector that describes the location of photon detection. We show here that if the expectation of U is the object's location θ scaled by the magnification $M > 0$ of the imaging system (i.e., if $\mathrm{E}[U] = [\mathrm{E}[X]\ \mathrm{E}[Y]] = [Mx_0\ My_0]$), then $\hat{\theta}_{CM}$ is an unbiased estimator of θ.

Note that although the fundamental data "image" technically consists of the detection time points of the N_0 photons in addition to their spatial coordinates, they play no role in this estimation problem and are therefore omitted from our problem formulation.

To show that $\hat{\theta}_{CM}$ is unbiased, we need to take its expectation and find that it equals θ. As mentioned in Section 16.2.1.1, when calculating the expectation of an estimator that operates on an image acquired under the fundamental data model, we need to account for the fact that the photon count N_0 is a realization of the Poisson random variable $\mathcal{N}(t)$ representing the number of photons detected during the acquisition time interval $[t_0, t]$, given that $\mathcal{N}(t) > 0$. To calculate the expectation of $\hat{\theta}_{CM}$, we will therefore condition it on the value of $\mathcal{N}(t)$ subject to the constraint $\mathcal{N}(t) > 0$. By this we mean that we will calculate, under the condition $\mathcal{N}(t) > 0$, the expected value of $\hat{\theta}_{CM}$ given each possible value of $\mathcal{N}(t)$, and take the mean of these conditional expected values weighted according to the probabilities with which $\mathcal{N}(t)$ takes on its possible values.

To determine the expectation of $\hat{\theta}_{CM}$, we therefore need to make use of the zero-truncated Poisson probability mass function of Eq. (15.6), which gives the probability that $\mathcal{N}(t)$ takes on a particular value N_0 given that $\mathcal{N}(t) > 0$. Using this mass function and applying the law of total expectation from probability theory, we obtain for the expectation of $\hat{\theta}_{CM}$

$$\begin{aligned}
\mathrm{E}\left[\hat{\theta}_{CM}\right] &= \sum_{N_0=1}^{\infty} \mathrm{E}\left[\hat{\theta}_{CM} \mid \mathcal{N}(t) = N_0\right] \cdot P\left(\mathcal{N}(t) = N_0 \mid \mathcal{N}(t) > 0\right) \\
&= \sum_{N_0=1}^{\infty} \mathrm{E}\left[\begin{bmatrix} \frac{1}{MN_0} \sum_{k=1}^{N_0} x_k & \frac{1}{MN_0} \sum_{k=1}^{N_0} y_k \end{bmatrix} \mid \mathcal{N}(t) = N_0\right] \cdot P\left(\mathcal{N}(t) = N_0 \mid \mathcal{N}(t) > 0\right) \\
&= \sum_{N_0=1}^{\infty} \mathrm{E}\left[\begin{bmatrix} \frac{1}{MN_0} \sum_{k=1}^{N_0} x_k & \frac{1}{MN_0} \sum_{k=1}^{N_0} y_k \end{bmatrix}\right] \cdot P\left(\mathcal{N}(t) = N_0 \mid \mathcal{N}(t) > 0\right) \\
&= \sum_{N_0=1}^{\infty} \begin{bmatrix} \frac{1}{MN_0} \sum_{k=1}^{N_0} \mathrm{E}\left[x_k\right] & \frac{1}{MN_0} \sum_{k=1}^{N_0} \mathrm{E}\left[y_k\right] \end{bmatrix} \cdot P\left(\mathcal{N}(t) = N_0 \mid \mathcal{N}(t) > 0\right) \\
&= \sum_{N_0=1}^{\infty} \begin{bmatrix} \frac{1}{MN_0} \sum_{k=1}^{N_0} Mx_0 & \frac{1}{MN_0} \sum_{k=1}^{N_0} My_0 \end{bmatrix} \cdot P\left(\mathcal{N}(t) = N_0 \mid \mathcal{N}(t) > 0\right) \\
&= \begin{bmatrix} x_0 & y_0 \end{bmatrix} \cdot \sum_{N_0=1}^{\infty} P\left(\mathcal{N}(t) = N_0 \mid \mathcal{N}(t) > 0\right) \\
&= \begin{bmatrix} x_0 & y_0 \end{bmatrix}.
\end{aligned}$$

Here we have used the fact that the expected location of a photon is the object location multiplied by the magnification, and the fact that a probability mass function must sum to one over all possible values of the random variable. We have thus shown that the expectation of $\hat{\theta}_{CM}$ equals the parameter $\theta = (x_0, y_0)$, and that $\hat{\theta}_{CM}$ is therefore unbiased.

Importantly, we can use this relatively general result to show that the maximum likelihood estimator obtained in Section 16.2.1.1 is unbiased. In that example, we showed that under the fundamental data model, the maximum likelihood estimate $\hat{\theta}_{MLE}$ of the location $\theta = (x_0, y_0) \in \mathbb{R}^2$ of a point source with a 2D Gaussian image profile is the magnification-corrected center of mass of the spatial coordinates of the detected photons (Eq. (16.13)). To show that $\hat{\theta}_{MLE}$ is unbiased, we then only need to prove that the expected location of a detected photon is the point source's location multiplied by the magnification. In this case, the random vector $U = [X \ Y]$ that describes the location of a detected photon has the 2D Gaussian photon distribution profile f_θ of Eq. (16.11) as its probability density function. Using f_X and f_Y to denote the probability densities of the component random variables X and Y, we have that

$$f_X(x) = \frac{1}{\sqrt{2\pi\sigma_g^2 M}} e^{-\frac{1}{2\sigma_g^2}\left(\frac{x}{M} - x_0\right)^2}, \ x \in \mathbb{R}, \text{ and } f_Y(y) = \frac{1}{\sqrt{2\pi\sigma_g^2 M}} e^{-\frac{1}{2\sigma_g^2}\left(\frac{y}{M} - y_0\right)^2}, \ y \in \mathbb{R}, \tag{16.18}$$

as it is easily verified that the 2D Gaussian density f_θ can be written as the product of two 1D Gaussian densities, one for X and one for Y. Using f_X and f_Y, the expectation of U is then simply obtained as

$$
\begin{aligned}
\mathrm{E}[U] &= \begin{bmatrix}\mathrm{E}[X] & \mathrm{E}[Y]\end{bmatrix} \\
&= \left[\int_{-\infty}^{\infty} x f_X(x)dx \quad \int_{-\infty}^{\infty} y f_Y(y)dy\right] \\
&= \left[\int_{-\infty}^{\infty} x\cdot\left(\frac{1}{M}\frac{1}{\sqrt{2\pi\sigma_g^2}}e^{-\frac{1}{2\sigma_g^2}\left(\frac{x}{M}-x_0\right)^2}\right)dx \quad \int_{-\infty}^{\infty} y\cdot\left(\frac{1}{M}\frac{1}{\sqrt{2\pi\sigma_g^2}}e^{-\frac{1}{2\sigma_g^2}\left(\frac{y}{M}-y_0\right)^2}\right)dy\right] \\
&= \left[\int_{-\infty}^{\infty} Mu\cdot\left(\frac{1}{M}\frac{1}{\sqrt{2\pi\sigma_g^2}}e^{-\frac{1}{2\sigma_g^2}(u-x_0)^2}\right)Mdu \quad \int_{-\infty}^{\infty} Mu\cdot\left(\frac{1}{M}\frac{1}{\sqrt{2\pi\sigma_g^2}}e^{-\frac{1}{2\sigma_g^2}(u-y_0)^2}\right)Mdu\right] \\
&= \left[M\cdot\int_{-\infty}^{\infty} u\cdot\left(\frac{1}{\sqrt{2\pi\sigma_g^2}}e^{-\frac{1}{2\sigma_g^2}(u-x_0)^2}\right)du \quad M\cdot\int_{-\infty}^{\infty} u\cdot\left(\frac{1}{\sqrt{2\pi\sigma_g^2}}e^{-\frac{1}{2\sigma_g^2}(u-y_0)^2}\right)du\right] \\
&= \begin{bmatrix}Mx_0 & My_0\end{bmatrix},
\end{aligned}
\tag{16.19}
$$

and therefore $\hat{\theta}_{MLE}$ is an unbiased estimator of $\theta = (x_0, y_0)$. The unbiasedness of this estimator can be seen in the example of Fig. 16.2(b) by the symmetric distribution of the 500 estimates around the point source location (x_0, y_0).

We have shown that the center of mass is an unbiased estimate of the object location under the technical condition that $\mathrm{E}[U] = [Mx_0 \ My_0]$, which states that the mean location of the detected photons is the magnification-adjusted location of the object. For a Gaussian image profile we have just shown that this condition is satisfied. In fact, we can show (Exercise 15.7) that this technical condition is satisfied when the object's image profile q is symmetric (i.e., for $(x, y) \in \mathbb{R}^2$, $q(x, y) = q(-x, y) = q(x, -y)$). This therefore implies that the center of mass provides an unbiased estimate of the object location if the spatial distribution of the photons detected from the object can be described by a symmetric image profile.

16.6 Variance of an estimator

We have investigated the unbiasedness of an estimator as a criterion for the quality of an estimator. The bias of an estimator is an indication of how close, on average, an estimate is to the true value of the parameter. A further important criterion for the quality of an estimator relates to the "spread", or variance, of the estimator. We clearly prefer an estimator that has a low variance over an estimator that has a large variance, provided the estimator does not have significant bias. Investigating this criterion for an estimator will be the topic of this section. An extensive analysis of how small the variance of an estimator can be, given the probabilistic characteristics of the data, will be provided in Chapter 17.

As is the case with the question of the bias of an estimator, the variance is in general difficult to determine analytically. However, we can find the variance of the maximum likelihood estimators for which we determined the bias in Sections 16.5.1 and 16.5.3. We will postpone the investigation of the variance of the maximum likelihood estimator for the various practical image data models to Chapter 18.

16.6.1 Example 1: mean of a Poisson and a Gaussian random variable

The maximum likelihood estimator $\hat{\theta}_{MLE}$ of the mean μ_k of a Poisson random variable, given N_{im} independent measurements $z_1, z_2, \ldots, z_{N_{im}}$, was derived in Section 16.1.1 as the sample mean. We now obtain its variance as

$$\operatorname{Var}\left(\hat{\theta}_{MLE}\right) = \operatorname{Var}\left(\frac{1}{N_{im}}\sum_{i=1}^{N_{im}} z_i\right) = \frac{1}{N_{im}^2}\sum_{i=1}^{N_{im}} \operatorname{Var}(z_i) = \frac{1}{N_{im}^2} \cdot N_{im} \cdot \mu_k = \frac{\mu_k}{N_{im}}. \quad (16.20)$$

For the maximum likelihood estimator $\hat{\theta}_{MLE}$ of the mean η_k of a Gaussian random variable with variance σ_k^2, derived in Section 16.1.2, the same simple calculation will show its variance to be $\frac{\sigma_k^2}{N_{im}}$.

These results clearly show that the variance of these two maximum likelihood estimators is inversely proportional to the number of measurements that are available. Increasing the number of data points is thus beneficial as it decreases the variance of these estimators.

Applying these results to the specific examples of Fig. 16.1, in which $N_{im} = 500$, $\mu_k = 5$ photons squared, and $\sigma_k^2 = 64$ electrons squared, the variance of the estimator of the Poisson mean is $\frac{\mu_k}{N_{im}} = 0.01$ photons squared, and the variance of the estimator of the Gaussian mean is $\frac{\sigma_k^2}{N_{im}} = 0.128$ electrons squared.

16.6.2 Example 2: center of mass as an object location estimator under the fundamental data model

We consider here the variance of the estimator $\hat{\theta}_{CM}$ of Section 16.5.3, which takes the location $\theta = (x_0, y_0) \in \mathbb{R}^2$ of a stationary object to be the magnification-corrected center of mass of the spatial coordinates $(x_k, y_k) \in \mathbb{R}^2$, $k = 1, \ldots, N_0$, of the N_0 photons that form the "image" of the object. The image of N_0 photons is described by the fundamental data model, and is recorded by an infinitely large detector over some acquisition time interval $[t_0, t]$. For convenience, we reproduce here the expression for the estimator, in which M denotes the magnification of the imaging system:

$$\hat{\theta}_{CM} = \begin{bmatrix}\hat{x}_0 & \hat{y}_0\end{bmatrix} = \begin{bmatrix}\frac{1}{MN_0}\sum_{k=1}^{N_0} x_k & \frac{1}{MN_0}\sum_{k=1}^{N_0} y_k\end{bmatrix}. \quad (16.21)$$

As our estimator here is a vector instead of a scalar, its "variance" is given by a *covariance matrix*. For a random vector U comprising scalar random variables $U_1, \ldots, U_n$, the covariance matrix $\operatorname{cov}(U)$ is an $n \times n$ matrix in which entry (i, j) is the *covariance* between U_i and U_j, defined as the quantity $\operatorname{E}\left[(U_i - \operatorname{E}[U_i])(U_j - \operatorname{E}[U_j])\right]$. The covariance provides a measure of the joint variability of a pair of random variables. Comparing its expression with the definition of the variance of a random variable (Eq. (16.16)), it is easy to see that the main diagonal entries of a covariance matrix are the variances of the component scalar random variables of the random vector.

Proceeding as with the calculation of the expectation of $\hat{\theta}_{CM}$ in Section 16.5.3, we condition the covariance matrix of $\hat{\theta}_{CM}$ on the value of $\mathcal{N}(t)$, the number of photons detected during the image acquisition time interval $[t_0, t]$, given that $\mathcal{N}(t) > 0$, and obtain

$$\begin{aligned}\operatorname{cov}\left(\hat{\theta}_{CM}\right) &= \sum_{N_0=1}^{\infty} \operatorname{cov}\left(\hat{\theta}_{CM} \mid \mathcal{N}(t) = N_0\right) \cdot P\left(\mathcal{N}(t) = N_0 \mid \mathcal{N}(t) > 0\right) \\ &= \sum_{N_0=1}^{\infty} \operatorname{cov}\left(\begin{bmatrix}\hat{x}_0 & \hat{y}_0\end{bmatrix} \mid \mathcal{N}(t) = N_0\right) \cdot P\left(\mathcal{N}(t) = N_0 \mid \mathcal{N}(t) > 0\right) \\ &= \sum_{N_0=1}^{\infty} \begin{bmatrix}C_{1,1} & C_{1,2} \\ C_{2,1} & C_{2,2}\end{bmatrix} \cdot P\left(\mathcal{N}(t) = N_0 \mid \mathcal{N}(t) > 0\right), \end{aligned} \quad (16.22)$$

where

$$C_{1,1} = \mathrm{E}\left[(\hat{x}_0 - \mathrm{E}[\hat{x}_0 \mid \mathcal{N}(t) = N_0])^2 \mid \mathcal{N}(t) = N_0\right],$$
$$C_{1,2} = C_{2,1} = \mathrm{E}\left[(\hat{x}_0 - \mathrm{E}[\hat{x}_0 \mid \mathcal{N}(t) = N_0])(\hat{y}_0 - \mathrm{E}[\hat{y}_0 \mid \mathcal{N}(t) = N_0]) \mid \mathcal{N}(t) = N_0\right],$$
$$C_{2,2} = \mathrm{E}\left[(\hat{y}_0 - \mathrm{E}[\hat{y}_0 \mid \mathcal{N}(t) = N_0])^2 \mid \mathcal{N}(t) = N_0\right].$$

From Eq. (16.21), we can see that the expectation of any function of $\hat{x}_0$ and/or $\hat{y}_0$, given that $\mathcal{N}(t) = N_0$, is just the expectation of that function. The entries of the conditional covariance matrix therefore reduce to

$$C_{1,1} = \mathrm{E}\left[(\hat{x}_0 - \mathrm{E}[\hat{x}_0])^2 \mid \mathcal{N}(t) = N_0\right] = \mathrm{E}\left[(\hat{x}_0 - \mathrm{E}[\hat{x}_0])^2\right] = \mathrm{Var}(\hat{x}_0),$$
$$C_{1,2} = C_{2,1} = \mathrm{E}\left[(\hat{x}_0 - \mathrm{E}[\hat{x}_0])(\hat{y}_0 - \mathrm{E}[\hat{y}_0]) \mid \mathcal{N}(t) = N_0\right] = \mathrm{E}\left[(\hat{x}_0 - \mathrm{E}[\hat{x}_0])(\hat{y}_0 - \mathrm{E}[\hat{y}_0])\right],$$
$$C_{2,2} = \mathrm{E}\left[(\hat{y}_0 - \mathrm{E}[\hat{y}_0])^2 \mid \mathcal{N}(t) = N_0\right] = \mathrm{E}\left[(\hat{y}_0 - \mathrm{E}[\hat{y}_0])^2\right] = \mathrm{Var}(\hat{y}_0).$$

Plugging these entries into Eq. (16.22), we obtain

$$\mathrm{cov}\left(\hat{\theta}_{CM}\right) = \sum_{N_0=1}^{\infty} \begin{bmatrix} \mathrm{Var}(\hat{x}_0) & \mathrm{E}\left[(\hat{x}_0 - \mathrm{E}[\hat{x}_0])(\hat{y}_0 - \mathrm{E}[\hat{y}_0])\right] \\ \mathrm{E}\left[(\hat{x}_0 - \mathrm{E}[\hat{x}_0])(\hat{y}_0 - \mathrm{E}[\hat{y}_0])\right] & \mathrm{Var}(\hat{y}_0) \end{bmatrix} \times P(\mathcal{N}(t) = N_0 \mid \mathcal{N}(t) > 0). \tag{16.23}$$

This gives us a general expression for the covariance matrix of $\hat{\theta}_{CM}$ that can be used with any probability density function that describes the spatial distribution of the photons detected from the object (i.e., any photon distribution profile).

We next derive the covariance matrix for a more specfic scenario. Suppose, as we did in Section 16.5.3, that the expected value of the location of a detected photon is the object's location multiplied by the magnification M. More precisely, we impose the condition that $\mathrm{E}[[X\ Y]] = [Mx_0\ My_0]$, where $[X\ Y]$ is the random vector that models the coordinates of a detected photon. Given this assumption, we showed in Section 16.5.3 that $\hat{\theta}_{CM}$ is an unbiased estimator of $\theta = (x_0, y_0)$, i.e., $\mathrm{E}\left[\hat{\theta}_{CM}\right] = \mathrm{E}[[\hat{x}_0\ \hat{y}_0]] = [x_0\ y_0]$. Additionally, we assume here that X and Y are uncorrelated, meaning that $\mathrm{E}[XY] = \mathrm{E}[X] \cdot \mathrm{E}[Y]$.

Plugging in the expression for $\hat{x}_0$, entry $C_{1,1}$ now evaluates to

$$C_{1,1} = \left(\frac{1}{MN_0}\right)^2 \sum_{k=1}^{N_0} \mathrm{Var}(x_k) = \left(\frac{1}{MN_0}\right)^2 N_0 \left(\mathrm{E}\left[X^2\right] - (\mathrm{E}[X])^2\right)$$
$$= \frac{1}{M^2 N_0}\left(\mathrm{E}\left[X^2\right] - (Mx_0)^2\right) = \frac{\mathrm{E}\left[X^2\right]}{M^2 N_0} - \frac{x_0^2}{N_0}.$$

Here we have used the mutual independence of the locations of different photons to write the variance of $\hat{x}_0$ as the sum of the variances of x_k, $k = 1, \ldots, N_0$. It is easy to see that an analogous evaluation of entry $C_{2,2}$ yields

$$C_{2,2} = \frac{\mathrm{E}\left[Y^2\right]}{M^2 N_0} - \frac{y_0^2}{N_0}.$$

For entries $C_{1,2}$ and $C_{2,1}$, we use the unbiasedness of $\hat{\theta}_{CM}$ to obtain

$$C_{1,2} = C_{2,1} = \mathrm{E}\left[(\hat{x}_0 - x_0)(\hat{y}_0 - y_0)\right] = \mathrm{E}[\hat{x}_0 \hat{y}_0] - y_0 \mathrm{E}[\hat{x}_0] - x_0 \mathrm{E}[\hat{y}_0] + x_0 y_0$$
$$= \mathrm{E}[\hat{x}_0]\,\mathrm{E}[\hat{y}_0] - x_0 y_0 - x_0 y_0 + x_0 y_0 = 0.$$

Note here that it is the fact that X and Y are uncorrelated that gives us the equality $\mathrm{E}[\hat{x}_0\hat{y}_0] = \mathrm{E}[\hat{x}_0]\,\mathrm{E}[\hat{y}_0]$. This means that when the x and y coordinates of a detected photon are uncorrelated, the components $\hat{x}_0$ and $\hat{y}_0$ of the estimator $\hat{\theta}_{CM}$ are also uncorrelated.

Replacing the matrix entries in Eq. (16.23) with these results yields the more specific expression

$$\begin{aligned}
\mathrm{cov}\left(\hat{\theta}_{CM}\right) &= \begin{bmatrix} \frac{1}{M^2}\mathrm{E}\left[X^2\right] - x_0^2 & 0 \\ 0 & \frac{1}{M^2}\mathrm{E}\left[Y^2\right] - y_0^2 \end{bmatrix} \cdot \sum_{N_0=1}^{\infty} \frac{1}{N_0} \cdot P\left(\mathcal{N}(t) = N_0 \mid \mathcal{N}(t) > 0\right) \\
&= \begin{bmatrix} \frac{1}{M^2}\mathrm{E}\left[X^2\right] - x_0^2 & 0 \\ 0 & \frac{1}{M^2}\mathrm{E}\left[Y^2\right] - y_0^2 \end{bmatrix} \cdot \mathrm{E}\left[\frac{1}{\mathcal{N}(t)} \mid \mathcal{N}(t) > 0\right]. \qquad (16.24)
\end{aligned}$$

We see here that the covariance matrix is now the product of two distinct parts. The first part is a matrix that, through $\mathrm{E}\left[X^2\right]$ and $\mathrm{E}\left[Y^2\right]$, depends on the spatial distribution of the detected photons. The second part is a scalar conditional expectation that depends on the temporal aspect of photon detection, as it is a function of the rate of photon detection Λ that characterizes the Poisson process $\mathcal{N}(\tau)$, $\tau \geq t_0$, which models the arrival of photons to the detector (see Eq. (15.6) for the dependence of this term on Λ). In terms of the spatial aspect, use of this expression in a concrete way requires only that $E\left[X^2\right]$ and $E\left[Y^2\right]$ be evaluated with the given photon distribution profile, which of course must satisfy the conditions under which this expression is valid.

For example, we consider the case where the photons detected from the object are distributed according to a 2D Gaussian density function. Here the photon distribution profile is given by Eq. (16.11), and we know from Section 16.5.3 that for this profile, $\mathrm{E}[[X\ Y]] = [Mx_0\ My_0]$. We also know that in this case X and Y are uncorrelated, as a quick inspection of Eq. (16.11) tells us that X and Y are independently distributed with the 1D Gaussian densities of Eq. (16.18). To find the covariance matrix of $\hat{\theta}_{CM}$, which as shown in Section 16.2.1.1 is in fact the maximum likelihood estimator for this specific scenario, we simply need to calculate $\mathrm{E}\left[X^2\right]$ and $\mathrm{E}\left[Y^2\right]$ and substitute the results into Eq. (16.24). Using the density f_X from Eq. (16.18), for $\mathrm{E}\left[X^2\right]$ we have

$$\begin{aligned}
\mathrm{E}\left[X^2\right] &= \int_{-\infty}^{\infty} x^2 f_X(x)dx = \frac{1}{M\sqrt{2\pi\sigma_g^2}} \int_{-\infty}^{\infty} x^2 e^{-\frac{1}{2\sigma_g^2}\left(\frac{x}{M} - x_0\right)^2} dx \\
&= \frac{M^2}{\sqrt{2\pi\sigma_g^2}} \int_{-\infty}^{\infty} u^2 e^{-\frac{1}{2\sigma_g^2}(u - x_0)^2} du = M^2\left(\sigma_g^2 + x_0^2\right),
\end{aligned}$$

where we have used the fact that the second non-central moment of a Gaussian random variable is the sum of its variance and the square of its mean. The term $\mathrm{E}\left[Y^2\right] = M^2\left(\sigma_g^2 + y_0^2\right)$ can be obtained analogously using the density f_Y from Eq. (16.18). The covariance matrix of $\hat{\theta}_{CM}$ in the case of a 2D Gaussian image profile is therefore given by

$$\mathrm{cov}\left(\hat{\theta}_{CM}\right) = \begin{bmatrix} \sigma_g^2 & 0 \\ 0 & \sigma_g^2 \end{bmatrix} \cdot \mathrm{E}\left[\frac{1}{\mathcal{N}(t)} \mid \mathcal{N}(t) > 0\right]. \qquad (16.25)$$

This expression will be analyzed in Section 18.7.1, where the efficiency of location estimators is investigated.

16.6.3 Example 3: center of mass as an object location estimator under a fixed photon count data model

In this section, we consider, as in Section 16.6.2, an estimator $\hat{\theta}_{CM} = [\hat{x}_0\ \hat{y}_0]$ of the location $\theta = (x_0, y_0) \in \mathbb{R}^2$ of a stationary object that yields as its estimate the magnification-corrected center of mass of the coordinates of the photons detected from the object. The difference here is the underlying data model that describes the image acquired of the object. The data model here is closely related to the fundamental data model, as an "image" of the object also consists of the mutually independent spatial coordinates $(x_k, y_k) \in \mathbb{R}^2$, $k = 1, \ldots, N_0$, of the N_0 photons detected by an unpixelated and, as assumed in Section 16.6.2, infinitely large, detector $C = \mathbb{R}^2$. This data model, however, is different from the fundamental data model in a subtle but important way. In the fundamental data model the photons forming an image are detected over a fixed time interval $[t_0, t]$. As the temporal component of photon detection is a Poisson process $\mathcal{N}(\tau)$, $\tau \geq t_0$, the number of photons detected in the fixed time interval is a random number modeled by the Poisson random variable $\mathcal{N}(t)$. Repeat acquisitions under the fundamental data model will therefore produce images composed of different numbers of photons (i.e., different realizations N_0 of $\mathcal{N}(t)$, given that $\mathcal{N}(t) > 0$). In contrast, with the current data model we consider a situation in which a fixed number of photons N_0 are detected in an image. Repeat acquisitions under this model will therefore yield images comprising the same number of photons (Note 16.5).

The estimator $\hat{\theta}_{CM}$ considered here is identical in form to Eq. (16.21). In analyzing its covariance matrix, we will take the same steps as in Section 16.6.2. Noting that the photon count N_0 here is not the realization of a random variable, the general expression for the covariance matrix is simply obtained, by the definition of a covariance matrix, as

$$\operatorname{cov}\left(\hat{\theta}_{CM}\right) = \begin{bmatrix} \operatorname{Var}(\hat{x}_0) & \mathrm{E}\left[(\hat{x}_0 - \mathrm{E}[\hat{x}_0])(\hat{y}_0 - \mathrm{E}[\hat{y}_0])\right] \\ \mathrm{E}\left[(\hat{x}_0 - \mathrm{E}[\hat{x}_0])(\hat{y}_0 - \mathrm{E}[\hat{y}_0])\right] & \operatorname{Var}(\hat{y}_0) \end{bmatrix}.$$

Under the assumptions that $\mathrm{E}[[X\ Y]] = [Mx_0\ My_0]$ and $\mathrm{E}[XY] = \mathrm{E}[X] \cdot \mathrm{E}[Y]$, where $[X\ Y]$ is the random vector that models the spatial coordinates of a detected photon, the same calculations as those shown in Section 16.6.2 give us the nearly completely specified covariance matrix

$$\operatorname{cov}\left(\hat{\theta}_{CM}\right) = \begin{bmatrix} \frac{1}{M^2}\mathrm{E}\left[X^2\right] - x_0^2 & 0 \\ 0 & \frac{1}{M^2}\mathrm{E}\left[Y^2\right] - y_0^2 \end{bmatrix} \cdot \frac{1}{N_0}.$$

Note here that the assumption $\mathrm{E}[[X\ Y]] = [Mx_0\ My_0]$ means that $\hat{\theta}_{CM}$ is also unbiased under the current data model (Exercise 16.3), and that therefore the off-diagonal entries of the matrix evaluate to 0. Note also the simplicity of this covariance matrix compared to the covariance matrix for the analogous estimator under the fundamental data model (Eq. (16.24)). The two expressions share the same matrix portion, but instead of the conditional expectation of the inverse of the random photon count which needs to be computed, the scalar portion for the expression here is just the inverse of the fixed photon count N_0.

Finally, taking the case of a 2D Gaussian image profile as a specific example, plugging in the results for $\mathrm{E}\left[X^2\right]$ and $\mathrm{E}\left[Y^2\right]$ from Section 16.6.2 yields the very simple covariance matrix

$$\operatorname{cov}\left(\hat{\theta}_{CM}\right) = \begin{bmatrix} \sigma_g^2 & 0 \\ 0 & \sigma_g^2 \end{bmatrix} \cdot \frac{1}{N_0}. \tag{16.26}$$

With this data model, it is easy to see from the main diagonal elements of the covariance matrix that the variance of the component estimators $\hat{x}_0$ and $\hat{y}_0$ decreases with increasing numbers of data points (i.e., photons).

17

Fisher Information and Cramér-Rao Lower Bound

In the prior chapter, we have considered parameter estimation problems such as the determination of the mean of a random variable and the localization of an object. We have also discussed approaches to solving such estimation problems, and in particular, the maximum likelihood estimator. Furthermore, for assessing the quality of an estimation approach, we have introduced the notion of the bias and the variance of an estimator. Here, it is the variance or standard deviation of an estimator that we consider in more detail. Provided that an estimator is not biased, we would want to use an estimator with the smallest variance in order to have a small "spread" of the estimates in the case of repeat estimates. Importantly, the variance of an unbiased estimator cannot be arbitrarily small. There is a bound, the Cramér-Rao lower bound, that provides a lower limit for the variance of an unbiased estimator. The Cramér-Rao lower bound, which is the inverse of a quantity known as the Fisher information matrix, is the subject of this chapter. Besides providing a benchmark against which the performance of an estimator can be assessed, we will demonstrate in later chapters that the Cramér-Rao lower bound is a useful tool for experimental design.

17.1 Cramér-Rao inequality

In this section, we will begin by stating the Cramér-Rao inequality and sketching its derivation for estimation problems involving a scalar parameter. We will then extend the result to the more general case of the estimation of a vector parameter.

Let θ be the scalar parameter that we want to estimate from the observed data x, which is characterized by the probability density function $p_\theta(x)$. The function $p_\theta(x)$ can instead be a probability mass function, provided that x takes on values described by such a distribution. Then the variance of any unbiased estimator $\hat{\theta}$ of θ satisfies the Cramér-Rao inequality

$$\mathrm{Var}\left(\hat{\theta}\right) \geq \mathbf{I}^{-1}(\theta), \tag{17.1}$$

where

$$\mathbf{I}(\theta) = \mathrm{E}\left[\left(\frac{\partial}{\partial\theta}\ln p_\theta(x)\right)^2\right] = -\mathrm{E}\left[\frac{\partial^2}{\partial\theta^2}\ln p_\theta(x)\right] \tag{17.2}$$

is called the *Fisher information matrix*. In our current situation, the matrix is in fact a scalar as it contains only a single element due to our assumption of a scalar θ. Note that $\ln p_\theta(x)$ inside the expectation is the log-likelihood function for the parameter θ (Section 16.1).

The Cramér-Rao inequality shows that the variance of an unbiased estimator is always at least as large as the inverse of the Fisher information. This result is very powerful as it is independent of the particulars of a specific estimation approach, the only stipulation being that it is unbiased. The bound $\mathbf{I}^{-1}(\theta)$ is called the *Cramér-Rao lower bound*, and it depends on the probabilistic description of the data through the probability density function p_θ. The

Fisher information $\mathbf{I}(\theta)$ can therefore be seen as a measure of the quality of the data that is available for the estimation process.

An important question is whether the equality in Eq. (17.1) is in fact attained, i.e., whether there is an unbiased estimator $\hat{\theta}$ whose variance equals the inverse of the Fisher information matrix. Such an unbiased estimator is called *efficient*. In fact, in the examples of Sections 17.1.3 and 17.1.4, we will see that the maximum likelihood estimator is in effect efficient for the specific problems considered. In general, it is known that the maximum likelihood estimator is asymptotically efficient, which means that as the amount of data increases to infinity, the variance of the maximum likelihood estimator approaches the inverse of the Fisher information matrix. We will also show in subsequent chapters, using simulated image data, that for many of the parameter estimation problems of interest to us, the variance of the maximum likelihood estimator is well described by the Cramér-Rao lower bound.

17.1.1 Sketch of derivation of Cramér-Rao lower bound

We will show in this subsection how the Cramér-Rao inequality for the case of a scalar parameter is obtained. The sketch of the derivation is necessarily technical in nature, and is intended for readers who are interested in seeing how one can arrive at the result. A reader can omit this subsection without losing continuity with the content of the rest of this chapter.

Before proceeding with the derivation, we should note that the Cramér-Rao inequality is not generally true as it depends on certain so-called regularity conditions to be satisfied. In particular, we need the technical condition that

$$\mathrm{E}\left[\frac{\partial}{\partial\theta}\ln p_\theta(x)\right] = 0 \tag{17.3}$$

is satisfied for all $\theta \in \Theta$, where Θ is the parameter space, i.e., the set of all allowable values for θ. We should also note that the derivation will assume that the observed data is a realization of a continuous random variable distributed according to a probability density function. A derivation for the case where the data takes on discrete values distributed according to a probability mass function would be analogous.

We now begin with the fact that since the estimator $\hat{\theta}$ is unbiased, we can write, for $\theta \in \Theta$,

$$\theta = \mathrm{E}\left[\hat{\theta}\right] = \int_{-\infty}^{\infty} \hat{\theta}(x) p_\theta(x) dx.$$

By differentiating both sides of the equation with respect to θ, we obtain

$$\begin{aligned}
1 = \frac{\partial}{\partial\theta}\theta &= \frac{\partial}{\partial\theta}\int_{-\infty}^{\infty} \hat{\theta}(x) p_\theta(x) dx = \int_{-\infty}^{\infty} \frac{\partial}{\partial\theta}\left(\hat{\theta}(x) p_\theta(x)\right) dx \\
&= \int_{-\infty}^{\infty} \left(\frac{\partial}{\partial\theta}\hat{\theta}(x)\right) p_\theta(x) dx + \int_{-\infty}^{\infty} \hat{\theta}(x) \left(\frac{\partial}{\partial\theta} p_\theta(x)\right) dx \\
&= \int_{-\infty}^{\infty} \hat{\theta}(x) \left(\frac{\partial}{\partial\theta} p_\theta(x)\right) dx = \int_{-\infty}^{\infty} \hat{\theta}(x) p_\theta(x) \left(\frac{\partial}{\partial\theta}\ln p_\theta(x)\right) dx.
\end{aligned} \tag{17.4}$$

By the regularity condition of Eq. (17.3), we have

$$0 = \int_{-\infty}^{\infty} p_\theta(x) \left(\frac{\partial}{\partial\theta}\ln p_\theta(x)\right) dx, \tag{17.5}$$

and therefore that

$$0 = 0 \cdot \theta = \int_{-\infty}^{\infty} \theta \; p_\theta(x) \left(\frac{\partial}{\partial \theta} \ln p_\theta(x) \right) dx. \tag{17.6}$$

Subtracting Eq. (17.6) from Eq. (17.4), we obtain

$$\begin{aligned} 1 &= \int_{-\infty}^{\infty} \left(\hat{\theta}(x) - \theta \right) p_\theta(x) \left(\frac{\partial}{\partial \theta} \ln p_\theta(x) \right) dx \\ &= \int_{-\infty}^{\infty} \left(\hat{\theta}(x) - \theta \right) \sqrt{p_\theta(x)} \sqrt{p_\theta(x)} \left(\frac{\partial}{\partial \theta} \ln p_\theta(x) \right) dx. \end{aligned}$$

Squaring both sides yields

$$1^2 = \left(\int_{-\infty}^{\infty} \left(\hat{\theta}(x) - \theta \right) \sqrt{p_\theta(x)} \sqrt{p_\theta(x)} \left(\frac{\partial}{\partial \theta} \ln p_\theta(x) \right) dx \right)^2,$$

at which point application of the Cauchy-Schwarz inequality gives us

$$\begin{aligned} 1^2 &= \left(\int_{-\infty}^{\infty} \left(\hat{\theta}(x) - \theta \right) \sqrt{p_\theta(x)} \sqrt{p_\theta(x)} \left(\frac{\partial}{\partial \theta} \ln p_\theta(x) \right) dx \right)^2 \\ &\leq \left(\int_{-\infty}^{\infty} \left(\left(\hat{\theta}(x) - \theta \right) \sqrt{p_\theta(x)} \right)^2 dx \right) \left(\int_{-\infty}^{\infty} \left(\sqrt{p_\theta(x)} \left(\frac{\partial}{\partial \theta} \ln p_\theta(x) \right) \right)^2 dx \right) \\ &= \left(\int_{-\infty}^{\infty} \left(\hat{\theta}(x) - \theta \right)^2 p_\theta(x) dx \right) \left(\int_{-\infty}^{\infty} \left(\frac{\partial}{\partial \theta} \ln p_\theta(x) \right)^2 p_\theta(x) dx \right) \\ &= \mathrm{E}\left[\left(\hat{\theta}(x) - \theta \right)^2 \right] \mathrm{E}\left[\left(\frac{\partial}{\partial \theta} \ln p_\theta(x) \right)^2 \right] \\ &= \mathrm{Var}\left(\hat{\theta} \right) \mathbf{I}(\theta), \end{aligned}$$

and hence the Cramér-Rao inequality $\mathrm{Var}\left(\hat{\theta} \right) \geq \mathbf{I}^{-1}(\theta)$. To show the equality of the two expressions in Eq. (17.2) for the Fisher information matrix, we differentiate both sides of Eq. (17.5) to obtain

$$\begin{aligned} 0 = \frac{\partial}{\partial \theta} 0 &= \frac{\partial}{\partial \theta} \int_{-\infty}^{\infty} \left(\frac{\partial}{\partial \theta} \ln p_\theta(x) \right) p_\theta(x) dx \\ &= \int_{-\infty}^{\infty} \frac{\partial}{\partial \theta} \left(\left(\frac{\partial}{\partial \theta} \ln p_\theta(x) \right) p_\theta(x) \right) dx \\ &= \int_{-\infty}^{\infty} \left(\frac{\partial}{\partial \theta} \left(\frac{\partial}{\partial \theta} \ln p_\theta(x) \right) \right) p_\theta(x) dx + \int_{-\infty}^{\infty} \left(\frac{\partial}{\partial \theta} \ln p_\theta(x) \right) \left(\frac{\partial}{\partial \theta} p_\theta(x) \right) dx \\ &= \int_{-\infty}^{\infty} \left(\frac{\partial^2}{\partial \theta^2} \ln p_\theta(x) \right) p_\theta(x) dx + \int_{-\infty}^{\infty} \left(\frac{\partial}{\partial \theta} \ln p_\theta(x) \right) \left(\frac{\partial}{\partial \theta} p_\theta(x) \right) dx \\ &= \mathrm{E}\left[\frac{\partial^2}{\partial \theta^2} \ln p_\theta(x) \right] + \int_{-\infty}^{\infty} \left(\frac{\partial}{\partial \theta} \ln p_\theta(x) \right) \left(\frac{\partial}{\partial \theta} p_\theta(x) \right) dx. \end{aligned}$$

Therefore,

$$\begin{aligned}
-\mathrm{E}\left[\frac{\partial^2}{\partial\theta^2}\ln p_\theta(x)\right] &= \int_{-\infty}^{\infty}\left(\frac{\partial}{\partial\theta}\ln p_\theta(x)\right)\left(\frac{\partial}{\partial\theta}p_\theta(x)\right)dx \\
&= \int_{-\infty}^{\infty}\left(\frac{\partial}{\partial\theta}\ln p_\theta(x)\right)\left(p_\theta(x)\left(\frac{\partial}{\partial\theta}\ln p_\theta(x)\right)\right)dx \\
&= \int_{-\infty}^{\infty}\left(\frac{\partial}{\partial\theta}\ln p_\theta(x)\right)^2 p_\theta(x)dx \\
&= \mathrm{E}\left[\left(\frac{\partial}{\partial\theta}\ln p_\theta(x)\right)^2\right] = \mathbf{I}(\theta),
\end{aligned}$$

which shows that the two expressions for the Fisher information are in fact identical.

17.1.2 Multivariate Cramér-Rao lower bound

The Cramér-Rao lower bound for a scalar parameter can be extended to a vector parameter $\theta = (\theta_1, \theta_2, \ldots, \theta_N) \in \Theta$, in which case we have the analogous relationship

$$\operatorname{cov}\left(\hat{\theta}\right) \geq \mathbf{I}^{-1}(\theta), \tag{17.7}$$

where $\operatorname{cov}\left(\hat{\theta}\right)$ denotes the $N \times N$ covariance matrix of $\hat{\theta}$, with element (i,j) given by the covariance between the estimators $\hat{\theta}_i$ and $\hat{\theta}_j$, and where the Fisher information matrix

$$\mathbf{I}(\theta) = \mathrm{E}\left[\left(\frac{\partial}{\partial\theta}\ln p_\theta(x)\right)^T\left(\frac{\partial}{\partial\theta}\ln p_\theta(x)\right)\right] \tag{17.8}$$

is an $N \times N$ matrix, with element (i,j) given by

$$[\mathbf{I}(\theta)]_{i,j} = \mathrm{E}\left[\frac{\partial}{\partial\theta_i}\ln p_\theta(x)\frac{\partial}{\partial\theta_j}\ln p_\theta(x)\right] = -\mathrm{E}\left[\frac{\partial^2}{\partial\theta_i\partial\theta_j}\ln p_\theta(x)\right]. \tag{17.9}$$

With the expression $A \geq B$ for two symmetric matrices $A, B \in \mathbb{R}^N$, we mean that $\langle Ax, x\rangle \geq \langle Bx, x\rangle$ for all vectors $x \in \mathbb{R}^N$. Therefore, from $\operatorname{cov}\left(\hat{\theta}\right) \geq \mathbf{I}^{-1}(\theta)$, we have, by taking $x = e_i \in \mathbb{R}^N$ where e_i is the vector having a 1 for its ith element and zeros for all other elements, that

$$\operatorname{Var}\left(\hat{\theta}_i\right) \geq \left[\mathbf{I}^{-1}(\theta)\right]_{i,i}. \tag{17.10}$$

This shows that the variance of the estimate $\hat{\theta}_i$ of parameter θ_i (i.e., the ith entry of θ) is bounded from below by the ith main diagonal entry of the inverse Fisher information matrix.

It is important to note that the Cramér-Rao lower bound cannot be stated when the Fisher information matrix $\mathbf{I}(\theta)$ is not invertible. We will not investigate the invertibility of the Fisher information matrix further, but we want to mention that a non-invertible Fisher information matrix is typically an indicator that the associated parameter estimation problem is badly posed, such as when a parameter cannot be estimated from the acquired data, or when there is a linear relationship between the parameters. In fact, in our discussion of the depth discrimination problem in the context of 3D location estimation (Section 19.4) and in our investigation of the resolution problem (Section 20.3), we will encounter important situations in which the Fisher information matrix is not invertible.

As mentioned in the derivation of the result for the scalar case, we note again that the Cramér-Rao lower bound depends on some technical assumptions to be satisfied. These assumptions are primarily "regularity" conditions that guarantee that certain derivatives and integrals can be computed. These conditions are typically satisfied for the type of problems we investigate in this book. However, the reader does need to be aware that, especially in the case where expressions are, for example, not differentiable, the Cramér-Rao lower bound may not hold (Note 17.1).

17.1.3 Example 1: mean of a Poisson random variable

We revisit the example of Section 16.1.1, where we derived the maximum likelihood estimator for the mean $\theta = \mu_k$ of a Poisson random variable $S_{\theta,k}$, given N_{im} independent measurements of $S_{\theta,k}$, denoted by $z_1, \ldots, z_{N_{im}}$. In that example, it was shown that the log-likelihood function is given by

$$\ln p_{\theta,joint}(z_1, \ldots, z_{N_{im}}) = \sum_{i=1}^{N_{im}} \ln p_\theta(z_i) = -\theta N_{im} + \left(\sum_{i=1}^{N_{im}} z_i\right) \cdot \ln\theta - \sum_{i=1}^{N_{im}} \ln z_i!,$$

and that its first derivative with respect to θ is

$$\frac{\partial}{\partial\theta} \ln p_{\theta,joint}(z_1, \ldots, z_{N_{im}}) = -N_{im} + \frac{1}{\theta}\sum_{i=1}^{N_{im}} z_i.$$

Differentiating again with respect to θ, we obtain the second derivative

$$\frac{\partial^2}{\partial\theta^2} \ln p_{\theta,joint}(z_1, \ldots, z_{N_{im}}) = -\frac{1}{\theta^2}\sum_{i=1}^{N_{im}} z_i.$$

Using Eq. (17.2), the Fisher information matrix for the estimation of the mean $\theta = \mu_k$ is then given by

$$\begin{aligned}\mathbf{I}(\theta) &= -\mathrm{E}\left[\frac{\partial^2}{\partial\theta^2} \ln p_{\theta,joint}(z_1, \ldots, z_{N_{im}})\right] \\ &= -\mathrm{E}\left[-\frac{1}{\theta^2}\sum_{i=1}^{N_{im}} z_i\right] = \frac{1}{\theta^2}\sum_{i=1}^{N_{im}} \mathrm{E}[z_i] = \frac{1}{\theta^2} \cdot N_{im} \cdot \theta = \frac{N_{im}}{\theta},\end{aligned}$$

and the Cramér-Rao lower bound is its inverse, given by $\frac{\theta}{N_{im}} = \frac{\mu_k}{N_{im}}$. We therefore have that for any unbiased estimator $\hat{\theta} = \hat{\mu}_k$ of μ_k,

$$\mathrm{Var}\left(\hat{\theta}\right) = \mathrm{Var}\left(\hat{\mu}_k\right) \geq \frac{\mu_k}{N_{im}}.$$

From Section 16.5.1, we know that the maximum likelihood estimator for this estimation problem is unbiased. Moreover, in Section 16.6.1 we have shown that the variance of this maximum likelihood estimator is given by $\mathrm{Var}\left(\hat{\theta}_{MLE}\right) = \frac{\mu_k}{N_{im}}$. This means that in the current situation the Cramér-Rao lower bound is attained for the maximum likelihood estimator, i.e., $\mathrm{Var}\left(\hat{\theta}_{MLE}\right) = \mathbf{I}^{-1}(\theta)$. Therefore in this case the maximum likelihood estimator is efficient.

17.1.4 Example 2: mean of a Gaussian random variable

From the example of Section 16.1.2 where we derived the maximum likelihood estimator for the mean $\theta = \eta_k$ of a Gaussian random variable $W_{\theta,k}$, we have that the log-likelihood function, given N_{im} independent measurements of $W_{\theta,k}$ denoted by $z_1, \ldots, z_{N_{im}}$, is given by

$$\ln p_{\theta,joint}(z_1, \ldots, z_{N_{im}}) = \sum_{i=1}^{N_{im}} \ln p_\theta(z_i) = -N_{im} \cdot \ln \sqrt{2\pi\sigma_k^2} + \sum_{i=1}^{N_{im}} \left[-\frac{1}{2\sigma_k^2}(\theta - z_i)^2 \right],$$

where σ_k^2 is the variance of $W_{\theta,k}$. To obtain the Fisher information matrix for the estimation of the mean $\theta = \eta_k$, we start by calculating the first derivative of the log-likelihood function with respect to θ:

$$\begin{aligned}\frac{\partial}{\partial\theta} \ln p_{\theta,joint}(z_1, \ldots, z_{N_{im}}) &= \frac{\partial}{\partial\theta}\left(-N_{im} \cdot \ln \sqrt{2\pi\sigma_k^2} + \sum_{i=1}^{N_{im}} \left[-\frac{1}{2\sigma_k^2}(\theta - z_i)^2 \right] \right) \\ &= \sum_{i=1}^{N_{im}} \left[-\frac{1}{2\sigma_k^2} \cdot 2 \cdot (\theta - z_i) \right] = \sum_{i=1}^{N_{im}} \left[-\frac{1}{\sigma_k^2}(\theta - z_i) \right].\end{aligned}$$

The second derivative is then obtained by differentiating again with respect to θ:

$$\frac{\partial^2}{\partial\theta^2} \ln p_{\theta,joint}(z_1, \ldots, z_{N_{im}}) = \frac{\partial}{\partial\theta} \sum_{i=1}^{N_{im}} \left[-\frac{1}{\sigma_k^2}(\theta - z_i) \right] = \sum_{i=1}^{N_{im}} \left[-\frac{1}{\sigma_k^2} \right] = -\frac{N_{im}}{\sigma_k^2}.$$

By Eq. (17.2), the Fisher information matrix is then

$$\mathbf{I}(\theta) = -\mathrm{E}\left[\frac{\partial^2}{\partial\theta^2} \ln p_{\theta,joint}(z_1, \ldots, z_{N_{im}}) \right] = -\mathrm{E}\left[-\frac{N_{im}}{\sigma_k^2} \right] = \frac{N_{im}}{\sigma_k^2},$$

and its inverse gives the Cramér-Rao lower bound $\frac{\sigma_k^2}{N_{im}}$.

As we know from Section 16.5.1, the maximum likelihood estimator $\hat{\theta}_{MLE}$ for this problem is unbiased. We also know from Section 16.6.1 that the variance of $\hat{\theta}_{MLE}$ equals the Cramér-Rao lower bound $\frac{\sigma_k^2}{N_{im}}$. The maximum likelihood estimator for this problem is therefore also an efficient estimator.

17.2 Fisher information for the fundamental data model

In this section, we apply the theory of the Fisher information to the estimation of parameters from image data described by the fundamental data model. We will present a general expression for the Fisher information matrix corresponding to a fundamental image, in the sense that it is an expression that can be made much more explicit by applying it to a more concrete estimation problem, as we will see in Chapters 18, 19, and 20 where we consider the estimation of the location of an object and the distance between two objects.

Although the fundamental data model is theoretical in nature and not applicable in practice, the Cramér-Rao lower bound on the variance of any unbiased estimator, calculated under its assumptions, importantly provides a benchmark for what can be achieved in practice. This benchmark, despite being possibly unattainable in an experimental setting, gives us an idea of how close a practically attainable lower bound (Section 17.3) is to the bound corresponding to an idealized imaging setting. This importantly provides an indication of how much more the practical bound might be further improved by reducing the deteriorative effects of the practical experimental setup, such as minimizing the detector readout noise level and the effective pixel size.

Suppose, as described in Section 15.1, an image is captured, under the fundamental data model, over the time interval $[t_0, t]$ by an idealized detector $C \subseteq \mathbb{R}^2$ that records both the time and the location at which each photon is detected. The acquired image is composed of photons that are detected temporally according to an inhomogeneous Poisson process $\mathcal{N}(\tau)$ with rate $\Lambda_\theta(\tau)$, $\tau \geq t_0$, and spatially at locations described by the family of photon distribution profiles $\{f_{\theta,\tau}(x,y)\}_{\tau \geq t_0}$, $(x,y) \in C$. Then, as shown in Chapter E of the Online Appendix, by evaluating the general Fisher information matrix expression of Eq. (17.8) with the log-likelihood function of Eq. (16.10), the Fisher information matrix $\mathbf{I}(\theta)$ corresponding to the acquired image is given by

$$
\begin{aligned}
\mathbf{I}(\theta) &= \frac{1}{1 - e^{-\int_{t_0}^{t} \Lambda_\theta(\tau)d\tau}} \int_{t_0}^{t} \int_C \frac{\Lambda_\theta(\tau)}{f_{\theta,\tau}(x,y)} \left(\frac{\partial f_{\theta,\tau}(x,y)}{\partial\theta}\right)^T \frac{\partial f_{\theta,\tau}(x,y)}{\partial\theta} dxdyd\tau \\
&\quad + \frac{1}{1 - e^{-\int_{t_0}^{t} \Lambda_\theta(\tau)d\tau}} \int_{t_0}^{t} \frac{1}{\Lambda_\theta(\tau)} \left(\frac{\partial \Lambda_\theta(\tau)}{\partial\theta}\right)^T \frac{\partial \Lambda_\theta(\tau)}{\partial\theta} d\tau \\
&\quad - \frac{e^{-\int_{t_0}^{t} \Lambda_\theta(\tau)d\tau}}{\left(1 - e^{-\int_{t_0}^{t} \Lambda_\theta(\tau)d\tau}\right)^2} \int_{t_0}^{t} \left(\frac{\partial \Lambda_\theta(\tau)}{\partial\theta}\right)^T d\tau \cdot \int_{t_0}^{t} \frac{\partial \Lambda_\theta(\tau)}{\partial\theta} d\tau \\
&= \frac{1}{1 - e^{-\int_{t_0}^{t} \Lambda_\theta(\tau)d\tau}} \int_{t_0}^{t} \int_C \frac{1}{\Lambda_\theta(\tau) f_{\theta,\tau}(x,y)} \\
&\quad \times \left(\frac{\partial \left[\Lambda_\theta(\tau) f_{\theta,\tau}(x,y)\right]}{\partial\theta}\right)^T \frac{\partial \left[\Lambda_\theta(\tau) f_{\theta,\tau}(x,y)\right]}{\partial\theta} dxdyd\tau \\
&\quad - \frac{e^{-\int_{t_0}^{t} \Lambda_\theta(\tau)d\tau}}{\left(1 - e^{-\int_{t_0}^{t} \Lambda_\theta(\tau)d\tau}\right)^2} \int_{t_0}^{t} \left(\frac{\partial \Lambda_\theta(\tau)}{\partial\theta}\right)^T d\tau \cdot \int_{t_0}^{t} \frac{\partial \Lambda_\theta(\tau)}{\partial\theta} d\tau. \qquad (17.11)
\end{aligned}
$$

Note that the term $e^{-\int_{t_0}^{t} \Lambda_\theta(\tau)d\tau}$ approaches zero quickly with increasing values of $\mathrm{E}[\mathcal{N}(t)] = \int_{t_0}^{t} \Lambda_\theta(\tau)d\tau$, the mean number of photons detected during the acquisition time interval $[t_0, t]$. Its value, for example, is already just $\sim$4.5400 $\times$ 10^{-5} at the low mean photon count of $\mathrm{E}[\mathcal{N}(t)] = 10$. Accordingly, the factors $\left(1 - e^{-\int_{t_0}^{t} \Lambda_\theta(\tau)d\tau}\right)^{-1}$ and $e^{-\int_{t_0}^{t} \Lambda_\theta(\tau)d\tau}\left(1 - e^{-\int_{t_0}^{t} \Lambda_\theta(\tau)d\tau}\right)^{-2}$ quickly approach one and zero, respectively, with increasing values of $\mathrm{E}[\mathcal{N}(t)]$. Therefore, provided that the mean photon count $\mathrm{E}[\mathcal{N}(t)]$ is not very small, the Fisher information matrix for an acquired image is well approximated by

$$
\begin{aligned}
\mathbf{I}(\theta) &\approx \int_{t_0}^{t} \int_C \frac{1}{\Lambda_\theta(\tau) f_{\theta,\tau}(x,y)} \left(\frac{\partial \left[\Lambda_\theta(\tau) f_{\theta,\tau}(x,y)\right]}{\partial\theta}\right)^T \frac{\partial \left[\Lambda_\theta(\tau) f_{\theta,\tau}(x,y)\right]}{\partial\theta} dxdyd\tau \\
&= \int_{t_0}^{t} \int_C \frac{\Lambda_\theta(\tau)}{f_{\theta,\tau}(x,y)} \left(\frac{\partial f_{\theta,\tau}(x,y)}{\partial\theta}\right)^T \frac{\partial f_{\theta,\tau}(x,y)}{\partial\theta} dxdyd\tau \\
&\quad + \int_{t_0}^{t} \frac{1}{\Lambda_\theta(\tau)} \left(\frac{\partial \Lambda_\theta(\tau)}{\partial\theta}\right)^T \frac{\partial \Lambda_\theta(\tau)}{\partial\theta} d\tau. \qquad (17.12)
\end{aligned}
$$

As the data produced by a fluorescence microscopy experiment typically satisfies the photon count requirement, this approximation can often be applied when using Eq. (17.11) to calculate benchmark Cramér-Rao lower bounds for practical scenarios.

The first expression of Eq. (17.11) is a summation of three terms, but can be viewed as the sum of two parts. The first part consists of the first term and relates primarily to the spatial aspect of photon detection, but includes the photon detection rate as a weight. The second part consists of the sum of the second and third terms and relates to the temporal aspect of photon detection. Using this expression, we next consider two examples to illustrate the general Fisher information matrix corresponding to the fundamental data model.

17.2.1 Example: known photon detection rate

As a first example, we investigate here the estimation of parameters pertaining to a stationary object (e.g., the object's location) from an image acquired during the time interval $[t_0, t]$. We assume the photon detection rate Λ is known and therefore does not depend on the vector θ of unknown parameters. Accordingly, $\frac{\partial \Lambda(\tau)}{\partial \theta}$, $\tau \geq t_0$, is a vector of N zeros given that θ comprises N parameters to be estimated. Furthermore, since the object is stationary, we have $f_{\theta,\tau} =: f_\theta$, $\tau \geq t_0$. Then, using the first expression of Eq. (17.11), the Fisher information matrix is given by

$$\begin{aligned}
\mathbf{I}(\theta) &= \frac{1}{1 - e^{-\int_{t_0}^t \Lambda_\theta(\tau)d\tau}} \int_{t_0}^t \int_C \frac{\Lambda(\tau)}{f_\theta(x,y)} \left(\frac{\partial f_\theta(x,y)}{\partial \theta}\right)^T \frac{\partial f_\theta(x,y)}{\partial \theta} dxdyd\tau + 0_{N,N} - 0_{N,N} \\
&= \frac{\int_{t_0}^t \Lambda(\tau)d\tau}{1 - e^{-\int_{t_0}^t \Lambda(\tau)d\tau}} \int_C \frac{1}{f_\theta(x,y)} \left(\frac{\partial f_\theta(x,y)}{\partial \theta}\right)^T \frac{\partial f_\theta(x,y)}{\partial \theta} dxdy \\
&= N_{photons} \int_C \frac{1}{f_\theta(x,y)} \left(\frac{\partial f_\theta(x,y)}{\partial \theta}\right)^T \frac{\partial f_\theta(x,y)}{\partial \theta} dxdy,
\end{aligned}$$

where $0_{N,N}$ is the $N \times N$ zero matrix and $N_{photons} := \int_{t_0}^t \Lambda(\tau)d\tau \cdot \left(1 - e^{-\int_{t_0}^t \Lambda(\tau)d\tau}\right)^{-1}$ is the mean photon count of an acquired image, as discussed in Section 15.1.1.1.

The Cramér-Rao lower bound is then given by the inverse of $\mathbf{I}(\theta)$, i.e.,

$$\mathbf{I}^{-1}(\theta) = \frac{1}{N_{photons}} \left(\int_C \frac{1}{f_\theta(x,y)} \left(\frac{\partial f_\theta(x,y)}{\partial \theta}\right)^T \left(\frac{\partial f_\theta(x,y)}{\partial \theta}\right) dxdy \right)^{-1}. \tag{17.13}$$

The inverse dependence on $N_{photons}$ shows that the lower bounds on the variances for estimating parameters, such as the object's positional coordinates x_0 and y_0, decrease with increasing numbers of detected photons. This corroborates the intuition that the greater the amount of data (i.e., the greater the number of photons) acquired, the better (i.e., the smaller) the variance for estimating a parameter.

While this result shows how the Cramér-Rao lower bound depends on the mean number of detected photons, it is not completely satisfactory, as the integral expression is not very explicit. This will be addressed, for example in Section 18.5, where explicit expressions for the Fisher information matrix will be obtained in the case of the estimation of an object's location parameters. It is, however, instructive to note that the integrand is a weighted square of the derivative of the image profile with respect to the parameter. This suggests that the larger the dependence of the image profile on changes in the estimated parameter, the larger the Fisher information and therefore the lower the Cramér-Rao lower bound. As a result, we can expect that a parameter can be estimated better, i.e., with lower variance, if the image changes significantly with changes in the parameter.

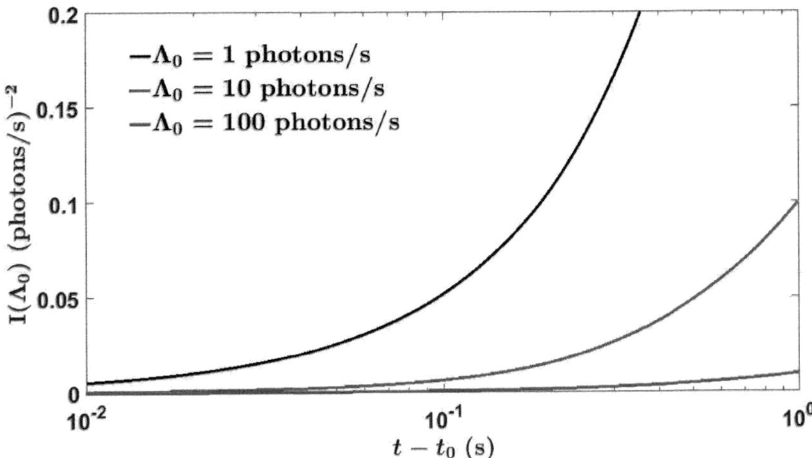

FIGURE 17.1
Fisher information $\mathbf{I}(\Lambda_0)$ of Eq. (17.14) as a function of the image acquisition time $t - t_0$, plotted for three different values of the photon detection rate Λ_0.

17.2.2 Example: known photon distribution profile

As a second example, we investigate here the estimation of the photon detection rate assuming the spatial distribution of the detected photons is known. Assume that the rate at which photons are detected from the object of interest, over the time interval $[t_0, t]$, is a constant. We therefore have $\Lambda_\theta(\tau) = \Lambda_0$, $\tau \geq t_0$, where $\Lambda_0 > 0$. Consider the estimation of the rate $\theta = \Lambda_0$ from the image acquired during $[t_0, t]$. As we assume the photon distribution profile to be known, we have that $\frac{\partial f_\tau(x,y)}{\partial \theta} = \frac{\partial f_\tau(x,y)}{\partial \Lambda_0} = 0$, $\tau \geq t_0$, $(x, y) \in C$. Then, according to the first expression of Eq. (17.11), the Fisher information matrix is given by

$$\begin{aligned}\mathbf{I}(\Lambda_0) &= 0 + \frac{1}{1 - e^{-\int_{t_0}^{t} \Lambda_0 d\tau}} \int_{t_0}^{t} \frac{1}{\Lambda_0} \left(\frac{\partial \Lambda_0}{\partial \Lambda_0} \right)^2 d\tau - \frac{e^{-\int_{t_0}^{t} \Lambda_0 d\tau}}{\left(1 - e^{-\int_{t_0}^{t} \Lambda_0 d\tau}\right)^2} \left(\int_{t_0}^{t} \frac{\partial \Lambda_0}{\partial \Lambda_0} d\tau \right)^2 \\ &= \frac{1}{1 - e^{-\Lambda_0(t-t_0)}} \cdot \frac{t - t_0}{\Lambda_0} - \frac{e^{-\Lambda_0(t-t_0)}}{\left(1 - e^{-\Lambda_0(t-t_0)}\right)^2} \cdot (t - t_0)^2, \end{aligned} \tag{17.14}$$

and it follows that the Cramér-Rao lower bound is just

$$\mathbf{I}^{-1}(\Lambda_0) = \frac{1}{\frac{1}{1-e^{-\Lambda_0(t-t_0)}} \cdot \frac{t-t_0}{\Lambda_0} - \frac{e^{-\Lambda_0(t-t_0)}}{\left(1-e^{-\Lambda_0(t-t_0)}\right)^2} \cdot (t - t_0)^2}. \tag{17.15}$$

By plotting Eq. (17.14) as a function of $t - t_0$, we can verify that the Fisher information $\mathbf{I}(\Lambda_0)$ increases with increasing values of $t - t_0$ (Fig. 17.1). This shows that $\mathbf{I}^{-1}(\Lambda_0)$ of Eq. (17.15), the lower bound on the variance with which Λ_0 can be estimated, can be improved (i.e., decreased) by increasing the image acquisition time. We can see this more easily by considering the scenario in which Λ_0 is large enough such that for the range of acquisition times $t - t_0$ considered, $e^{-\Lambda_0(t-t_0)} \approx 0$. In this case the Fisher information is approximated by $\mathbf{I}(\Lambda_0) \approx \frac{t-t_0}{\Lambda_0}$ and the Cramér-Rao lower bound by $\mathbf{I}^{-1}(\Lambda_0) \approx \frac{\Lambda_0}{t-t_0}$. It is then straightforward to see that the lower bound improves with increasing values of $t - t_0$.

As increasing the acquisition time increases the mean number of detected photons, this result again demonstrates that the variance of parameter estimation can be improved by acquiring more data. This is, of course, provided that an estimator exists that is efficient, or at least has a variance that is close to the Cramér-Rao lower bound. Note also that the expression for the Fisher information is independent of the photon distribution profile f_τ. This is not surprising as here the spatial profile of the image is in no way connected to the rate of photon detection.

17.3 Fisher information for the practical data models

We consider in this section the Fisher information matrix corresponding to the estimation of parameters from image data described by the various practical data models presented in Section 15.2. Here we will again present only relatively high-level expressions for the Fisher information matrix. Concrete illustrations of these expressions will be provided in the next three chapters, where they will be applied to specific estimation problems.

We begin by summarizing the settings that underlie a practical data model. In a practical scenario, we assume that an image is captured over the time interval $[t_0, t]$ by a pixelated detector. The detector can introduce noise of different types to the signals detected in its pixels. Exactly as in the fundamental case, the acquired image is composed of photons detected temporally according to an inhomogeneous Poisson process with rate $\Lambda_\theta(\tau)$, $\tau \geq t_0$, and spatially at locations described by the family of photon distribution profiles $\{f_{\theta,\tau}\}_{\tau \geq t_0}$ defined on a fundamental detector C. What makes the practical image different, however, is that the time point at which each photon is detected is only imprecisely known to fall within the acquisition time interval $[t_0, t]$, and that all photons impacting the same pixel are represented by a single photon count. For the kth pixel, we denote the mean of this photon count by $\nu_{\theta,k}$. Based on Eqs. (15.18) and (15.15), and adding the subscript θ to indicate the dependence of a function on the parameter θ, $\nu_{\theta,k}$ is generally given by

$$\nu_{\theta,k} = \int_{t_0}^{t} \Lambda_\theta(\tau) \int_{C_k} f_{\theta,\tau}(x, y)dxdyd\tau + \beta_{\theta,k}, \tag{17.16}$$

where $C_k \subseteq C$ is the region in C that is occupied by the kth pixel, and $\beta_{\theta,k}$ is the mean of the background signal detected in the kth pixel, which can itself be expressed in terms of integrals of a photon detection rate Λ_θ^b and a photon distribution profile $f_{\theta,\tau}^b$.

As we will see directly, for an N_p-pixel practical image, the general Fisher information matrix expressions will be given in terms of $\nu_{\theta,k}$, $k = 1, \ldots, N_p$, and its partial derivatives with respect to the parameters in θ. Specific realizations of these high-level expressions, which correspond to concretely defined estimation problems, are obtained by computing $\nu_{\theta,k}$ and its partial derivative $\frac{\partial \nu_{\theta,k}}{\partial \theta}$ with problem-specific expressions for θ, Λ_θ, $f_{\theta,\tau}$, and $\beta_{\theta,k}$, $k = 1, \ldots, N_p$. For example, as we will see in Chapter 18, for a localization problem θ will usually comprise the positional coordinates of an object of interest and the photon distribution profile $f_{\theta,\tau}$ will typically be specified in terms of a single image function q that describes the object's image, as in Eq. (15.8). For a resolution problem, we will see in Chapter 20 that θ will usually include the distance that separates a pair of objects of interest and $f_{\theta,\tau}$ will typically be given as a weighted sum of two image functions, one for each object. Additional details such as the specific expression for the image function will further differentiate one concrete estimation problem from another.

In many cases, the data captured in the pixels of a practical detector can be assumed to be stochastically independent. Given this assumption, the log-likelihood function corresponding to an N_p-pixel image is the sum of the log-likelihood functions $\ln p_{\theta,1}, \ldots, \ln p_{\theta,N_p}$

for the pixel data $z_1, \ldots, z_{N_p}$, and is given by Eq. (16.14), which we repeat here for convenience:

$$\mathcal{L}\left(\theta \mid z_1, \ldots, z_{N_p}\right) = \sum_{k=1}^{N_p} \ln p_{\theta,k}(z_k). \tag{17.17}$$

Given such a log-likelihood function, the Fisher information matrix $\mathbf{I}(\theta)$ is equal to the sum of the Fisher information matrices corresponding to the individual pixels (see Online Appendix Section A.3 for the additivity of Fisher information matrices), i.e.,

$$\mathbf{I}(\theta) = \sum_{k=1}^{N_p} \mathbf{I}_k(\theta),$$

where $\mathbf{I}_k(\theta)$ is the Fisher information matrix for the kth pixel. To obtain the Fisher information matrix corresponding to an entire image, the key then lies in calculating the Fisher information matrix for each pixel of the image.

We first derive the Fisher information for a somewhat general scenario that will apply to all the specific cases that will be considered here. As pointed out in Section 16.2.2, it is typically the case that information about θ, the parameter of interest, is contained only in the pixel photon counts that arise from the detection of photons from the imaged sample. Given this assumption, for every practical data model presented in Section 15.2, with the exception of the simple form of the deterministic data model, the probability distribution $p_{\theta,k}$ for the data in the kth image pixel depends on θ only through the mean $\nu_{\theta,k}$ of the Poisson-distributed signal in the kth pixel. We will not consider here the Fisher information for the simple form of the deterministic data model, but its derivation is provided in Online Appendix Section F.1.

17.3.1 Noise coefficient and the Fisher information

Under the above assumption, and applying the definition of Eq. (17.8), the Fisher information matrix $\mathbf{I}_k(\theta)$ corresponding to the kth image pixel is given by

$$\begin{aligned}
\mathbf{I}_k(\theta) &= \mathrm{E}\left[\left(\frac{\partial \mathcal{L}(\theta \mid z_k)}{\partial \theta}\right)^T \frac{\partial \mathcal{L}(\theta \mid z_k)}{\partial \theta}\right] = \mathrm{E}\left[\left(\frac{\partial \ln p_{\theta,k}(z_k)}{\partial \theta}\right)^T \frac{\partial \ln p_{\theta,k}(z_k)}{\partial \theta}\right] \\
&= \mathrm{E}\left[\left(\frac{1}{p_{\theta,k}(z_k)} \cdot \frac{\partial p_{\theta,k}(z_k)}{\partial \nu_{\theta,k}} \cdot \frac{\partial \nu_{\theta,k}}{\partial \theta}\right)^T \frac{1}{p_{\theta,k}(z_k)} \cdot \frac{\partial p_{\theta,k}(z_k)}{\partial \nu_{\theta,k}} \cdot \frac{\partial \nu_{\theta,k}}{\partial \theta}\right] \\
&= \left(\frac{\partial \nu_{\theta,k}}{\partial \theta}\right)^T \frac{\partial \nu_{\theta,k}}{\partial \theta} \cdot \mathrm{E}\left[\left(\frac{1}{p_{\theta,k}(z_k)} \cdot \frac{\partial p_{\theta,k}(z_k)}{\partial \nu_{\theta,k}}\right)^2\right] \\
&= \left(\frac{\partial \nu_{\theta,k}}{\partial \theta}\right)^T \frac{\partial \nu_{\theta,k}}{\partial \theta} \cdot \mathrm{E}\left[\left(\frac{\partial \ln p_{\theta,k}(z_k)}{\partial \nu_{\theta,k}}\right)^2\right].
\end{aligned} \tag{17.18}$$

This expression shows that the Fisher information matrix for the kth pixel is the product of two parts. The first part is a matrix of products of partial derivatives of $\nu_{\theta,k}$ with respect to the elements of the parameter vector θ. This matrix is a function of the specific estimation problem, as it depends on how $\nu_{\theta,k}$ is defined and how θ is defined. It is typically not dependent on the particular detector or type of detector that is used for the image acquisition, and therefore, for a given estimation problem, will be the same regardless of the detector that is employed. In contrast, the second part of the product is a scalar expectation term that typically depends on the particular detector and detector type through the probability distribution $p_{\theta,k}$, which accounts for the corruption of the detected signal by noise sources due to the detector. The scalar expectation term is not dependent on the specific estimation problem, in the sense that while it is a function of $\nu_{\theta,k}$, it is not impacted by how θ is defined and how it parameterizes $\nu_{\theta,k}$. In other words, for a given $p_{\theta,k}$, the same value for $\nu_{\theta,k}$ will produce the same value for the expectation term, regardless of the specific definition of $\nu_{\theta,k}$.

It is important to note that the scalar expectation term is in fact the Fisher information, with respect to $\nu_{\theta,k}$, for the kth image pixel. More practically, for a given value of $\nu_{\theta,k}$, it can be viewed as a detector-dependent scaling factor for the matrix portion of $\mathbf{I}_k(\theta)$ that indicates to what extent the signal in the kth pixel is corrupted by noise due to the detector that is used to capture the image. The smaller the degree of corruption, the larger the value of the expectation term, and hence the greater the amount of information that the data in the kth pixel carries about θ, the parameter we wish to estimate. A larger expectation term therefore means a smaller lower bound on the variance with which θ can be estimated based solely on the data in the kth pixel.

For a given pixel signal level $\nu_{\theta,k}$, the detector-dependent expectation term thus provides a useful means for comparing the parameter estimation performance of different detectors and detector types. To facilitate such comparisons, we introduce a normalized version of the expectation term which we call the *noise coefficient*. Denoted by γ_k, the noise coefficient for the kth image pixel is given by

$$\gamma_k = \frac{\mathrm{E}\left[\left(\frac{\partial \ln p_{\theta,k}(z_k)}{\partial \nu_{\theta,k}}\right)^2\right]}{\frac{1}{\nu_{\theta,k}}} = \nu_{\theta,k} \cdot \mathrm{E}\left[\left(\frac{\partial \ln p_{\theta,k}(z_k)}{\partial \nu_{\theta,k}}\right)^2\right], \tag{17.19}$$

which is just the expectation term for the kth pixel normalized with respect to the inverse of the signal level $\nu_{\theta,k}$ in the kth pixel. The reason for using this specific normalization will be made clear in Section 17.4, where we demonstrate the use of the noise coefficient for the comparison of detectors and detector types in terms of parameter estimation performance.

Expressed using the noise coefficient, the Fisher information matrix $\mathbf{I}_k(\theta)$ for the data in the kth image pixel is given by

$$\begin{aligned}\mathbf{I}_k(\theta) &= \left(\frac{\partial \nu_{\theta,k}}{\partial \theta}\right)^T \frac{\partial \nu_{\theta,k}}{\partial \theta} \cdot \frac{1}{\nu_{\theta,k}} \cdot \left(\nu_{\theta,k} \cdot \mathrm{E}\left[\left(\frac{\partial \ln p_{\theta,k}(z_k)}{\partial \nu_{\theta,k}}\right)^2\right]\right) \\ &= \left(\frac{\partial \nu_{\theta,k}}{\partial \theta}\right)^T \frac{\partial \nu_{\theta,k}}{\partial \theta} \cdot \frac{\gamma_k}{\nu_{\theta,k}}.\end{aligned} \tag{17.20}$$

Given the expression for $\mathbf{I}_k(\theta)$, the Fisher information matrix $\mathbf{I}(\theta)$ for an N_p-pixel image is obtained easily, by applying the additivity property of the Fisher information matrix, as

$$\mathbf{I}(\theta) = \sum_{k=1}^{N_p} \mathbf{I}_k(\theta) = \sum_{k=1}^{N_p} \left(\frac{\partial \nu_{\theta,k}}{\partial \theta}\right)^T \left(\frac{\partial \nu_{\theta,k}}{\partial \theta}\right) \cdot \frac{\gamma_k}{\nu_{\theta,k}}. \tag{17.21}$$

For a particular specification of θ and $\nu_{\theta,k}$ (i.e., for a given parameter estimation problem), the practical data models of Section 15.2 differ through the probability distribution $p_{\theta,k}$, which accounts for detector-dependent noise and therefore provides a stochastic description of the data acquired in the kth image pixel. It is thus easy to see, from the relatively general expression of Eq. (17.21), that the Fisher information matrix expression for a particular practical data model is obtained by evaluating the noise coefficient γ_k using the data model's definition of $p_{\theta,k}$. The evaluation of γ_k actually requires the logarithm of $p_{\theta,k}$ (i.e., the log-likelihood function for the data in the kth pixel), and this is given for each practical data model in Table 16.1.

General expression for the Fisher information matrix corresponding to an N_p-pixel practical image:	$\mathbf{I}(\theta)=\sum_{k=1}^{N_p}\mathbf{I}_k(\theta)=\sum_{k=1}^{N_p}\left(\frac{\partial\nu_{\theta,k}}{\partial\theta}\right)^T\left(\frac{\partial\nu_{\theta,k}}{\partial\theta}\right)\cdot\frac{\gamma_k}{\nu_{\theta,k}}$	
Data model	**Noise coefficient γ_k**	**Notes**
Poisson	1	
CCD/CMOS	$\nu_{\theta,k}\cdot\left(\int_{-\infty}^{\infty}\frac{[s_{\theta,k}(z)]^2}{p_{\theta,k}(z)}dz-1\right)$	$s_{\theta,k}(z)=\frac{e^{-\nu_{\theta,k}}}{\sqrt{2\pi}\sigma_k}\sum_{l=1}^{\infty}\frac{\nu_{\theta,k}^{l-1}}{(l-1)!}e^{-\frac{1}{2}\left(\frac{z-l-\eta_k}{\sigma_k}\right)^2}$
CCD/CMOS Gaussian approximation	$\nu_{\theta,k}\cdot\left(\frac{1}{\nu_{\theta,k}+\sigma_k^2}+\frac{1}{2\left(\nu_{\theta,k}+\sigma_k^2\right)^2}\right)$	
EMCCD	$\nu_{\theta,k}\cdot\left(\int_{-\infty}^{\infty}\frac{[s_{\theta,k}(z)]^2}{p_{\theta,k}(z)}dz-1\right)$	$s_{\theta,k}(z)=\frac{e^{-\nu_{\theta,k}}}{\sqrt{2\pi}\sigma_k}\sum_{l=1}^{\infty}e^{-\left(\frac{z-l-\eta_k}{\sqrt{2}\sigma_k}\right)^2}\sum_{j=0}^{l-1}\frac{\binom{l-1}{j}\left(1-\frac{1}{g}\right)^{l-j-1}\nu_{\theta,k}^j}{j!g^{j+1}}$
EMCCD high gain approximation	$\nu_{\theta,k}\cdot\left(\int_{-\infty}^{\infty}\frac{[s_{\theta,k}(z)]^2}{p_{\theta,k}(z)}dz-1\right)$	$s_{\theta,k}(z)=\frac{e^{-\nu_{\theta,k}}}{\sqrt{2\pi}\sigma_k}\int_0^{\infty}\frac{e^{-\left(\frac{z-u-\eta_k}{\sqrt{2}\sigma_k}\right)^2-\frac{u}{g}}}{g}I_0\left(2\sqrt{\frac{\nu_{\theta,k}u}{g}}\right)du$
EMCCD Gaussian approximation	$\nu_{\theta,k}\cdot\left(\frac{g^2}{(2g^2-g)\nu_{\theta,k}+\sigma_k^2}+\frac{(2g^2-g)^2}{2\left((2g^2-g)\nu_{\theta,k}+\sigma_k^2\right)^2}\right)$	

TABLE 17.1
Fisher information matrix corresponding to an N_p-pixel image under various practical data models. The data models share the same general expression for the Fisher information matrix and differ by the noise coefficient γ_k which depends on the probability distribution $p_{\theta,k}$ for the data in the kth image pixel. For each data model, $p_{\theta,k}$ is given in Table 15.1. The log-likelihood function corresponding to each probability distribution, which is used in the calculation of the corresponding noise coefficient, is given in Table 16.1.

In what follows, we will only show the evaluation of the noise coefficient for the Poisson and CCD/CMOS data models, as the differences between the evaluations for the different data models lie only in the mathematical details associated with the particular probability distribution $p_{\theta,k}$. For the interested reader, evaluations of the noise coefficients for the Gaussian approximations for the CCD/CMOS and EMCCD data models are shown in Online Appendix Sections F.2 and F.3, respectively. For the EMCCD data model and its high gain approximation, the noise coefficient evaluations are given in Online Appendix Sections G.3 and G.4, respectively. For a listing of the noise coefficients of the various practical data models, see Table 17.1.

Poisson data model. For an N_p-pixel image captured by a hypothetical pixelated detector that does not introduce readout noise, the log-likelihood function $\ln p_{\theta,k}$ for the data z_k in the kth pixel, $k=1,\ldots,N_p$, is the expression shown for the Poisson data model in Table 16.1. To calculate the noise coefficient for the Poisson data model, we simply substitute this log-likelihood expression into Eq. (17.19) and obtain

$$\begin{aligned}\gamma_k &= \nu_{\theta_k}\cdot\mathrm{E}\left[\left(\frac{\partial}{\partial\nu_{\theta,k}}\ln p_{\theta,k}(z_k)\right)^2\right]=\nu_{\theta_k}\cdot\mathrm{E}\left[\left(\frac{\partial}{\partial\nu_{\theta,k}}\left(z_k\ln(\nu_{\theta,k})-\nu_{\theta,k}-\ln(z_k!)\right)\right)^2\right]\\ &= \nu_{\theta_k}\cdot\mathrm{E}\left[\left(\frac{z_k}{\nu_{\theta,k}}-1\right)^2\right]=\nu_{\theta_k}\cdot\mathrm{E}\left[\frac{z_k^2}{\nu_{\theta,k}^2}-\frac{2z_k}{\nu_{\theta,k}}+1\right]=\nu_{\theta_k}\cdot\left(\frac{1}{\nu_{\theta,k}^2}\mathrm{E}\left[z_k^2\right]-\frac{2\mathrm{E}\left[z_k\right]}{\nu_{\theta,k}}+1\right)\\ &= \nu_{\theta_k}\cdot\left(\frac{1}{\nu_{\theta,k}^2}\left[\nu_{\theta,k}+\nu_{\theta,k}^2\right]-\frac{2\nu_{\theta,k}}{\nu_{\theta,k}}+1\right)=\nu_{\theta_k}\cdot\left(\frac{1}{\nu_{\theta,k}}+1-2+1\right)=1. \end{aligned} \tag{17.22}$$

Therefore, the Fisher information matrix for an N_p-pixel image based on the Poisson data model is, according to Eq. (17.21), given by

$$\mathbf{I}(\theta) = \sum_{k=1}^{N_p} \left(\frac{\partial \nu_{\theta,k}}{\partial \theta}\right)^T \left(\frac{\partial \nu_{\theta,k}}{\partial \theta}\right) \cdot \frac{1}{\nu_{\theta,k}}. \tag{17.23}$$

CCD/CMOS data model. For an N_p-pixel image acquired using a CCD or CMOS camera, using in Eq. (17.19) the expression for $\ln p_{\theta,k}(z_k)$ as shown for the CCD/CMOS data model in Table 16.1 yields the noise coefficient

$$\begin{aligned}
\gamma_k &= \nu_{\theta,k} \cdot \mathrm{E}\left[\left(\frac{\partial \ln p_{\theta,k}(z_k)}{\partial \nu_{\theta,k}}\right)^2\right] \\
&= \nu_{\theta,k} \cdot \mathrm{E}\left[\left(\frac{\partial}{\partial \nu_{\theta,k}}\left(-\nu_{\theta,k} - \ln\left(\sqrt{2\pi}\sigma_k\right) + \ln\left(\sum_{l=0}^{\infty} \frac{\nu_{\theta,k}^l}{l!} e^{-\frac{1}{2}\frac{(z_k - l - \eta_k)^2}{\sigma_k^2}}\right)\right)\right)^2\right] \\
&= \nu_{\theta,k} \cdot \mathrm{E}\left[\left(-1 + \frac{1}{\sum_{l=0}^{\infty} \frac{\nu_{\theta,k}^l}{l!} e^{-\frac{1}{2}\frac{(z_k - l - \eta_k)^2}{\sigma_k^2}}} \sum_{l=0}^{\infty} \frac{l \nu_{\theta,k}^{l-1}}{l!} e^{-\frac{1}{2}\frac{(z_k - l - \eta_k)^2}{\sigma_k^2}}\right)^2\right] \\
&= \nu_{\theta,k} \cdot \left(\mathrm{E}\left[\left(\frac{e^{-\nu_{\theta,k}}}{\sqrt{2\pi}\sigma_k \cdot p_{\theta,k}(z_k)} \sum_{l=1}^{\infty} \frac{\nu_{\theta,k}^{l-1}}{(l-1)!} e^{-\frac{1}{2}\frac{(z_k - l - \eta_k)^2}{\sigma_k^2}}\right)^2\right]\right. \\
&\qquad \left. -2 \cdot \mathrm{E}\left[\frac{e^{-\nu_{\theta,k}}}{\sqrt{2\pi}\sigma_k \cdot p_{\theta,k}(z_k)} \sum_{l=1}^{\infty} \frac{\nu_{\theta,k}^{l-1}}{(l-1)!} e^{-\frac{1}{2}\frac{(z_k - l - \eta_k)^2}{\sigma_k^2}}\right] + 1\right) \\
&= \nu_{\theta,k} \cdot \left(\int_{-\infty}^{\infty} \frac{1}{p_{\theta,k}(z)} \left(\frac{e^{-\nu_{\theta,k}}}{\sqrt{2\pi}\sigma_k} \sum_{l=1}^{\infty} \frac{\nu_{\theta,k}^{l-1}}{(l-1)!} e^{-\frac{1}{2}\frac{(z - l - \eta_k)^2}{\sigma_k^2}}\right)^2 dz\right. \\
&\qquad \left. -2 \cdot \sum_{l=1}^{\infty} \frac{e^{-\nu_{\theta,k}} \nu_{\theta,k}^{l-1}}{(l-1)!} \int_{-\infty}^{\infty} \frac{1}{\sqrt{2\pi}\sigma_k} e^{-\frac{1}{2}\frac{(z - l - \eta_k)^2}{\sigma_k^2}} dz + 1\right) \\
&= \nu_{\theta,k} \cdot \left(\int_{-\infty}^{\infty} \frac{1}{p_{\theta,k}(z)} \left(\frac{e^{-\nu_{\theta,k}}}{\sqrt{2\pi}\sigma_k} \sum_{l=1}^{\infty} \frac{\nu_{\theta,k}^{l-1}}{(l-1)!} e^{-\frac{1}{2}\frac{(z - l - \eta_k)^2}{\sigma_k^2}}\right)^2 dz - 2 \cdot 1 + 1\right) \\
&= \nu_{\theta,k} \cdot \left(\int_{-\infty}^{\infty} \frac{1}{p_{\theta,k}(z)} \left(\frac{e^{-\nu_{\theta,k}}}{\sqrt{2\pi}\sigma_k} \sum_{l=1}^{\infty} \frac{\nu_{\theta,k}^{l-1}}{(l-1)!} e^{-\frac{1}{2}\frac{(z - l - \eta_k)^2}{\sigma_k^2}}\right)^2 dz - 1\right) \\
&= \nu_{\theta,k} \cdot \left(\int_{-\infty}^{\infty} \frac{[s_{\theta,k}(z)]^2}{p_{\theta,k}(z)} dz - 1\right),
\end{aligned} \tag{17.24}$$

where

$$s_{\theta,k}(z) = \frac{e^{-\nu_{\theta,k}}}{\sqrt{2\pi}\sigma_k} \sum_{l=1}^{\infty} \frac{\nu_{\theta,k}^{l-1}}{(l-1)!} e^{-\frac{1}{2}\left(\frac{z - l - \eta_k}{\sigma_k}\right)^2},$$

and $p_{\theta,k}$ is as given in Table 15.1 for the CCD/CMOS data model. By Eq. (17.21), the Fisher information matrix for an N_p-pixel CCD/CMOS image is then given by

$$\mathbf{I}(\theta) = \sum_{k=1}^{N_p} \left(\frac{\partial \nu_{\theta,k}}{\partial \theta}\right)^T \left(\frac{\partial \nu_{\theta,k}}{\partial \theta}\right) \cdot \left(\int_{-\infty}^{\infty} \frac{[s_{\theta,k}(z)]^2}{p_{\theta,k}(z)} dz - 1\right). \tag{17.25}$$

17.4 Noise coefficient analysis of the pixel signal level

We saw in the previous section that for a given mean signal $\nu_{\theta,k}$, the noise coefficient γ_k represents the part of Eq. (17.20) that accounts for detector-dependent differences in the amount of information that the data in the kth pixel contains about the parameter of interest θ. As such, it determines the detector-dependent differences in the lower bound on the variance for estimating θ based on the data in the kth pixel. The noise coefficient therefore provides a natural means for selecting the best detector type and noise settings to use from the perspective of information content and estimation performance. Specifically, we can use it to make comparisons of different detectors at different signal levels by calculating its value for different combinations of detector type, detector noise settings, and $\nu_{\theta,k}$. This will be illustrated in the subsections that follow, but we will first explain the normalization that is used in its definition to facilitate such comparisons.

17.4.1 Noise coefficient — an in-depth look

For a given signal level $\nu_{\theta,k}$, the noise coefficient as defined in Eq. (17.19) is quite simply the ratio of the scalar expectation term for the given detector to the scalar expectation term $\frac{1}{\nu_{\theta,k}}$ for the corresponding Poisson data model (Eq. (17.22)). As the Poisson data model describes the ideal scenario in which the detector does not contribute readout noise to the acquired signal, with this definition the noise coefficient incorporates a meaningful point of reference that makes it a useful tool for comparing the performance of detectors in terms of parameter estimation.

By its definition, the noise coefficient is equal to 1 for the Poisson data model (Eq. (17.22)) for any given signal level $\nu_{\theta,k}$. This means that for a data model that has as its basis a Poisson signal with mean $\nu_{\theta,k}$, the noise coefficient will have a value between 0 and 1 (Note 17.2). The minimum value is 0 because the noise coefficient is the product of $\nu_{\theta,k} > 0$ and the expected value of a squared expression. Intuitively, it can be seen that the maximum value is 1 because it must be the case that data that entails the corruption of the detected Poisson signal will contain less information about the parameter θ than data that consists of the uncorrupted Poisson signal. In other words, because the Poisson data model represents the best-case scenario, its corresponding expectation term in the denominator of the noise coefficient can never be smaller than the expectation term in the numerator, which corresponds to a data model that represents a scenario in which there is corruption of the signal by detector noise. For a given signal level $\nu_{\theta,k}$, the problem of selecting the best detector from a set of candidate detectors is therefore reduced to a comparison of noise coefficient values between 0 and 1, with values close to 0 indicating that the data for the given detector carries little information about θ, and values close to 1 indicating that the data contains nearly the same amount of information as the uncorrupted Poisson signal.

In the next subsections, we will use the noise coefficient to investigate and compare the information content, and hence the parameter estimation performance, of different detectors and detector types at different pixel signal levels. As the analyses will be in reference to a generic pixel, the signal level and the noise coefficient will simply be notated as ν_θ and γ, respectively, without the subscript k.

17.4.2 Noise coefficient for CCD/CMOS detectors

Figure 17.2 provides an illustration of the comparison of different detectors using the noise coefficient. Shown in the figure are the noise coefficients γ for two CCD and two EMCCD

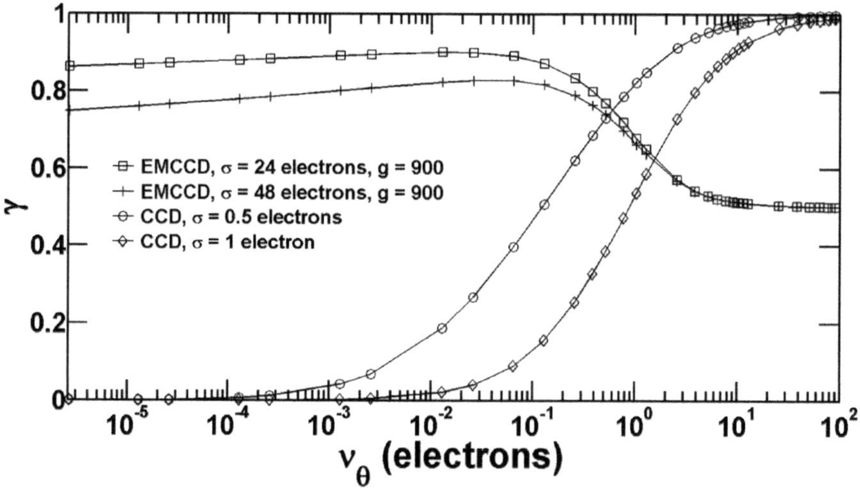

FIGURE 17.2
CCD and EMCCD noise coefficients as functions of the pixel photon count. The noise coefficient γ is shown for EMCCD detectors used with an EM gain of $g = 900$ and having readout noise with standard deviations of $\sigma = 24$ electrons ($\square$) and $\sigma = 48$ electrons ($+$), and for CCD detectors having readout noise with standard deviations of $\sigma = 0.5$ electrons ($\circ$) and $\sigma = 1$ electron ($\diamond$). In all cases, the readout noise is assumed to have a mean of $\eta = 0$ electrons.

detectors, plotted as functions of the mean photon signal ν_θ in a given pixel. Clear differences can be seen between the curves for the two different detector types. The two CCD noise coefficient curves both approach the maximum value of 1 at large values of ν_θ, indicating that when the number of photons detected in a pixel is large enough to render the readout noise insignificant, the minimally corrupted data in the pixel carries nearly the same amount of information about θ as the pure Poisson signal that comprises the data under the Poisson data model. In contrast, at values of ν_θ significantly less than one photon, the same two curves both approach the minimum value of 0, demonstrating that when the number of photons detected in a pixel is so small that the data in the given pixel is dominated by the CCD detector's readout noise, the severely corrupted data carries almost no information about θ. From the perspective of being able to estimate θ with high precision, the behavior of these curves affirms that a CCD or CMOS detector should be used when a reasonably good level of fluorescence can be expected from the imaged object. A comparison of the two CCD curves, which differ by their readout noise standard deviations, further corroborates the intuition that a lower level of readout noise will yield data that contains more information about θ. The curves show that this is particularly the case when ν_θ takes on values that fall between the convergence of the curves to 0 and the convergence of the curves to 1.

17.4.3 EMCCD detectors as low-light detectors

Unlike CCD noise coefficient curves, Fig. 17.2 shows that EMCCD noise coefficient curves converge to the value 0.5 at large values of ν_θ, provided that a relatively high EM gain is used. The value of 0.5 indicates that the data in the EMCCD pixel carries half of the maximum amount of information the corresponding pure Poisson signal carries about θ. This halving of the maximum information content has been shown to be consistent with the excess noise factor (Note 17.3) associated with stochastic electron multiplication, based on which it is commonly assumed that the lower bound on the variance with which a parameter can be estimated from an EMCCD image can be no smaller than twice the lower bound on the variance with which it can be estimated from a corresponding Poisson data image. This is indeed approximately the case when relatively large numbers of photons are detected in the pixels of an EMCCD image, but the EMCCD noise coefficient curves demonstrate that the factor of 2 penalty does not generally apply for all EMCCD images. In fact, Fig. 17.2 shows that at values of ν_θ less than one photon, both EMCCD curves attain values significantly greater than 0.5, meaning that the data in the given EMCCD pixel has an information content that is substantially greater than half of the maximum. Therefore, when very low photon counts are expected in the pixels of an EMCCD detector, the Cramér-Rao lower bound for estimating θ is potentially substantially closer to the Cramér-Rao lower bound for estimating θ from a corresponding Poisson data image. Indeed, based on this idea, a microscopy modality has been proposed that uses an EMCCD detector in conjunction with high magnification and/or a small exposure time to ensure very low pixel photon counts, thereby producing information-rich image data that allows for parameter estimation with minimal variance (Note 17.4). For a given small light level ν_θ, comparison of the two EMCCD noise coefficient curves in Fig. 17.2 further illustrates that the information content of the data in a given pixel can be brought closer to the maximum by reducing the EMCCD detector's readout noise standard deviation. It has also been shown, using the noise coefficient, that the same result can be achieved by increasing the EM gain of the EMCCD detector.

The different behaviors exhibited by the EMCCD noise coefficient at relatively high and at very low mean pixel photon counts can be explained by the fact that signal amplification by the EMCCD detector, being stochastic in nature, is itself a source of noise and should be used only when it is beneficial to do so. When the mean pixel photon count is high enough such that readout noise represents only a small component of the data in the pixel, electron multiplication becomes unnecessary and using it will actually introduce more stochasticity to the data, leading to the undesirable halving of the data's maximum information content. On the other hand, when the mean pixel photon count is so small that readout noise becomes the dominant component of the data in the pixel, electron multiplication has the beneficial effect of significantly amplifying the photon signal, thereby rendering the readout noise small in comparison and affording an information content for the data that is close to the maximum.

17.4.4 Comparison of CCD/CMOS and EMCCD detectors

We will see in Section 18.6.2 that the different ways in which CCD and EMCCD noise coefficients depend on the mean pixel photon count are consistent with the patterns exhibited by CCD-based and EMCCD-based Cramér-Rao lower bounds as a function of the effective pixel size of the image. There, in the context of the object localization problem, we will see that whereas the CCD-based lower bound on the standard deviation of location estimation is superior (i.e., smaller) at large effective pixel sizes which correspond to relatively large mean pixel photon counts, the EMCCD-based lower bound on the standard deviation of location estimation is superior at relatively small effective pixel sizes which correspond to relatively small mean pixel photon counts. We will also see in Section 18.6.2 that just as the CCD noise coefficient approaches the Poisson noise coefficient of 1 at high mean pixel photon counts, the CCD-based lower bound on the standard deviation of location estimation approaches the Poisson-based lower bound on the standard deviation of location estimation at large effective pixel sizes. Similarly, we will see that just as the EMCCD noise coefficient comes close to the Poisson noise coefficient of 1 at very small mean pixel photon counts, the EMCCD-based lower bound on the standard deviation of location estimation comes close to the Poisson-based lower bound on the standard deviation of location estimation at small effective pixel sizes.

17.5 Fisher information for multi-image data

In Sections 17.2 and 17.3, the Fisher information matrix expressions presented pertain to the estimation of parameters of interest from a single image. In many situations, however, a parameter of interest is estimated by using the information contained in more than one image. One example is the estimation of parameters associated with a moving object's trajectory. In this case, parameters such as the speed of the object are estimated from a time sequence of images that capture the object's trajectory. Another example, which relates to the resolution problem as introduced in Chapter 20, is the determination of the distance that separates two closely spaced objects from a combination of images, each containing at least one of the objects. In this case, the use of an image that contains just one of the objects allows the localization of that object in the absence of overlapping signal from the other object, and helps to improve the Cramér-Rao lower bound on the variance with which the distance of separation can be estimated.

Provided that the images from which the vector of parameters θ is to be estimated are the results of stochastically independent photon detection processes, the Fisher information matrix corresponding to the set of images is just the sum of the Fisher information matrices corresponding to the individual images (Online Appendix Section A.3). For a set of N_{im} images with Fisher information matrices $\mathbf{I}_k(\theta)$, $k = 1, \ldots, N_{im}$, the Fisher information matrix $\mathbf{I}_{set}(\theta)$ is thus given by

$$\mathbf{I}_{set}(\theta) = \sum_{k=1}^{N_{im}} \mathbf{I}_k(\theta). \tag{17.26}$$

The summing of Fisher information matrices shows that more information can be gained about the parameters in θ from multiple images, meaning that we can expect the use of additional images in an estimation to improve the variance of the estimation. For example, in the simplest case where we have acquired multiple images of a stationary object under unchanging conditions, the Fisher information matrix $\mathbf{I}_{set}(\theta)$ corresponding to N_{im} of these repeat images, given that $\mathbf{I}_{single}(\theta)$ is the Fisher information matrix corresponding to a single image, is $\mathbf{I}_{set}(\theta) = \sum_{k=1}^{N_{im}} \mathbf{I}_k(\theta) = N_{im} \cdot \mathbf{I}_{single}(\theta)$. This implies that the Cramér-Rao lower bound on the variance of any unbiased estimator of θ that utilizes the N_{im} images is improved by a factor of $\frac{1}{N_{im}}$ relative to the bound when only one image is used.

Using Eq. (17.26), we will demonstrate in Section 19.6.4 how the poor depth discrimination of the optical microscope can be overcome by estimating the z_0 coordinate of a point source from the multiple images acquired using a MUM setup. The images in this case are those captured simultaneously from different focal planes within the sample, and together they enable the axial localization of a near-focus object with reasonable variance by providing views of the object from different distances along the optical axis.

18

Localizing Objects and Single Molecules in Two Dimensions

Having introduced the necessary background in the prior chapters, we can now discuss one of the central topics of this part of the book, the estimation of the location of objects with single molecules being an important example. The main focus of this chapter is the estimation of the (x, y) location, meaning the position of the object relative to the image coordinate system. The more general situation entailing the estimation of all three spatial coordinates of the object will be discussed in the next chapter.

18.1 Object localization as a parameter estimation problem

As discussed in Chapter 16, parameter estimation refers to the determination of quantities of interest from some acquired data. In this chapter, we are particularly interested in the determination of the location of an object from image data acquired using a microscopy setup. This will generally involve the use of a mathematical model that provides a description of the image data, and that is parameterized by the positional coordinates of the object. The goal of the object localization problem is to find values for the positional coordinates that yield the instance of this model that best fits the image data according to some criterion. Typically, it is not possible to analytically determine these values. Therefore, the solution to the object localization problem will generally also involve the use of a numerical optimization algorithm that finds the best-fit values through an iterative process. A high-level illustration of the optimization process is given in Fig. 18.1, where the object to be localized is a fluorescent molecule and the model used to fit the acquired image is an Airy profile. We see that the optimization starts with an initial guess for the values of the molecule's positional coordinates and the instance of the Airy profile that corresponds to those values. It then proceeds by iteratively comparing the current instance of the Airy profile with the image and generating a new set of values for the coordinates based on the comparison. The routine terminates when a set of coordinate values is found that produces the instance of the Airy profile that best fits the image.

An important aspect of parameter estimation is the criterion used to determine how well a given instance of the model fits the data. This criterion represents a critical way by which one parameter estimation method differs from another. In Chapter 16, for example, we introduced maximum likelihood estimation which uses the criterion of how likely the data resulted from the current set of parameter values, and least squares estimation which makes the determination based on the square of the difference between the current instance of the model and the data. In this chapter, we will consider the maximum likelihood method of estimation for the object localization problem.

In more precise terms, the object localization problem can be formulated as follows. An imaging experiment yields a N_p-pixel image of the object. We then estimate the parameter

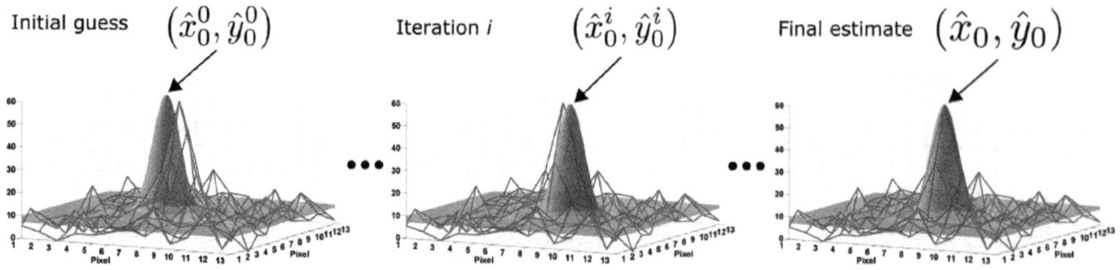

FIGURE 18.1
Localization as a parameter estimation problem. The positional coordinates x_0 and y_0 of a molecule are estimated from an image (red mesh) of the molecule via an optimization algorithm that iteratively tries different sets of values for x_0 and y_0 and compares the resulting instances of the model (gray surface) with the image. The optimization starts with the initial guess $(\hat{x}_0^0, \hat{y}_0^0)$ for the position of the molecule and iterates until arriving at the estimate $(\hat{x}_0, \hat{y}_0)$, which yields an instance of the model that best fits the image according to some given criterion. The vertices of the red mesh represent the pixel values of the image, and the gray surface is an Airy profile that models the image. The Airy profile is parameterized by x_0 and y_0, which correspond to the coordinates of its center. It should be noted that only a conceptual illustration is provided here, and that in practice the model that is compared to the image in each iteration is a pixelated and differently scaled version of the gray surface that is more accurately represented by a mesh similar to the image (see, e.g., Fig. 18.2(b)).

of interest $\theta \in \Theta$, namely the (x, y) location of the object, from the measured pixel values $z_1, \ldots, z_{N_p}$. Here $\Theta \subseteq \mathbb{R}^2$ is the parameter space of candidate locations for the object. We employ maximum likelihood estimation, wherein given the measured data $z_1, \ldots, z_{N_p}$, we maximize the log-likelihood function of Eq. (16.14) with respect to θ. In more concrete terms, we would like to obtain, for the location of an object, the maximum likelihood estimate

$$\hat{\theta}_{MLE} = \arg\max_{\theta \in \Theta} \mathcal{L}\left(\theta \mid z_1, \ldots, z_{N_p}\right) = \arg\max_{\theta \in \Theta} \sum_{k=1}^{N_p} \ln p_{\theta,k}(z_k), \tag{18.1}$$

where $p_{\theta,k}$ is a function that characterizes the distribution of the value of the data z_k in the kth pixel, and is given as one of the probability density functions or mass functions from Table 15.1, depending on the type of detector that was used to capture the image. The function $p_{\theta,k}$ is itself a function of the mean photon count $\nu_{\theta,k}$ in the kth pixel which, reproduced here from Eq. (17.16), is given by

$$\nu_{\theta,k} = \int_{t_0}^{t} \Lambda_\theta(\tau) \int_{C_k} f_{\theta,\tau}(x, y)dxdyd\tau + \beta_{\theta,k}. \tag{18.2}$$

In this general expression, the function Λ_θ describes the rate at which photons are detected from the object, the function $f_{\theta,\tau}$ describes the spatial distribution of the photons detected from the object at time τ, and the function $\beta_{\theta,k}$ gives the mean number of photons in the pixel that are due to the background component. This general formulation allows one to define and estimate the location of a moving object, but in the example that we will present directly, we will customize this expression for the localization of an object that is, or can be assumed to be, stationary.

From our description we can see that determining an expression for the probability distribution $p_{\theta,k}$, $k = 1, \ldots, N_p$, is key to the maximum likelihood estimation of the location of an object of interest. As it depends on both the signal detected from the object and the characteristics of the detector that was used to acquire the image, in the modeling aspect we have to make at least two important decisions. One is the model for the image of the object, and the other is the practical data model that accounts for how the specific detector used affects the signal detected from the object.

18.2 Example: estimating the location of a single molecule

As an example, we analyze a location estimation problem as it arises in single molecule microscopy. We consider the case where a stationary molecule is to be localized from an acquired image. The fact that the object of interest is immobile allows us to remove the time dependence of the photon distribution profile in Eq. (18.2). By writing $f_{\theta,\tau} = f_\theta$, $\tau \geq t_0$, the double integral in Eq. (18.2) becomes a product of the spatial integral $\int_{C_k} f_\theta(x,y)dxdy$ and the mean number of photons $N_{photons} = \int_{t_0}^{t} \Lambda_\theta(\tau)d\tau$ that are detected from the object over the image acquisition time interval $[t_0, t]$.

For our example, we will further make the common assumption that the imaging system is translation-invariant. Then based on Eq. (15.8), the photon distribution profile f_θ can be expressed more explicitly via an image function q as

$$f_\theta(x,y) = \frac{1}{M^2} q\left(\frac{x}{M} - x_0, \frac{y}{M} - y_0\right), \quad (x,y) \in \mathbb{R}^2, \tag{18.3}$$

where x_0 and y_0 are the object's positional coordinates, M is the magnification of the optics used to image the object, and θ is the vector of parameters to be estimated, which for a localization problem would typically include x_0 and y_0.

Suppose our stationary molecule is in focus, and that its image is described by an Airy profile. We can therefore write the photon distribution profile of Eq. (18.3) explicitly by substituting q with the Airy image function of Eq. (15.9), obtaining

$$f_\theta(x,y) = \frac{1}{M^2} \frac{J_1^2\left(\alpha\sqrt{\left(\frac{x}{M} - x_0\right)^2 + \left(\frac{y}{M} - y_0\right)^2}\right)}{\pi\left(\left(\frac{x}{M} - x_0\right)^2 + \left(\frac{y}{M} - y_0\right)^2\right)}, \quad (x,y) \in \mathbb{R}^2. \tag{18.4}$$

Furthermore, suppose that the rate at which photons are detected from the molecule is constant. This allows us to write the photon detection rate in Eq. (18.2) as $\Lambda_\theta(\tau) = \Lambda_0 > 0$, $\tau \geq t_0$, and the mean number of photons detected from the object reduces to $N_{photons} = \Lambda_0(t - t_0)$. In terms of the background component, we assume a uniform distribution of noise over the array of N_p pixels from which we will estimate the location of the molecule. Accordingly, we can set the mean number of background photons to be the same in all N_p pixels, i.e., we can write $\beta_{\theta,k} = \beta_0$, $k = 1, \ldots, N_p$.

We have now arrived at a complete description of the model for the image of our object of interest. In other words, provided that values are specified for parameters such as the magnification M, the parameter α of the Airy image function, and the molecule's positional coordinates x_0 and y_0, we can now compute $\nu_{\theta,k}$ of Eq. (18.2) for all pixels k that make up the pixel array from which we will localize the molecule. Of the various parameters, some will be considered "known" while others need to be estimated. For our current problem, we will assume that we know the values for M, α, β_0, the image acquisition time $t - t_0$, the photon detection rate Λ_0, and the integration limits corresponding to each 2D region C_k

which represents the kth pixel in the pixel array. On the other hand, we wish to estimate x_0 and y_0. We therefore define $\theta = (x_0, y_0) \in \mathbb{R}^2$, for which we will obtain the maximum likelihood estimate $\hat{\theta}_{MLE}$ as described in Eq. (18.1).

Before we can determine $\hat{\theta}_{MLE}$, however, we still need an expression for the probability distribution $p_{\theta,k}$ for the data z_k in each pixel k of the N_p-pixel array. As mentioned above, this distribution is dependent on the type of detector used to record the image of, in our current example, the molecule. We will assume that our molecule is imaged using a CCD camera, and therefore use the probability density function for the CCD/CMOS data model, which for $k = 1, \ldots, N_p$ is given by

$$p_{\theta,k}(z) = \frac{e^{-\nu_{\theta,k}}}{\sqrt{2\pi}\sigma_k} \sum_{l=0}^{\infty} \frac{\nu_{\theta,k}^l}{l!} e^{-\frac{1}{2}\frac{(z-l-\eta_k)^2}{\sigma_k^2}}, \quad z \in \mathbb{R}. \tag{18.5}$$

This density function is reproduced from Eq. (15.21), with the subscript θ added and ν_k replaced by $\nu_{\theta,k}$. We see that it is a function of the mean photon count $\nu_{\theta,k}$ in the kth pixel, which we have completely specified above. It is also a function of the mean η_k and standard deviation σ_k of the CCD camera's readout noise at the kth pixel. As the readout noise for all pixels of a CCD camera can be assumed to be statistically identical, we have that $\eta_k = \eta_0$ and $\sigma_k = \sigma_0 > 0$, where η_0 and σ_0 are constants, for all pixels k. Provided that we know the values of η_0 and σ_0, we have now fully specified Eq. (18.1) and can proceed to determine the maximum likelihood estimate $\hat{\theta}_{MLE}$. Note that if we had assumed the use of a CMOS camera, then the readout noise level would depend on the specific pixel. In that case η_k and σ_k would generally differ from pixel to pixel and we would need to know their values on the basis of individual pixels.

Given an array of pixel values $z_1, \ldots, z_{N_p}$ that contain an image of our molecule of interest, we can now use a standard numerical optimization method to find $\hat{\theta}_{MLE}$ by maximizing the sum of the logarithms of $p_{\theta,k}$, $k = 1, \ldots, N_p$. The logarithm of $p_{\theta,k}$ is in fact the log-likelihood function for the data in the kth pixel, and is as given in Table 16.1 for the CCD/CMOS data model. Starting with an initial guess for the value of $\theta = (x_0, y_0)$, the optimization algorithm will iteratively change the values of x_0 and y_0 until the combination $\hat{\theta}_{MLE}$ is found that maximizes the sum of log-likelihood functions, or more precisely, until a termination criterion for the algorithm is satisfied. An example of this maximum likelihood estimation is given in Fig. 18.2. A simulated image of the molecule is shown in part (a) of the figure, and comparisons of this image with the pixelated Airy profile that has been fitted to it through the maximization are shown in part (b). The image has been simulated by computing, for the kth pixel, $\nu_{\theta,k}$ as described above using the true value $\theta = (x_0, y_0) = (688.00 \text{ nm}, 656.00 \text{ nm})$, and by using the fact that $\nu_{\theta,k}$ is the mean of a Poisson random variable and η_k and σ_k are the mean and standard deviation of a Gaussian random variable (Section 15.2.2). The Airy profile that has been fitted to the image is given by $\nu_{\theta,k}$ that has been computed for the kth pixel using the obtained maximum likelihood estimate $\hat{\theta}_{MLE} = (689.95 \text{ nm}, 654.48 \text{ nm})$.

In Section 18.10, we will give a detailed example of the localization of a dye molecule from experimentally acquired image data. There, we will discuss issues that arise in a practical setting, such as the determination of the values of the various "known" parameters. In that example, we will also touch on the determination of an initial guess for the value of the parameter of interest.

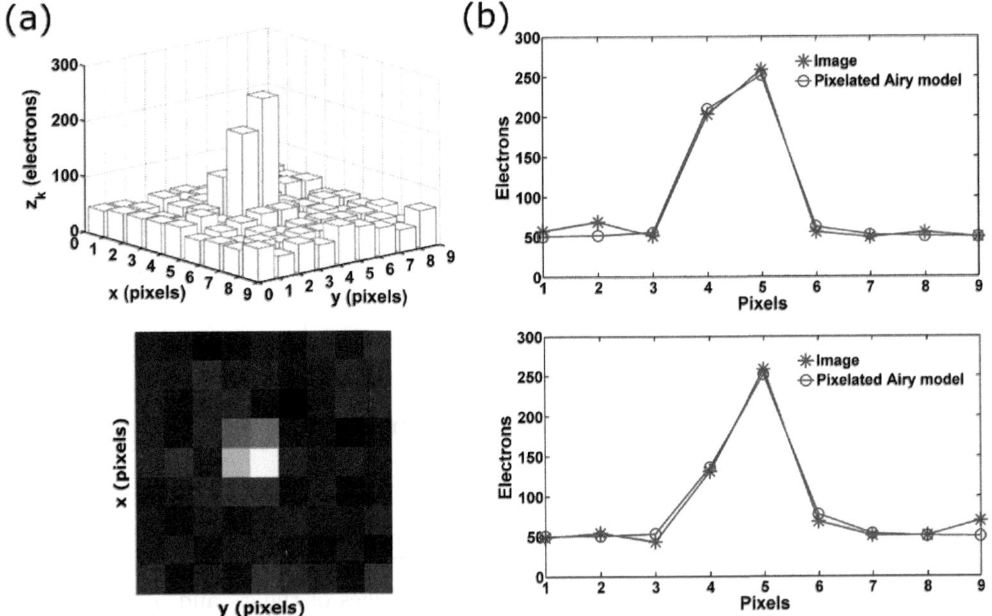

FIGURE 18.2
Maximum likelihood localization of a stationary and in-focus molecule from a simulated image. (a) Bar graph (top) and intensity (bottom) representations of an image of a stationary and in-focus molecule captured with a CCD detector. (b) Overlay of the middle row (top) and column (bottom) of the image from (a) with the corresponding row and column of the pixelated Airy profile that has been fitted to the image through maximum likelihood localization.

18.3 How well can the location of an object be estimated?

In the above sections, we presented a formulation for the maximum likelihood localization of an object and provided a descriptive example where we obtained estimates for the positional coordinates of a molecule. Here we are interested in assessing how good such estimates are, or put another way, how much we can rely on such estimates to be representative of the true location of the object. In Sections 16.5 and 16.6, we introduced two main criteria for the quality of an estimator, namely its bias and variance.

18.3.1 Bias of location estimation

In general the question of bias for a location estimator is intractable from a theoretical standpoint, although we encountered an exception in Section 16.5.3. There we showed that under the fundamental data scenario in which the detector is infinitely large, and provided that a certain condition is satisfied by the photon distribution profile, the estimator that calculates the center of mass of the photons detected from a stationary object is an unbiased estimator of the location of the object. This class of estimators includes the maximum likelihood estimator for the case where the image of the object is described by a 2D Gaussian image function.

In a practical situation, pixelation of the data makes the determination of whether

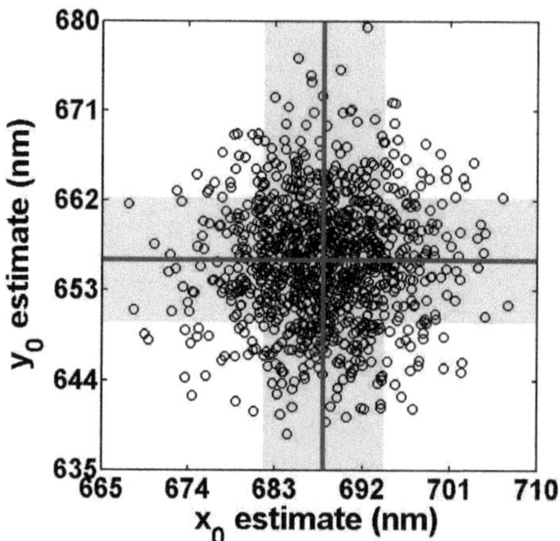

FIGURE 18.3
Maximum likelihood estimates of the location (x_0, y_0) of a stationary and in-focus point source determined from simulated CCD image data. The estimates ($\circ$) shown are obtained from 1000 repeat images of the point source. The averages of the x_0 and y_0 estimates are indicated with blue lines, and the yellow region about each average delineates one standard deviation above and below the average. The values of the averages and standard deviations are as given in Table 18.1 for the data set simulated with the CCD/CMOS data model and the Airy image function. See the table caption for details relating to the image data and the estimation.

an estimator is biased a more complex problem, and we need to resort to verification by simulation. As an example, we simulated 1000 repeat images of the stationary and in-focus molecule from Section 18.2 and carried out the maximum likelihood localization described there on each image. The obtained location estimates are plotted in Fig. 18.3 along with the average estimate for each coordinate. From the first row of Table 18.1, we see that the averages are within a tenth of a nanometer from their respective true values. Although they are themselves only estimates of the mean of the estimator, the closeness of these averages to the true values suggests that the maximum likelihood estimator is effectively unbiased under the specific conditions assumed for this data set.

Table 18.1 further shows the results for maximum likelihood localizations carried out on five additional data sets involving the same molecule, but simulated with different combinations of practical data model and image function. The localizations for these data sets follow the same descriptions given in Section 18.2, but with $\ln p_{\theta,k}$, the log-likelihood function for the data in the kth pixel, replaced by that shown in Table 16.1 for the appropriate data model. Similarly, the image function q in the photon distribution profile of Eq. (18.3) is replaced by the 2D Gaussian image function as needed. For each of these data sets, we again see a closeness between the average of the 1000 estimates for each coordinate and the coordinate's true value. The largest differences we see are within 1 nm. As with the first data set, we can therefore consider the maximum likelihood location estimator to be unbiased, or to have very small bias, under the specific conditions assumed for these data sets.

Based on simulation studies similar to the one presented here, we can assume that in many cases of practical relevance, the maximum likelihood estimator is unbiased or has very small bias (Note 18.1). This assumption is especially valid in cases where estimation under the corresponding fundamental data model is unbiased, and where the effective pixel

Data model	Image function	x_0 (nm)	Mean of x_0 estimates (nm)	δ_{x_0} (nm)	SD of x_0 estimates (nm)	y_0 (nm)	Mean of y_0 estimates (nm)	δ_{y_0} (nm)	SD of y_0 estimates (nm)
CCD/CMOS	Airy	688.00	687.99	6.41	6.18	656.00	656.07	6.35	6.27
Poisson	Airy	688.00	688.07	5.59	5.39	656.00	655.99	5.71	5.60
EMCCD†	Airy	688.00	688.15	16.64	17.02	656.00	656.52	17.71	18.13
CCD/CMOS	2D Gaussian	688.00	687.77	6.10	5.93	656.00	656.11	5.90	5.94
Poisson	2D Gaussian	688.00	688.03	5.36	5.53	656.00	655.92	5.31	5.33
EMCCD	2D Gaussian	688.00	687.79	16.22	16.61	656.00	656.81	16.40	16.53

† Results based on estimates from 991 instead of the entire set of 1000 images, as estimates from 9 images were discarded for placing the point source outside of the 9×9-pixel region.

TABLE 18.1
Maximum likelihood localization of an in-focus and stationary point source from simulated CCD, Poisson, and EMCCD image data. Results for six data sets are shown. Each data set comprises 1000 repeat images of the point source, and is simulated by assuming either an Airy profile or a 2D Gaussian profile for the image of the point source. For each data set, the mean and standard deviation of the maximum likelihood estimates of the coordinates x_0 and y_0 are reported alongside the true values x_0 and y_0 and the bounds δ_{x_0} and δ_{y_0} on the standard deviations for estimating x_0 and y_0. In all data sets, an image comprises a 9×9-pixel region with a pixel size of 16 μm × 16 μm, and is assumed to have been acquired over an exposure time of $t-t_0 = 0.1$ s. Furthermore, the point source is imaged at a magnification of $M = 100$ and is positioned such that the center of its image is located at 0.3 pixels in the x direction and 0.1 pixels in the y direction with respect to the upper left corner of the center pixel of the 9×9-pixel region. For the data sets simulated under the CCD/CMOS data model, the CCD detector adds readout noise with mean $\eta_k = 0$ electrons and standard deviation $\sigma_k = 6$ electrons to each pixel k. For the data sets simulated under the EMCCD data model, the EMCCD detector amplifies the photon signal in each pixel with an EM gain of $g = 800$ and adds readout noise with mean $\eta_k = 0$ electrons and standard deviation $\sigma_k = 20$ electrons to each pixel k. For the data sets simulated under the CCD/CMOS and Poisson data models, photons from the point source are assumed to have been detected at a rate of $\Lambda_0 = 7000$ photons/s, and a background component is assumed such that each pixel k detects, on average, $\beta_k = \beta_0 = 50$ background photons. For the data sets simulated under the EMCCD data model, photons from the point source are detected at a rate of $\Lambda_0 = 1000$ photons/s, and each pixel k detects, on average, $\beta_k = \beta_0 = 2$ background photons. For the data sets simulated with the Airy image function, it is assumed that photons of wavelength $\lambda = 573$ nm are collected by an objective with numerical aperture $n_a = 1.4$. For the data sets simulated with the 2D Gaussian image function, the width of the Gaussian profile is set to $\sigma_g = 86.18$ nm. For each data set, the initial values provided to the estimator for the parameters x_0 and y_0 were randomly generated, for each image, to be within 20% of their true values.

size of the image data is sufficiently small to reasonably approximate the corresponding unpixelated image produced by the idealized detector.

18.3.1.1 Bias of the center of mass as a location estimator under the practical data models

We have seen in Section 16.2.1.1 that the center of mass of the spatial coordinates of the detected photons provides a simple approach to estimating the location of an object in the case of the fundamental data model. We will now examine this estimator in the context of a practical data model. Suppose the image is an $N_x \times N_y$ array of pixels, where N_x is the number of columns and N_y is the number of rows. Then in the x direction, the center of mass, in units of pixels, is just the sum of the column indices of all pixels, each weighted by the value of the pixel. The center of mass in the y direction is likewise the weighted sum of the row indices of all pixels. Denoting by $z(k,j)$ the value of the pixel in the kth column and jth row, the center of mass $(\hat{x}_i, \hat{y}_i)$ of the image space positional coordinates (x_i, y_i) of the object, in units of pixels, is given by

$$\hat{x}_i = \frac{1}{z_{total}} \sum_{k=1}^{N_x} \sum_{j=1}^{N_y} k \cdot z(k,j) - 0.5, \qquad \hat{y}_i = \frac{1}{z_{total}} \sum_{k=1}^{N_x} \sum_{j=1}^{N_y} j \cdot z(k,j) - 0.5, \tag{18.6}$$

where $z_{total} = \sum_{k=1}^{N_x} \sum_{j=1}^{N_y} z(k,j)$ normalizes the values of the pixels. The adjustment by 0.5 is necessary as $(0,0)$ is taken to be the upper left corner of the $N_x \times N_y$ image. Note that to translate this estimator to the object space coordinate system which we normally work with, we simply need to divide by the magnification, i.e., $(\hat{x}_0, \hat{y}_0) = \left(\frac{\hat{x}_i}{M}, \frac{\hat{y}_i}{M}\right)$. As we will see directly, however, image space coordinates and units of pixels are better suited for our discussion here.

While this estimator is simple to implement, it is very sensitive to the background component in the image and can become significantly biased as a result. If background noise is uniformly distributed throughout the image, then the center of mass will tend to the center of the image with increasing levels of the background, assuming the object signal level remains the same. Intuitively, one can see that increasing the uniform background while maintaining the same object signal level causes the image to become increasingly flat. Therefore it is to be expected that the center of mass will converge to the geometric center of the image.

The effect of background on the center of mass is demonstrated in Fig. 18.4. It can be seen in the plot shown that the average center of mass, determined from 1000 repeat images of a stationary and in-focus point source, is reasonably close to the point source's true location when the background level is low. However, with increasing background level, the average center of mass deviates towards the center of the pixel array used for the estimation, illustrating the potential for severe bias. Note that even in the absence of background noise, the center of mass exhibits bias.

While we have shown that the center of mass is an unbiased location estimator for the fundamental data model given that a mild technical assumption is satisfied, we have seen here that in a practical scenario the center of mass can exhibit very significant bias. This is one of the main reasons why the center of mass is rarely used in practice.

18.3.2 Variance of location estimation

An analysis of the second criterion, the variance of an estimator, is again very difficult in general, especially in the context of practical data models. We saw an example in Section 16.6.2 where we derived analytically the variance of the maximum likelihood location estimator of Section 16.2.1.1. By and large, however, similar to the assessment of bias the variance of an estimator needs to be evaluated using simulations. For example, looking again at the maximum likelihood localization results in Table 18.1, for each of the six scenarios we

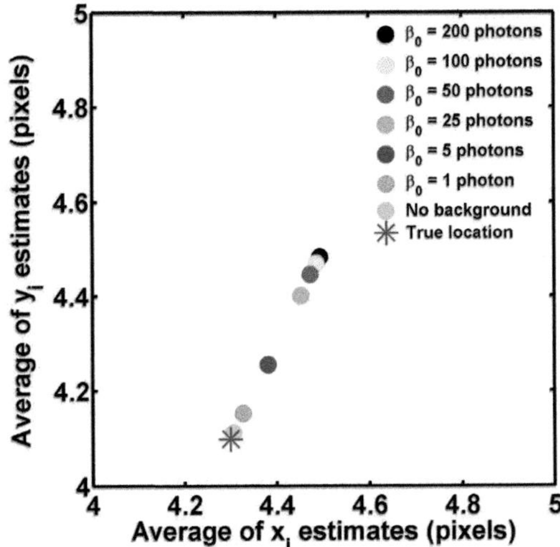

FIGURE 18.4
Effect of background noise level on center of mass location estimation. Each data point (filled circle) represents the average of 1000 center of mass estimates of the image space positional coordinates (x_i, y_i) of a stationary and in-focus point source. The estimates are obtained from 1000 repeat images of the point source, simulated according to the Poisson data model. The data sets for the different data points differ only in the uniform per-pixel background level β_0. The true image space location (red star) of the point source is $(x_i, y_i) = (4.3, 4.1)$ in units of pixels, and the limits of the plot map exactly to the center pixel of a 9×9 pixel array, from which each location estimate is obtained. Bias of the center of mass as a location estimator can be seen to increase as the average of estimates gravitates towards the center of the pixel array with increasing levels of background noise. Details on the simulation of the data sets are as described in Table 18.1 for the data set simulated with the Poisson data model and the Airy image function, except the per-pixel background levels β_0 are as given in the plot legend.

can take the reported standard deviations (i.e., square roots of the variances) of the x_0 and y_0 coordinate estimates as an estimate of the variance of the particular estimator used. As an illustration of the standard deviations of the coordinate estimates, the plot of Fig. 18.3 uses yellow regions to delineate one standard deviation above and below the averages of the x_0 and y_0 estimates obtained for the first data set of Table 18.1.

Since the variance provides a measure of the spread of an estimator, we can use it as a criterion for choosing between estimators. Given a choice of estimators for a particular problem, one would naturally select the estimator with the smallest variance, provided that all candidates perform similarly in terms of bias. While knowing the variance importantly allows us to compare estimators from a set of candidates, what we still don't know is how good the smallest variance from that set really is. The question remains as to how much smaller the variance of an estimator can be for the given problem, and hence whether it might make sense to look for another estimator that can potentially provide a significant improvement in terms of the variance. From Chapter 17, we know that the variance of an unbiased estimator for a given problem cannot be arbitrarily small, and that the Cramér-Rao lower bound gives a limit on how small the variance can be. The Cramér-Rao lower bound thus provides an answer to our question, and will in fact be the focus of much of the rest of this chapter as we look at it in detail in the context of the localization problem. In Section 18.7, we will revisit its use as a benchmark for assessing the performance of an estimator.

18.4 Estimation of other parameters

In the above sections, we have been concerned with the estimation of only the location of an object. In general, however, other parameters of interest can also be estimated. For example, together with the object location, we could have estimated parameters such as the α parameter for the Airy image function and the photon detection rate Λ_0, assuming their values were also unknown. The question arises, however, as to why we should not always simultaneously estimate these other parameters and the location coordinates. There are often disadvantages to estimating more parameters. First, the computational complexity is higher, leading to increased computational time. Then, if the estimators of the parameters are correlated, the "quality" of the estimates might be lower due to increased standard deviations of the estimates. In a worst-case situation, bias in the estimate of one parameter might also lead to bias in the estimates of other parameters. Therefore, if for example through independent experiments a parameter can be determined with a high level of certainty, it might be appropriate to treat it as a known parameter instead of estimating it simultaneously with the location coordinates. To independently ascertain the value of the parameter α of an Airy profile, for example, one could estimate it from samples of the fluorophore that have been separately imaged using the same experimental setup. Ideally, these samples should produce high signal and low noise levels so that α can be estimated with a small variance about the average value (i.e., with a high level of certainty).

18.5 Cramér-Rao lower bound for location estimation — fundamental data model

In this and the next few sections, we will consider in detail the calculation of Cramér-Rao lower bounds pertaining to object localization in two dimensions. Besides providing a measure against which the variance of a location estimator can be assessed, the lower bound expressions we obtain importantly afford a means by which we can make informed decisions concerning the design of experiments and the analysis of the resulting data. Specifically, we will see that the lower bound expressions will depend on practically customizable parameters such as wavelength and magnification, and that we can therefore use them to determine settings that yield an acceptable level of uncertainty (i.e., variance) for estimating the location of the object of interest. A significant portion of our presentation will accordingly be devoted to the discussion and illustration of the type of practically useful insights that can be gained by examining and computing the lower bound expressions. We will begin by considering in this section Cramér-Rao lower bounds under the fundamental data model.

In order to calculate the Cramér-Rao lower bound on the variance with which the location of an object can be estimated from an image of the object, we need to determine the Fisher information matrix corresponding to the underlying image data. Suppose the image is captured over the time interval $[t_0, t]$ and by a detector with an infinitely large photon detection area $C = \mathbb{R}^2$. Assume that the object is stationary, such that the photon distribution profile $f_{\theta,\tau} = f_\theta$, $\tau \geq t_0$. Further assume that the rate at which photons are detected from the object is a constant, so that $\Lambda_\theta(\tau) = \Lambda_0$, $\tau \geq t_0$, where $\Lambda_0 > 0$. Making the necessary modifications in the general Fisher information matrix expression of Eq. (17.11), we begin with the Fisher information matrix

$$
\begin{aligned}
\mathbf{I}(\theta) &= \frac{1}{1-e^{-\Lambda_0(t-t_0)}}\int_{t_0}^{t}\int_{\mathbb{R}^2}\frac{1}{\Lambda_0 f_\theta(x,y)}\left(\frac{\partial[\Lambda_0 f_\theta(x,y)]}{\partial\theta}\right)^T\frac{\partial[\Lambda_0 f_\theta(x,y)]}{\partial\theta}dxdyd\tau \\
&\quad - \frac{e^{-\Lambda_0(t-t_0)}}{\left(1-e^{-\Lambda_0(t-t_0)}\right)^2}\int_{t_0}^{t}\left(\frac{\partial\Lambda_0}{\partial\theta}\right)^T d\tau\cdot\int_{t_0}^{t}\frac{\partial\Lambda_0}{\partial\theta}d\tau \\
&= \frac{1}{1-e^{-\Lambda_0(t-t_0)}}\cdot\frac{t-t_0}{\Lambda_0}\int_{\mathbb{R}^2}\frac{1}{f_\theta(x,y)}\left(\frac{\partial[\Lambda_0 f_\theta(x,y)]}{\partial\theta}\right)^T\frac{\partial[\Lambda_0 f_\theta(x,y)]}{\partial\theta}dxdy \\
&\quad - \frac{e^{-\Lambda_0(t-t_0)}}{\left(1-e^{-\Lambda_0(t-t_0)}\right)^2}\int_{t_0}^{t}\left(\frac{\partial\Lambda_0}{\partial\theta}\right)^T d\tau\cdot\int_{t_0}^{t}\frac{\partial\Lambda_0}{\partial\theta}d\tau
\end{aligned}
\tag{18.7}
$$

for a general parameter estimation problem.

We consider the estimation of the positional coordinates x_0 and y_0 of a stationary object, as well as the photon detection rate Λ_0, which we also assume to be unknown. We therefore have $\theta = (x_0, y_0, \Lambda_0)$. Substituting Eq. (18.3), which expresses f_θ in terms of an image function q, into Eq. (18.7), and letting $u = \frac{x}{M} - x_0$ and $v = \frac{y}{M} - y_0$, we obtain

$$
\begin{aligned}
\mathbf{I}(\theta) &= \frac{1}{1-e^{-\Lambda_0(t-t_0)}}\cdot\frac{t-t_0}{M^2\Lambda_0}\int_{\mathbb{R}^2}\frac{1}{q(u,v)}\left(\frac{\partial[\Lambda_0 q(u,v)]}{\partial\theta}\right)^T\frac{\partial[\Lambda_0 q(u,v)]}{\partial\theta}dxdy \\
&\quad - \frac{e^{-\Lambda_0(t-t_0)}}{\left(1-e^{-\Lambda_0(t-t_0)}\right)^2}\int_{t_0}^{t}\left(\frac{\partial\Lambda_0}{\partial\theta}\right)^T d\tau\cdot\int_{t_0}^{t}\frac{\partial\Lambda_0}{\partial\theta}d\tau \\
&= \frac{1}{1-e^{-\Lambda_0(t-t_0)}}\cdot\frac{t-t_0}{M^2\Lambda_0}
\begin{bmatrix}
\Lambda_0^2\int_{\mathbb{R}^2}\frac{1}{q(u,v)}\left(\frac{\partial q(u,v)}{\partial x_0}\right)^2dxdy & \Lambda_0^2\int_{\mathbb{R}^2}\frac{1}{q(u,v)}\frac{\partial q(u,v)}{\partial x_0}\frac{\partial q(u,v)}{\partial y_0}dxdy & \Lambda_0\int_{\mathbb{R}^2}\frac{\partial q(u,v)}{\partial x_0}\frac{\partial\Lambda_0}{\partial\Lambda_0}dxdy \\
\Lambda_0^2\int_{\mathbb{R}^2}\frac{1}{q(u,v)}\frac{\partial q(u,v)}{\partial x_0}\frac{\partial q(u,v)}{\partial y_0}dxdy & \Lambda_0^2\int_{\mathbb{R}^2}\frac{1}{q(u,v)}\left(\frac{\partial q(u,v)}{\partial y_0}\right)^2dxdy & \Lambda_0\int_{\mathbb{R}^2}\frac{\partial q(u,v)}{\partial y_0}\frac{\partial\Lambda_0}{\partial\Lambda_0}dxdy \\
\Lambda_0\int_{\mathbb{R}^2}\frac{\partial q(u,v)}{\partial x_0}\frac{\partial\Lambda_0}{\partial\Lambda_0}dxdy & \Lambda_0\int_{\mathbb{R}^2}\frac{\partial q(u,v)}{\partial y_0}\frac{\partial\Lambda_0}{\partial\Lambda_0}dxdy & \int_{\mathbb{R}^2}q(u,v)\left(\frac{\partial\Lambda_0}{\partial\Lambda_0}\right)^2dxdy
\end{bmatrix}
- \begin{bmatrix}
0 & 0 & 0 \\
0 & 0 & 0 \\
0 & 0 & \frac{(t-t_0)^2e^{-\Lambda_0(t-t_0)}}{\left(1-e^{-\Lambda_0(t-t_0)}\right)^2}
\end{bmatrix} \\
&= \frac{1}{1-e^{-\Lambda_0(t-t_0)}}\cdot\frac{t-t_0}{M^2}
\begin{bmatrix}
\Lambda_0\int_{\mathbb{R}^2}\frac{1}{q(u,v)}\left(\frac{\partial q(u,v)}{\partial x_0}\right)^2dxdy & \Lambda_0\int_{\mathbb{R}^2}\frac{1}{q(u,v)}\frac{\partial q(u,v)}{\partial x_0}\frac{\partial q(u,v)}{\partial y_0}dxdy & \int_{\mathbb{R}^2}\frac{\partial q(u,v)}{\partial x_0}dxdy \\
\Lambda_0\int_{\mathbb{R}^2}\frac{1}{q(u,v)}\frac{\partial q(u,v)}{\partial x_0}\frac{\partial q(u,v)}{\partial y_0}dxdy & \Lambda_0\int_{\mathbb{R}^2}\frac{1}{q(u,v)}\left(\frac{\partial q(u,v)}{\partial y_0}\right)^2dxdy & \int_{\mathbb{R}^2}\frac{\partial q(u,v)}{\partial y_0}dxdy \\
\int_{\mathbb{R}^2}\frac{\partial q(u,v)}{\partial x_0}dxdy & \int_{\mathbb{R}^2}\frac{\partial q(u,v)}{\partial y_0}dxdy & \frac{1}{\Lambda_0}\int_{\mathbb{R}^2}q(u,v)\,dxdy - \frac{M^2(t-t_0)e^{-\Lambda_0(t-t_0)}}{1-e^{-\Lambda_0(t-t_0)}}
\end{bmatrix}.
\end{aligned}
\tag{18.8}
$$

We next perform a change of variables to express all integrals and partial derivatives in terms of u and v, obtaining

$$
\begin{aligned}
\mathbf{I}(\theta) &= \frac{1}{1-e^{-\Lambda_0(t-t_0)}} \cdot \frac{t-t_0}{M^2} \\
&\quad \times \begin{bmatrix}
\Lambda_0 \int_{\mathbb{R}^2} \frac{M^2 dudv}{q(u,v)} \left(\frac{\partial q(u,v)}{\partial u}\frac{\partial u}{\partial x_0}\right)^2 & \Lambda_0 \int_{\mathbb{R}^2} \frac{M^2 dudv}{q(u,v)} \left(\frac{\partial q(u,v)}{\partial u}\frac{\partial u}{\partial x_0}\right)\left(\frac{\partial q(u,v)}{\partial v}\frac{\partial v}{\partial y_0}\right) & \int_{\mathbb{R}^2} \frac{\partial q(u,v)}{\partial u}\frac{\partial u}{\partial x_0} M^2 dudv \\
\Lambda_0 \int_{\mathbb{R}^2} \frac{M^2 dudv}{q(u,v)} \left(\frac{\partial q(u,v)}{\partial u}\frac{\partial u}{\partial x_0}\right)\left(\frac{\partial q(u,v)}{\partial v}\frac{\partial v}{\partial y_0}\right) & \Lambda_0 \int_{\mathbb{R}^2} \frac{M^2 dudv}{q(u,v)} \left(\frac{\partial q(u,v)}{\partial v}\frac{\partial v}{\partial y_0}\right)^2 & \int_{\mathbb{R}^2} \frac{\partial q(u,v)}{\partial v}\frac{\partial v}{\partial y_0} M^2 dudv \\
\int_{\mathbb{R}^2} \frac{\partial q(u,v)}{\partial u}\frac{\partial u}{\partial x_0} M^2 dudv & \int_{\mathbb{R}^2} \frac{\partial q(u,v)}{\partial v}\frac{\partial v}{\partial y_0} M^2 dudv & \frac{1}{\Lambda_0}\int_{\mathbb{R}^2} q(u,v) M^2 dudv - \frac{M^2(t-t_0)e^{-\Lambda_0(t-t_0)}}{1-e^{-\Lambda_0(t-t_0)}}
\end{bmatrix} \\
&= \frac{t-t_0}{1-e^{-\Lambda_0(t-t_0)}} \begin{bmatrix}
\Lambda_0 \int_{\mathbb{R}^2} \frac{1}{q(u,v)} \left(\frac{\partial q(u,v)}{\partial u}\right)^2 dudv & \Lambda_0 \int_{\mathbb{R}^2} \frac{1}{q(u,v)} \frac{\partial q(u,v)}{\partial u}\frac{\partial q(u,v)}{\partial v} dudv & -\int_{\mathbb{R}^2} \frac{\partial q(u,v)}{\partial u} dudv \\
\Lambda_0 \int_{\mathbb{R}^2} \frac{1}{q(u,v)} \frac{\partial q(u,v)}{\partial u}\frac{\partial q(u,v)}{\partial v} dudv & \Lambda_0 \int_{\mathbb{R}^2} \frac{1}{q(u,v)} \left(\frac{\partial q(u,v)}{\partial v}\right)^2 dudv & -\int_{\mathbb{R}^2} \frac{\partial q(u,v)}{\partial v} dudv \\
-\int_{\mathbb{R}^2} \frac{\partial q(u,v)}{\partial u} dudv & -\int_{\mathbb{R}^2} \frac{\partial q(u,v)}{\partial v} dudv & \frac{1}{\Lambda_0} - \frac{(t-t_0)e^{-\Lambda_0(t-t_0)}}{1-e^{-\Lambda_0(t-t_0)}}
\end{bmatrix},
\end{aligned}
\tag{18.9}
$$

where in entry $(3,3)$ we have made use of the fact that an image function integrates to 1 over $\mathbb{R}^2$. This expression shows that when the detector is infinitely large, the Fisher information matrix for the 2D localization problem under the fundamental data model is independent of the magnification and the positional coordinates x_0 and y_0 of the object of interest.

For the special but useful case where the image function q is symmetric, meaning that $q(x,y) = q(-x,y) = q(x,-y)$, $(x,y) \in \mathbb{R}^2$, this Fisher information matrix can be simplified further and in fact reduces to a diagonal matrix. Given the symmetry, we can use the fact that the derivative of an even function is an odd function to show that the off-diagonal entries of $\mathbf{I}(\theta)$ are all zeros. We demonstrate this with entry $(1,2)$ of the matrix, using the notations $w_1(u,v) := \frac{\partial q(u,v)}{\partial u}$, $(u,v) \in \mathbb{R}^2$, and $w_2(u,v) := \frac{\partial q(u,v)}{\partial v}$, $(u,v) \in \mathbb{R}^2$, to represent the explicitly evaluated form of the respective partial derivative expressions. From Eq. (18.9), we have for entry $(1,2)$

$$
\begin{aligned}
[\mathbf{I}(\theta)]_{1,2} &= N_{photons} \int_{\mathbb{R}^2} \frac{w_1(u,v)w_2(u,v)}{q(u,v)} dudv \\
&= N_{photons} \int_{-\infty}^{\infty} \left[\int_0^{\infty} \frac{w_1(u,v)w_2(u,v)}{q(u,v)} du + \int_{-\infty}^{0} \frac{w_1(u,v)w_2(u,v)}{q(u,v)} du \right] dv \\
&= N_{photons} \int_{-\infty}^{\infty} \left[\int_0^{\infty} \frac{w_1(u,v)w_2(u,v)}{q(u,v)} du + \int_{0}^{\infty} \frac{w_1(-u,v)w_2(-u,v)}{q(-u,v)} du \right] dv \\
&= N_{photons} \int_{-\infty}^{\infty} \left[\int_0^{\infty} \frac{w_1(u,v)w_2(u,v)}{q(u,v)} du + \int_{0}^{\infty} \frac{(-w_1(u,v))\, w_2(u,v)}{q(u,v)} du \right] dv \\
&= 0,
\end{aligned}
$$

where $N_{photons} := \Lambda_0(t-t_0)/\left(1-e^{-\Lambda_0(t-t_0)}\right)$ is the expected number of photons (Eq. (15.7)), and where we have used the fact that for a given value of v, $q(u,v)$ and $w_2(u,v)$ are even functions with respect to u, but $w_1(u,v)$ is an odd function with respect to u. The same approach can be taken to show that entries $(1,3)$ and $(2,3)$ are also zero, and the remaining three off-diagonal entries are thereby also zero as they are identical to entries $(1,2)$, $(1,3)$ and $(2,3)$. For a symmetric image function q, we therefore have the diagonal Fisher information matrix

$$\mathbf{I}(\theta) = \begin{bmatrix} N_{photons} \int_{\mathbb{R}^2} \frac{\left(\frac{\partial q(u,v)}{\partial u}\right)^2}{q(u,v)} dudv & 0 & 0 \\ 0 & N_{photons} \int_{\mathbb{R}^2} \frac{\left(\frac{\partial q(u,v)}{\partial v}\right)^2}{q(u,v)} dudv & 0 \\ 0 & 0 & \frac{N_{photons}}{\Lambda_0^2} - \frac{(t-t_0)^2 e^{-\Lambda_0(t-t_0)}}{\left(1-e^{-\Lambda_0(t-t_0)}\right)^2} \end{bmatrix}. \tag{18.10}$$

Diagonality of the Fisher information matrix has important consequences as it implies that the Cramér-Rao lower bound for the estimation of one parameter does not influence the Cramér-Rao lower bound for the estimation of any of the other parameters. We will see in our discussion of estimation under the practical data models that we have to assume that the Fisher information is not diagonal, which implies that the estimation of one parameter can negatively influence the estimation of other parameters by increasing the variance of the estimation, or that the estimations are correlated. Knowing that the Fisher information is diagonal for the fundamental data model does, however, indicate that these effects are likely small in cases where the practical data model is well approximated by the fundamental one.

By the Cramér-Rao inequality, we have that for any unbiased estimator $\hat{\theta} = (\hat{x}_0, \hat{y}_0, \hat{\Lambda}_0)$ of the parameter $\theta = (x_0, y_0, \Lambda_0)$, $\text{cov}\left(\hat{\theta}\right) \geq \mathbf{I}^{-1}(\theta)$. Therefore, given the diagonality of the matrix of Eq. (18.10), we have for the components $\hat{x}_0$, $\hat{y}_0$, and $\hat{\Lambda}_0$ that

$$\text{Var}\left(\hat{x}_0\right) \geq \frac{1}{[\mathbf{I}(\theta)]_{1,1}} = \frac{1}{N_{photons}} \cdot \frac{1}{\int_{\mathbb{R}^2} \frac{1}{q(u,v)} \left(\frac{\partial q(u,v)}{\partial u}\right)^2 dudv}, \tag{18.11}$$

$$\text{Var}\left(\hat{y}_0\right) \geq \frac{1}{[\mathbf{I}(\theta)]_{2,2}} = \frac{1}{N_{photons}} \cdot \frac{1}{\int_{\mathbb{R}^2} \frac{1}{q(u,v)} \left(\frac{\partial q(u,v)}{\partial v}\right)^2 dudv}, \tag{18.12}$$

$$\text{Var}\left(\hat{\Lambda}_0\right) \geq \frac{1}{[\mathbf{I}(\theta)]_{3,3}} = \frac{1}{\frac{N_{photons}}{\Lambda_0^2} - \frac{(t-t_0)^2 e^{-\Lambda_0(t-t_0)}}{\left(1-e^{-\Lambda_0(t-t_0)}\right)^2}}. \tag{18.13}$$

If the image function q is not only symmetric as defined above, but circularly symmetric, then the integrals over $\mathbb{R}^2$ in Eqs. (18.11) and (18.12) will evaluate to the same value, and the expressions for the lower bounds for the x_0 and y_0 coordinates will be identical. This is intuitively clear as in this situation there is nothing that distinguishes the two coordinates. We will use this result in the two subsections that follow, in which the specific image functions considered are in fact circularly symmetric.

Equations (18.11) and (18.12) show that the bounds on the variance for estimating the object's positional coordinates decrease inverse proportionally with the expected number of detected photons. This importantly means that, assuming an efficient estimator exists, the coordinates can be estimated arbitrarily precisely, provided that sufficiently many photons have been detected. The same can be said about the estimation of the photon detection rate based on Eq. (18.13), which is identical to Eq. (17.15) and therefore discussed in Section 17.2.2.

To obtain explicit expressions for the Fisher information matrix, we next consider two scenarios in which the imaged object is a point-like object such as a fluorescent molecule. In one case, the image function q that is used to model the image of the object is the Airy image function, and in the other case, it is the 2D Gaussian image function (Section 15.1.2.1).

18.5.1 Cramér-Rao lower bound for the Airy image function

By writing the Airy image function of Eq. (15.9) in the form

$$q(u,v) = \frac{\left(\frac{2\pi n_a}{\lambda}\right)^2}{\pi}\left(\frac{J_1\left(\frac{2\pi n_a}{\lambda}\sqrt{u^2+v^2}\right)}{\frac{2\pi n_a}{\lambda}\sqrt{u^2+v^2}}\right)^2, \quad (u,v)\in\mathbb{R}^2, \tag{18.14}$$

its partial derivative with respect to u can be shown, using the Bessel function identity $\frac{\partial}{\partial z}\left[z^{-n}J_n(z)\right] = -z^{-n}J_{n+1}(z)$ with $n=1$, to be given by

$$\frac{\partial q(u,v)}{\partial u} = -\frac{4n_a}{\lambda}\cdot u\cdot\frac{J_1\left(\frac{2\pi n_a}{\lambda}\sqrt{u^2+v^2}\right)}{\sqrt{u^2+v^2}}\cdot\frac{J_2\left(\frac{2\pi n_a}{\lambda}\sqrt{u^2+v^2}\right)}{u^2+v^2}, \quad (u,v)\in\mathbb{R}^2. \tag{18.15}$$

Since the Airy image function is symmetric, the Cramér-Rao lower bounds on the variances for estimating the positional coordinates x_0 and y_0 and the photon detection rate Λ_0 can be obtained by evaluating Eqs. (18.11), (18.12), and (18.13). Using Eqs. (18.14) and (18.15) in the denominator of Eq. (18.11), we obtain

$$\begin{aligned}
[\mathbf{I}(\theta)]_{1,1} &= N_{photons}\int_{\mathbb{R}^2}\frac{1}{q(u,v)}\left(\frac{\partial q(u,v)}{\partial u}\right)^2 dudv \\
&= N_{photons}\cdot\frac{16\pi n_a^2}{\lambda^2}\cdot\int_{\mathbb{R}^2} u^2\cdot\frac{J_2^2\left(\frac{2\pi n_a}{\lambda}\sqrt{u^2+v^2}\right)}{(u^2+v^2)^2}dudv \\
&= \frac{16N_{photons}\pi n_a^2}{\lambda^2}\int_0^{2\pi}\int_0^\infty r^2\cos^2(\phi)\cdot\frac{J_2^2\left(\frac{2\pi n_a}{\lambda}r\right)}{r^4}r\,drd\phi \\
&= \frac{16N_{photons}\pi n_a^2}{\lambda^2}\int_0^{2\pi}\cos^2(\phi)d\phi\cdot\int_0^\infty\frac{J_2^2\left(\frac{2\pi n_a}{\lambda}r\right)}{r}dr \\
&= \frac{16N_{photons}\pi n_a^2}{\lambda^2}\cdot\pi\cdot\int_0^\infty\frac{J_2^2(z)}{z}dz \\
&= \frac{4N_{photons}\pi^2 n_a^2}{\lambda^2},
\end{aligned} \tag{18.16}$$

where we have applied the polar transformation $u = r\cos\phi$ and $v = r\sin\phi$ and made the substitution $z = \frac{2\pi n_a}{\lambda}r$ before using the Bessel function identity $\int_0^\infty (J_n^2(z)/z)dz = 1/(2n)$ with $n=2$.

Due to the circular symmetry of the Airy image function q, we know from the above discussion that $[\mathbf{I}(\theta)]_{2,2} = [\mathbf{I}(\theta)]_{1,1}$. Given the Airy image function, the Cramér-Rao lower bounds $\delta_{x_0}^2$, $\delta_{y_0}^2$, and $\delta_{\Lambda_0}^2$ on the variances for estimating x_0, y_0, and Λ_0 are therefore given by

$$\delta_{x_0}^2 = \delta_{y_0}^2 = \frac{\lambda^2}{4N_{photons}\pi^2 n_a^2},$$

$$\delta_{\Lambda_0}^2 = \frac{1}{\frac{N_{photons}}{\Lambda_0^2} - \frac{(t-t_0)^2 e^{-\Lambda_0(t-t_0)}}{\left(1-e^{-\Lambda_0(t-t_0)}\right)^2}}.$$

Accordingly, the square roots of the bounds are given by

$$\delta_{x_0} = \delta_{y_0} = \frac{\lambda}{2\pi n_a\sqrt{N_{photons}}}, \tag{18.17}$$

$$\delta_{\Lambda_0} = \frac{1}{\sqrt{\frac{N_{photons}}{\Lambda_0^2} - \frac{(t-t_0)^2 e^{-\Lambda_0(t-t_0)}}{\left(1-e^{-\Lambda_0(t-t_0)}\right)^2}}}. \tag{18.18}$$

The square root of the Cramér-Rao lower bound gives us a lower bound on the standard deviation for estimating the parameter. Rather than working with the Cramér-Rao lower bound directly, we work with its square root because as a bound on the standard deviation instead of the variance of any unbiased estimator of the parameter, it allows us to express the measure of variability in estimating the parameter in the same unit as the parameter itself.

Dependence on optical parameters. Equations (18.17) and (18.18) demonstrate that lower bounds on the standard deviations for estimating parameters of interest are dependent on the particular experimental setting. Here, we see that experimental parameters such as the wavelength λ of the detected photons, the expected number $N_{photons}$ of detected photons, and the numerical aperture n_a of the objective can all play a role in determining the bound on the standard deviation with which a parameter can be estimated from an image. Bounds on the standard deviation can therefore be used as a tool for experimental design, as they can be calculated for different sets of experimental parameters to determine an appropriate parameter combination that yields the desired level of estimator performance in terms of the spread of the estimates given repeat images of an object of interest.

Independence from location and magnification. As they derive from Eq. (18.10), δ_{x_0} and δ_{y_0} do not depend on the positional coordinates x_0 and y_0 of the object or the magnification M of the microscope system. This is again a consequence of the infinitely large and unpixelated detection region of the detector in the fundamental data model scenario considered here. In the case of a practical data model, however, the analogous bounds will be dependent on x_0, y_0, and M. See, for example, Figs. 18.7 and 18.8 and the associated discussions in Sections 18.6.2 and 18.6.3 for illustrations of how a practical bound on the standard deviation is affected by the magnification and the position of the object in a pixel array.

Dependence on photon count and profile width. As is to be expected from our discussion of the general results above, we see from the expression for δ_{x_0} and δ_{y_0} an inverse square root dependence of the bounds on the expected number of detected photons $N_{photons}$. This inverse dependence again shows that the more photons that are detected from the object, the smaller the bound on the standard deviation with which the object's positional coordinates can be estimated. The expression also shows a dependence on the width parameter $\frac{2\pi n_a}{\lambda}$ of the Airy profile. Increasing the width parameter (by either increasing the numerical aperture n_a or decreasing the wavelength λ of the detected photons) narrows the profile, and improves the bound on the standard deviation for estimating x_0 or y_0. Both the dependence on the photon count and the dependence on the Airy profile width are in accordance with intuition, as one would expect the collection of more data (i.e., more photons) and a sharper image profile to improve the standard deviation with which the object can be localized.

From Fig. 17.1 and the discussion in Section 17.2.2, we know that the denominator in the expression for δ_{Λ_0} increases with the image acquisition time $t - t_0$, and therefore with $N_{photons}$. The bound on the standard deviation for estimating the photon detection rate Λ_0 can thus also be improved by increasing the mean number of detected photons. Since $N_{photons} = \Lambda_0(t - t_0)/\left(1 - e^{-\Lambda_0(t-t_0)}\right)$ is itself a function of Λ_0, the improvement of δ_{Λ_0} for a given value of Λ_0 is achieved by increasing the image acquisition time $t - t_0$.

Examples. By plotting δ_{x_0} and δ_{Λ_0} as functions of the expected number of detected photons $N_{photons}$, Figs. 18.5(a) and (b) demonstrate the improvement in both bounds as the expected photon count is increased. For example, the plots show that when $N_{photons} = 660$ photons are, on average, detected from a point-like object such as a single molecule, the lower bounds on the standard deviations for estimating, from the resulting image, the object's x_0 coordinate and the rate at which photons are detected from the object, are $\delta_{x_0} = 2.30$ nm and $\delta_{\Lambda_0} = 2569$ photons/s, respectively. When the expected photon count is increased

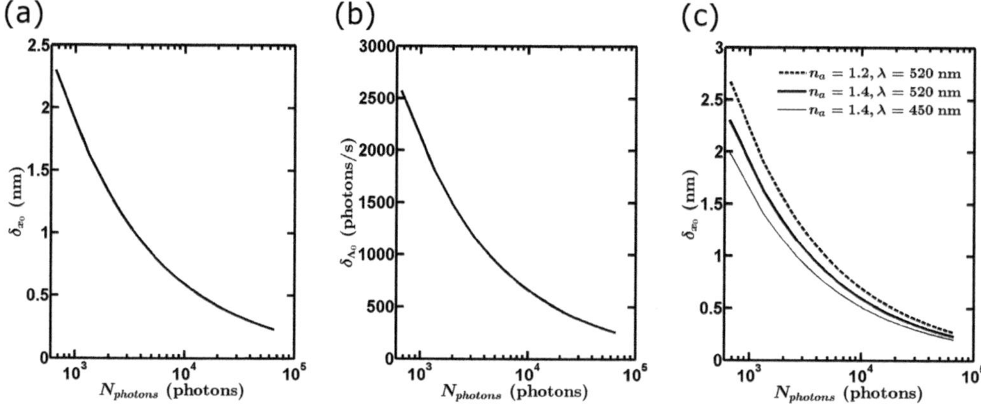

FIGURE 18.5
Dependence of lower bounds on optical parameters. The bounds shown are computed for a stationary and in-focus point-like object whose image is described by an Airy profile and acquired under the fundamental data scenario in which the detector has an infinitely large photon detection region. They are shown as a function of the expected number of photons $N_{photons}$ detected from the object. (a) Lower bound δ_{x_0} on the standard deviation for estimating the positional coordinate x_0 of the object. The wavelength λ of the detected photons is set to 520 nm and the numerical aperture n_a of the objective is set to 1.4. (b) Lower bound δ_{Λ_0} on the standard deviation for estimating the rate Λ_0 at which photons are detected from the object. The value of Λ_0 is set to 66000 photons/s. (c) Lower bound δ_{x_0} for three combinations of the numerical aperture n_a and the wavelength λ.

tenfold to $N_{photons} = 6600$, however, the bounds are improved by a factor of $1/\sqrt{10}$ to $\delta_{x_0} = 0.73$ nm and $\delta_{\Lambda_0} = 812$ photons/s. These values are specific to the experimental setting in which photons of wavelength $\lambda = 520$ nm are detected from the point-like object at a rate of $\Lambda_0 = 66000$ photons/s, and are collected by an objective with a numerical aperture of $n_a = 1.4$. Note that the improvement factor of $1/\sqrt{10}$ for δ_{Λ_0} in this example is due to Λ_0 being large enough for the acquisition times $t - t_0$ considered such that $e^{-\Lambda_0(t-t_0)} \approx 0$ and therefore $\delta_{\Lambda_0} \approx \Lambda_0/\sqrt{N_{photons}}$.

In Fig. 18.5(c), the effect that the wavelength and the numerical aperture have on the lower bound δ_{x_0} is illustrated. As discussed above, both of these parameters affect the width of the Airy profile. Taking $\lambda = 520$ nm and $n_a = 1.4$ as the reference combination, we see from the plot that for any given expected photon count $N_{photons}$, a shorter wavelength of $\lambda = 450$ nm, which narrows the Airy profile, lowers the value of δ_{x_0}. On the other hand, a smaller numerical aperture of $n_a = 1.2$ broadens the Airy profile and is seen to increase the value of δ_{x_0}. When the expected photon count is $N_{photons} = 660$, for example, δ_{x_0} improves from 2.30 nm to 1.99 nm when the wavelength λ is reduced from 520 nm to 450 nm, and worsens from 2.30 nm to 2.68 nm when the numerical aperture n_a is reduced from 1.4 to 1.2.

18.5.2 Cramér-Rao lower bound for the 2D Gaussian image function

Having determined Cramér-Rao lower bounds under the fundamental data model in the case of an Airy profile, we now investigate the same question for a 2D Gaussian profile. Writing the 2D Gaussian image function of Eq. (15.10) in terms of u and v, such that we have

$$q(u,v) = \frac{1}{2\pi\sigma_g^2} e^{-\frac{1}{2\sigma_g^2}(u^2+v^2)}, \quad (u,v) \in \mathbb{R}^2, \tag{18.19}$$

we follow the Airy image function example and first obtain its partial derivative with respect to u as

$$\frac{\partial q_\theta(u,v)}{\partial u} = -\frac{1}{2\pi\sigma_g^4} \cdot u \cdot e^{-\frac{1}{2\sigma_g^2}(u^2+v^2)}, \quad (u,v) \in \mathbb{R}^2. \tag{18.20}$$

As the 2D Gaussian function is symmetric, we use Eqs. (18.19) and (18.20) in the denominator of Eq. (18.11) and obtain

$$\begin{aligned}
[\mathbf{I}(\theta)]_{1,1} &= N_{photons} \int_{\mathbb{R}^2} \frac{1}{q(u,v)} \left(\frac{\partial q(u,v)}{\partial u} \right)^2 dudv \\
&= N_{photons} \cdot \frac{1}{2\pi\sigma_g^6} \int_{\mathbb{R}^2} u^2 e^{-\frac{1}{2\sigma_g^2}(u^2+v^2)} dudv \\
&= N_{photons} \cdot \frac{1}{2\pi\sigma_g^6} \int_{\mathbb{R}} u^2 e^{-\frac{1}{2\sigma_g^2}u^2} du \cdot \int_{\mathbb{R}} e^{-\frac{1}{2\sigma_g^2}v^2} dv \\
&= N_{photons} \cdot \frac{1}{2\pi\sigma_g^6} \cdot \frac{1}{2}\sqrt{8\pi\sigma_g^6} \cdot \sqrt{2\pi}\sigma_g \\
&= \frac{N_{photons}}{\sigma_g^2},
\end{aligned} \tag{18.21}$$

where we have made use of the identity $\int_0^\infty z^2 e^{-az^2} dz = \frac{1}{4}\sqrt{\frac{\pi}{a^3}}$ for $a > 0$. Since the 2D Gaussian profile is not just symmetric, but circularly symmetric, again we know that analogous calculations will lead to the same result for $[\mathbf{I}(\theta)]_{2,2}$ of Eq. (18.12). Given the 2D Gaussian image function, we therefore have the Cramér-Rao lower bounds

$$\begin{aligned}
\delta_{x_0}^2 &= \delta_{y_0}^2 = \frac{\sigma_g^2}{N_{photons}}, \\
\delta_{\Lambda_0}^2 &= \frac{1}{\frac{N_{photons}}{\Lambda_0^2} - \frac{(t-t_0)^2 e^{-\Lambda_0(t-t_0)}}{\left(1-e^{-\Lambda_0(t-t_0)}\right)^2}},
\end{aligned}$$

and the corresponding lower bounds δ_{x_0}, δ_{y_0}, and δ_{Λ_0} on the standard deviations for estimating x_0, y_0, and Λ_0 are then given by

$$\delta_{x_0} = \delta_{y_0} = \frac{\sigma_g}{\sqrt{N_{photons}}}, \tag{18.22}$$

$$\delta_{\Lambda_0} = \frac{1}{\sqrt{\frac{N_{photons}}{\Lambda_0^2} - \frac{(t-t_0)^2 e^{-\Lambda_0(t-t_0)}}{\left(1-e^{-\Lambda_0(t-t_0)}\right)^2}}}. \tag{18.23}$$

From the expressions for δ_{x_0} and δ_{y_0}, we see the same dependence of the bounds on the mean detected photon count $N_{photons}$ and the width of the object's image, in this case the width σ_g of a 2D Gaussian profile. Once again, we see that the bounds can be improved by increasing the number of detected photons or decreasing the width of the object's image profile.

18.5.3 Extensions to further experimental situations

We have here investigated the Cramér-Rao lower bound for the estimation of the location of an object such as a single molecule. The basic assumption for our derivation consisted of a photon distribution profile $f_{\theta,\tau}$ that is constant with respect to time (i.e., $f_{\theta,\tau} = f_\theta$, $\tau \geq t_0$), as well as a photon detection rate $\Lambda_\theta(\tau)$ that is not only constant with respect to time (i.e., $\Lambda_\theta(\tau) = \Lambda_0$, $\tau \geq t_0$), but importantly does not depend on the positional coordinates to be estimated. We know from the first expression of Eq. (17.11) that the Fisher information matrix is the sum of two matrices, one that relates to the spatial aspect of photon detection, and one that relates to the temporal aspect. The temporal matrix is a function of only the image acquisition time interval $[t_0, t]$, the photon detection rate Λ_θ, and the derivative of Λ_θ with respect to θ. Given that our photon detection rate Λ_0 does not depend on the coordinates x_0 and y_0 of the object, we have that $\frac{\partial \Lambda_0}{\partial x_0} = \frac{\partial \Lambda_0}{\partial y_0} = 0$, and the temporal matrix makes no contribution to the overall matrix $\mathbf{I}(\theta)$ in terms of information about x_0 and y_0.

Clearly, using an experimental setup that results in a photon detection rate that depends on parameters of interest such as the object location can increase the Fisher information pertaining to those parameters via the temporal matrix, as the contribution of the temporal matrix to $\mathbf{I}(\theta)$ will be positive. This positive contribution can be verified in relatively straightforward fashion using the Fisher information matrix of Eq. (17.12) (Exercise 18.5). A photon detection rate that depends on parameters such as the object location can be achieved, for example, through nonuniform and possibly time-varying illumination of the sample (Note 18.4). In the context of 3D location estimation, which we will discuss in the next chapter, nonuniform illumination can be achieved in different ways. One example is total internal reflection illumination, which we encountered in Section 11.4.

The Cramér-Rao lower bound can also be derived for situations where the object of interest is not stationary. Estimation problems can be analyzed in the context of the fundamental data model for deterministic and stochastic object trajectories, leading to generalizations of the Fisher information results obtained in this section (Note 18.5).

18.6 Cramér-Rao lower bound for location estimation — practical data model

Having analyzed examples of Cramér-Rao lower bounds under the fundamental data model, we now come to analyze the situation for the practical data models. In this case, the pixelation of the detector renders the mathematical expressions less transparent, and we will not be able to obtain very explicit results. The lack of explicit expressions in the practical case is one of the justifications for the detailed examination of fundamental scenarios as they represent important limiting cases that often have relatively explicit and easy-to-analyze expressions. Instead, we will describe how the Cramér-Rao lower bound is calculated, and will demonstrate the usefulness of the result as a tool for experimental design, by virtue of its dependence on experimental settings such as the magnification and the detector type. For our illustrations, we will start by using the Poisson data model to demonstrate how the major characteristics of a practical data model affect localization performance. We will then proceed to compare the lower bounds for the CCD/CMOS and EMCCD data models before closing the section with an investigation of how the location of the object impacts how well it can be determined.

We again start with the Fisher information matrix corresponding to the underlying image data from which we want to estimate the location of an object of interest. The general expression for the Fisher information for a detector with N_p pixels is given by Eq. (17.21), where for $k = 1, \ldots, N_p$, the mean photon count $\nu_{\theta,k}$ in the kth pixel is given generally by Eq. (18.2). To obtain an expression for $\nu_{\theta,k}$ that is specific to our localization problem, we make the same assumption, as in the fundamental case, of a constant rate at which photons are detected from the object, so that $\Lambda_\theta(\tau) = \Lambda_0 > 0$, $\tau > t_0$. Moreover, given that the object is stationary, for the photon distribution profile we have $f_{\theta,\tau} = f_\theta$, $\tau \geq t_0$. The function f_θ can again be expressed through an image function, as in Eq. (18.3). If we further assume that photons from the background component are uniformly distributed over the N_p pixels of the image, such that the mean background photon count in the kth pixel, $k = 1, \ldots, N_p$, is given by $\beta_{\theta,k} = \beta_0 > 0$, then Eq. (18.2) becomes, for $k = 1, \ldots, N_p$,

$$\nu_{\theta,k} = N_{photons} \cdot \int_{C_k} f_\theta(x, y)dxdy + \beta_0, \tag{18.24}$$

where $N_{photons} = \Lambda_0(t - t_0)$. Note that this description of $\nu_{\theta,k}$ is the same as that used in the localization example of Section 18.2. For the illustrations that follow, we will mirror that example further by assuming a point-like object whose in-focus image is described by the Airy image function. With this assumption, f_θ in Eq. (18.24) is explicitly given by Eq. (18.4). Unlike in Section 18.5, we consider here the case where only the positional coordinates of the object are estimated, and therefore define the vector of unknown parameters to be $\theta = (x_0, y_0)$. Going forward we will present and discuss only the results obtained for the x_0 coordinate, as the results for the y_0 coordinate are similar.

18.6.1 Poisson data model — effects of pixelation, finite image size, and background noise

We begin our illustrations by considering the case where the image of the object is captured with a hypothetical pixelated detector that introduces no readout noise to the image. Consideration of this hypothetical case allows us to take a significant step towards imaging with a realistic detector, and at the same time enables a straightforward analysis that highlights the primary differences between the fundamental data model and a practical data model. To obtain lower bounds corresponding to imaging with a pixelated detector that adds no readout noise to the acquired data, we need the Fisher information matrix for the Poisson data model, which in Section 17.3.1 we derived from the general Fisher information expression of Eq. (17.21). The result is given by Eq. (17.23), and we simply evaluate the expression with $\nu_{\theta,k}$ as described, take the inverse of the computed matrix, and obtain δ_{x_0} and δ_{y_0}, the lower bounds on the standard deviations for estimating x_0 and y_0, as the square roots of entries $(1, 1)$ and $(2, 2)$ of the inverse matrix, respectively.

We compute the bounds under two scenarios — with and without a uniformly distributed background component in the image. For the scenario with background noise, the mean photon count $\nu_{\theta,k}$ in the kth pixel, $k = 1, \ldots, N_P$, is as given by Eq. (18.24). For the scenario without background noise, $\nu_{\theta,k}$, $k = 1, \ldots, N_P$, is given by the same expression, but without the constant background term β_0.

The bound δ_{x_0} computed under each scenario is shown in Fig. 18.6. In this figure, δ_{x_0} is shown as a function of the image's effective pixel size, and is compared with δ_{x_0} for the corresponding fundamental data scenario, given by Eq. (18.17). The computations performed are such that the different effective pixel sizes are achieved by varying the magnification M (which appears in the photon distribution profile of Eq. (18.4)) while keeping the actual pixel size of the detector constant (i.e., the region of integration C_k, $k = 1, \ldots, N_p$, in Eq. (18.24) does not change). To allow a fair assessment of how the effective pixel size affects the value of δ_{x_0}, the size of a given image (i.e., the number of pixels N_p comprising the pixel array used for the object localization) is scaled proportionally to ensure that the image captures, on average, the same number of photons from the object regardless of its effective pixel size. Note that δ_{x_0} for the corresponding fundamental data scenario is rep-

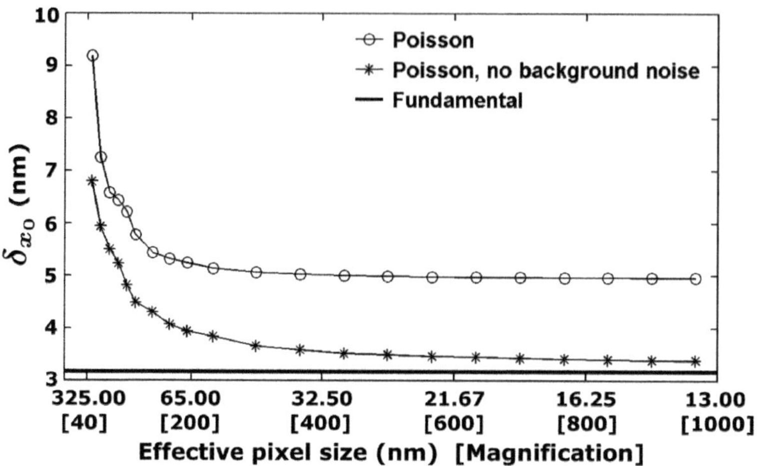

FIGURE 18.6
Lower bound for point source localization under the Poisson data model. The lower bound δ_{x_0} on the standard deviation for estimating the positional coordinate x_0 of a stationary and in-focus point-like object is calculated under the Poisson data model and shown as a function of the effective pixel size of the underlying image. Two scenarios, one with background noise (∘) and one without (∗), are shown along with the corresponding lower bound (Eq. (18.17)) under the fundamental data model (thick line). For the Poisson scenario without background noise, no background component is modeled and the background term β_0 in the expression for the mean photon count in the kth pixel (Eq. (18.24)) is dropped. In all data scenarios, photons are detected from the object at a constant rate of $\Lambda_0 = 7000$ photons/s over an acquisition time of $t - t_0 = 0.05$ s, such that $N_{photons} = \Lambda_0(t - t_0) = 350$ photons are on average detected from the object over the detector plane. Moreover, it is assumed that the image of the object is given by the Airy image function. For both Poisson data scenarios, the physical pixel size is 13 μm × 13 μm, and the different effective pixel sizes are obtained by varying the magnification from $M = 46.67$ to $M = 966.67$. The image is a 15×15 pixel array at $M = 100$, and a proportionately scaled pixel array at each of the other magnifications. The object is positioned at $(x_0, y_0) = (916.5 \text{ nm}, 942.5 \text{ nm})$, such that at $M = 100$, the center of its image (i.e., the center of the Airy image function), is located at 0.05 pixels in the x direction and 0.25 pixels in the y direction with respect to the upper left corner of the center pixel of the 15×15 pixel array. For the Poisson data scenario with background noise, a background component is assumed that is uniformly distributed over the pixel array. The level of the background noise is such that at $M = 100$, $\beta_k = \beta_0 = 8$ photons for each pixel k. For each of the other magnifications, the same background level per pixel is retained by a proportionate scaling. In all cases, all photons are of wavelength of $\lambda = 520$ nm, and are assumed to be collected by an objective with numerical aperture $n_a = 1.4$.

resented by a horizontal line, as the effective pixel size does not apply to the fundamental data model.

Our investigation of δ_{x_0} as a function of the effective pixel size provides an example of how one can compute lower bounds for different values of one or more parameters to determine the effect of those parameters on the standard deviation with which a quantity of interest can be extracted from the acquired image data. Such calculations are particularly useful for the design of imaging experiments, as they allow the identification of settings that optimize the standard deviation for estimating the quantity of interest. In addition to the effective pixel size, one can investigate the effect of, for example, the expected number of object photons that are detected in an image, the wavelength of those photons, and the numerical aperture of the objective used.

Finer pixelation improves localization performance in the absence of detector readout noise. From Fig. 18.6, we see that for any given effective pixel size, δ_{x_0} under the Poisson data model, with or without background noise, is larger in value, and therefore represents a worse lower bound, than δ_{x_0} under the fundamental data model. This is due, in large part, to the fact that while the unpixelated detector of the fundamental data model records the locations of photon detection with arbitrarily high precision, pixelation of the photon detection area under the Poisson data model reduces the precision with which the location of a detected photon is known. To obtain better bounds under the Poisson data model, we therefore need to increase the precision with which photon locations are recorded. Indeed, we see that increasing this precision by decreasing the effective pixel size improves the bounds and moves them closer to the bound for the fundamental data model. Note that in practice, detectors introduce readout noise and as a result, reducing the effective pixel size does not always improve the lower bound as in the case of the Poisson data model. This will be discussed in Section 18.6.2.

Finite image size deteriorates localization performance. In our current example, the fact that δ_{x_0} under the Poisson data model is always worse than δ_{x_0} under the fundamental data model is also partly attributable to the finite physical extent of an image under the Poisson data scenarios considered here. The finiteness of the image ensures that photons falling outside the image will contribute no spatial information about the object. In contrast, the fundamental data scenario used for comparison here assumes an image of infinite size, so that no spatial information is lost due to photons not being available. (See Section 18.9.1 for more on the importance of choosing, for localization analysis, a sufficiently large array of pixels that captures a large percentage of the photons from the object.) Consequently, a larger lower bound on the standard deviation for localizing the object is to be expected for the Poisson data scenarios, even when the effect of pixelation is minimized. This is demonstrated by δ_{x_0} for the Poisson data scenario without background noise. We see in Fig. 18.6 that this bound comes quite close to δ_{x_0} for the fundamental data scenario at very small effective pixel sizes. However, the fact that it levels off as the effective pixel size reduction continues beyond a certain point shows that it will never actually attain the bound for the fundamental data scenario. The small gap that will not be closed is due to the finite image size of the Poisson data scenario.

Background noise degrades localization performance. In general, noise of any kind can be expected to make it more difficult to estimate parameters with precision. Noise due to the background component is no exception. Figure 18.6 shows that while δ_{x_0} for the Poisson data scenario without background noise is able to attain values approaching δ_{x_0} for the fundamental data scenario, the presence of background noise prevents δ_{x_0} for the other Poisson data scenario from coming close. The background noise level assumed here is such that, at any given effective pixel size, the lower bound is nontrivally larger than that for the scenario without background noise.

18.6.2 Localizing objects from CCD/CMOS and EMCCD images

Continuing with the same localization example, with the specific settings specified in Fig. 18.6, we consider here the lower bound δ_{x_0} when the image of the object is captured with a realistic detector. We consider two scenarios, one in which the image is recorded with a CCD or CMOS detector, and one in which the image is acquired with an EMCCD detector. To obtain the lower bound in either case, we calculate the same mean photon

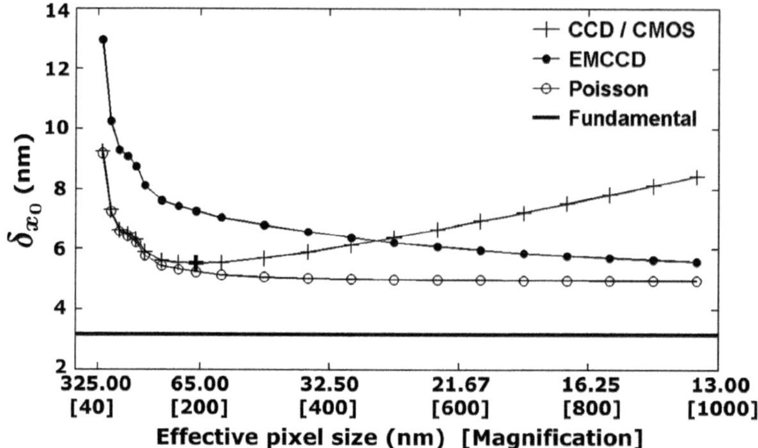

FIGURE 18.7
Lower bounds for point source localization under the CCD/CMOS and EMCCD data models. The lower bound δ_{x_0} on the standard deviation for estimating the positional coordinate x_0 of a stationary and in-focus point-like object is calculated under the CCD/CMOS (+) and EMCCD (•) data models and shown as a function of the effective pixel size of the underlying image. For δ_{x_0} calculated under the CCD/CMOS data model, the thick marker identifies the smallest value out of all the data points shown. For comparison, corresponding lower bounds calculated under the Poisson data model with background noise (○) and the fundamental data model (thick line) are reproduced from Fig. 18.6. For both the CCD/CMOS and the EMCCD data models, the mean of the readout noise in each pixel k is $\eta_k = 0$ electrons. For the CCD/CMOS data model, the standard deviation of the readout noise in each pixel k is $\sigma_k = 1$ electron, and for the EMCCD data model, it is $\sigma_k = 24$ electrons. An EM gain of $g = 900$ is used for the EMCCD data model. All other details are as given in Fig. 18.6.

counts $\nu_{\theta,k}$, $k = 1, \ldots, N_p$, as before, but use them to compute the Fisher information matrix for the CCD/CMOS data model (Eq. (17.25)) or the EMCCD data model (Table 17.1). The computed bounds are shown in Fig. 18.7 as a function of the same effective pixel sizes from Fig. 18.6. For purposes of comparison, also included in the plot are the bounds for the fundamental data scenario and the Poisson data scenario, with background noise, from Fig. 18.6. Note that in our illustrations, the readout noise standard deviation is assumed to be the same for every CCD/CMOS image pixel. For a CMOS detector, we know from Chapter 12 that this assumption is typically not valid. However, with this assumption we can regard the modeled CMOS detector as one that is very well calibrated, so much so that the individual readout circuitries for the pixels produce a uniform level of readout noise.

Choice of appropriate detector type based on pixel photon count levels. Compared with δ_{x_0} for the CCD/CMOS data scenario, Fig. 18.7 shows that δ_{x_0} for the EMCCD data scenario is inferior at relatively large effective pixel sizes, but superior at smaller effective pixel sizes. This can be explained by considering the magnitudes of the photon signals detected in the pixels. When the effective pixel size is relatively large, each pixel is allocated a greater number of the photons detected from the object. The detector's readout noise therefore represents a relatively minor component of the data in each pixel, and there is no need to amplify the photon signal. In fact, use of electron multiplication in such a scenario is counter-productive, as the net effect will be to add more stochasticity to the data acquired in each pixel, and to make the estimation of parameters of interest a more difficult task. This is reflected in the higher δ_{x_0} values for the EMCCD data scenario when the pixelation is relatively coarse.

A small effective pixel size, on the other hand, means that fewer object photons are allocated to each pixel of an image. Readout noise thus becomes a significant component of the data acquired in a pixel, and amplification of the photon signal, despite being stochastic in nature, is desirable. In such a scenario, the overall effect of electron multiplication is a reduced corruption of the object photon signal by detector noise, and hence we can expect a better lower bound on the standard deviation with which a parameter of interest can be estimated. This is seen in the smaller δ_{x_0} values for the EMCCD data scenario when the image is finely pixelated. For more on how the performance of CCD/CMOS detectors differs from the performance of EMCCD detectors in terms of the Fisher information content of the data in a pixel under different signal levels, see Section 17.4.

Figure 18.7 thus illustrates the use of lower bounds on the standard deviation of parameter estimation to compare the performance of different detector types under a given set of conditions. This again showcases the Fisher information-based lower bound as a tool for experimental design, specifically demonstrating that an appropriate detector type can be chosen based on the photon signal levels that are expected in the pixels of the acquired image. In addition to comparing the performance of different detector types, the lower bounds can be used to compare the performance of a given detector type under different noise settings. Lower bounds corresponding to both CCD/CMOS and EMCCD imaging, for example, can be computed for different readout noise standard deviations to assess the extent to which the performance of the detector is affected by changes to the readout noise level for a given experimental scenario.

Choice of effective pixel size for imaging with a CCD or CMOS detector. Figure 18.7 demonstrates that for a given set of experimental conditions, there is an optimal effective pixel size when it comes to the lower bound on the standard deviation for estimating the location of an object from a CCD or CMOS image. For the settings detailed in the figure, we see for the CCD/CMOS data scenario that the smallest value for δ_{x_0} is attained at an effective pixel size of approximately 65 nm (which corresponds to a magnification of $M = 200$). As the effective pixel size is increased or decreased relative to this optimal size, δ_{x_0} exhibits a deteriorating trend. On the one hand, increasing the effective pixel size worsens the bound because a coarser pixelation means a lower precision with which the location of each detected photon is known. Imagine, for example, the extreme scenario where the effective pixel size is so large that a vast majority of the photons distributed according to an Airy profile are captured within one pixel. In this case, the shape of the Airy profile is largely "lost" inside the one pixel because all photons detected in that pixel are spatially indistinguishable from one another. This loss of spatial information makes the location of the Airy profile very difficult to estimate with a reasonable standard deviation. On the other hand, decreasing the effective pixel size worsens the bound because a finer pixelation means that a smaller signal from the object is detected in each pixel, and that the detector's readout noise becomes a more significant component of the data in each pixel. As an example, imagine the opposite extreme where the effective pixel size is so small that the signal in each pixel is effectively buried in the readout noise. In this case, the position of the Airy profile is also very difficult to estimate with an acceptable standard deviation.

Poisson data model provides a more realistic benchmark than fundamental data model. From Fig. 18.7, we see that at any given effective pixel size, δ_{x_0} for the Poisson data scenario is smaller than δ_{x_0} for the CCD/CMOS and EMCCD data scenarios. This is to be expected because an image captured by a CCD, CMOS, or EMCCD detector is the result of more forms of signal loss or degradation compared to an image recorded by the corresponding hypothetical detector under the Poisson data model. Besides being pixelated, finite in size, and corrupted by background noise, just as in the case of an image under the

Poisson data model, a CCD or CMOS image is deteriorated by detector readout noise, and an EMCCD image is deteriorated by the stochastic electron multiplication process that attempts to drown out the detector readout noise by amplifying the detected photon signal.

Importantly, Fig. 18.7 shows that compared to δ_{x_0} for the fundamental data scenario, δ_{x_0} for the Poisson data scenario represents a significantly tighter lower bound to δ_{x_0} for the CCD/CMOS and EMCCD data scenarios. This means that δ_{x_0} for the Poisson data scenario provides a more realistic benchmark against which δ_{x_0} for the CCD/CMOS and EMCCD data scenarios can be assessed, and makes for a more practical target to which the latter can come close. This of course should not be surprising, as by virtue of accounting for the same finite and pixelated photon detection area and the same background component, the Poisson data model is much more similar to the CCD/CMOS and EMCCD data models than the fundamental data model. In fact, we see that at the largest effective pixel sizes shown, δ_{x_0} corresponding to the CCD/CMOS detector comes very close to δ_{x_0} for the Poisson data scenario. In these cases, the photon counts in the pixels of the CCD/CMOS detector are high enough that the readout noise becomes relatively insignificant, making the Poisson data model a reasonable description of the CCD/CMOS data. We also see that at the smallest effective pixel sizes shown, δ_{x_0} corresponding to the EMCCD detector approaches δ_{x_0} for the Poisson data scenario. This is a result of the stochastic amplification of the photon signals in the pixels of the EMCCD detector, a phenomenon that, as discussed above, is most effectively utilized to drown out the detector readout noise when the mean photon counts in the pixels are small.

18.6.3 Object location makes a difference

We have seen from Eq. (18.9) and the more specific results of Eqs. (18.17) and (18.22) that in the ideal scenario in which the image of a point source is captured with an infinitely large and unpixelated detector, the lower bounds δ_{x_0} and δ_{y_0} do not depend on the point source's x_0 and y_0 positional coordinates. This result makes intuitive sense, as the lack of an image boundary and the absence of pixels to partition the detected photons mean that effectively the same image is produced regardless of the location of the object. In the practical scenario of a detector with a finite and pixelated detection surface, however, δ_{x_0} and δ_{y_0} will necessarily depend on x_0 and y_0 as different images will be produced depending on the object's location, and will carry different amounts of information about that particular location. One can easily see, for example, that placing a point source at the intersection of four pixels as opposed to the center of a pixel will result in different pixelated versions of its Airy profile.

Using δ_{x_0}, calculated for the case in which the x_0 and y_0 positional coordinates are the only parameters to be estimated (i.e., for the case of $\theta = (x_0, y_0)$), the object location dependence of the lower bound on the standard deviation for estimating an object's positional coordinates is illustrated in Fig. 18.8. In this figure, the object is a point source with an Airy profile as its image. Reflecting the common practice of selecting a region of interest that is approximately centered about the location of the object to be localized, the x_0 coordinate of this point source is incrementally varied over an interval that corresponds, in image space, to a 48-μm interval centered in a 240 μm $\times$ 240 μm pixel array. Four different pixel sizes are considered, and in the case of the largest size of 24 μm, the 48-μm interval corresponds exactly to the middle two pixels of the array. From the plots shown, we see that the value of δ_{x_0} oscillates as the point source is moved over the 48-μm interval in the center of the pixel array. Importantly, from the plots for the two largest pixel sizes considered, we see that δ_{x_0} can take on significantly different values depending on the point source's location within a relatively large pixel. In the case of the 24-μm pixel size, for instance, within the middle two pixels of the pixel array δ_{x_0} attains its largest (i.e., worst) value when the point source is

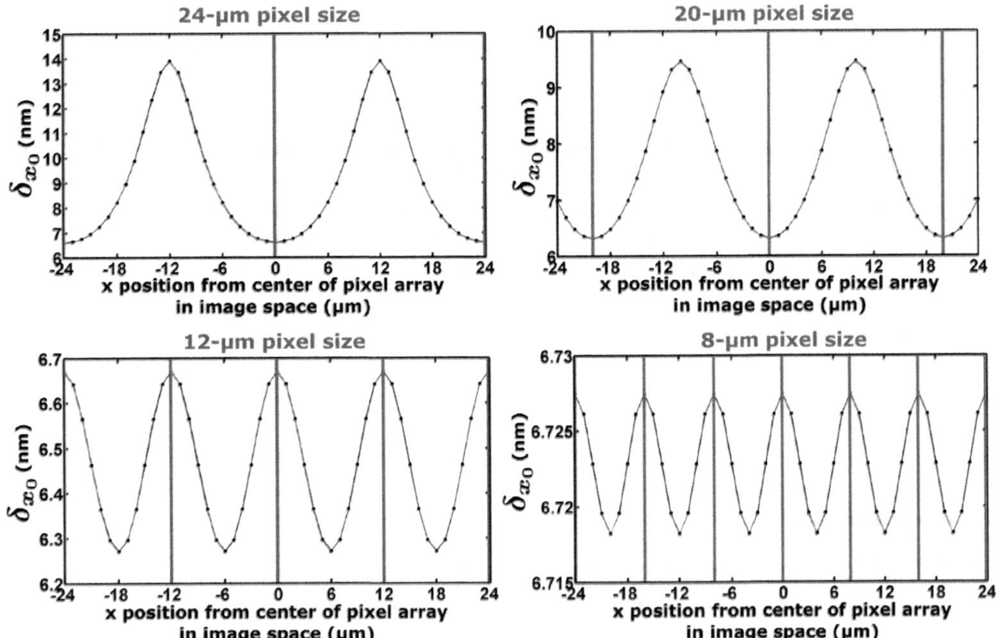

FIGURE 18.8
Dependence of the lower bound for point source localization on the point source location itself, given a pixelated detector. The lower bound δ_{x_0} on the standard deviation for estimating the positional coordinate x_0 of a stationary and in-focus point-like object is shown as a function of x_0 itself in the case of a pixelated detector. The lower bound is calculated, under the CCD/CMOS data model, for four detectors with pixel sizes of 24 μm $\times$ 24 μm, 20 μm $\times$ 20 μm, 12 μm $\times$ 12 μm, and 8 μm $\times$ 8 μm. A separate plot is shown for each pixel size, and in each case x_0 is given in the image space coordinate system, with respect to the center of a 240 μm $\times$ 240 μm pixel array. The vertical magenta lines in each plot demarcate the boundaries between pixels in the x direction. In all cases the x_0 coordinate of the object is varied in increments of 1 μm, in image space, across a 48-μm interval centered in the pixel array. As the magnification is set to $M = 100$, this is equivalent to changing the object's x position in increments of 10 nm over a 480-nm interval in object space. In the y direction, the object is positioned at the center of the pixel array. The image of the object is given by the Airy image function. Photons are detected from the object at a constant rate of $\Lambda_0 = 7000$ photons/s over an acquisition time of $t - t_0 = 0.05$ s, such that an average of $N_{photons} = \Lambda_0(t - t_0) = 350$ photons are detected from the object over the detector plane. A background component is assumed that is uniformly distributed over the pixel array. The level of this component is such that $\beta_k = \beta_0 = 18$ photons for each pixel k in the case of the 24-μm pixel size. The same level is retained by a proportionate scaling for the other pixel sizes. In all cases, the mean and standard deviation of the readout noise in each pixel k are $\eta_k = 0$ electrons and $\sigma_k = 3$ electrons. It is assumed that the object and background photons are of wavelength $\lambda = 520$ nm, and that they are collected by an objective with numerical aperture $n_a = 1.4$.

located at the center of either pixel. This largest value is more than twice its smallest (i.e., best) value, which it attains when the point source is located at an edge of either pixel.

On the other hand, with the smaller pixel sizes of 12 μm and 8 μm, we see that the oscillating value of δ_{x_0} is confined to much smaller ranges. This is to be expected, as smaller pixel sizes lead to less pronounced differences between the differently pixelated Airy profiles that result from the positioning of the point source at closely spaced locations. In the case of the 8-μm pixel size, for example, the oscillations are contained within a very small range of less than 0.015 nm. In such scenarios, the Airy profile is pixelated at a sufficiently fine level as to allow δ_{x_0} to reasonably be considered, within the confines of a small region, to be independent of the point source location. Note that due to the finiteness of the pixel array, even in the case of the smallest of pixel sizes the approximate independence of δ_{x_0} from the point source location cannot be extended to the whole of the pixel array. This is due to the fact that the detected image can be expected to be very different when the point source is located near the periphery of the pixel array. When this is the case, a significant portion of the Airy profile will not be recorded in the pixel array, and this loss of information will significantly increase the value of δ_{x_0}. Indeed, the same result can be expected even if the finite region corresponding to the pixel array is unpixelated.

18.7 Efficiency of estimators: how well is the behavior of estimators described by the Cramér-Rao lower bound?

We have introduced the Cramér-Rao lower bound as a means to analyze the variance or standard deviation of parameter estimation approaches. We know that the variance of any unbiased estimator is at least as large as the Cramér-Rao lower bound, the inverse of the Fisher information matrix. If we can establish that the variance of an unbiased estimator matches closely the Cramér-Rao lower bound, we can compute the Cramér-Rao lower bound to predict the variance of such an estimator. In Chapter 17 we introduced the notion of the efficiency of an estimator, i.e., the property of an unbiased estimator whose variance equals the Cramér-Rao lower bound. We will now investigate, under the fundamental and the practical data models, the question of whether a specific location estimator is efficient, or at least has a variance that closely matches the Cramér-Rao lower bound.

18.7.1 Fundamental data model

In Chapter 16, we extensively studied, under the fundamental data model, the center of mass as an object location estimator as it allowed us to analytically show that it is unbiased, provided that a technical condition is satisfied. In that chapter, we also obtained analytical expressions for the variance of this estimator. As it is the maximum likelihood estimator in the case where the image of the object is described by a 2D Gaussian profile, it is a particularly interesting estimator. Specifically, for the fundamental data model with a Gaussian image profile, we showed that the center of mass of the spatial coordinates of the photons detected in an image has a covariance matrix given by (Eq. (16.25))

$$\operatorname{cov}\left(\hat{\theta}_{CM}\right) = \mathrm{E}\left[\frac{1}{\mathcal{N}(t)} \mid \mathcal{N}(t) > 0\right] \cdot \begin{bmatrix} \sigma_g^2 & 0 \\ 0 & \sigma_g^2 \end{bmatrix},$$

where σ_g is the width of the Gaussian profile, and $\mathcal{N}(t)$ is the Poisson-distributed number of photons that are detected in an image acquired over the time interval $[t_0, t]$. In Section 18.5.2, we showed that the Cramér-Rao lower bound pertaining to this particular localization problem is given by

$$\mathbf{I}^{-1}(\theta) = \frac{1}{N_{photons}} \cdot \begin{bmatrix} \sigma_g^2 & 0 \\ 0 & \sigma_g^2 \end{bmatrix},$$

where $N_{photons} = \mathrm{E}[\mathcal{N}(t) \mid \mathcal{N}(t) > 0] = \mathrm{E}[\mathcal{N}(t)] / \left(1 - e^{-\mathrm{E}[\mathcal{N}(t)]}\right)$. By the Cramér-Rao inequality we know that

$$\operatorname{cov}\left(\hat{\theta}_{CM}\right) = \mathrm{E}\left[\frac{1}{\mathcal{N}(t)} \mid \mathcal{N}(t) > 0\right] \cdot \begin{bmatrix} \sigma_g^2 & 0 \\ 0 & \sigma_g^2 \end{bmatrix} \geq \frac{1}{N_{photons}} \cdot \begin{bmatrix} \sigma_g^2 & 0 \\ 0 & \sigma_g^2 \end{bmatrix} = \mathbf{I}^{-1}(\theta).$$

This is indeed the case as $\mathrm{E}\left[\frac{1}{\mathcal{N}(t)} \mid \mathcal{N}(t) > 0\right] \geq \frac{1}{N_{photons}}$, as shown in Fig. 18.9. In this figure, the expectation on the left-hand side of the inequality, given more explicitly (using the zero-truncated Poisson probability mass function of Eq. (15.6)) by

$$\begin{aligned} \mathrm{E}\left[\frac{1}{\mathcal{N}(t)} \mid \mathcal{N}(t) > 0\right] &= \sum_{N_0=1}^{\infty} \frac{1}{N_0} \cdot P\left(\mathcal{N}(t) = N_0 \mid \mathcal{N}(t) > 0\right) \\ &= \frac{e^{-\mathrm{E}[\mathcal{N}(t)]}}{1 - e^{-\mathrm{E}[\mathcal{N}(t)]}} \sum_{N_0=1}^{\infty} \frac{\left(\mathrm{E}[\mathcal{N}(t)]\right)^{N_0}}{N_0 \cdot N_0!}, \end{aligned} \tag{18.25}$$

and the expression $\frac{1}{N_{photons}}$ on the right-hand side of the inequality are compared as a function of $N_{photons}$, the mean number of photons detected in an image over the acquisition interval $[t_0, t]$. We can see that these two quantities are very close when $N_{photons}$ is close to 1, quite far apart for values of $N_{photons}$ up to 10 or so, and again very close for values of $N_{photons}$ greater than 10 or so. The left-hand side expression, however, is always greater than the right-hand side expression, as required by the Cramér-Rao inequality. Importantly, we see that as $N_{photons}$ increases, the difference between the two expressions decreases to zero. This shows that asymptotically, under the stated assumptions for this localization problem, the center of mass is an efficient location estimator.

In Section 16.6.3 we investigated a modification of the fundamental data model wherein we assumed that a fixed number, N_0, of photons are acquired in each repeat image. In the case where the image of the object is described by a 2D Gaussian profile with width σ_g, we showed that the center of mass of the spatial coordinates of the N_0 photons has a covariance matrix given by

$$\operatorname{cov}\left(\hat{\theta}_{CM}\right) = \frac{1}{N_0} \cdot \begin{bmatrix} \sigma_g^2 & 0 \\ 0 & \sigma_g^2 \end{bmatrix}.$$

In contrast to the above analysis we therefore have that in this case the center of mass is an efficient location estimator as its covariance matrix equals the inverse Fisher information matrix $\mathbf{I}^{-1}(\theta)$, which can be shown to be given by the same expression (Exercise 17.4).

The analysis of these two center of mass location estimators shows that subtle differences in the data model can lead to different results. Here it is particularly interesting that it is the variability in the number of photons that can be acquired in an experiment that leads to different characterizations of the efficiency of an estimator.

18.7.2 Practical data models

While we could analytically determine the efficiency of at least one maximum likelihood estimator under the fundamental data model, in the case of a practical data model we can only resort to simulations to investigate efficiency of the maximum likelihood estimator.

In Section 18.6 we have seen how the Cramér-Rao lower bound can be computed for the location estimation problem under the Poisson, CCD/CMOS, and EMCCD data models. Here we use the lower bounds computed in the manner described to assess the variances of the maximum likelihood estimators that produced the localization results presented in Table 18.1. From the table we see that for each of the six data sets, the standard deviations

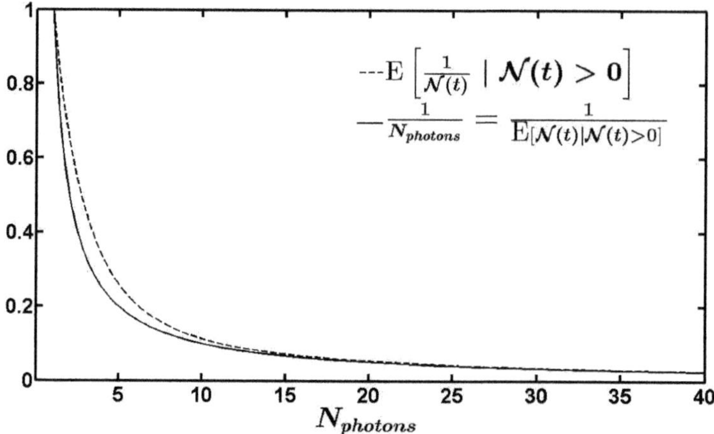

FIGURE 18.9
Comparing the mean of the inverse with the inverse of the mean of a zero-truncated Poisson random variable. The expectation of the inverse of $\mathcal{N}(t)$, the number of photons detected during the acquisition time interval $[t_0, t]$ according to the Poisson process $\mathcal{N}(\tau)$, $\tau \geq t_0$, given that $\mathcal{N}(t)$ is positive (i.e., Eq. (18.25)), is plotted together with the inverse of $N_{photons}$, the expectation of $\mathcal{N}(t)$ given that $\mathcal{N}(t)$ is positive. The two quantities are plotted as a function of $N_{photons}$, which is just the mean number of photons in an image acquired over the interval $[t_0, t]$.

of the x_0 and y_0 estimates are within 5% of their corresponding bounds δ_{x_0} and δ_{y_0}. Based on this observation, we can reasonably conclude that under the conditions assumed for these data sets, the variance of the maximum likelihood estimator is well described by the Cramér-Rao lower bound. As the latter is a lower bound on the variance of any unbiased estimator, this shows that the maximum likelihood estimator can be considered to be optimal with regard to its variance, and that no other unbiased estimator will have superior performance in terms of a lower variance.

While this analysis has only considered a small number of cases, it confirms other studies in the literature that indicate that the maximum likelihood estimator can be assumed to have little, if any bias, and has a variance that is well characterized by the Cramér-Rao lower bound (Note 18.1). This is also consistent with the general theoretical result that the maximum likelihood estimator is asymptotically efficient, which means that for large data sets the variance approximates the Cramér-Rao lower bound.

18.8 Approximations

In the discussion of Fig. 18.7, we pointed out that the Poisson data model makes a good substitute for the CCD/CMOS data model when lower bounds on the standard deviation of location estimation need to be computed for CCD or CMOS images that have high photon signals in their pixels. In this section, we will present additional approximations that may be used to calculate the Cramér-Rao lower bound for estimating parameters from a CCD, CMOS, or EMCCD image. Like the Poisson data model, these approximations provide computationally faster alternatives to the CCD/CMOS and EMCCD data models.

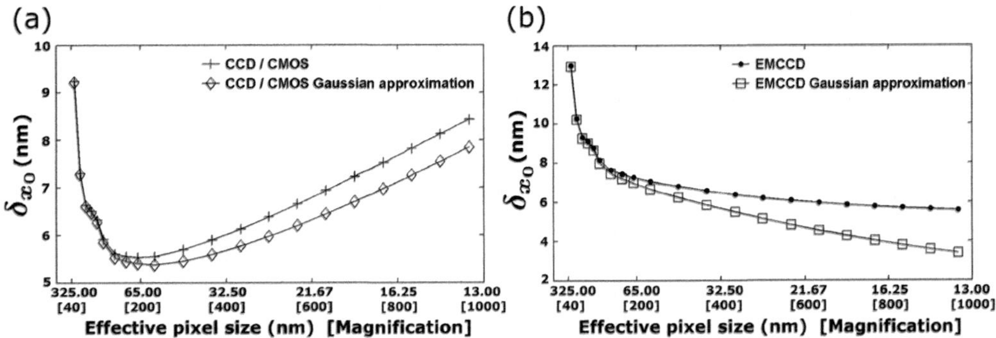

FIGURE 18.10
Lower bounds for point source localization under the Gaussian approximations for the CCD/CMOS and EMCCD data models. The lower bound δ_{x_0} on the standard deviation for estimating the positional coordinate x_0 of a stationary and in-focus point-like object is calculated under the Gaussian approximation for (a) the CCD/CMOS data model and (b) the EMCCD data model, and shown as a function of the effective pixel size of the underlying image. The lower bounds are calculated with parameters as specified in Fig. 18.7 for the CCD and EMCCD scenarios considered there. In (a), δ_{x_0} for the Gaussian approximation ($\diamond$) is compared with δ_{x_0} for the CCD/CMOS data model (+) from Fig. 18.7. In (b), δ_{x_0} for the Gaussian approximation ($\square$) is compared with δ_{x_0} for the EMCCD data model ($\bullet$) from Fig. 18.7.

However, also as in the case of the Poisson data model, they should be used with care as they are only appropriate under certain conditions.

18.8.1 Gaussian approximations for the CCD/CMOS and EMCCD data models

In Sections 15.2.3.1 and 15.2.4.2, we presented the Gaussian approximations for the CCD/CMOS and EMCCD data models, both of which are appropriate when the photon signal levels in the pixels of the image are high. Here we use the Fisher information matrices for these approximations (Table 17.1) to compute lower bounds in the context of the 2D point source localization problem, and compare the obtained bounds with those calculated for the CCD/CMOS and EMCCD data models.

We first look at the lower bound δ_{x_0} calculated using the Gaussian approximation for the CCD/CMOS data model for the same settings used for the CCD/CMOS data scenario in Fig. 18.7. It is plotted in Fig. 18.10(a) together with δ_{x_0} for the CCD/CMOS data model, which is reproduced from Fig. 18.7. Comparing the two curves, we see that their values match well at the largest effective pixel sizes shown, but diverge with decreasing effective pixel size.

An analogous result is seen in Fig. 18.10(b) for δ_{x_0} calculated using the Gaussian approximation for the EMCCD data model for the same settings used for the EMCCD data scenario in Fig. 18.7. Compared with δ_{x_0} for the EMCCD data model, which is replicated from Fig. 18.7, we see that δ_{x_0} for the Gaussian approximation approximates well the lower bounds for the EMCCD data model at the largest effective pixel sizes considered.

As large effective pixel sizes correspond to relatively large mean pixel photon counts, these results are consistent with the Gaussian approximations being predicated on relatively large photon signals having been detected in the pixels of the image.

18.8.2 Inverse square root approximation of the dependence on the photon count

It is sometimes stated in literature that the standard deviation of location estimation has an inverse square root dependence on the number of photons detected from the object of interest. One should be aware, however, that the relationship referred to is technically only an approximate one. The statement that the standard deviation of location estimation changes with the inverse of the square root of the detected object photon count is really only true in the special and unrealistic situation where images of the object are acquired in the absence of any extraneous noise. From the perspective of the Cramér-Rao lower bound, we saw an example of this in the fundamental data example of Section 18.5. By taking the square root of the right-hand side of the relatively general expressions of Eqs. (18.11) and (18.12), one can easily see that the lower bounds on the standard deviations for estimating the object's x_0 and y_0 coordinates improve with increasing photon count by a factor that is exactly equal to the inverse of the square root of the expected photon count $N_{photons}$. This can further be seen explicitly in the specific results of Eq. (18.17) and Eq. (18.22), which assume the object to be a point source whose image is given by the Airy and 2D Gaussian image functions, respectively.

In practical situations, background noise and detector noise will be present during image acquisition, and as a result, the inverse square root relationship will, at best, be a good approximation when the signal detected from the object is large enough to render the extraneous noise negligible in comparison. In the other extreme, it will make a rather poor approximation when the extraneous noise is a significant component of the acquired image. This is illustrated in Fig. 18.11 using δ_{x_0} for the case of a point-like object imaged with a CCD detector, in which the only parameters to be estimated are the object's positional coordinates (i.e., $\theta = (x_0, y_0)$). In the figure, δ_{x_0} is shown as a function of the expected object photon count and is compared with its approximation as per the inverse square root assumption. We see that while the approximation works well when the object photon count is within a relatively high range, it increasingly underestimates the deterioration of the bound due to the extraneous noise as the object photon count decreases. The discrepancy between the actual and approximate values of δ_{x_0} can become quite significant, as we can see from the substantially wider separation of the two curves at the lowest object photon counts shown.

18.9 Lower bound as a tool for the design of data analysis

We saw in Section 18.6 how the Fisher information-based lower bound can be used to inform the design of microscopy experiments. In particular, we saw that by computing the lower bound δ_{x_0} for different effective pixel sizes and different detector types, we can determine the detector and magnification to use for a particular experiment. In this section, we give two examples that illustrate how, in a similar way, the Fisher information-based lower bound can be a useful tool for the design of the analysis of the data generated by an experiment. Specifically, we will use δ_{x_0} to help with the selection of an appropriately sized pixel array for a localization analysis, and to investigate the performance advantage that we can gain by adding repeat images acquired of an object of interest.

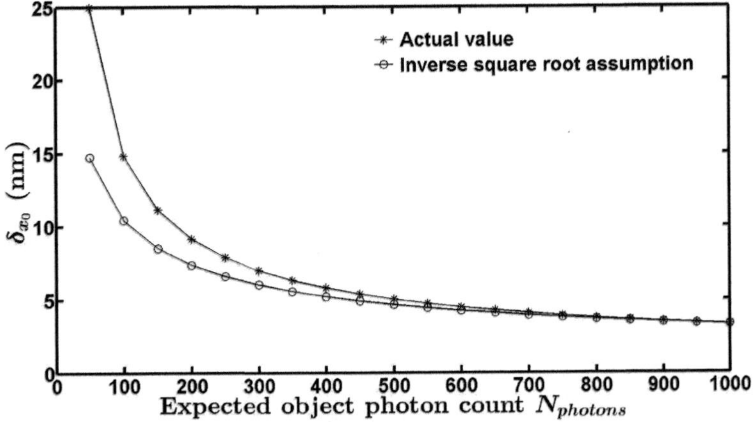

FIGURE 18.11
Inverse square root approximation of the dependence of the lower bound for location estimation on the expected photon count. The lower bound δ_{x_0} ($*$) on the standard deviation for estimating the positional coordinate x_0 of a stationary and in-focus point-like object is shown as a function of the expected number of photons $N_{photons}$ that are detected from the object over the detector plane, and is compared with δ_{x_0} ($\circ$) obtained by applying the inverse square root assumption. For a given expected photon count $N_{photons}$, the δ_{x_0} value under the inverse square root assumption is calculated by dividing the actual δ_{x_0} value for the highest expected photon count of 1000 by the square root of the photon count ratio $N_{photons}/1000$. The actual δ_{x_0} values are calculated for a CCD detector with 13 μm $\times$ 13 μm pixels. The object is positioned at $(x_0, y_0) = (916.5 \text{ nm}, 942.5 \text{ nm})$, such that using a magnification of $M = 100$, the center of its image, given by the Airy image function, is located at 0.05 pixels in the x direction and 0.25 pixels in the y direction with respect to the upper left corner of the center pixel of the 15$\times$15 pixel array considered here. Photons are detected from the object at a constant rate of $\Lambda_0 = 5000$ photons/s, and a background component is assumed that is uniformly distributed over the pixel array at a level of $\beta_k = \beta_0 = 8$ photons for each pixel k. The mean and standard deviation of the readout noise in each pixel k are $\eta_k = 0$ electrons and $\sigma_k = 1$ electron, respectively. The detected photons are of wavelength $\lambda = 520$ nm, and are collected by an objective with numerical aperture $n_a = 1.4$.

18.9.1 Choosing the region of interest

A question that arises in setting up a localization analysis or, more generally, a parameter estimation analysis, is the size of the pixel array, or region of interest, that should be used. A rule of thumb is that a significant portion of the image of the object of interest should be included in the pixel array, so that enough spatial information is provided to the estimator. Using the lower bounds δ_{x_0} and δ_{y_0} for the localization of a stationary and point-like object, we demonstrate here that choosing a sufficiently large region of interest can be important for ensuring a good standard deviation for the estimation.

In Fig. 18.12, results are shown for δ_{x_0} and δ_{y_0} calculated for regions of interest that range in size from a 3$\times$3 to a 31$\times$31 pixel array. Two scenarios are considered that differ in the size of a pixel. In both cases, we see that a 3$\times$3 pixel array is perhaps too small, and that the lower bounds can be improved significantly by using a slightly larger 5$\times$5 pixel array. This is especially the case for the smaller pixel size of 6.45 μm $\times$ 6.45 μm, as a 3$\times$3 pixel array in that scenario accounts for a substantially smaller portion of the Airy

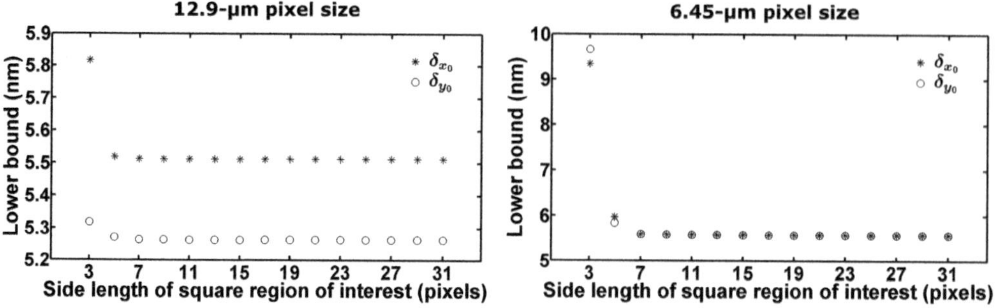

FIGURE 18.12
Dependence of location estimation lower bounds on the size of the region of interest. Lower bounds δ_{x_0} and δ_{y_0} for estimating the x_0 and y_0 coordinates of a stationary and in-focus point-like object are shown as functions of the side length of the square pixel array on which the estimation is carried out. The lower bounds are calculated under two scenarios, one in which the size of a pixel is 12.9 μm $\times$ 12.9 μm, and the other in which the size of a pixel is 6.45 μm $\times$ 6.45 μm. In both scenarios, the size of the pixel array is varied from 3×3 pixels to 31×31 pixels. The image of the object is assumed to be captured with an objective with a numerical aperture of $n_a = 1.4$ and a magnification of $M = 100$, and to be described by an Airy profile. The detector used is assumed to be a CCD detector with readout noise mean and standard deviation given by $\eta_k = 0$ electrons and $\sigma_k = 3$ electrons for each pixel k. In image space the object is located at 0.05 pixels in the x direction and 0.25 pixels in the y direction with respect to the upper left corner of the center pixel of the given pixel array. Photons are detected from the object at a constant rate of $\Lambda_0 = 10000$ photons/s, and for each pixel k, background levels of $\beta_k = \beta_0 = 8$ photons and $\beta_k = \beta_0 = 2$ photons are assumed in the 12.9-μm and 6.45-μm scenarios, respectively. The exposure time of the image is set to $t - t_0 = 0.05$ s, and the wavelength of the detected photons is assumed to be $\lambda = 520$ nm.

profile that describes the image of the object than its counterpart in the other scenario. More precisely, summing the mean number of detected object photons $\mu_{\theta,k}$ for all pixels k comprising the 3×3 pixel array and dividing the sum by the expected number of object photons $N_{photons} = 500$ that are detected over the entire detector plane, we find that while a 3×3 pixel array already accounts for 81.5% of the 500 photons when the pixel size is 12.9 μm $\times$ 12.9 μm, it only captures 52.6% of those photons when the pixel size is 6.45 μm $\times$ 6.45 μm. Increasing the size of the pixel array to 5×5 boosts these percentages to 89.0% and 79.8%, respectively, allowing the estimator to "see" significantly more of the Airy profile.

At the same time, we see from Fig. 18.12 that continuing to increase the size of the region of interest does not yield improvement of much consequence, although in the 6.45-μm-pixel scenario we do obtain a noticeable improvement by going to a 7×7 pixel array. This indicates that the larger-sized arrays do not contain significantly more photons from the object, and therefore do not provide much more spatial information to the estimator.

18.9.2 Improving estimation performance by adding images

In Section 18.10, we will give an example of object localization that is carried out on experimental image data. In that example, we will add multiple repeat images of a stationary dye molecule and use the resulting sum image to estimate some parameters we need. The reason for using the sum image is to determine the values of those parameters with higher certainty. By calculating the lower bounds on the standard deviations for estimating the object location and other parameters, we demonstrate here that a sum image can indeed provide more information about the parameters and enable estimation with a significantly improved standard deviation.

For our illustration, we consider the molecule localization problem exactly as presented in Section 18.10. To calculate the lower bounds corresponding to a single image, we use the Fisher information matrix for the EMCCD data model (Table 17.1), with $\theta = (x_0, y_0, \alpha, \Lambda_0, \beta_0)$ and the values of these five parameters given by the estimated values reported in Section 18.10 (the values for x_0 and y_0 are given by the averages of the x_0 and y_0 estimates obtained from the data set of 301 repeat images). To calculate the lower bounds corresponding to a sum image formed from N_{im} repeat images, we use instead the Fisher information matrix for the Gaussian approximation for the EMCCD data model (Table 17.1). This approximation is appropriate as the photon signals in the pixels of the sum image are large (Section 18.8.1), and it has the advantage of being fast to compute. For the sum image the parameters are given by $\theta = (x_0, y_0, \alpha, \Lambda_{0,sum}, \beta_{0,sum})$, where the values for x_0, y_0, and α are the same as those used in the case of a single image, while the values for $\Lambda_{0,sum}$ and $\beta_{0,sum}$ are equal to $\Lambda_0 \times N_{im}$ and $\beta_0 \times N_{im}$, respectively. Additionally, due to the adding of readout noise from its constituent repeat images, for the sum image the readout noise standard deviation is given by $\sigma_{k,sum} = \sqrt{\sigma_k^2 \times N_{im}}$ for each pixel k.

The lower bounds calculated for all five parameters of interest are shown in Fig. 18.13, each plotted as a function of the number of repeat images used to form the sum image. For each parameter, we see that compared with a single image, a sum image formed by adding just 10 repeat images already provides a substantial improvement in terms of the lower bound on the standard deviation for its estimation. As more repeat images are used to form the sum image, each lower bound continues to improve, though the rate of improvement does drop significantly. Since the values of $\Lambda_{0,sum}$ and $\beta_{0,sum}$ change with the number of images summed, their respective lower bounds are shown as a percentage of their values.

The advantage gained with the use of a sum image can be understood from the fact that we are exploiting information contained in not one but multiple images about the parameters of interest. Note, however, that while this is conceptually similar to the use of a set of images as discussed in Section 17.5, the Fisher information matrix is not calculated the same way. As here we are adding the multiple images and using the resulting sum image rather than keeping the images separate but using them simultaneously, the Fisher information matrix is not the sum of the Fisher information matrices for the individual images. Instead, as described above, we have calculated the matrix the same way that we calculate it for a regular image, but simply with appropriately scaled-up values for the photon detection rate, background level, and readout noise standard deviation. The scaling-up of these values is equivalent to modeling the data in the pixels of the sum image as sum random variables obtained by adding the random variables that model the data in the corresponding pixels of the constituent images. Note that scenarios do exist in which the Fisher information matrix for a set of images is the same as that for the sum image formed by adding the images. For an example, see Exercise 18.7.

18.10 Example: single molecule localization from experimentally acquired images

To close out this chapter, we present the analysis of a location estimation problem as it arises in single molecule microscopy. Figure 18.14 shows a unit-converted version of the image of

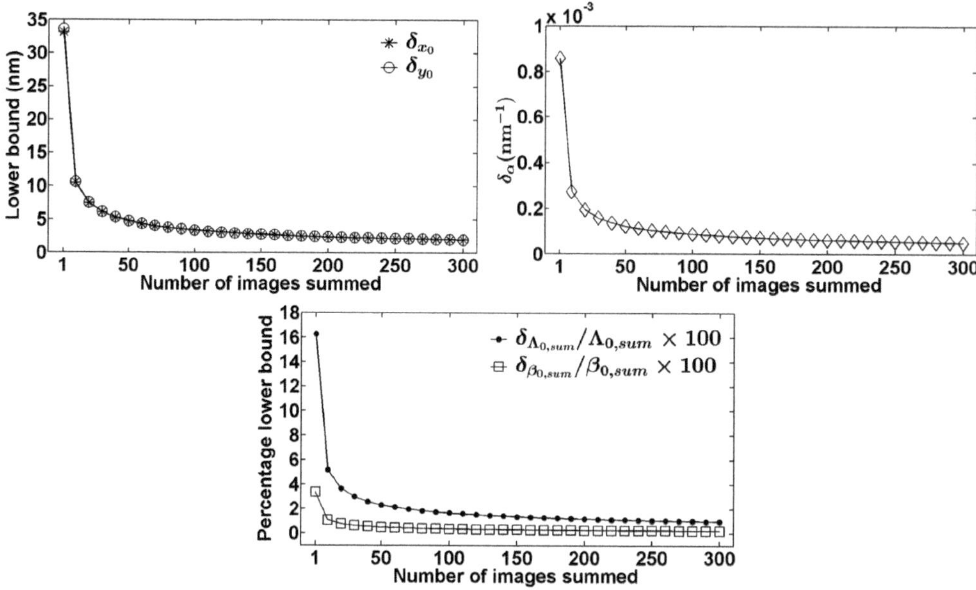

FIGURE 18.13
Dependence of lower bounds on the number of images used to form the sum image on which parameter estimation is carried out. Lower bounds on the standard deviation are shown, as functions of the number of images that form the sum image, for the estimation of the positional coordinates x_0 and y_0 of a stationary and in-focus point-like object, the width parameter α of the Airy profile that describes the image of the object, the object photon detection rate $\Lambda_{0,sum}$, and the mean per-pixel background photon count $\beta_{0,sum}$. The bounds for $\Lambda_{0,sum}$ and $\beta_{0,sum}$ are shown as percentages of their respective parameter values. The bounds are calculated for an EMCCD detector with a pixel size of 16 μm $\times$ 16 μm and an EM gain setting of $g = 950$, and the region of interest considered is a 13×13 pixel array. The object is located at $x_0 = 1635.26$ nm and $y_0 = 1529.01$ nm, which correspond in image space to a position within the center pixel of the pixel array, given that the magnification is $M = 63$ and the origin of the coordinate system coincides with the upper left corner of the pixel array. The width parameter is set to $\alpha = 7.08\ \mu\text{m}^{-1}$, and for a single image, the object photon detection rate is set to $\Lambda_0 = 4716$ photons/s and the mean per-pixel background photon count is set to $\beta_0 = 11.51$ photons. The exposure time for a single image is set to $t - t_0 = 0.04$ s. Assuming that the object and background photon detection processes remain stable over time, for a sum image formed from N_{im} constituent images, the parameters $\Lambda_{0,sum}$ and $\beta_{0,sum}$ are respectively given by $\Lambda_0 \times N_{im}$ and $\beta_0 \times N_{im}$. For each pixel k, the readout noise standard deviation is set to $\sigma_k = 40.9$ electrons for a single image, and $\sigma_{k,sum} = \sqrt{\sigma_k^2 \times N_{im}}$ for a sum image obtained from N_{im} single images. In all cases, the readout noise mean is set to $\eta_k = 0$ electrons for each pixel k.

a single in-focus and stationary ATTO 647N fluorescent dye molecule. From this image, we would like to determine the molecule's location using a maximum likelihood estimator. The raw image of the molecule was acquired and unit-converted as follows. The molecule was excited with a 635-nm laser, and fluorescence of wavelength $\lambda = 700$ nm was captured by an objective with a magnification of $M = 63$ and a nominal numerical aperture of $n_a = 1.4$, and detected by a grayscale EMCCD camera that recorded images with an exposure time of 40 ms. The camera had pixels of size 16 μm $\times$ 16 μm, and the region of interest shown in

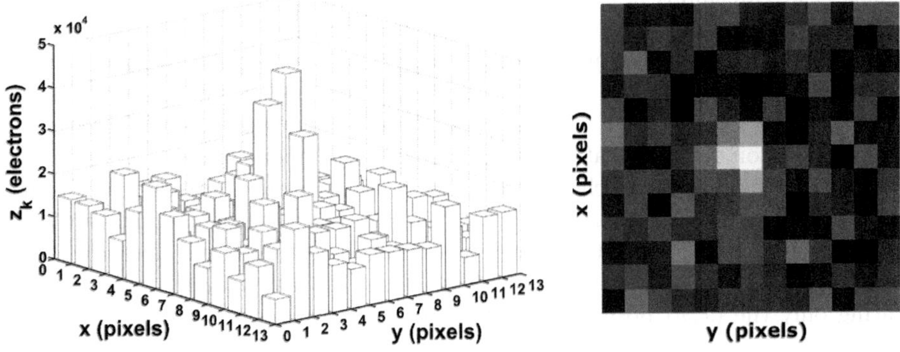

FIGURE 18.14
Bar graph and intensity representations of an image of a stationary and in-focus ATTO 647N molecule captured with an EMCCD detector.

the figure (and to be used for the localization of the dye molecule) is a 13×13 array of these pixels. The camera was operated in a readout mode that produced 14-bit values, in units of DU, in the image pixels. According to the camera's spec sheet, the electron-count-to-DU conversion factor associated with this readout mode is 10.8 electrons per DU. The data shown in the figure is in units of electrons, and was obtained by converting the acquired pixel values from units of DU to units of electrons using this factor. This conversion is necessary as our data modeling, which is required for the localization analysis, assumes units of photons or, equivalently, electrons. Prior to the conversion, a constant camera offset of 98.72 DU, determined as the mean of the readout noise using the method of Section 12.7.2.3, was subtracted from each pixel.

18.10.1 Choice of data model based on the detector used

As the image of Fig. 18.14 was acquired using an EMCCD detector, we choose to describe the data using the high gain approximation for the EMCCD data model (Section 15.2.4.1). More specifically, for the location estimation, we will maximize the log-likelihood function $\mathcal{L}$ of Eq. (18.1) with the $N_p = 169$ measured electron counts from the 13×13 pixel array, and with the log-likelihood function $\ln p_{\theta,k}$ for the kth pixel, $k = 1, \ldots, 169$, as given in Table 16.1 for the high gain approximation. The choice of the high gain approximation, which enables faster computation than the EMCCD data model, is justified by the fact that a high EM gain of $g = 950$ was used to acquire the image. In the expression for $\ln p_{\theta,k}$, we set, for $k = 1, \ldots, 169$, the readout noise mean to $\eta_k = 0$ electrons because the data $z_1, \ldots, z_{169}$ are already the offset-subtracted version of the original pixel values. Also for $k = 1, \ldots, 169$, we set the readout noise standard deviation σ_k to 40.9 electrons, the value that was obtained along with the offset using the method of Section 12.7.2.3.

18.10.2 Modeling the image of the molecule and the background component

What remains to be specified in the expression for $\ln p_{\theta,k}$ is a model for the acquired image as defined through the mean photon counts in its pixels, i.e., through $\nu_{\theta,k}$, $k = 1, \ldots, 169$. Here we have a dye molecule that is stationary and in focus. Furthermore, as for our purposes we need only know the expected number of photons that are detected from the molecule in a given image, we can simply assume that the photons emitted by the molecule were detected at a constant rate. In terms of the background component in the image, we will assume a uniform distribution of background photons based on the observation of an approximately constant background level in our 13×13 pixel array. For this description of the image, $\nu_{\theta,k}$, $k = 1, \ldots, 169$, is as given by Eq. (18.24), in which f_θ is the photon distribution profile, Λ_0 is the constant photon detection rate, and β_0 is the per-pixel mean of the uniform background. As the molecule is in focus, we choose to model its image using the Airy image function, so that f_θ is given by Eq. (18.4). Lastly, in Eq. (18.24), the region of integration C_k is simply defined by the x and y coordinates of the corners of the kth pixel, and the term $t - t_0$ equals 0.04 s as the exposure time for the image was 40 ms. In this example, we are interested in estimating only the (x, y) positional coordinates of the molecule, and we therefore define $\theta = (x_0, y_0) \in \mathbb{R}^2$.

Before continuing with our example, we first state a couple of assumptions that we will utilize shortly. Based on repeat acquisitions, the photon count due to the dye molecule is observed to have remained relatively stable over time. We will therefore assume the same constant rate of photon detection Λ_0 for all repeat images of the molecule. Similarly, the background level is observed to have remained relatively stable throughout the acquisition of repeat images. We will therefore also assume the same per-pixel background level of β_0 for all repeat images of the molecule.

18.10.3 Determining the "known" parameters

As we are only estimating x_0 and y_0, the photon detection rate Λ_0, the mean per-pixel background photon count β_0, and the parameter α in Eq. (18.4) which specifies the width of the Airy profile are considered "known" parameters for which we need to provide values. Although we could estimate these quantities together with x_0 and y_0, as we pointed out in Section 18.4 there are disadvantages to doing so. For Λ_0 and β_0, we could use simple but crude estimates of their values. For example, the parameter β_0 could be calculated by taking the average of our 13×13 pixel array's edge pixels, which contain mostly background noise, and dividing by 950 to correct for the EM gain. The parameter Λ_0 could then be obtained by determining what remains in the pixels after "removal" of the background component. Specifically, it could be computed by adding all pixels, dividing the sum by 950 to correct for the EM gain, subtracting 169 times the estimated value of β_0 from the gain-adjusted value to remove the background contribution, and dividing the resulting estimate of the molecule photon count by 0.04 s to obtain the desired rate of photon detection. Crude estimates obtained with this approach from repeat images of the molecule were in fact used to assess the stability of the molecule and background photon counts noted in the prior subsection.

For the parameter α, we could use its theoretical value of $\frac{2\pi n_a}{\lambda}$, but this gives a poor estimate of its true value and the resulting Airy profile will not match well with the image of the molecule in terms of the width. This is due to the fact that the actual numerical aperture of the objective used here is significantly smaller than the nominal numerical aperture of 1.4.

Since Λ_0, β_0, and α are taken to be "known" parameters, in principle it is best to determine their values with a high level of certainty. Instead of using crude estimates of their values, we determined their values by carrying out a separate maximum likelihood estimation. Specifically, we performed this estimation on a sum image formed by adding 301 repeat images of the stationary ATTO 647N molecule (see Section 18.9.2 for an illustration of the advantage of estimating parameters from a sum image). In this estimation, we used the Gaussian approximation for the EMCCD data model, which was appropriate given the large signals in the pixels of the sum image (Note 18.6). Specifically, for $k = 1, \ldots, 169$, the log-likelihood function $\ln p_{\theta,k}$ for the kth pixel of the sum image was given by the expression shown in Table 16.1 for the Gaussian approximation for the EMCCD data model, with the mean photon count $\nu_{\theta,k}$ in the kth image pixel replaced by $\nu_{\theta,k,sum}$ and the readout noise mean and standard deviation η_k and σ_k replaced by $\eta_{k,sum}$ and $\sigma_{k,sum}$. As the summing of repeat images changes only the amount and not the intrinsic description of the data, the definition of $\nu_{\theta,k,sum}$ is identical to the definition of $\nu_{\theta,k}$ as described above, except that Λ_0 and β_0 are replaced by $\Lambda_{0,sum}$ and $\beta_{0,sum}$, the photon detection rate and background level for the sum image. The readout noise mean $\eta_{k,sum}$ is the sum of the readout noise means for the individual images, and was accordingly set to 0 electrons. The readout noise standard deviation was set to $\sigma_{k,sum} = \sqrt{40.9^2 \times 301}$ electrons, as the variance of the sum of independent random variables is the sum of their individual variances. Using $\ln p_{\theta,k}$ as described, we maximized the log-likelihood function $\mathcal{L}$ of Eq. (18.1) (using a standard numerical optimization method) with respect to $\Lambda_{0,sum}$, $\beta_{0,sum}$, and α, as well as the positional coordinates x_0 and y_0 of the molecule, as their values were also unknown. However, while they were simultaneously estimated with the three parameters of interest, the estimated values of x_0 and y_0 were discarded as our ultimate goal is to determine x_0 and y_0 from an individual image of the molecule rather than from the sum image.

The maximum likelihood estimation on the sum image produced the parameter values $\alpha = 7.08\ \mu\text{m}^{-1}$, $\Lambda_0 = 4716$ photons/s, and $\beta_0 = 11.51$ photons, where Λ_0 and β_0, which pertain to an individual acquired image rather than the sum image, were obtained, based on the aforementioned stability of the molecule and background photon counts over time, by simply dividing the estimated values of $\Lambda_{0,sum}$ and $\beta_{0,sum}$, respectively, by 301.

18.10.4 Location estimates

Now that we have fully specified our log-likelihood function for an individual image of the ATTO 647N molecule, we take the measured electron counts $z_1, \ldots, z_{169}$ from the image shown in Fig. 18.14, and we maximize the log-likelihood function with respect to $\theta = (x_0, y_0)$ using a standard numerical optimization routine. The resulting maximum likelihood estimates of x_0 and y_0 are 1652.33 nm and 1523.61 nm, respectively, assuming that the origin of the coordinate system coincides with the upper left corner of the 13×13 pixel array. A drift-corrected version (Section 18.10.6) of this result is plotted in Fig. 18.15(a), along with the drift-corrected estimates obtained from the other 300 images of the data set which we used for the sum image analysis in Section 18.10.3. In Fig. 18.15(b), comparisons are given of the image from Fig. 18.14 and the pixelated Airy profile that is fitted to it through the maximization of the log-likelihood function. More precisely, the pixelated Airy profile comprises $\nu_{\theta,k}$, $k = 1, \ldots, 169$, as described above in this example, computed using the obtained maximum likelihood estimates $\hat{x}_0 = 1652.33$ nm and $\hat{y}_0 = 1523.61$ nm, and scaled up by an EM gain of 950. The two plots of Fig. 18.15(b) show the overlay of the middle row and column of the 13×13-pixel image with the middle row and column of the pixelated Airy profile.

18.10.5 Initial values

For both the estimation on the sum image and the estimation on an individual acquired image, we should note that initial values for the unknown parameters were required by the numerical optimization routines used. An initial value should be reasonably close to the true value of the parameter, and can often be obtained in a simple way. For example, for the estimation on an individual image, we chose the initial values for x_0 and y_0 to coincide with the center of the 13×13 pixel array. This is because, by the location of the peak of the image profile of the dye molecule (Fig. 18.14), we know that the molecule is located close to the center of the pixel array. Likewise, crude estimates for the photon detection

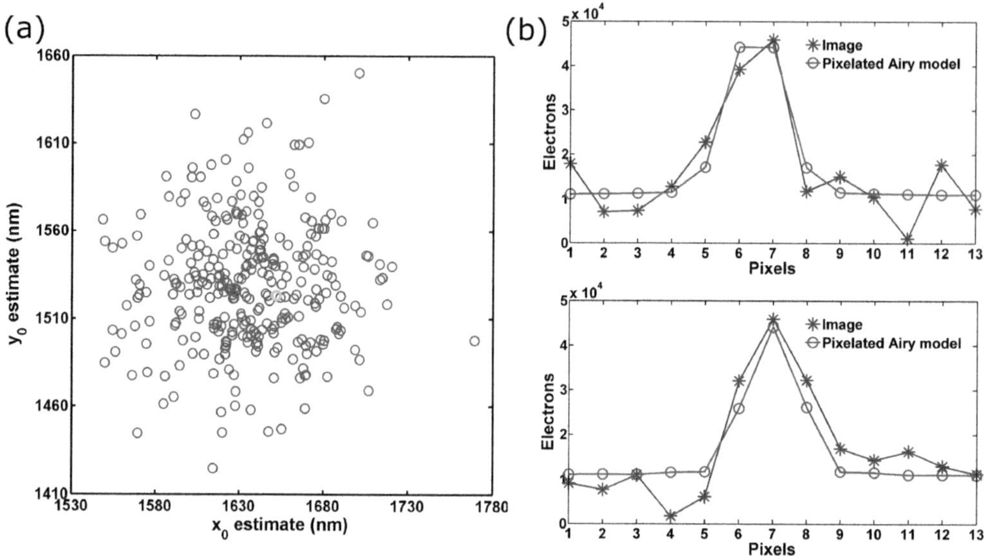

FIGURE 18.15
Maximum likelihood localization of a stationary and in-focus ATTO 647N molecule. (a) Drift-corrected estimates of the x_0 and y_0 coordinates of the molecule from 301 repeat images recorded with an EMCCD detector. The point corresponding to the estimates from the image shown in Fig. 18.14 is highlighted in green. (b) Overlay of the middle row (top) and column (bottom) of the image from Fig. 18.14 with the corresponding row and column of the pixelated Airy profile that has been fitted to the image through the maximum likelihood localization.

rate and background level, obtained using a simple approach similar to the one described in Section 18.10.3, were used as initial values for $\Lambda_{0,sum}$ and $\beta_{0,sum}$ in the estimation on the sum image.

18.10.6 Assessing the standard deviation of the localization

By carrying out maximum likelihood localization on each of the 301 repeat images of the ATTO 647N molecule used to form the sum image in Section 18.10.3, and after correcting for lateral drift of the sample during the image acquisition, we obtain, for the x_0 coordinate, an average value of 1635.26 nm and a standard deviation of 36.52 nm, and for the y_0 coordinate, an average value of 1529.01 nm and a standard deviation of 35.75 nm. (The 301 pairs of drift-corrected estimates are shown in Fig. 18.15(a).) Lateral drift refers to movement of the sample in the xy-plane with respect to the optics of the microscope during image acquisition. Lateral drift can have a variety of causes, including temperature fluctuations in the environment of, and within, the imaging setup, and vibrations due to the motion of components of the imaging setup and the operation of equipment in the vicinity of the setup. To correct for drift in our analysis, a relatively simple method was used by which drift-corrected estimates are obtained by subtracting from the original estimates a line that best fits the original estimates. This method was applied independently to the set of 301 x_0 estimates and the set of 301 y_0 estimates returned by the maximum likelihood estimator. Note that due to a multitude of factors, drift can also occur in the axial direction and cause shifts in the plane of focus during imaging. In our current example, however, visual inspection of the data set suggests that a negligible amount of axial drift occurred during acquisition.

Given the obtained statistics, a question that arises naturally is how good the standard deviations are. We can answer this question by computing, based on the given data, the Cramér-Rao lower bounds on the variances for estimating x_0 and y_0, and comparing the square root of these bounds with the respective standard deviations. Closeness to the square root of the bounds, which we denote by δ_{x_0} and δ_{y_0}, will indicate that the standard deviations are close to the best values that can be expected. Importantly, whatever one may designate as close enough or not close enough, the difference between a standard deviation and its corresponding bound will indicate the amount by which the standard deviation can potentially be improved. The bounds for our current example can be calculated using the Fisher information matrix shown in Table 17.1 for the EMCCD data model. Specifically, we do so by setting $\theta = (x_0, y_0)$ and the EM gain g and the readout noise mean η_k and standard deviation σ_k, $k = 1, \ldots, 169$, to the values given in the example, and by computing $\nu_{\theta,k}$, $k = 1, \ldots, 169$, and its partial derivatives with respect to θ, according to the descriptions above. For the values of x_0 and y_0, which are required for the evaluation of $\nu_{\theta,k}$, $k = 1, \ldots, 169$, and its partial derivatives, we use the average of the drift-corrected maximum likelihood estimates reported above. The bounds δ_{x_0} and δ_{y_0} computed in this way are 33.45 nm and 33.91 nm, respectively. The standard deviations we have obtained are therefore larger than their respective lower bounds by less than 10%.

19

Localizing Objects and Single Molecules in Three Dimensions

In the prior chapter, we investigated the problem of estimating the (x, y) location of an object in the object plane. We now turn to the more general problem of estimating the location of an object in three dimensions or, more particularly, along the optical axis. This problem presents particular challenges due to the characteristics of image formation in a 3D context.

19.1 Parameter estimation for object localization in three dimensions

The formulation of the problem of estimating the 3D location of an object is essentially identical to that discussed in the context of 2D localization, differing only in the model that is used to describe the image of the object. For example, if we have images of a molecule whose (x, y, z) position is not known and needs to be estimated, we would proceed in effectively the same manner as described in Sections 18.1 and 18.2, but using an image function corresponding to a 3D point spread function that appropriately models the image of the molecule. Such an image function would be dependent on the axial location of the molecule, and therefore allow its estimation.

Just as in the 2D scenario, in a 3D localization problem we are given an N_p-pixel image of the object of interest, with measured values $z_1, \ldots, z_{N_p}$. From this image, we likewise want to determine the location of the object, but here the parameter θ will include the axial coordinate of the object. As before we would like to find the maximum likelihood estimate of θ, denoted by $\hat{\theta}_{MLE}$. The general expression for $\hat{\theta}_{MLE}$ is as given in Eq. (18.1), which shows that it is nothing more than the value of θ that maximizes the sum of the log-likelihood functions $\ln p_{\theta,k}$, $k = 1, \ldots, N_p$, for the values measured in the N_p pixels of the image. The function $p_{\theta,k}$ for the kth pixel depends on the mean photon count $\nu_{\theta,k}$ in the pixel, and as can be seen in Eq. (18.2), $\nu_{\theta,k}$ is itself a function of the detection rate Λ_θ and spatial distribution profile $f_{\theta,\tau}$ of the photons that form the image.

What differentiates the 3D localization problem from its 2D analogue is the fact that in the 3D case the image that we obtain of the object is dependent on the object's positional coordinate in the axial dimension. As mentioned above, this is expressed mathematically through an image function that depends on the object's z coordinate, an example of which is given in Eq. (15.11). This means that in the 3D case the photon distribution profile, for an imaging system that can be assumed to be translation-invariant, is given by

$$f_{\theta,\tau}(x, y) = \frac{1}{M^2} q_{z_{0,\tau}}\left(\frac{x}{M} - x_{0,\tau}, \frac{y}{M} - y_{0,\tau}\right), \quad (x, y) \in \mathbb{R}^2, \quad \tau \geq t_0. \tag{19.1}$$

Compared to the general photon distribution profile of Eq. (15.8), here we have simply replaced the image function q with an image function $q_{z_{0,\tau}}$ that depends on $z_{0,\tau}$, the axial coordinate of the point source at time τ. With this modification, and provided that an appropriate expression for $q_{z_{0,\tau}}$ is used, the maximum likelihood estimation of the 3D location of an object can be carried out as detailed in Section 18.2, with θ defined to include the object's axial location.

As the formulation and implementation of 3D localization are essentially no different from what has been presented for the 2D problem, in this chapter we will concentrate on analyzing the Cramér-Rao lower bound in the 3D context. However, we will present and discuss a few examples of 3D localization carried out on simulated image data using maximum likelihood estimation. A major topic of this chapter will be the optical microscope's poor depth discrimination capability and how multifocal plane microscopy and an associated 3D localization algorithm can be used to overcome it.

19.2 Cramér-Rao lower bound for 3D location estimation — fundamental data

We consider here the localization of a possibly out-of-focus object, which entails the estimation of the object's (x, y, z) location. Just as we did in Chapter 18, we will assume here a stationary object. As in the case of the 2D localization problem, for the 3D case the Fisher information matrix under the fundamental data model, assuming that the image is captured over the time interval $[t_0, t]$ by an infinitely large detector $C = \mathbb{R}^2$, and that photons are detected from the object at a constant rate $\Lambda_0 > 0$, is given by Eq. (18.7). Here, however, to account for the fact that the object may be out of focus, the photon distribution profile is given by

$$f_\theta(x, y) = \frac{1}{M^2} q_{z_0}\left(\frac{x}{M} - x_0, \frac{y}{M} - y_0\right), \quad (x, y) \in \mathbb{R}^2, \tag{19.2}$$

where $z_0 \in \mathbb{R}$. This is just Eq. (19.1) with the dependence on τ removed, as here we have confined ourselves to the scenario of an immobile object.

Also as in the 2D examples of Section 18.5, we consider here the simultaneous estimation of the object's positional coordinates and the constant photon detection rate Λ_0. Therefore, in the 3D case we have $\theta = (x_0, y_0, \Lambda_0, z_0)$, which clearly shows that the 3D problem is just the 2D problem with the axial coordinate z_0 as an additional parameter that needs to be estimated. Using this fact, and by substituting the photon distribution profile f_θ of Eq. (19.2) into Eq. (18.7), we obtain the Fisher information matrix for the 3D case as

$$\begin{aligned}
\mathbf{I}(\theta) &= \frac{1}{1 - e^{-\Lambda_0(t-t_0)}} \cdot \frac{t - t_0}{\Lambda_0} \int_{\mathbb{R}^2} \frac{1}{f_\theta(x, y)} \left(\frac{\partial[\Lambda_0 f_\theta(x, y)]}{\partial\theta}\right)^T \left(\frac{\partial[\Lambda_0 f_\theta(x, y)]}{\partial\theta}\right) dxdy \\
&\quad - \frac{e^{-\Lambda_0(t-t_0)}}{\left(1 - e^{-\Lambda_0(t-t_0)}\right)^2} \int_{t_0}^{t} \left(\frac{\partial\Lambda_0}{\partial\theta}\right)^T d\tau \cdot \int_{t_0}^{t} \frac{\partial\Lambda_0}{\partial\theta} d\tau \\
&= \frac{1}{1 - e^{-\Lambda_0(t-t_0)}} \cdot \frac{t - t_0}{M^2\Lambda_0} \int_{\mathbb{R}^2} \frac{1}{q_{z_0}(u, v)} \left(\frac{\partial[\Lambda_0 q_{z_0}(u, v)]}{\partial\theta}\right)^T \left(\frac{\partial[\Lambda_0 q_{z_0}(u, v)]}{\partial\theta}\right) dxdy \\
&\quad - \begin{bmatrix} 0 & 0 & 0 & 0 \\ 0 & 0 & 0 & 0 \\ 0 & 0 & \frac{(t-t_0)^2 e^{-\Lambda_0(t-t_0)}}{\left(1-e^{-\Lambda_0(t-t_0)}\right)^2} & 0 \\ 0 & 0 & 0 & 0 \end{bmatrix}
\end{aligned}$$

$$= \begin{bmatrix} & \mathbf{I}_{2D}(\theta) & & \begin{matrix}[\mathbf{I}(\theta)]_{1,4} \\ [\mathbf{I}(\theta)]_{2,4} \\ [\mathbf{I}(\theta)]_{3,4}\end{matrix} \\ [\mathbf{I}(\theta)]_{4,1} & [\mathbf{I}(\theta)]_{4,2} & [\mathbf{I}(\theta)]_{4,3} & [\mathbf{I}(\theta)]_{4,4} \end{bmatrix},$$

where we have again used the abbreviations $u = \frac{x}{M} - x_0$ and $v = \frac{y}{M} - y_0$, and where $\mathbf{I}_{2D}(\theta)$ is the 3×3 matrix $\mathbf{I}(\theta)$ of Eq. (18.8) with q replaced by q_{z_0}, and

$$[\mathbf{I}(\theta)]_{1,4} = [\mathbf{I}(\theta)]_{4,1} = \frac{\Lambda_0(t-t_0)}{\left(1-e^{-\Lambda_0(t-t_0)}\right)M^2}\int_{\mathbb{R}^2}\frac{1}{q_{z_0}(u,v)}\frac{\partial q_{z_0}(u,v)}{\partial x_0}\frac{\partial q_{z_0}(u,v)}{\partial z_0}dxdy,$$

$$[\mathbf{I}(\theta)]_{2,4} = [\mathbf{I}(\theta)]_{4,2} = \frac{\Lambda_0(t-t_0)}{\left(1-e^{-\Lambda_0(t-t_0)}\right)M^2}\int_{\mathbb{R}^2}\frac{1}{q_{z_0}(u,v)}\frac{\partial q_{z_0}(u,v)}{\partial y_0}\frac{\partial q_{z_0}(u,v)}{\partial z_0}dxdy,$$

$$[\mathbf{I}(\theta)]_{3,4} = [\mathbf{I}(\theta)]_{4,3} = \frac{t-t_0}{\left(1-e^{-\Lambda_0(t-t_0)}\right)M^2}\int_{\mathbb{R}^2}\frac{\partial q_{z_0}(u,v)}{\partial z_0}dxdy,$$

$$[\mathbf{I}(\theta)]_{4,4} = \frac{\Lambda_0(t-t_0)}{\left(1-e^{-\Lambda_0(t-t_0)}\right)M^2}\int_{\mathbb{R}^2}\frac{1}{q_{z_0}(u,v)}\left(\frac{\partial q_{z_0}(u,v)}{\partial z_0}\right)^2 dxdy.$$

Applying a change of variables to express all integrals and partial derivatives with respect to u and v then yields

$$\mathbf{I}(\theta) = \begin{bmatrix} & \mathbf{I}_{2D}(\theta) & & \begin{matrix}[\mathbf{I}(\theta)]_{1,4} \\ [\mathbf{I}(\theta)]_{2,4} \\ [\mathbf{I}(\theta)]_{3,4}\end{matrix} \\ [\mathbf{I}(\theta)]_{4,1} & [\mathbf{I}(\theta)]_{4,2} & [\mathbf{I}(\theta)]_{4,3} & [\mathbf{I}(\theta)]_{4,4} \end{bmatrix},$$

where $\mathbf{I}_{2D}(\theta)$ is the 3×3 matrix $\mathbf{I}(\theta)$ of Eq. (18.9) with q replaced by q_{z_0}, and

$$[\mathbf{I}(\theta)]_{1,4} = [\mathbf{I}(\theta)]_{4,1} = \frac{-\Lambda_0(t-t_0)}{1-e^{-\Lambda_0(t-t_0)}}\int_{\mathbb{R}^2}\frac{1}{q_{z_0}(u,v)}\frac{\partial q_{z_0}(u,v)}{\partial u}\frac{\partial q_{z_0}(u,v)}{\partial z_0}dudv,$$

$$[\mathbf{I}(\theta)]_{2,4} = [\mathbf{I}(\theta)]_{4,2} = \frac{-\Lambda_0(t-t_0)}{1-e^{-\Lambda_0(t-t_0)}}\int_{\mathbb{R}^2}\frac{1}{q_{z_0}(u,v)}\frac{\partial q_{z_0}(u,v)}{\partial v}\frac{\partial q_{z_0}(u,v)}{\partial z_0}dudv,$$

$$[\mathbf{I}(\theta)]_{3,4} = [\mathbf{I}(\theta)]_{4,3} = \frac{t-t_0}{1-e^{-\Lambda_0(t-t_0)}}\int_{\mathbb{R}^2}\frac{\partial q_{z_0}(u,v)}{\partial z_0}dudv,$$

$$[\mathbf{I}(\theta)]_{4,4} = \frac{\Lambda_0(t-t_0)}{1-e^{-\Lambda_0(t-t_0)}}\int_{\mathbb{R}^2}\frac{1}{q_{z_0}(u,v)}\left(\frac{\partial q_{z_0}(u,v)}{\partial z_0}\right)^2 dudv.$$

Note that the expression for $[\mathbf{I}(\theta)]_{3,4}$ and $[\mathbf{I}(\theta)]_{4,3}$ evaluates to zero, as the partial derivative with respect to z_0 can be swapped with the integral with respect to u and v. By first integrating q_{z_0} over $\mathbb{R}^2$, we obtain the value 1 as per the definition of an image function. Taking the derivative of the constant 1 then yields zero.

If we next assume the image function to be symmetric, such that $q_{z_0}(x,y) = q_{z_0}(-x,y) = q_{z_0}(x,-y)$, $(x,y) \in \mathbb{R}^2$, we know that $\mathbf{I}_{2D}(\theta)$ reduces to the diagonal matrix of Eq. (18.10), with q replaced by q_{z_0}. Furthermore, since $\frac{\partial q_{z_0}(u,v)}{\partial u}$ is then an odd function with respect to u and $\frac{\partial q_{z_0}(u,v)}{\partial v}$ is an odd function with respect to v, their integrals over infinity with respect to u and v, respectively, evaluate to zero. This can be used to show that the expressions for $[\mathbf{I}(\theta)]_{1,4}$, $[\mathbf{I}(\theta)]_{4,1}$, $[\mathbf{I}(\theta)]_{2,4}$, and $[\mathbf{I}(\theta)]_{4,2}$ are zero when q_{z_0} is symmetric. Therefore, just as in the 2D case, the Fisher information matrix is a diagonal matrix when the image function is symmetric. Writing out the main diagonal elements of $\mathbf{I}_{2D}(\theta)$ explicitly, we have

$$\mathbf{I}(\theta) = \begin{bmatrix} N_{photons}\int_{\mathbb{R}^2}\frac{\left(\frac{\partial q_{z_0}(u,v)}{\partial u}\right)^2}{q_{z_0}(u,v)}dudv & 0 & 0 & 0 \\ 0 & N_{photons}\int_{\mathbb{R}^2}\frac{\left(\frac{\partial q_{z_0}(u,v)}{\partial v}\right)^2}{q_{z_0}(u,v)}dudv & 0 & 0 \\ 0 & 0 & \frac{N_{photons}}{\Lambda_0^2} - \frac{(t-t_0)^2 e^{-\Lambda_0(t-t_0)}}{\left(1-e^{-\Lambda_0(t-t_0)}\right)^2} & 0 \\ 0 & 0 & 0 & N_{photons}\int_{\mathbb{R}^2}\frac{\left(\frac{\partial q_{z_0}(u,v)}{\partial z_0}\right)^2}{q_{z_0}(u,v)}dudv \end{bmatrix}. \quad (19.3)$$

This gives us an expression for the Fisher information matrix for the 3D localization problem under the fundamental data model. Importantly, this expression is in terms of a general image function q_{z_0} and can therefore be used to analyze different situations arising from different choices of 3D image profiles. In the next subsection, we will consider the scenario in which the object to be localized is a point source whose image profile is based on the 3D point spread function that we obtained in Section 14.5.5.

19.2.1 3D localization of a point source

To obtain a more explicit expression of the Fisher information matrix, we proceed by evaluating the first, second, and fourth main diagonal entries of Eq. (19.3) using a specific representation of the image function q_{z_0}. As in the 2D localization examples of Section 18.5, we will consider the case where we have a point-like object such as a fluorescent molecule. Under diffraction theory as presented in Chapter 14, the image function q_{z_0} for such an object is symmetric, and can therefore be used in Eq. (19.3). It takes on the form exemplified by the square of the modulus of the 3D point spread function of Eq. (14.15), but here we state an expression that replaces the exponential term with a generic function W_{z_0} to allow for the modeling of the point spread function under more general situations. This image function is given by

$$\begin{aligned} q_{z_0}(x,y) &= \frac{1}{C_{z_0}}\left|\int_0^1 J_0\left(\frac{2\pi n_a}{\lambda}\sqrt{x^2+y^2}\rho\right)e^{-iW_{z_0}(\rho)}\rho d\rho\right|^2 \\ &= \frac{1}{C_{z_0}}\left(U_{z_0}^2(x,y)+V_{z_0}^2(x,y)\right), \quad (x,y)\in\mathbb{R}^2, \quad z_0\in\mathbb{R}, \end{aligned} \quad (19.4)$$

where

$$\begin{aligned} U_{z_0}(x,y) &= \int_0^1 J_0\left(\frac{2\pi n_a}{\lambda}\sqrt{x^2+y^2}\rho\right)\cos\left(W_{z_0}(\rho)\right)\rho d\rho, \\ V_{z_0}(x,y) &= \int_0^1 J_0\left(\frac{2\pi n_a}{\lambda}\sqrt{x^2+y^2}\rho\right)\sin\left(W_{z_0}(\rho)\right)\rho d\rho, \\ C_{z_0} &= \int_{\mathbb{R}^2}\left(U_{z_0}^2(x,y)+V_{z_0}^2(x,y)\right)dxdy. \end{aligned}$$

The term C_{z_0} is a normalization term which ensures that q_{z_0} integrates to 1 over $\mathbb{R}^2$. The generic phase function W_{z_0} is importantly the part of the expression that is dependent on the axial coordinate z_0 of the point source. Its precise definition differs from one model of the

3D point spread function to another, depending on how one models the specific conditions under which the image is formed. For example, the derivation of the point spread function of Eq. (14.15) assumes that the object space consists of a single medium with refractive index n_o. In that case, the phase function is given by

$$W_{z_0}(\rho) = \frac{\pi n_a^2}{\lambda}\left(\frac{z_i}{M^2 n_i} - \frac{z_0}{n_o}\right)\rho^2, \quad 0 \le \rho \le 1, \quad z_0 \in \mathbb{R}. \tag{19.5}$$

With reference to the notation of Eq. (14.15), here we have replaced the point source's axial coordinate z_o with z_0, the wave number k_{vac} with $2\pi/\lambda$, and the magnification M_0 with M. However, if one were to take into account the fact that the object space comprises media of different refractive indices (i.e., the sample, the cover glass, and the immersion medium for the objective), then one would arrive at a more complex expression for the phase function W_{z_0} (Note 14.5).

To evaluate the first main diagonal entry of the Fisher information matrix, we need to calculate the partial derivative of q_{z_0} with respect to u. For q_{z_0} of Eq. (19.4), we obtain this partial derivative as

$$\begin{aligned}
\frac{\partial q_{z_0}(u,v)}{\partial u} &= \frac{2}{C_{z_0}}\left(U_{z_0}(u,v)\frac{\partial U_{z_0}(u,v)}{\partial u} + V_{z_0}(u,v)\frac{\partial V_{z_0}(u,v)}{\partial u}\right) \\
&= \frac{4\pi n_a}{C_{z_0}\lambda}\frac{u}{\sqrt{u^2+v^2}}\left[U_{z_0}(u,v)\left(-\int_0^1 J_1\left(\frac{2\pi n_a \rho}{\lambda}\sqrt{u^2+v^2}\right)\cos(W_{z_0}(\rho))\rho^2 d\rho\right)\right. \\
&\qquad \left. + V_{z_0}(u,v)\left(-\int_0^1 J_1\left(\frac{2\pi n_a \rho}{\lambda}\sqrt{u^2+v^2}\right)\sin(W_{z_0}(\rho))\rho^2 d\rho\right)\right],
\end{aligned}$$

where the Bessel function identity $\frac{\partial}{\partial z}\left[z^{-n}J_n(z)\right] = -z^{-n}J_{n+1}(z)$ with $n=0$ has been used.

The first main diagonal entry of $\mathbf{I}(\theta)$ of Eq. (19.3) is then given by

$$\begin{aligned}
[\mathbf{I}(\theta)]_{1,1} &= N_{photons}\int_{\mathbb{R}^2}\frac{1}{q_{z_0}(u,v)}\left(\frac{\partial q_{z_0}(u,v)}{\partial u}\right)^2 dudv \\
&= N_{photons}\cdot\left(\frac{4\pi n_a}{C_{z_0}\lambda}\right)^2\int_{\mathbb{R}^2}\frac{1}{q_{z_0}(u,v)}\cdot\frac{u^2}{u^2+v^2} \\
&\qquad \left[U_{z_0}(u,v)\left(-\int_0^1 J_1\left(\frac{2\pi n_a \rho}{\lambda}\sqrt{u^2+v^2}\right)\cos(W_{z_0}(\rho))\rho^2 d\rho\right)\right. \\
&\qquad \left. + V_{z_0}(u,v)\left(-\int_0^1 J_1\left(\frac{2\pi n_a \rho}{\lambda}\sqrt{u^2+v^2}\right)\sin(W_{z_0}(\rho))\rho^2 d\rho\right)\right]^2 dudv \\
&= N_{photons}\cdot\frac{16\pi^2 n_a^2}{\tilde{C}_{z_0}^2\lambda^2}\int_0^\infty\frac{1}{\tilde{q}_{z_0}(r)}\left(\int_0^{2\pi}\cos^2\phi\, d\phi\right) \\
&\qquad \left[\tilde{U}_{z_0}(r)\left(\int_0^1 J_1\left(\frac{2\pi n_a \rho}{\lambda}r\right)\cos(W_{z_0}(\rho))\rho^2 d\rho\right)\right. \\
&\qquad \left. + \tilde{V}_{z_0}(r)\left(\int_0^1 J_1\left(\frac{2\pi n_a \rho}{\lambda}r\right)\sin(W_{z_0}(\rho))\rho^2 d\rho\right)\right]^2 rdr \\
&= N_{photons}\cdot\frac{16\pi^3 n_a^2}{\tilde{C}_{z_0}^2\lambda^2}\int_0^\infty\frac{1}{\tilde{q}_{z_0}(r)}\left[\tilde{U}_{z_0}(r)\left(\int_0^1 J_1\left(\frac{2\pi n_a \rho}{\lambda}r\right)\cos(W_{z_0}(\rho))\rho^2 d\rho\right)\right. \\
&\qquad \left. + \tilde{V}_{z_0}(r)\left(\int_0^1 J_1\left(\frac{2\pi n_a \rho}{\lambda}r\right)\sin(W_{z_0}(\rho))\rho^2 d\rho\right)\right]^2 rdr,
\end{aligned}$$

where we have applied the polar transformation $u = r\cos\phi$ and $v = r\sin\phi$. The various functions and the normalization term in polar form are, for $r \geq 0$, given by

$$\tilde{U}_{z_0}(r) = \int_0^1 J_0\left(\frac{2\pi n_a \rho}{\lambda} r\right)\cos(W_{z_0}(\rho))\rho d\rho, \quad \tilde{V}_{z_0}(r) = \int_0^1 J_0\left(\frac{2\pi n_a \rho}{\lambda} r\right)\sin(W_{z_0}(\rho))\rho d\rho,$$

$$\tilde{q}_{z_0}(r) = \frac{1}{\tilde{C}_{z_0}}\left(\tilde{U}^2_{z_0}(r) + \tilde{V}^2_{z_0}(r)\right), \qquad \tilde{C}_{z_0} = 2\pi\int_0^\infty \left(\tilde{U}^2_{z_0}(r) + \tilde{V}^2_{z_0}(r)\right) r dr.$$

As the image function q_{z_0} is not only symmetric, but circularly symmetric, we know from the discussion of Eqs. (18.11) and (18.12) that the second main diagonal entry of the Fisher information matrix is identical to the first main diagonal entry, i.e., $[\mathbf{I}(\theta)]_{2,2} = [\mathbf{I}(\theta)]_{1,1}$.

For the fourth main diagonal entry of the Fisher information matrix, we again start by calculating the partial derivative, in this case with respect to z_0, and obtain

$$\begin{aligned}
\frac{\partial q_{z_0}(u,v)}{\partial z_0} &= \frac{1}{C_{z_0}}\left[2\left(U_{z_0}(u,v)\frac{\partial U_{z_0}(u,v)}{\partial z_0} + V_{z_0}(u,v)\frac{\partial V_{z_0}(u,v)}{\partial z_0}\right) - q_{z_0}(u,v)\cdot\frac{\partial C_{z_0}}{\partial z_0}\right] \\
&= \frac{-2}{C_{z_0}}\left[U_{z_0}(u,v)\int_0^1 J_0\left(\frac{2\pi n_a\rho}{\lambda}\sqrt{u^2+v^2}\right)\sin(W_{z_0}(\rho))\frac{\partial W_{z_0}(\rho)}{\partial z_0}\rho d\rho\right. \\
&\quad - V_{z_0}(u,v)\int_0^1 J_0\left(\frac{2\pi n_a\rho}{\lambda}\sqrt{u^2+v^2}\right)\cos(W_{z_0}(\rho))\frac{\partial W_{z_0}(\rho)}{\partial z_0}\rho d\rho \\
&\quad - q_{z_0}(u,v)\int_{\mathbb{R}^2}\left(U_{z_0}(\xi,\zeta)\int_0^1 J_0\left(\frac{2\pi n_a\rho}{\lambda}\sqrt{\xi^2+\zeta^2}\right)\sin(W_{z_0}(\rho))\frac{\partial W_{z_0}(\rho)}{\partial z_0}\rho d\rho\right. \\
&\quad \left.\left. - V_{z_0}(\xi,\zeta)\int_0^1 J_0\left(\frac{2\pi n_a\rho}{\lambda}\sqrt{\xi^2+\zeta^2}\right)\cos(W_{z_0}(\rho))\frac{\partial W_{z_0}(\rho)}{\partial z_0}\rho d\rho\right) d\xi d\zeta\right].
\end{aligned}$$

Using this expression, the fourth main diagonal entry of Eq. (19.3) is then

$$\begin{aligned}
[\mathbf{I}(\theta)]_{4,4} &= N_{photons}\int_{\mathbb{R}^2}\frac{1}{q_{z_0}(u,v)}\left(\frac{\partial q_{z_0}(u,v)}{\partial z_0}\right)^2 dudv \\
&= N_{photons}\cdot\frac{4}{C^2_{z_0}}\int_{\mathbb{R}^2}\frac{1}{q_{z_0}(u,v)} \\
&\quad \left[U_{z_0}(u,v)\int_0^1 J_0\left(\frac{2\pi n_a\rho}{\lambda}\sqrt{u^2+v^2}\right)\sin(W_{z_0}(\rho))\frac{\partial W_{z_0}(\rho)}{\partial z_0}\rho d\rho\right. \\
&\quad - V_{z_0}(u,v)\int_0^1 J_0\left(\frac{2\pi n_a\rho}{\lambda}\sqrt{u^2+v^2}\right)\cos(W_{z_0}(\rho))\frac{\partial W_{z_0}(\rho)}{\partial z_0}\rho d\rho \\
&\quad - q_{z_0}(u,v)\int_{\mathbb{R}^2}\left(U_{z_0}(\xi,\zeta)\int_0^1 J_0\left(\frac{2\pi n_a\rho}{\lambda}\sqrt{\xi^2+\zeta^2}\right)\sin(W_{z_0}(\rho))\frac{\partial W_{z_0}(\rho)}{\partial z_0}\rho d\rho\right. \\
&\quad \left.\left. - V_{z_0}(\xi,\zeta)\int_0^1 J_0\left(\frac{2\pi n_a\rho}{\lambda}\sqrt{\xi^2+\zeta^2}\right)\cos(W_{z_0}(\rho))\frac{\partial W_{z_0}(\rho)}{\partial z_0}\rho d\rho\right) d\xi d\zeta\right]^2 dudv \\
&= N_{photons}\cdot\frac{8\pi}{\tilde{C}^2_{z_0}}\int_0^\infty\frac{1}{\tilde{q}_{z_0}(r)}\left[\tilde{U}_{z_0}(r)\int_0^1 J_0\left(\frac{2\pi n_a\rho}{\lambda}r\right)\sin(W_{z_0}(\rho))\frac{\partial W_{z_0}(\rho)}{\partial z_0}\rho d\rho\right. \\
&\quad - \tilde{V}_{z_0}(r)\int_0^1 J_0\left(\frac{2\pi n_a\rho}{\lambda}r\right)\cos(W_{z_0}(\rho))\frac{\partial W_{z_0}(\rho)}{\partial z_0}\rho d\rho \\
&\quad - 2\pi\tilde{q}_{z_0}(r)\int_0^\infty\left(\tilde{U}_{z_0}(\gamma)\int_0^1 J_0\left(\frac{2\pi n_a\rho}{\lambda}\gamma\right)\sin(W_{z_0}(\rho))\frac{\partial W_{z_0}(\rho)}{\partial z_0}\rho d\rho\right. \\
&\quad \left.\left. - \tilde{V}_{z_0}(\gamma)\int_0^1 J_0\left(\frac{2\pi n_a\rho}{\lambda}\gamma\right)\cos(W_{z_0}(\rho))\frac{\partial W_{z_0}(\rho)}{\partial z_0}\rho d\rho\right)\gamma d\gamma\right]^2 r dr, \quad (19.6)
\end{aligned}$$

where we have applied the same polar transformation of u and v as in the calculation of the first main diagonal entry. A similar polar transformation has also been applied to the double integral over $\mathbb{R}^2$ inside the partial derivative expression.

The square roots of the Cramér-Rao lower bounds on the variances for estimating the positional coordinates x_0, y_0, z_0, denoted by δ_{x_0}, δ_{y_0}, δ_{z_0}, respectively, and the square root of the Cramér-Rao lower bound on the variance for estimating the photon detection rate Λ_0, denoted by δ_{Λ_0}, are thus given explicitly by

$$\begin{aligned}\delta_{x_0} = \delta_{y_0} &= \sqrt{[\mathbf{I}^{-1}(\theta)]_{1,1}} \\ &= \frac{\tilde{C}_{z_0}\lambda}{4\pi^{\frac{3}{2}} n_a \sqrt{N_{photons}}} \left\{ \int_0^\infty \frac{1}{\tilde{q}_{z_0}(r)} \left[\tilde{U}_{z_0}(r) \left(\int_0^1 J_1\left(\frac{2\pi n_a \rho}{\lambda} r\right) \cos(W_{z_0}(\rho))\rho^2 d\rho \right) \right.\right. \\ &\qquad \left.\left. + \tilde{V}_{z_0}(r) \left(\int_0^1 J_1\left(\frac{2\pi n_a \rho}{\lambda} r\right) \sin(W_{z_0}(\rho))\rho^2 d\rho \right) \right]^2 r dr \right\}^{-\frac{1}{2}}, \end{aligned} \tag{19.7}$$

$$\begin{aligned}\delta_{z_0} &= \sqrt{[\mathbf{I}^{-1}(\theta)]_{4,4}} \\ &= \frac{\tilde{C}_{z_0}}{2\sqrt{2\pi N_{photons}}} \left\{ \int_0^\infty \frac{1}{\tilde{q}_{z_0}(r)} \left[\tilde{U}_{z_0}(r) \int_0^1 J_0\left(\frac{2\pi n_a \rho}{\lambda} r\right) \sin(W_{z_0}(\rho)) \frac{\partial W_{z_0}(\rho)}{\partial z_0} \rho d\rho \right.\right. \\ &\quad - \tilde{V}_{z_0}(r) \int_0^1 J_0\left(\frac{2\pi n_a \rho}{\lambda} r\right) \cos(W_{z_0}(\rho)) \frac{\partial W_{z_0}(\rho)}{\partial z_0} \rho d\rho \\ &\quad - 2\pi \tilde{q}_{z_0}(r) \int_0^\infty \left(\tilde{U}_{z_0}(\gamma) \int_0^1 J_0\left(\frac{2\pi n_a \rho}{\lambda} \gamma\right) \sin(W_{z_0}(\rho)) \frac{\partial W_{z_0}(\rho)}{\partial z_0} \rho d\rho \right. \\ &\quad \left.\left.\left. - \tilde{V}_{z_0}(\gamma) \int_0^1 J_0\left(\frac{2\pi n_a \rho}{\lambda} \gamma\right) \cos(W_{z_0}(\rho)) \frac{\partial W_{z_0}(\rho)}{\partial z_0} \rho d\rho \right) \gamma d\gamma \right]^2 r dr \right\}^{-\frac{1}{2}}, \end{aligned} \tag{19.8}$$

$$\delta_{\Lambda_0} = \sqrt{[\mathbf{I}^{-1}(\theta)]_{3,3}} = \frac{1}{\sqrt{\frac{N_{photons}}{\Lambda_0^2} - \frac{(t-t_0)^2 e^{-\Lambda_0(t-t_0)}}{\left(1-e^{-\Lambda_0(t-t_0)}\right)^2}}}. \tag{19.9}$$

From the expressions for δ_{x_0}, δ_{y_0}, and δ_{z_0}, we again see an inverse square root dependence of the lower bounds on the expected number of detected photons $N_{photons}$, demonstrating that the collection of more data helps to improve the bounds on the standard deviations for estimating an object's positional coordinates. The lower bound δ_{Λ_0} remains the same as that for the 2D localization problem. The fact that it is unchanged is not unexpected, as the problem formulation here differs only in the description of the spatial distribution of the detected photons.

The obtained lower bounds apply generally to all point spread functions having the form of Eq. (19.4). As mentioned above, different point spread function models differ in terms of the phase function W_{z_0}, which has been left unspecified in our calculations. By using an explicit definition of W_{z_0} and the associated derivative $\frac{\partial W_{z_0}(\rho)}{\partial z_0}$ in the expressions for the lower bounds, one can apply the results to a particular point spread function model. For example, to calculate the lower bounds based on the 3D point spread function of Eq. (14.15), we would use W_{z_0} as given in Eq. (19.5). In fact, for our illustrations in this chapter, we will use this phase function under the assumption that the detector is located in the design image plane. For this specific scenario, setting $z_i = 0$ in Eq. (19.5) yields

$$W_{z_0}(\rho) = \frac{-\pi n_a^2 z_0}{n_o \lambda}\rho^2 \quad \text{and} \quad \frac{\partial W_{z_0}(\rho)}{\partial z_0} = \frac{-\pi n_a^2}{n_o \lambda}\rho^2, \quad 0 \le \rho \le 1, \quad z_0 \in \mathbb{R}. \tag{19.10}$$

For an expression of the image function q_{z_0} that is written out using this phase function, see Eq. (15.11). It should be noted that through numerical calculations, the normalization term for this particular point spread function model has been found to be well approximated by

$$C_{z_0} = \tilde{C}_{z_0} \approx \frac{\lambda^2}{4\pi n_a^2}.$$

Under this approximation, C_{z_0} does not depend on the axial coordinate z_0, and the expression for δ_{z_0} simplifies. This can be seen right from the calculation of $\frac{\partial q_{z_0}(u,v)}{\partial z_0}$, where we can set $\frac{\partial C_{z_0}}{\partial z_0} = 0$. This approximation is used in all our calculations involving this specific point spread function.

It is instructive to evaluate the expression for δ_{x_0} and δ_{y_0} in the special case of $z_0 = 0$, meaning that the point source is in focus. In this case, if the phase function W_{z_0} is that of Eq. (19.10), then $W_0(\rho) = 0$, $0 \leq \rho \leq 1$, and Eq. (19.7) can be shown to reduce to Eq. (18.17), the expression for δ_{x_0} and δ_{y_0} when the image of the point source is given by the Airy image function (Exercise 19.2). This is in fact to be expected, as the image function described here, i.e., Eq. (15.11), reduces to the Airy image function when $z_0 = 0$ (Exercise 19.1).

19.3 Cramér-Rao lower bound for 3D location estimation — practical data

As with any parameter estimation problem where the underlying image data comprises a finite array of pixel values, Cramér-Rao lower bounds in the context of the 3D object localization problem can be computed with proper utilization of the general Fisher information matrix expression of Eq. (17.21). As we saw in the fundamental data scenario above, the formulation of the 3D localization problem is identical to that of the 2D localization problem, the only essential difference being that the image function used needs to be one that can model the image of an out-of-focus object. For the 3D localization problem in the context of practical data, we can therefore proceed from the general matrix of Eq. (17.21) in the same manner as we did for the 2D localization problem in Section 18.6, making only a straightforward adjustment in the specification of the photon distribution profile. Specifically, we write the mean photon count $\nu_{\theta,k}$ in the kth pixel in terms of the constant photon detection rate $\Lambda_0 > 0$ and the photon distribution profile of Eq. (19.2). Moreover, as in the 2D localization examples, we assume, for the computations carried out here, the existence of a background component that is uniformly distributed over the N_p pixels of the image. Accordingly, we have that the mean background photon count in the kth pixel, $k = 1, \ldots, N_p$, is given by $\beta_{\theta,k} = \beta_0 > 0$. Therefore, $\nu_{\theta,k}$ is as given in Eq. (18.24), except f_θ is given by Eq. (19.2). Using $\nu_{\theta,k}$ as described, we can, as we did in the case of the 2D localization problem, utilize any of the more specific Fisher information expressions derived from Eq. (17.21) in Section 17.3 to compute Cramér-Rao lower bounds for the 3D localization problem under the various practical data models.

For our illustrations in this chapter, we assume that a CCD detector is used to acquire the image data, and will hence make use of the Fisher information matrix of Eq. (17.25). As in the case of the above illustration for the fundamental data scenario, we consider the localization of a stationary and out-of-focus point-like object whose image is described by the 3D point spread function with phase function W_{z_0} as given in Eq. (19.10), such that q_{z_0} in Eq. (19.2) is the image function of Eq. (15.11). Unlike in the above illustration, however, we assume here that the rate Λ_0 at which photons are detected from the object is known, and consider here the estimation of only the positional coordinates of the object, so that $\theta = (x_0, y_0, z_0)$.

19.4 Depth discrimination problem

We now have the tools available to investigate the estimation of the z position of an object and, in particular, a point source. In doing so, we will discover that the determination of the axial position has characteristics that are very different from the estimation of the (x, y) location.

Figure 19.1 shows δ_{z_0} of Eq. (19.8), the lower bound on the standard deviation for estimating the axial coordinate z_0 of a point source, which has been computed using the phase function of Eq. (19.10), and therefore the image function of Eq. (15.11). The plot shows how δ_{z_0} depends on the coordinate z_0 itself, and that it exhibits a behavior that may at first be surprising. We see that when the point source is located in the plane of focus $z_0 = 0$ nm, δ_{z_0} is not defined as the plot exhibits a singularity at this point. In particular, close to the in-focus position δ_{z_0} attains very large values, indicating that the z position of the point source cannot be estimated with a reasonable standard deviation. As the point source moves away from the in-focus position, however, δ_{z_0} improves and attains its lowest

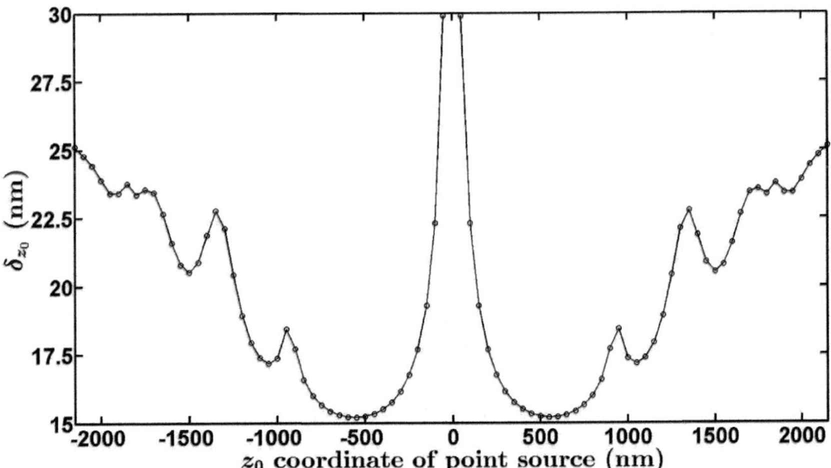

FIGURE 19.1
Dependence of axial location estimation on the axial location. The lower bound δ_{z_0} on the standard deviation for estimating the z_0 coordinate of a stationary and out-of-focus point-like object is shown as a function of z_0. The lower bound is computed under the fundamental data scenario in which the detector has an infinitely large photon detection area. The image of the object is assumed to be described by the 3D point spread function of Eq. (19.4), with the phase function as given in Eq. (19.10). The wavelength λ of the detected photons is set to 520 nm, the expected number of detected photons $N_{photons}$ is set to 500, the numerical aperture n_a of the objective is set to 1.3, and the immersion medium refractive index n_o is set to 1.515.

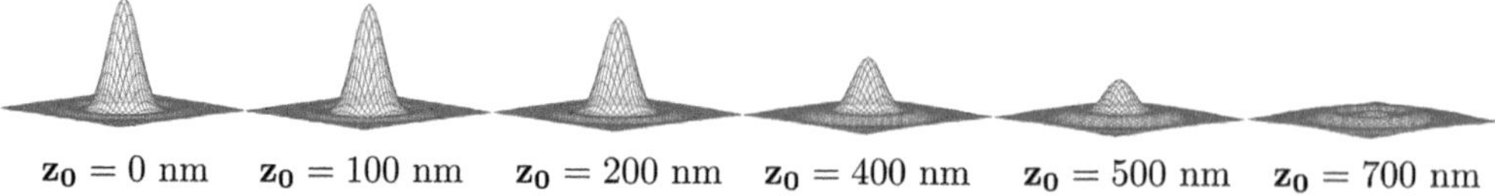

FIGURE 19.2
Dependence of a 3D point spread function on the axial position of a point source. The 3D point spread function shown here is given by the image function of Eq. (19.4), and is computed with the phase function of Eq. (19.10) for different values of the axial position z_0 of the point source. The similarity of the first three profiles shows that the image of a point source changes little when the point source is located near the focal plane ($z_0 = 0$ nm).

value at around $z_0 = 500$ nm above and below the focal plane. For larger defocus values, δ_{z_0} increases in non-monotonic fashion, exhibiting several local maxima within the axial range shown. The general increasing trend for large defocus values corresponds to the expectation that the localization task should become more difficult as the point source moves farther away from the plane of focus. What may not seem as intuitive, however, are the extremely large values of δ_{z_0} when the point source is close to the in-focus position.

We refer to the phenomenon that the axial position of a point source is essentially impossible to determine precisely when the point source is in focus, or very close to being in focus, as the *depth discrimination problem*. The depth discrimination problem is one of the main reasons that it is generally difficult to carry out precise quantitative single molecule experiments in three dimensions. Analogous difficulties arise in the tracking of vesicles or organelles in live cell microscopy experiments as it is not possible to obtain very precise estimates of the z position of a vesicle or organelle when it is in proximity to the focal plane.

It can at first be surprising that the focus position is the worst position to estimate the z location of a point source. However, remembering the discussion on the depth of field (Section 14.5.5.2) may help to make it clearer. There we discussed that around the focus the image of an object, and in particular the point spread function, does not change much. This means that there are no appreciable differences between the images of a point source when it is located at different near-focus z positions (Fig. 19.2), and that these images therefore carry little information about the point source's z position. Hence, it is very difficult to estimate the z position from these images, and this difficulty is reflected in the large values for the lower bound δ_{z_0} on the standard deviation of the estimates.

Mathematically, the singularity at $z_0 = 0$ nm can be explained as follows. We see that the phase function of Eq. (19.10) is zero when $z_0 = 0$ nm. This means that the sine of the phase function is also zero, and using this fact one can easily verify that Eq. (19.6), the entry of the Fisher information matrix corresponding to z_0, evaluates to zero. This entry is therefore not invertible, and the Cramér-Rao lower bound on the variance for estimating z_0 is accordingly undefined, when $z_0 = 0$ nm. In this scenario, the Cramér-Rao lower bound can be interpreted as being infinitely large, and hence the variance of any unbiased estimator of z_0 is infinitely large. We should point out that this is a heuristic interpretation as the Cramér-Rao lower bound is not defined when the Fisher information is zero, but it is consistent with the large values for δ_{z_0} that we see in Fig. 19.1 for small values of z_0 around $z_0 = 0$ nm.

The depth discrimination problem is also a phenomenon that a microscopist knows. When a sample is first placed on the microscope it is not straightforward to determine whether the sample is in focus. In fact it is much easier to ascertain whether the sample is out of focus. Even if a sample is in focus, a microscopist often changes the focus position to several out-of-focus positions before returning to the starting position, convinced only then that it is indeed an in-focus position.

Note that in Fig. 19.1, the symmetry of δ_{z_0} with respect to $z_0 = 0$ nm results from the symmetry about the focal plane of the particular point spread function considered here, and is not generally expected for other point spread function models. Also, it should be noted that the δ_{z_0} values shown correspond to the specific scenario in which photons of wavelength $\lambda = 520$ nm are collected by an objective with a numerical aperture of $n_a = 1.3$ and an immersion medium of refractive index $n_o = 1.515$, and in which an average of $N_{photons} = 500$ photons are detected in the image.

In Section 19.6, we will see how a MUM setup (Section 10.2.2), which can simultaneously acquire multiple images of the object of interest from different focal perspectives, can be used to overcome the depth discrimination problem of a conventional microscopy setup.

19.5 Dependence of lateral location estimation on the axial position

In Chapter 18, we studied extensively the square roots of the Cramér-Rao lower bounds pertaining to the estimation of the spatial coordinates x_0 and y_0 of a point-like object. Here we want to briefly investigate how these lower bounds, namely δ_{x_0} and δ_{y_0}, change as a function of the axial position z_0 of the object. As we can see from Fig. 19.2, as the object's distance from the in-focus position $z_0 = 0$ increases, the point spread function broadens and becomes increasingly more diffuse. As a result we expect the task of estimating x_0 and y_0 to become more difficult, and hence the values of the lower bounds δ_{x_0} and δ_{y_0} to worsen, with increasing defocus, i.e., as $|z_0|$ increases. This is indeed what we see in Fig. 19.3 for the x_0 coordinate of a point-like object imaged under the fundamental data scenario in which the detector is infinitely large. We see, however, that the increase in the value of δ_{x_0} is not monotonic. This is due to the changing appearance of the defocused image of the point source as different diffraction patterns arise with increasing defocus.

We next examine the dependence of lateral localization on an object's axial position under the practical scenario in which a CCD detector is used for the imaging. For this we follow the description given in Section 19.3 to compute the lower bound δ_{x_0}. In Fig. 19.4, the computed δ_{x_0} is plotted as a function of a point source's z_0 coordinate for two different levels of detector readout noise — one characterized by a standard deviation of $\sigma_0 = 3$ electrons, the other by a higher standard deviation of $\sigma_0 = 6$ electrons — in each pixel of the pixel array from which x_0 is estimated. In both cases, due to the broadening of the point spread function with increasing defocus, deterioration of δ_{x_0} is seen with increasing values of $|z_0|$, just as in the case of the fundamental data scenario considered above. The curve of Fig. 19.3 is in fact reproduced here for comparison, and as can be expected from what we learned in Section 18.6.1, the two δ_{x_0} curves corresponding to CCD imaging are larger in value because of the pixelation and the finiteness of the image used for the localization, and because of the presence of background noise as well as readout noise. Additionally, δ_{x_0} is shown for the corresponding Poisson data scenario, in which imaging is carried out with a hypothetical pixelated detector that adds no readout noise to the acquired data. We can see that this curve sits below the CCD curves because of the absence of readout noise, but above the fundamental curve due to the pixelation and finiteness of the image.

As one would expect, in Fig. 19.4 we see that for any given value of z_0, δ_{x_0} corresponding to the higher level of readout noise is larger than δ_{x_0} corresponding to the lower level of readout noise. What might not be immediately intuitive, however, is the widening of the

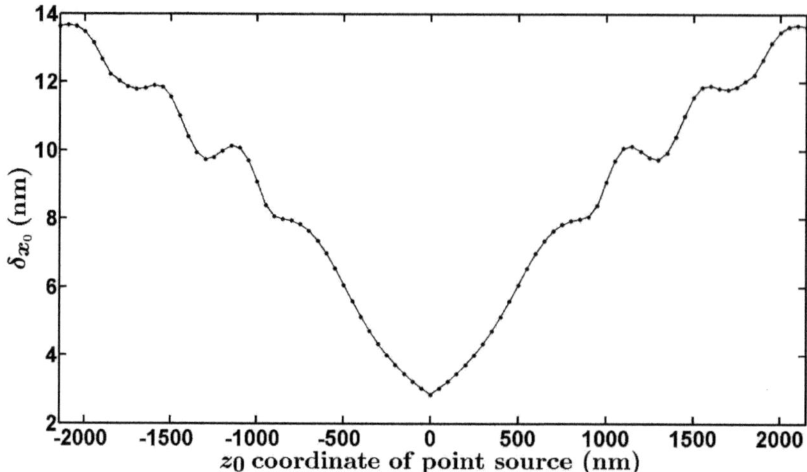

FIGURE 19.3
Dependence of lateral location estimation on the axial location. The lower bound δ_{x_0} on the standard deviation for estimating the x_0 coordinate of a stationary and out-of-focus point-like object is shown as a function of the object's axial coordinate z_0. The lower bound is computed under the fundamental data model with an infinitely large detector, using the same experimental setting assumed in Fig. 19.1.

gap between the two curves as the defocus $|z_0|$ increases. As the only difference between the two curves is the readout noise level, the widening of the gap is attributable to the photon signal in the pixels becoming increasingly weaker as the point spread function becomes more diffuse with increasing defocus (Fig. 19.2). When the point source is near focus and the point spread function is characterized by a strong lobe, the difference between the effects of the two readout noise levels is somewhat suppressed by the high photon signals in the pixels corresponding to the lobe. However, as the lobe weakens with increasing defocus, the readout noise becomes more and more of a major component of the data in each pixel. As a result the difference between the effects of the two readout noise levels becomes increasingly apparent. Indeed, depending on what one determines to be an acceptable value for δ_{x_0}, it may be that neither readout noise level is tolerable beyond a certain defocus value. In that case one needs to find a way to increase the photon signal from the object so that the signals detected in the pixels are not overwhelmed by the readout noise. Note that even though we have used the lateral location in our example here, the discussion here applies generally to the estimation of other parameters of interest such as the axial coordinate z_0.

19.6 Multifocal plane microscopy

In this section, we will revisit the technique of MUM, which we discussed in Sections 9.4.3, 10.2.2, and 11.7. In those sections, we saw that the applications of MUM include the acquisition of precise 3D views of cellular structures and the imaging of the 3D itineraries of moving objects. In contrast to conventional microscopy where a single focal plane inside the sample is imaged, MUM entails the simultaneous imaging of distinct focal planes within the

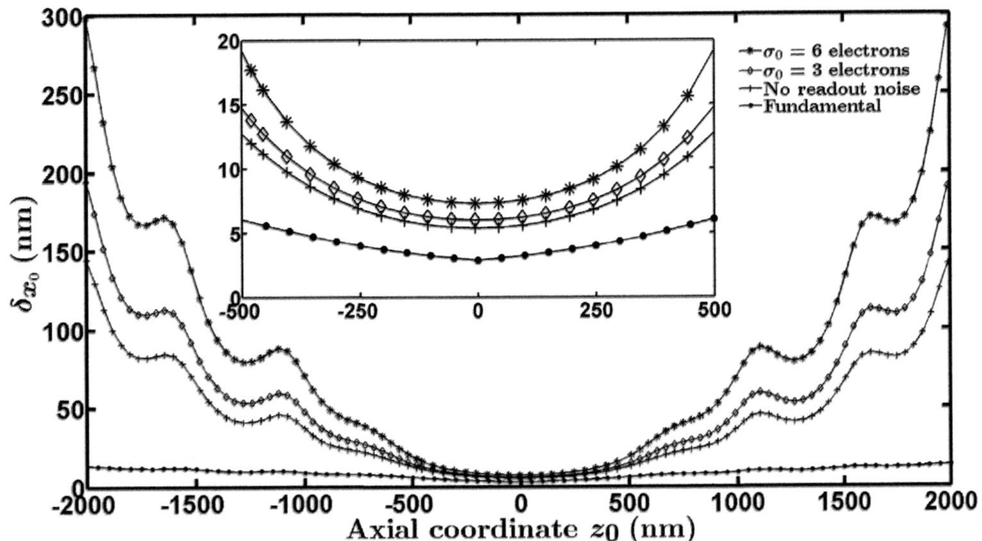

FIGURE 19.4
Effect of readout noise on lateral location estimation. For two different levels of readout noise, the lower bound δ_{x_0} on the standard deviation for estimating the x_0 coordinate of a stationary and out-of-focus point-like object is computed under the CCD/CMOS data model and shown as a function of the object's axial coordinate z_0. As references for comparison, the corresponding lower bounds computed under the Poisson and fundamental data models are also shown. The image of the object is described by the 3D point spread function of Eq. (19.4), with phase function as given in Eq. (19.10). For δ_{x_0} under the CCD/CMOS data model, an image comprises an 11×11 pixel array that is acquired by a CCD detector with pixels of size 13 μm $\times$ 13 μm. To each pixel k, the CCD detector adds readout noise with mean $\eta_k = 0$ electrons. The readout noise standard deviation is $\sigma_k = 3$ electrons in one case, and $\sigma_k = 6$ electrons in the other. Additionally, a background component is assumed that is uniformly distributed over the pixel array. The mean of this component in each pixel k is $\beta_k = \beta_0 = 10$ photons. The object is positioned such that the center of its image is located at 0.05 pixels in the x direction and 0.25 pixels in the y direction with respect to the upper left corner of the center pixel of the 11×11 pixel array. Photons are detected from the object at a constant rate of $\Lambda_0 = 5000$ photons/s over an acquisition time of $t - t_0 = 0.1$ s, such that $N_{photons} = 500$ object photons are on average detected over the detector plane. Photons of wavelength $\lambda = 520$ nm are collected by an objective with a magnification of $M = 100$, a numerical aperture of $n_a = 1.3$, and an immersion medium with refractive index $n_o = 1.515$. For δ_{x_0} under the Poisson data model, the conditions are as described, except the detector is one that adds no readout noise. For δ_{x_0} under the fundamental data model, the curve shown is that of Fig. 19.3.

sample so that at a given acquisition time point, multiple images are available that provide views of the sample from distinct focal perspectives. It is this aspect of MUM that makes it superior to conventional microscopy in terms of depth discrimination when the object of interest is in focus or near focus.

As we can see by the similarity of the first three image profiles in Fig. 19.2, with a conventional setup a point source's image (i.e., the point spread function) changes very little as a function of the point source's axial position in the near-focus region. Based on a conventionally acquired image it is therefore very difficult to determine the z position of

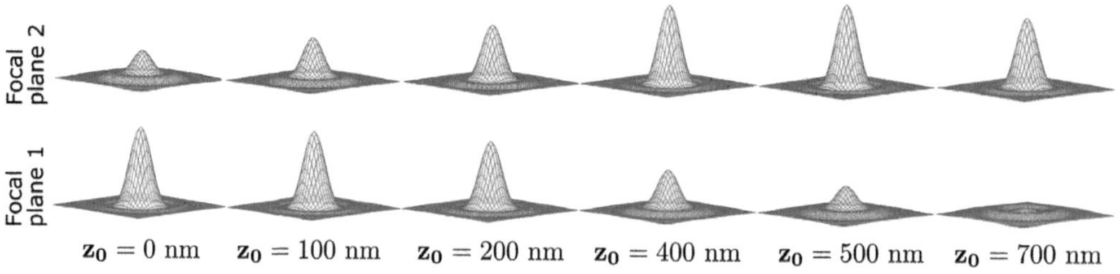

FIGURE 19.5
MUM overcomes the depth discrimination problem with the simultaneous imaging of multiple focal planes. The illustration here shows the advantage gained by using a two-plane MUM setup to image a point source. In contrast to the conventional situation depicted in Fig. 19.2, here the similarity of the first three image profiles is effectively "broken" when they are considered in conjunction with the significantly different profiles obtained from the imaging of an additional focal plane. Therefore, with MUM the near-focus positions of the point source can be determined with a high level of certainty. In this example, focal plane 1 is located at $z_0 = 0$ nm and focal plane 2 is located at $z_0 = 500$ nm.

a near-focus point source with an acceptable level of uncertainty. With a two-plane MUM setup, however, Fig. 19.5 shows that each of the three similar image profiles is now paired with an image profile that has been acquired with respect to a second focal plane. As this second focal plane is located farther from the point source than the first focal plane, the resulting image profiles for the same three near-focus axial positions differ significantly from one another. In the near-focus region the two-plane MUM modality therefore produces sufficiently different pairs of images from which the point source's z position can be estimated with a much higher level of certainty. Along with a suitable localization algorithm for analyzing the generated data, MUM can thus be used to address the depth discrimination problem.

In what follows, we will describe an algorithm for estimating the 3D location of an object from MUM data. Furthermore, from the perspective of Fisher information, we will discuss how MUM data allows us to estimate the axial location of a near-focus, or even in-focus, object with a high level of certainty.

19.6.1 Estimating the axial location from MUM data

When MUM is used to track the movement of an object in 3D space, the key task that forms the basis of the data analysis is the estimation of the object's 3D location from a set of MUM images corresponding to a time point in the captured itinerary. One method for carrying out this estimation is the *MUM localization algorithm* (MUMLA), which has been used to construct the 3D itinerary of a QD-labeled protein molecule in the examples of Sections 9.4.3, 11.7.2, and 19.6.2 below. MUMLA determines the 3D positional coordinates of the object by simultaneously fitting a model to the set of MUM images using, for example, maximum likelihood estimation or least squares estimation. In the case of an out-of-focus point-like object, for instance, an implementation of MUMLA might describe each image in the set with the same 3D point spread function model, specified appropriately in terms of parameters of interest that include the object's axial coordinate z_0, and globally fit these instances of the model to their respective images to determine an estimate for z_0 and the other parameters. In this way, information contained in views of the object from different focal perspectives is used in aggregate to estimate the 3D location of the object. It is this availability and utilization of image data acquired from one or more additional focal planes that enables the use of MUM and an algorithm like MUMLA to overcome the depth discrimination problem encountered in conventional microscopy.

For an example of MUMLA in more concrete terms, suppose we have a MUM setup that simultaneously acquires images of N_{plns} distinct focal planes within a sample. From a set of N_{plns} simultaneously acquired images, we want to estimate the positional coordinates (x_0, y_0, z_0) of a stationary fluorescent molecule using a maximum likelihood implementation of MUMLA. This requires that we have a log-likelihood function that is maximized with respect to the parameter vector θ, which we define to be (x_0, y_0, z_0). As the estimation of θ is to be done using all N_{plns} images, the log-likelihood function we need to maximize is the joint log-likelihood function $\mathcal{L}_{joint}$ for these images. Under the assumption that the simultaneous acquisitions of the distinct focal planes are stochastically independent photon detection processes, $\mathcal{L}_{joint}$ is just the sum of the log-likelihood functions for the individual images, given by

$$\begin{aligned}\mathcal{L}_{joint}&\left(\theta|z_{1,1},\ldots,z_{N_{p,1},1},z_{1,2},\ldots,z_{N_{p,2},2},\ldots,z_{1,N_{plns}},\ldots,z_{N_{p,N_{plns}},N_{plns}}\right)\\ &=\sum_{j=1}^{N_{plns}}\mathcal{L}_j\left(\theta|z_{1,j},\ldots,z_{N_{p,j},j}\right),\end{aligned}$$

where $\mathcal{L}_j$ is the log-likelihood function for the image of the jth focal plane, $z_{k,j}$ is the data in the kth pixel of the jth focal plane image, and $N_{p,j}$ is the number of pixels comprising the jth focal plane image. For the image acquired of the jth focal plane, the log-likelihood function $\mathcal{L}_j$ is given generally by

$$\mathcal{L}_j\left(\theta|z_{1,j},\ldots,z_{N_{p,j}}\right)=\sum_{k=1}^{N_{p,j}}\ln p_{\theta,k,j}(z_{k,j}),$$

which is just the log-likelihood function for a practical image as per Eq. (17.17), with $p_{\theta,k,j}$ denoting the probability distribution of the data in the kth pixel of the image of the jth focal plane. The distribution of the data in a pixel depends on the mean number of photons detected in the pixel, given generally by Eq. (18.2). For the kth pixel of the image of the jth focal plane, the mean photon count in the case of a stationary object can be written as

$$\nu_{\theta,k,j}=\int_{t_0}^{t}\Lambda_j(\tau)d\tau\cdot\int_{C_{k,j}}f_{\theta,j}(x,y)dxdy+\beta_{k,j}, \tag{19.11}$$

where Λ_j and $f_{\theta,j}$ are the object photon detection rate and distribution profile for the jth focal plane, $\beta_{k,j}$ is the mean background photon count in the pixel, and $C_{k,j}$ is the region of the detector plane corresponding to the pixel. As can be seen in Fig. 10.2, for a given focal plane the number of photons detected over an acquisition time interval is determined by the proportion of the emission that is diverted to its corresponding detector via beam splitters. Therefore, Λ_j is just a percentage of the total rate of detection of photons from the object. In terms of the photon distribution profile $f_{\theta,j}$, we know that it is of the form given by Eq. (19.2) if the imaging system can be assumed to be translation-invariant. However, we need to take care to specify the axial location of the object with respect to the jth focal plane. Further, by the result of Section 13.3.3, we need to use the lateral magnification that is associated with the jth focal plane. The profile $f_{\theta,j}$ is therefore given by

$$f_{\theta,j}(x,y)=\frac{1}{M_j^2}q_{z_0-\Delta z_{f,j}}\left(\frac{x}{M_j}-x_{0,j},\frac{y}{M_j}-y_{0,j}\right),\quad (x,y)\in\mathbb{R}^2, \tag{19.12}$$

where M_j is the lateral magnification for the jth focal plane, $\Delta z_{f,j}$ is the distance between the jth focal plane and the focal plane that is taken to be the reference focal plane, and $z_0 - \Delta z_{f,j}$ is accordingly the z coordinate of the molecule with respect to the jth focal plane. The magnification M_j can be measured, or it can be calculated using Eq. (13.14), with Δ_{i1} given by the distance between the microscope's image focal point and the position of the detector for the reference focal plane, and with δ_o given by the focal plane spacing $\Delta z_{f,j}$.

The expression for $p_{\theta,k,j}$ also depends on the detector that is used to image the jth focal plane. If it is a CCD detector, for example, then one might use the probability density function of Eq. (18.5) with appropriate values for the readout noise mean and variance. For an EMCCD detector, however, one might use the density function shown in Table 15.1 for the EMCCD data model, with parameters reflecting the readout and amplification settings used.

Having specified the log-likelihood function for the set of MUM images, the estimate of θ obtained with this implementation of MUMLA is just the maximum likelihood estimate

$$\hat{\theta}_{MLE} = (\hat{x}_0, \hat{y}_0, \hat{z}_0) = \arg\max_{\theta \in \Theta} \sum_{j=1}^{N_{plns}} \mathcal{L}_j \left(\theta | z_{1,j}, \ldots, z_{N_{p,j},j}\right).$$

19.6.2 Experimental example

An example of a two-plane MUM experiment is shown in Fig. 19.6. In part (a) of the figure, we see pairs of images of a QD-labeled IgG molecule that were simultaneously acquired from two focal planes spaced 500 nm apart in an endothelial cell. The pairs of images were recorded at different times and capture the movement of the IgG molecule over a nearly five-second time interval. From these images we can see that over this interval, the molecule moved from a near-focus position to a decidedly out-of-focus position with respect to one focal plane (i.e., the "membrane" plane), and from an obviously out-of-focus position to a near-focus position with respect to the other focal plane (i.e., the "top" plane). Based on this we can deduce that axially the IgG molecule traversed much of the space between the two focal planes over these few seconds. This conclusion is corroborated by the 3D trajectory of the molecule shown in part (b) of the figure. To construct this trajectory, the 3D location of the molecule at each available time point was estimated from the corresponding pair of MUM images using an implementation of MUMLA, and the obtained estimates were connected in order of time. In part (c) of the figure, for the same trajectory the estimated z coordinate of the molecule is plotted as a function of time. This trajectory reveals significant details regarding the movement of the IgG molecule, which may seem surprising considering that it is constructed from experimental image data.

19.6.3 Maximum likelihood localization with simulated data

To demonstrate the maximum likelihood implementation of MUMLA described in Section 19.6.1 and the advantage it provides in terms of localization performance, we present here the results of 3D localizations carried out on simulated images of stationary and out-of-focus point sources. Each of the three data sets we consider here has been simulated using the CCD/CMOS data model, with settings similar to those used in Fig. 19.4. The first data set provides an example of the scenario in which a point source imaged using a conventional microscope setup is located far enough from the focal plane to be axially localized with a reasonably good standard deviation. In this data set, the point source is located at $z_0 = 350$ nm above the focal plane, and Table 19.1 shows that in this case the lower bound δ_{z_0} pre-

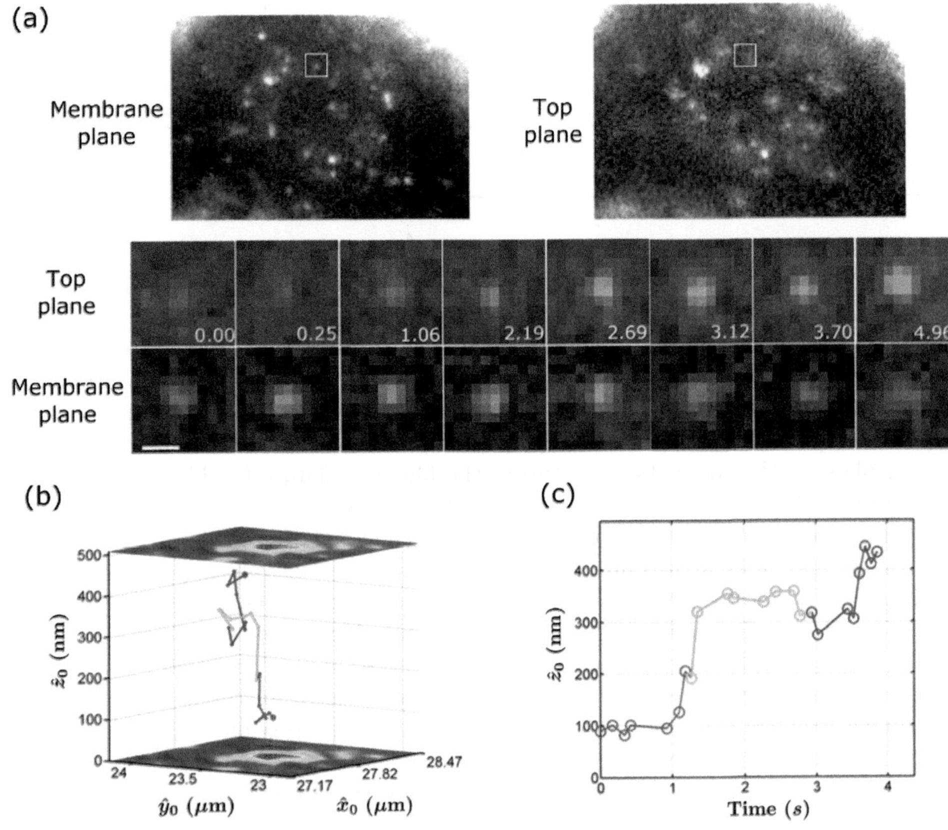

FIGURE 19.6
MUM and MUMLA for determining the trajectory of a molecule in 3D space. (a) The montage comprises pairs of images of a QD-labeled IgG molecule simultaneously acquired from two distinct focal planes inside an HMEC-1 cell. The "membrane" plane is positioned to coincide with the plasma membrane of the cell, and the "top" plane is positioned 500 nm above the membrane plane. The numbers shown in white indicate the time, in seconds, at which each pair of images were acquired relative to the first pair. The scale bar indicates a distance of 0.5 μm. The images above the montage provide a broad view of the HMEC-1 cell from each focal plane. In these images, the yellow boxes delineate the region that is shown in the images of the montage. (b) The 3D trajectory of the IgG molecule determined by estimating the molecule's 3D position (x_0, y_0, z_0) at each available time point from the corresponding pair of MUM images. The estimation of the 3D position was carried out using an implementation of MUMLA. The time information for the trajectory is encoded in color, with progression of time indicated by transition from red to green to blue. (c) Estimates of the IgG molecule's z_0 coordinate from the trajectory of (b), plotted as a function of time.

dicts that an unbiased estimator can at best determine the point source's z_0 coordinate with a standard deviation of 14.13 nm. The table further shows that the maximum likelihood estimates of z_0, obtained from the 1000 repeat images of the point source which comprise the data set, have an average value of 349.93 nm and a standard deviation of 14.16 nm, suggesting that under the conditions assumed for this data set, the maximum likelihood estimator is able to closely recover the true value of z_0 with a standard deviation that is

Imaging setup	x_0 (nm)	Mean of x_0 estimates (nm)	δ_{x_0} (nm)	SD of x_0 estimates (nm)	y_0 (nm)	Mean of y_0 estimates (nm)	δ_{y_0} (nm)	SD of y_0 estimates (nm)	z_0 (nm)	Mean of z_0 estimates (nm)	δ_{z_0} (nm)	SD of z_0 estimates (nm)
Conventional	656.50	656.31	5.10	5.20	682.50	682.60	4.98	4.91	350.00	349.93	14.13	14.16
Conventional	656.50	656.53	3.37	3.32	682.50	682.51	3.28	3.20	100.00	91.11	35.70	41.39
Two-plane MUM	656.50	656.58	4.27	4.33	682.50	682.62	4.17	4.15	100.00	100.43	18.41	18.35

TABLE 19.1
Maximum likelihood localization of an out-of-focus and stationary point source from simulated CCD image data. Results for three data sets are shown. The first two data sets simulate images acquired with a conventional microscopy setup, each consisting of 1000 repeat images of the point source. The third data set simulates images acquired with a two-plane MUM setup with a 500-nm plane spacing, and comprises 1000 pairs of repeat images of the point source. In the two-plane setup, focal plane 1 is taken to coincide with the focal plane of the conventional setup of the first two data sets. The detected photons are assumed to be divided equally between focal planes 1 and 2, and the distribution of background photons is assumed to be depth-independent. For each data set, the averages of the x_0, y_0, and z_0 estimates are reported alongside the true values of these parameters. Likewise, the standard deviations of the estimates are reported alongside the lower bounds δ_{x_0}, δ_{y_0}, and δ_{z_0} on the standard deviations for estimating x_0, y_0, and z_0. All settings and parameter values used in the simulation of the data sets are as given in Fig. 19.4, except for the value of z_0 which is indicated in the table itself, and except for the following. The acquisition time is changed to $t - t_0 = 0.3$ s, such that 1500 and 750 point source photons are on average detected over the detector plane in the conventional setup and each detector plane in the MUM setup, respectively, given the photon detection rate of $\Lambda_0 = 5000$ photons/s for the conventional setup. The mean of the uniformly distributed background component is changed to $\beta_k = \beta_0 = 30$ photons in each pixel k in the conventional setup, and accordingly, $\beta_{k,1} = \beta_{0,1} = 15$ photons for each pixel k in the image corresponding to focal plane 1 of the MUM setup. For the image corresponding to focal plane 2 of the MUM setup, $\beta_{0,2}$ is slightly larger than 15 photons as the image is scaled by the smaller magnification of $M_2 = 97.98$, determined using Eq. (13.14) with focal plane spacing $\delta_o = \Delta z_{f,2} = 500$ nm and distance $\Delta_{i1} = 160$ mm between the microscope's image focal point and the detector corresponding to focal plane 1. The center of the image of the object in the image corresponding to focal plane 2 is similarly adjusted with M_2.

very close to the lower bound δ_{z_0}. The same conclusion can be made concerning the x_0 and y_0 coordinates of the point source, which have been simultaneously estimated with z_0.

The second data set we consider here provides an illustration of the poor depth discrimination capability of a conventional microscope setup. In this data set, the point source is located substantially closer to the focal plane ($z_0 = 100$ nm), which makes the determination of its z_0 coordinate a significantly more challenging task. In Table 19.1, the increase in the difficulty of axial localization can be seen in the nontrivial worsening of the lower bound δ_{z_0} to 35.70 nm. Moreover, the maximum likelihood estimator is seen here to perform poorly, as based on the average of 1000 estimates (91.11 nm) it does not appear able to closely recover the true value of z_0. Note, however, that the estimator is able to closely recover the true values of x_0 and y_0 with standard deviations that are very comparable to the lower bounds δ_{x_0} and δ_{y_0}.

Lastly, our third data set demonstrates the MUM modality's ability to overcome the poor depth discrimination of the conventional microscopy setup. This data set assumes that the same point source from the second data set is imaged using a two-plane MUM setup with a 500-nm plane spacing, and it is analyzed using the maximum likelihood implementation of MUMLA described in Section 19.6.1. In Table 19.1, we see that the simultaneous imaging of two distinct focal planes enables the axial localization of the point source with a much improved standard deviation. Specifically, the lower bound of $\delta_{z_0} = 18.41$ nm for this third data set is nearly half of the value of δ_{z_0} for the second data set. For details on how lower bounds are calculated for a MUM setup, see Section 19.6.4. Furthermore, based on the estimates obtained from the 1000 pairs of repeat images of the point source which comprise the data set (each pair consisting of one image each of each focal plane), the maximum likelihood estimator appears able to closely recover the true value of z_0 with a standard deviation close to δ_{z_0}. As is the case with each of the other two data sets, the results here for the x_0 and y_0 coordinates suggest that this estimator performs optimally as well with respect to the lateral coordinates.

19.6.4 Overcoming the depth discrimination problem

We describe next the calculation of the Fisher information matrix for the 3D localization problem in the context of MUM, and demonstrate the advantage gained with MUM by comparing the resulting bound δ_{z_0} with that obtained for the corresponding conventional microscopy setup. We will use a two-plane MUM setup in our descriptions and illustration, but the method presented for the calculation of the Fisher information matrix can be extended to a MUM setup with any number of focal planes in straightforward fashion.

We begin by demonstrating again the depth discrimination problem from the information-theoretic perspective, but in this case for the scenario in which a CCD detector is used to acquire the image. Assuming the same point-like object from the example of Fig. 19.1, we compute δ_{z_0} as described in Section 19.3 for a range of values for the object's axial location z_0. We see in Fig. 19.7 that when the object is located away from the focal plane of a conventional microscopy setup (i.e., the focal plane located at $z_0 = 0$ nm), δ_{z_0} takes on reasonably small values. When the object position is moved towards the focal plane, however, we see that the value of δ_{z_0} increases dramatically, indicating that we lose the ability to estimate the object's z_0 coordinate with a reasonable level of certainty. This is as we saw in Fig. 19.1, and demonstrates the problem we encounter when the axial position of a near-focus object is to be determined from an image of the object acquired with a conventional microscopy setup.

Now assume that the same point-like object is imaged using a two-plane MUM setup in which the photons collected from the object are divided equally between two CCD detectors, one capturing images of focal plane 1 which coincides with the focal plane of the conventional setup, and the other capturing images of a focal plane 2 that is distinct from focal plane 1. As the photon detection processes corresponding to these two detectors can be assumed to be independent, the Fisher information matrix corresponding to two images, one acquired by each detector, is given by the sum of the Fisher information matrices corresponding to the two images individually (Section 17.5). For the image acquired with detector 1, which images focal plane 1, the Fisher information matrix $\mathbf{I}_1(\theta)$ is calculated as for the image acquired with the conventional setup, except the object and background photon detection rates are halved to reflect the equal partitioning of the detected photons between the two detectors. For the image acquired with detector 2, the Fisher information matrix $\mathbf{I}_2(\theta)$ is calculated using the same rates of photon detection as for detector 1, but with an adjusted lateral magnification, and with the object's axial position specified with respect to focal plane 2. Depending on the nature of the background component, a different background photon distribution profile may also need to be used when calculating $\mathbf{I}_2(\theta)$.

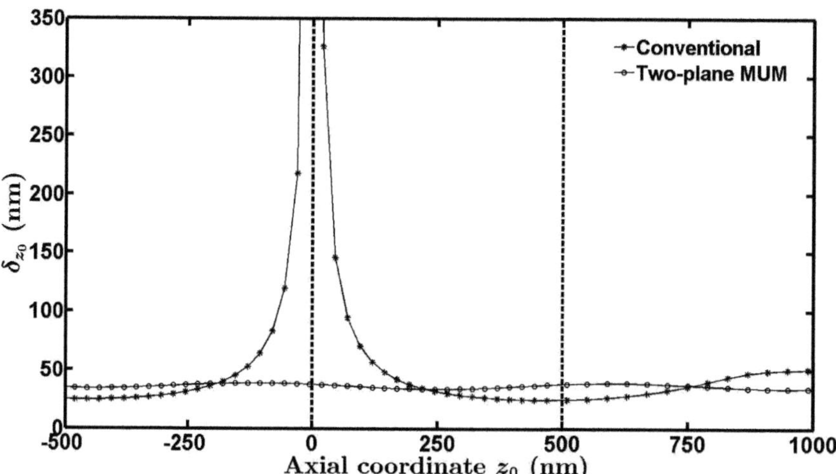

FIGURE 19.7
Overcoming the depth discrimination problem with a two-plane MUM setup. The lower bounds δ_{z_0} on the standard deviation for estimating the axial coordinate z_0 of a stationary and out-of-focus point-like object from a CCD image acquired with a conventional microscopy setup ($*$) and from a pair of CCD images simultaneously acquired with a two-plane MUM setup ($\circ$) are shown as a function of z_0. Focal planes 1 and 2 for the two-plane MUM setup, located at $z_0 = 0$ nm and $z_0 = 500$ nm, respectively, are marked with vertical dashed lines. The value of z_0 is given with respect to focal plane 1, which is also the focal plane for the conventional setup. For the conventional setup, the conditions under which δ_{z_0} is computed are as described in Fig. 19.4, with the CCD detector adding readout noise with standard deviation $\sigma_k = 3$ electrons to each pixel k. For the MUM setup, the same conditions are used, but the values for various parameters are adjusted based on the assumption that the object and background photons are divided equally between the detectors corresponding to the two focal planes. Specifically, the object photon detection rate is set to $\Lambda_0 = 2500$ photons/s per image, and the background level for the image of focal plane 1 is set to $\beta_{0,1} = 5$ photons per pixel. For the image of focal plane 2, $\beta_{0,2}$ is slightly larger as the image is scaled by the smaller magnification of $M_2 = 97.98$, calculated using Eq. (13.14) with focal plane spacing $\delta_o = \Delta z_{f,2} = 500$ nm and an assumed distance of $\Delta_{i1} = 160$ mm between the microscope's image focal point and the detector corresponding to focal plane 1. For this image, the center of the image of the object is likewise adjusted based on M_2.

More specifically, to compute $\mathbf{I}_1(\theta)$ and $\mathbf{I}_2(\theta)$, Eq. (19.11) for the mean photon count in the kth pixel is used with $j = 1$ and 2, respectively. Due to the equal splitting of the object photons between the two detectors, we have that $\Lambda_1(\tau) = \Lambda_2(\tau) = \Lambda(\tau)/2$ for $\tau \geq t_0$, where Λ is the photon detection rate for the corresponding conventional setup. Taking focal plane 1 as the reference focal plane, we have the focal plane spacings $\Delta z_{f,1} = 0$ and $\Delta z_{f,2} \neq 0$, the latter being equal to the distance of focal plane 2 from focal plane 1. Accordingly, we also have the magnifications $M_1 = M$, where M is the magnification for the corresponding conventional setup, and M_2 which can be calculated using Eq. (13.14), with $\delta_o = \Delta z_{f,2}$ and Δ_{i1} which is particular to the microscope used. Lastly, we have the background levels $\beta_{k,1} = \beta_k/2$, where β_k is the mean background photon count for the corresponding conventional setup, and $\beta_{k,2}$ which needs to be determined using magnification M_2 and the background photon distribution profile for focal plane 2.

For our example, we calculate M_2 with an assumed distance of $\Delta_{i1} = 160$ mm and an assumed focal plane spacing of $\Delta z_{f,2} = 500$ nm for our two-plane MUM setup. We also assume that the background photon distribution profile is not depth-dependent, and therefore we have the same uniform distribution for focal plane 2, but adjusted by the magnification M_2.

Calculating the two Fisher information matrices and taking the square root of entry $(3,3)$ of the inverse of their sum, we obtain the lower bound δ_{z_0} on the standard deviation with which we can estimate the axial coordinate z_0 of the object from the pair of images produced by the MUM setup. Plotted in Fig. 19.7 as a function of the object's axial location z_0, we see that this δ_{z_0} remains relatively constant throughout the range of axial locations shown. No sharp deterioration occurs near either focal plane, due to the fact that even though little information about the object's axial location is contained in the image of the focal plane closest to the object, it is compensated for by the relatively large amount of information contained in the image of the other focal plane. The MUM setup thus overcomes the depth discrimination problem, allowing a near-focus object to be localized with an acceptable standard deviation. Note that while the MUM setup does not provide a better lower bound than the conventional setup for all axial positions z_0, it has the important advantage of yielding a lower bound that is relatively constant and small over the entire axial range shown. This is a desirable property for applications such as the 3D tracking of single molecules, where one would ideally like to be able to estimate the axial location of the tracked object with approximately the same acceptable standard deviation over a large axial range.

19.6.5 Zero Fisher information and the depth discrimination problem

To more explicitly illustrate, in the context of MUM and the depth discrimination problem, how the relatively large amount of information contained in one image can compensate for the relatively small amount of information contained in another image, we return to the fundamental data model which affords us simpler mathematical expressions for analysis. Suppose we are only interested in estimating the z_0 coordinate of a point source. Then we have $\theta = z_0$, and for the image of focal plane 1, the scalar Fisher information matrix $\mathbf{I}_1(z_0)$ is simply given by the expression of Eq. (19.6). From the discussion in Section 19.4, we know that given the phase function of Eq. (19.10), $\mathbf{I}_1(z_0) = 0$ when the point source is in-focus with respect to focal plane 1, i.e., when $z_0 = 0$. This is again indicative of the depth discrimination problem, and it means that the image of focal plane 1 contains no information about the point source's z_0 coordinate. The lower bound δ_{z_0} is then infinitely large (as $\mathbf{I}_1(z_0)$ is non-invertible), and it would not be possible to determine z_0 precisely when this image is used by itself for the estimation of z_0. However, as focal plane 2 is a non-zero distance $\Delta z_{f,2}$ from focal plane 1, the point source is axially located at $z_0 - \Delta z_{f,2} = 0 - \Delta z_{f,2} = -\Delta z_{f,2}$ with respect to focal plane 2, and therefore not in-focus with respect to focal plane 2. Hence the scalar Fisher information matrix $\mathbf{I}_2(z_0)$ for the image of focal plane 2 is, unlike $\mathbf{I}_1(z_0)$, a positive quantity, indicating that this image carries some information about z_0. The scalar Fisher information matrix $\mathbf{I}_{MUM}(z_0)$ corresponding to both the image of focal plane 1 and the image of focal plane 2 is then also positive when $z_0 = 0$, as

$$\mathbf{I}_{MUM}(z_0) = \mathbf{I}_1(z_0) + \mathbf{I}_2(z_0) = 0 + \mathbf{I}_2(z_0) > 0.$$

From the perspective of Fisher information, this shows that MUM overcomes the depth discrimination problem by adding a second matrix that is invertible. Estimation using both images thus enables the precise determination of z_0 even when the point source is in-focus with respect to one of the focal planes.

For simplicity of argument in our illustration, we have considered the scenario where the point source is in-focus with respect to one of the focal planes, and therefore one of the scalar Fisher information matrices is zero. In general, neither $\mathbf{I}_1(z_0)$ nor $\mathbf{I}_2(z_0)$ is exactly zero, and the idea is that when one of these scalar matrices has a small value due to the depth discrimination problem, the other matrix compensates for that by having a sufficiently larger value. In practice, this can be achieved by choosing an appropriate spacing between the two focal planes. A spacing that is too small may mean that neither image will contain enough information about z_0 when the point source is located near both planes. A spacing that is too large may mean that when the point source is close to one plane, the other plane is so far away that its image does not contain enough information to adequately compensate for the little information contained in the image of the first plane. For an illustration of how the plane spacing can affect the performance of a two-plane MUM setup in terms of the estimation of z_0, see Section 19.6.6. We should also mention that the principle of the image of one plane compensating for the image of another plane applies more generally to MUM configurations with more than two focal planes. In the general case, $\mathbf{I}_{MUM}(z_0)$ is a sum of more than two Fisher information matrices, and the multiple spacings between the planes will need to be chosen appropriately to ensure that the depth discrimination problem at each focal plane is sufficiently overcome.

19.6.6 Experimental design: finding appropriate focal plane spacings

An important consideration in the design of a MUM setup is the positioning of the multiple focal planes within the sample. The locations of the focal planes, and therefore the distances between them, play a crucial role in determining how well one can localize a near-focus object in 3D space. In a two-plane MUM setup, for example, having too small a spacing between the two focal planes will mean that two very similar images are acquired of the object, and consequently leave the depth discrimination problem effectively unaddressed. On the other hand, a spacing that is too large will result in the second focal plane not receiving enough signal from the object to capture the necessary amount of information about its axial location. In this scenario, one also fails to adequately address the depth discrimination problem.

As an illustration, Fig. 19.8 uses the lower bound δ_{z_0} to demonstrate how the choice of the focal plane spacing in a two-plane MUM setup can affect the standard deviation with which the z_0 coordinate of a point source can be estimated. In doing so, it provides another example of how a Fisher information-based lower bound can be used to inform the design of an experiment. The figure compares δ_{z_0} as a function of the z_0 coordinate for three different plane spacings. The curve for the 500-nm spacing is the two-plane MUM curve from Fig. 19.7, but is shown here over a larger range of values for z_0. As we can see, this particular spacing yields a relatively constant and small value for δ_{z_0} over a relatively large range of z_0 values that span more than 1500 nm (i.e., from just below $z_0 = -500$ nm to just above $z_0 = 1000$ nm). With this spacing and an appropriate estimator, we can therefore expect to determine the axial location of a point source with a reasonably good standard deviation over a wide axial range.

When a smaller spacing of 250 nm is used, we see that compared to the 500-nm spacing, smaller values are obtained for δ_{z_0} over significant axial ranges both above and below the pair of planes. However, a major drawback of using this spacing is that δ_{z_0} takes on substantially larger values over the axial region between the two planes. As a result, a high level of uncertainty can be expected for the localization of a point source that is near-focus with respect to both focal planes, and therefore the 250-nm plane spacing does not fully overcome the depth discrimination problem. Lastly, for the largest spacing of 1000 nm, we see that over a range of a few hundred nanometers around the midpoint between the two planes, δ_{z_0}

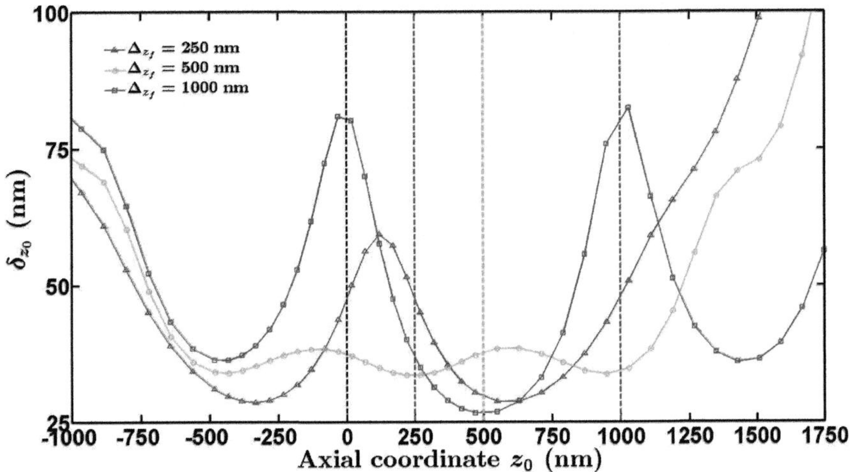

FIGURE 19.8
Effect of focal plane spacing on axial localization with a two-plane MUM setup. For three different values of the focal plane spacing Δz_f, the lower bound δ_{z_0} on the standard deviation for estimating the axial coordinate z_0 of a stationary and out-of-focus point source from a pair of CCD images simultaneously acquired with a two-plane MUM setup is shown as a function of z_0. In all three cases, focal plane 1 is located at $z_0 = 0$ nm. For $\Delta z_f = 250$ nm, $\Delta z_f = 500$ nm, and $\Delta z_f = 1000$ nm, focal plane 2 is located at $z_0 = 250$ nm, $z_0 = 500$ nm, and $z_0 = 1000$ nm, respectively. Except for the different plane spacings considered here and their respective magnifications for focal plane 2, all details required for the calculation of δ_{z_0} are as given in Fig. 19.7 for the two-plane MUM setup. The magnifications associated with focal plane 2 are $M_2 = 98.98$, $M_2 = 97.98$, and $M_2 = 96.04$ for $\Delta z_f = 250$ nm, $\Delta z_f = 500$ nm, and $\Delta z_f = 1000$ nm, respectively.

attains the smallest values in the entire plot. Moreover, beyond approximately $z_0 = 1250$ nm, this largest spacing yields considerably smaller values for δ_{z_0} than the other two spacings. The major disadvantage here, however, is that δ_{z_0} is very large when the point source is located near either focal plane. This again means that the depth discrimination problem is not adequately overcome, in this case due to the focal planes being too far apart to compensate for each other.

From Fig. 19.8, we can see that different focal spacings in a MUM setup will afford different advantages and present different drawbacks. The complex shapes of the δ_{z_0} curves in our example demonstrate that even for a MUM setup with just two focal planes, determining an appropriate plane spacing can be a challenging task. The example also shows, however, that the task can be facilitated by computing and comparing, for different focal plane spacings, the lower bounds δ_{z_0} for the expected range of axial locations over which object localization will be carried out during data analysis. We should mention that the best choice of focal plane spacings in a MUM setup depends on what one defines to be optimal for the experiment in question. In our illustration, for instance, the 500-nm plane spacing could be considered to be close to optimal if the primary goal is to be able to localize objects with a consistently low standard deviation over a continuous and relatively large axial range.

19.6.7 Further approaches to address the depth discrimination problem

Here we have seen how the simultaneous acquisition of two or more distinct focal planes can be used to address the depth discrimination problem. There are several other approaches that have been proposed, the most prominent ones being based on changing the point spread function at or near the focal plane so that different near-focus positions of a point source will yield images that are sufficiently different to allow the estimation of the point source's axial location with high certainty. In the astigmatic method, for example, a cylindrical lens is used to elongate the point spread function in one direction along the focal plane as a function of defocus. This point spread function is often modeled with an asymmetric Gaussian profile. A further approach, the double helix method, employs a complex design of a phase plate that is inserted in the optical light path to produce a point spread function that consists of two lobes. These two lobes capture defocus information by "rotating" around their center of gravity as the distance to the focal plane is changed.

20

Resolution

The question of the resolution of a microscope has appeared throughout this text starting with the introductory treatment in Part I, where we discussed Rayleigh's criterion. We have now acquired significant tools to revisit this problem using a more modern approach. In particular, we can now address one of the major shortcomings of Rayleigh's classical criterion, which is that it does not take into account the amount of data, or more precisely the number of photons, that are acquired during the acquisition of an image. Our approach considers resolution in the context of a parameter estimation problem, and the measure obtained is based on the Cramér-Rao lower bound for estimating the distance separating two point sources. Using this approach, we consider resolution in both two and three dimensions.

20.1 Resolution as a parameter estimation problem

As in Rayleigh's classical resolution criterion, we consider the problem that arises when imaging two identical objects that are closely spaced. Here we do, however, modify the resulting question from *whether* the two objects can be distinguished to *how well* they can be distinguished. This change in approach is justified for the following reason. Rayleigh's criterion was developed as a heuristic criterion at a time when images were investigated by eye. Today we have imaging sensors available that allow us to record the images for detailed analyses with possibly sophisticated algorithms. Certainly, if as assumed in the discussion of Rayleigh's criterion we had the precise deterministic image profiles of the two point sources available, it would not be a problem to determine the positions of the point sources or the distance that separates them, irrespective of how small the distance is. However, as we have discussed in previous chapters, what we have available is instead a stochastic realization of the point sources' image profiles in the form of photons detected from the point sources. Given such photon-based image data, it is reasonable to assume that the distance between two objects can be estimated, provided that sufficient data, or more precisely a sufficient number of photons, are available. We therefore rephrase the resolution problem in terms of a parameter estimation problem for the distance between the two objects, and using the theory of Fisher information we will evaluate how well, in terms of variance or standard deviation, this distance can be estimated.

20.2 Cramér-Rao lower bound for distance estimation — fundamental data

We consider here the problem of estimating the distance between two objects from image data captured under the fundamental data model. The image in this case is that of a pair of stationary objects, and the primary parameter of interest is the distance between them. Our goal is to calculate the Cramér-Rao lower bound on the variance for estimating this separation distance.

To derive the Cramér-Rao lower bound, we will proceed in the same fashion as we did in Chapters 18 and 19 for the localization problem. We again assume that the image is captured over the time interval $[t_0, t]$ by a detector with an infinitely large photon detection area $C = \mathbb{R}^2$, and start with the same general Fisher information matrix expression of Eq. (17.11). To properly define the photon detection rate Λ_θ and the photon distribution profile $f_{\theta,\tau}$, $\tau \geq t_0$, for the distance estimation problem, we need to take into account of the fact that the image is formed from the photons detected from two objects. We denote the photon detection rates of the two objects by Λ_1 and Λ_2, so that the total photon detection rate Λ_θ is given by $\Lambda_\theta(\tau) = \Lambda_1(\tau) + \Lambda_2(\tau)$, $\tau \geq t_0$. Similarly, we describe the photon distribution profile $f_{\theta,\tau}$ as the sum of two image functions q_1 and q_2 corresponding to the image profiles of the two objects, weighted by the objects' relative brightness. If x_{01} and y_{01} are the positional coordinates of one object and x_{02} and y_{02} are the positional coordinates of the other object, then $f_{\theta,\tau}$ is given by

$$f_{\theta,\tau}(x,y) = \frac{1}{M^2}\left[\frac{\Lambda_1(\tau)}{\Lambda_\theta(\tau)} q_1\left(\frac{x}{M} - x_{01}, \frac{y}{M} - y_{01}\right) + \frac{\Lambda_2(\tau)}{\Lambda_\theta(\tau)} q_2\left(\frac{x}{M} - x_{02}, \frac{y}{M} - y_{02}\right)\right], \tag{20.1}$$

$(x, y) \in \mathbb{R}^2$, where M is the magnification of the microscope system.

For the remainder of this chapter, we assume that photons are detected from each object at the same constant rate $\Lambda_0 > 0$, so that $\Lambda_1(\tau) = \Lambda_2(\tau) = \Lambda_0$, $\tau \geq t_0$. As we will see, this assumption enables a comparison of the obtained lower bounds with classical resolution criteria which consider the imaging of two identical point sources. With this assumption, the total photon detection rate Λ_θ is given by $\Lambda_\theta(\tau) = \Lambda_1(\tau) + \Lambda_2(\tau) = 2\Lambda_0$, $\tau \geq t_0$, and the photon distribution profile of Eq. (20.1) simplifies to

$$f_\theta(x,y) = \frac{1}{2M^2}\left[q_1\left(\frac{x}{M} - x_{01}, \frac{y}{M} - y_{01}\right) + q_2\left(\frac{x}{M} - x_{02}, \frac{y}{M} - y_{02}\right)\right], \quad (x,y) \in \mathbb{R}^2, \tag{20.2}$$

where the subscript τ has been dropped as the constant photon detection rate removes the distribution profile's dependence on time.

In addition, we assume that Λ_0 is known and therefore does not need to be estimated along with the distance between the two objects. With this further assumption, the derivative of the photon detection rate with respect to θ is zero, and the general Fisher information matrix expression of Eq. (17.11) reduces to

$$\begin{aligned} \mathbf{I}(\theta) &= \frac{1}{1 - e^{-\int_{t_0}^{t} 2\Lambda_0 d\tau}} \int_{t_0}^{t} \int_{\mathbb{R}^2} \frac{1}{2\Lambda_0 f_\theta(x,y)} \left(\frac{\partial\left[2\Lambda_0 f_\theta(x,y)\right]}{\partial\theta}\right)^T \frac{\partial\left[2\Lambda_0 f_\theta(x,y)\right]}{\partial\theta} dx dy d\tau \\ &= N_{photons} \int_{\mathbb{R}^2} \frac{1}{f_\theta(x,y)} \left(\frac{\partial f_\theta(x,y)}{\partial\theta}\right)^T \frac{\partial f_\theta(x,y)}{\partial\theta} dx dy, \end{aligned} \tag{20.3}$$

where f_θ is given by Eq. (20.2) and $N_{photons} := 2\Lambda_0(t - t_0)/\left(1 - e^{-2\Lambda_0(t-t_0)}\right)$, as per Eq. (15.7). In the next two sections, we will use Eq. (20.3) to derive example Cramér-Rao lower bounds pertaining to the estimation of the distance between two in-focus objects and the distance between two out-of-focus objects under the fundamental data model.

20.3 Two in-focus objects: an information-theoretic Rayleigh's criterion

For an example of the Cramér-Rao lower bound in the case where both objects are located in the focal plane of the microscope system, we consider the scenario where both objects lie on the x-axis, and are equidistant from the origin, of the object space coordinate system (Fig. 20.1(a)). For this specific situation, if we denote by $d > 0$ the distance separating the two objects, then the location of one object is given by $(x_{01}, y_{01}) = (-d/2, 0)$ and the location of the other object is given by $(x_{02}, y_{02}) = (d/2, 0)$. Using these values in the photon distribution profile of Eq. (20.2), we obtain

$$f_\theta(x, y) = \frac{1}{2M^2}\left[q_1\left(\frac{x}{M} + \frac{d}{2}, \frac{y}{M}\right) + q_2\left(\frac{x}{M} - \frac{d}{2}, \frac{y}{M}\right)\right], \quad (x, y) \in \mathbb{R}^2. \tag{20.4}$$

We next assume that the two objects are point-like and identical, and that the image of each object is given by the Airy profile (Fig. 20.1(b)). We therefore have, from Eq. (15.9), the image functions

$$q_1(x, y) = q_2(x, y) = \frac{J_1^2\left(\alpha\sqrt{x^2 + y^2}\right)}{\pi(x^2 + y^2)}, \quad (x, y) \in \mathbb{R}^2,$$

where $\alpha = \frac{2\pi n_a}{\lambda}$, with n_a the numerical aperture of the objective and λ the wavelength of the photons emitted by the objects. The photon distribution profile then becomes, for $(x, y) \in \mathbb{R}^2$,

$$f_\theta(x, y) = \frac{1}{2M^2}\left[\frac{J_1^2(\alpha\rho_{01})}{\pi\rho_{01}^2} + \frac{J_1^2(\alpha\rho_{02})}{\pi\rho_{02}^2}\right] = \frac{1}{2\pi M^2}\left[\frac{J_1^2(\alpha\rho_{01})}{\rho_{01}^2} + \frac{J_1^2(\alpha\rho_{02})}{\rho_{02}^2}\right], \tag{20.5}$$

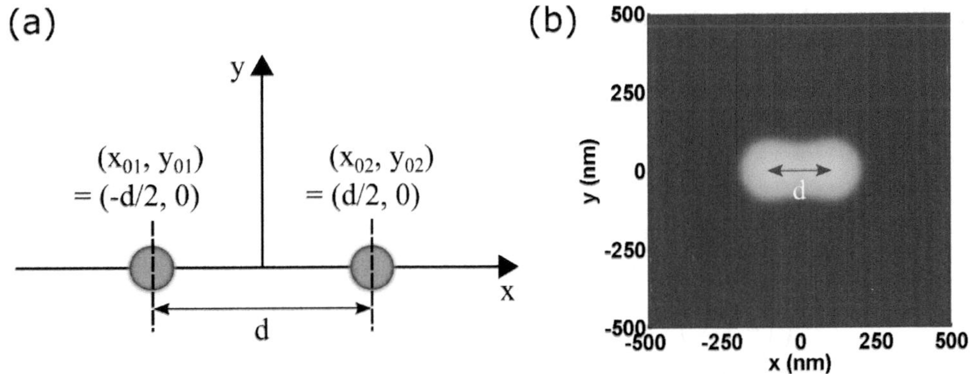

FIGURE 20.1
Resolution as a distance estimation problem. (a) Resolution is determined by how well the distance d between two closely spaced identical point sources can be estimated. In the scenario depicted here, the two point sources are located on the x-axis, equidistant from the origin of the object space coordinate system. (b) Model image of the two point sources from (a), assuming the point spread function of the microscope is the Airy profile. This image is computed using the photon distribution profile of Eq. (20.5), under unit magnification and with numerical aperture $n_a = 1.4$ and wavelength $\lambda = 520$ nm. The separation distance d is set to 200 nm.

where $\rho_{01} = \sqrt{\left(\frac{x}{M}+\frac{d}{2}\right)^2+\left(\frac{y}{M}\right)^2}$ and $\rho_{02} = \sqrt{\left(\frac{x}{M}-\frac{d}{2}\right)^2+\left(\frac{y}{M}\right)^2}$.

As we are interested in estimating the distance d, we let $\theta = d$. The partial derivative of f_θ with respect to θ can then be verified to be

$$\begin{aligned}\frac{\partial f_\theta(x,y)}{\partial \theta} &= \frac{1}{2M^2}\left[-\frac{\alpha}{\pi}\left(\frac{x}{M}+\frac{d}{2}\right)\frac{J_1(\alpha\rho_{01})}{\rho_{01}}\frac{J_2(\alpha\rho_{01})}{\rho_{01}^2}+\frac{\alpha}{\pi}\left(\frac{x}{M}-\frac{d}{2}\right)\frac{J_1(\alpha\rho_{02})}{\rho_{02}}\frac{J_2(\alpha\rho_{02})}{\rho_{02}^2}\right]\\ &= -\frac{\alpha}{2\pi M^2}\left[\left(\frac{x}{M}+\frac{d}{2}\right)\frac{J_1(\alpha\rho_{01})J_2(\alpha\rho_{01})}{\rho_{01}^3}-\left(\frac{x}{M}-\frac{d}{2}\right)\frac{J_1(\alpha\rho_{02})J_2(\alpha\rho_{02})}{\rho_{02}^3}\right].\end{aligned} \tag{20.6}$$

Substituting Eqs. (20.5) and (20.6) into Eq. (20.3), we obtain the Fisher information matrix

$$\begin{aligned}\mathbf{I}(d) = {} & \frac{N_{photons}\alpha^2}{2\pi M^2}\int_{\mathbb{R}^2}\frac{1}{\frac{J_1^2(\alpha\rho_{01})}{\rho_{01}^2}+\frac{J_1^2(\alpha\rho_{02})}{\rho_{02}^2}}\\ & \times\left(\left(\frac{x}{M}+\frac{d}{2}\right)\frac{J_1(\alpha\rho_{01})J_2(\alpha\rho_{01})}{\rho_{01}^3}-\left(\frac{x}{M}-\frac{d}{2}\right)\frac{J_1(\alpha\rho_{02})J_2(\alpha\rho_{02})}{\rho_{02}^3}\right)^2 dxdy.\end{aligned}$$

By making the change of variables $u = \frac{x}{M}$ and $v = \frac{y}{M}$, the magnification M cancels out, and we obtain, after renaming u and v back to x and y,

$$\begin{aligned}\mathbf{I}(d) = {} & \frac{N_{photons}\alpha^2}{2\pi}\int_{\mathbb{R}^2}\frac{1}{\frac{J_1^2(\alpha r_{01})}{r_{01}^2}+\frac{J_1^2(\alpha r_{02})}{r_{02}^2}}\\ & \times\left(\left(x+\frac{d}{2}\right)\frac{J_1(\alpha r_{01})J_2(\alpha r_{01})}{r_{01}^3}-\left(x-\frac{d}{2}\right)\frac{J_1(\alpha r_{02})J_2(\alpha r_{02})}{r_{02}^3}\right)^2 dxdy,\end{aligned} \tag{20.7}$$

where $r_{01} = \sqrt{\left(x+\frac{d}{2}\right)^2+y^2}$ and $r_{02} = \sqrt{\left(x-\frac{d}{2}\right)^2+y^2}$.

The lower bound δ_d on the standard deviation with which the distance d can be determined by an unbiased estimator is then just the square root of the inverse of $\mathbf{I}(d)$, given by

$$\delta_d = \frac{1}{\sqrt{2\pi N_{photons}\Gamma_0(d)}}\cdot\frac{\lambda}{n_a}, \tag{20.8}$$

where

$$\begin{aligned}\Gamma_0(d) = {} & \int_{\mathbb{R}^2}\frac{1}{\frac{J_1^2(\alpha r_{01})}{r_{01}^2}+\frac{J_1^2(\alpha r_{02})}{r_{02}^2}}\\ & \times\left(\left(x+\frac{d}{2}\right)\frac{J_1(\alpha r_{01})J_2(\alpha r_{01})}{r_{01}^3}-\left(x-\frac{d}{2}\right)\frac{J_1(\alpha r_{02})J_2(\alpha r_{02})}{r_{02}^3}\right)^2 dxdy.\end{aligned}$$

Having derived this result, we next look at what it can tell us in terms of the estimation of the distance between two closely spaced point-like objects.

Intuitively, we expect that smaller separation distances will be more difficult to determine than larger separation distances, simply due to the fact that a smaller separation means that there is more overlap between the images of the two objects. Calculated with $\lambda = 520$ nm for the wavelength of the detected photons, $n_a = 1.45$ for the numerical aperture of the objective, and $N_{photons} = 6000$ for the mean number of photons detected from the pair of point sources (i.e., 3000 from each point source), the lower bound δ_d of Eq. (20.8) is shown in Fig. 20.2(a) as a function of the separation distance d, ranging from 1 nm to 300 nm. The curve shown corroborates our intuition, as the lower bound is seen to improve (i.e., to

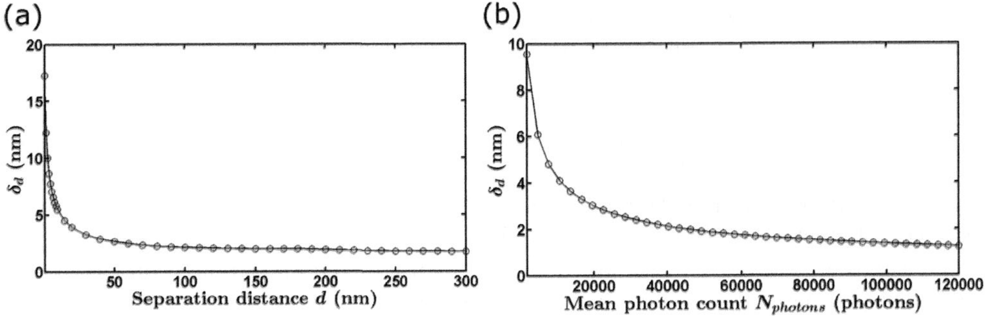

FIGURE 20.2
Dependence of separation distance estimation on the distance itself and the mean photon count. The lower bound δ_d on the standard deviation for estimating the distance d between two identical stationary and in-focus point-like objects is shown as a function of (a) the distance d, ranging from 1 nm to 300 nm, and (b) the mean number of photons $N_{photons}$ detected from the pair of objects, ranging from 2000 to 120000 (i.e., from 1000 to 60000 per object). The bound is calculated under the fundamental data scenario in which the detector has an infinitely large photon detection area, and the images of the two objects are each described by the Airy profile. In (a), the mean number of photons $N_{photons}$ detected from the object pair is 6000 (i.e., 3000 per object). In (b), the pair of objects are separated by a distance of $d = 10$ nm. In both (a) and (b), the wavelength λ of the photons is 520 nm, and the numerical aperture n_a of the objective is 1.45.

decrease in value) as the distance between the two point sources increases. The improvement is more substantial as the distance is increased from 1 nm to 50 nm, and slows significantly as the distance is increased further.

Importantly, the lower bound δ_d shows that separation distances smaller than the minimum distance specified by Rayleigh's criterion (Section 4.2) can in fact be determined. Whereas Rayleigh's criterion deems two point sources to be either distinguishable or indistinguishable depending on whether their separation is above or below the minimum distance, the Fisher information-based lower bound δ_d recasts the question of resolvability to one of how well the distance of separation, no matter how small, can be determined. For example, for $\lambda = 520$ nm and $n_a = 1.45$, Rayleigh's criterion specifies that two point sources are unresolvable if they are separated by a distance of less than $d_{min} = 0.61\lambda/n_a \approx 219$ nm. Figure 20.2(a) shows, however, that the much smaller separation distance of $d = 50$ nm, for instance, can potentially be estimated with a standard deviation as small as $\delta_d = 2.65$ nm when an average of 3000 photons are detected from each point source. Furthermore, as is the case for the estimation of any parameter (e.g., a positional coordinate, the photon detection rate; see Fig. 18.5), the lower bound on the standard deviation of the estimates can be improved by acquiring more photons. Similar to the lower bounds derived in Sections 18.5 and 19.2 for the estimation of positional coordinates, Eq. (20.8) shows an inverse square root dependence on the mean photon count $N_{photons}$ detected from the object of interest, in this case a pair of point sources. This means, for example, that if the mean photon count is doubled to $N_{photons} = 12000$ (i.e., 6000 per point source), then the lower bound for the 50-nm distance will improve by almost 30% to $\delta_d = 1.88$ nm. The dependence of δ_d on the mean photon count $N_{photons}$ is shown explicitly in Fig. 20.2(b) for the case where the distance separating the pair of point sources is just 10 nm. Under the conditions assumed, when $N_{photons} = 2000$ (i.e., 1000 per point source) we see that $\delta_d = 9.55$ nm, meaning that the lower bound on the standard deviation for estimating the 10-nm distance is nearly as large as the distance itself. However, the curve shows that by detecting about five times as many photons from the point source pair, δ_d can be significantly improved to just above 4 nm. To obtain a lower bound that is under 2 nm, the curve shows that a further fivefold increase in $N_{photons}$ would be needed.

It is also interesting to formally compare the lower bound δ_d with Rayleigh's criterion $0.61\lambda/n_a$. Both expressions have the term λ/n_a, indicating a dependence on the wavelength of the detected photons and the numerical aperture of the objective. However, the constant term 0.61 of Rayleigh's criterion is replaced by the term $1/\sqrt{2\pi N_{photons}\Gamma_0(d)}$ in δ_d. This term encapsulates the inverse square root dependence of the lower bound on the expected number of photons $N_{photons}$ detected from the pair of point sources. It also encapsulates the lower bound's dependence, in a complex fashion, on the distance d between the two point sources as discussed above.

In the discussion of the depth discrimination problem, we encountered a situation where the Fisher information is zero. In that case, it occurred in the estimation of the z position of a point source when the point source is in focus. Here in the resolution problem we have an analogous situation, i.e., that the Fisher information is zero when the distance between the two point sources is zero. One can easily verify this by setting $d = 0$ in the Fisher information expression of Eq. (20.7). This leads to the singularity of the Cramér-Rao lower bound at $d = 0$, with the accompanying sharp increase in its value as d decreases to zero (Fig. 20.2(a)). This is of course nothing but the mathematical description of the above observation that, for a given photon count, estimating the distance between the point sources becomes progressively more difficult with decreasing separation.

Experimentally, it has been shown that separation distances below the minimum distance according to Rayleigh's criterion can indeed be measured. Specifically, distances as small as 12 nm have been estimated between individual Cy5 molecules with standard deviations that are comparable with the lower bounds δ_d calculated based on the experimental image data (see Section 20.5 for a discussion and illustration of δ_d for practical data). The 12-nm distance is substantially smaller than the nearly 290-nm minimum resolvable distance predicted by Rayleigh's criterion for the experimental conditions used for the demonstration (Note 20.2).

20.4 Two objects in 3D space

We next investigate the distance estimation problem in the more general context in which at least one of the two objects is located outside the focal plane of the microscope system. For our example, we consider the specific case where both objects are located on the optical axis (i.e., the z-axis), such that their locations are given by $(x_{01}, y_{01}, z_{01}) = (0, 0, s_z + d/2)$ and $(x_{02}, y_{02}, z_{02}) = (0, 0, s_z - d/2)$, where s_z is the z coordinate of the midpoint between the objects and, as in the example above, d is the distance between the objects (Fig. 20.3(a)). Accordingly, the photon distribution profile of Eq. (20.2) becomes

$$f_\theta(x, y) = \frac{1}{2M^2}\left[q_{z_{01},1}\left(\frac{x}{M}, \frac{y}{M}\right) + q_{z_{02},2}\left(\frac{x}{M}, \frac{y}{M}\right)\right], \quad (x, y) \in \mathbb{R}^2, \tag{20.9}$$

where the subscripts z_{01} and z_{02} denote the dependence of the objects' image functions $q_{z_{01},1}$ and $q_{z_{02},2}$ on the objects' axial positions.

We again consider the situation of two identical point-like objects. Further, we assume the image of each object to be described by the 3D point spread function of Eq. (19.4), with phase function W_{z_0} as given in Eq. (19.10) (Figs. 20.3(b) and (c)). This means that

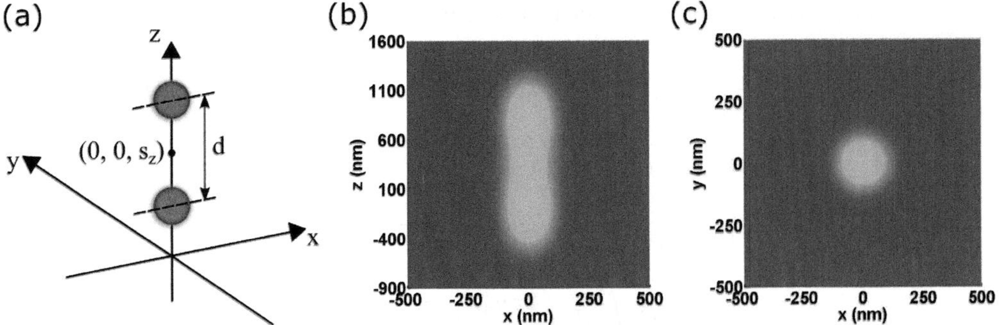

FIGURE 20.3
Distance estimation problem in 3D space. (a) In this example, the two point sources separated by a distance d are located on the optical axis. The midpoint between the point sources is located at $(0, 0, s_z)$. (b) Cross section $y = 0$ nm of the 3D model image of the two point sources from (a), when the point spread function of the microscope is the 3D point spread function of Eq. (19.4), with phase function W_{z_0} given by Eq. (19.10). The 3D model image is obtained using the photon distribution profile of Eq. (20.9), with magnification $M = 1$, axial coordinates $z_{01} = s_z + d/2$ and $z_{02} = s_z - d/2$ where $s_z = 350$ nm and $d = 900$ nm, and image functions $q_{z_{01},1}$ and $q_{z_{02},2}$ each given by Eq. (19.4) with the phase function of Eq. (19.10). The expressions for $q_{z_{01},1}$ and $q_{z_{02},2}$ are evaluated with numerical aperture $n_a = 1.45$, wavelength $\lambda = 655$ nm, and immersion medium refractive index $n_o = 1.515$. (c) Cross section $z = 0$ nm of the 3D model image. This cross section corresponds to the image that is typically acquired and used to estimate d. The view is such that one point source is situated directly in front of the other.

the normalization terms $C_{z_{01}}$ and $C_{z_{02}}$ can be assumed to be independent of z_{01} and z_{02}, respectively, and are each given by the constant $\frac{\lambda^2}{4\pi n_a^2}$. For $(x, y) \in \mathbb{R}^2$, the image functions $q_{z_{01},1}$ and $q_{z_{02},2}$ can therefore be written as

$$q_{z_{01},1}(x, y) = \frac{4\pi n_a^2}{\lambda^2}\left(U_{z_{01}}^2(x, y) + V_{z_{01}}^2(x, y)\right),$$
$$q_{z_{02},2}(x, y) = \frac{4\pi n_a^2}{\lambda^2}\left(U_{z_{02}}^2(x, y) + V_{z_{02}}^2(x, y)\right),$$

where

$$U_{z_{01}}(x, y) = \int_0^1 J_0\left(\frac{2\pi n_a}{\lambda}\sqrt{x^2 + y^2}\rho\right)\cos\left(W_{z_{01}}(\rho)\right)\rho d\rho,$$
$$V_{z_{01}}(x, y) = \int_0^1 J_0\left(\frac{2\pi n_a}{\lambda}\sqrt{x^2 + y^2}\rho\right)\sin\left(W_{z_{01}}(\rho)\right)\rho d\rho, \quad W_{z_{01}}(\rho) = \frac{-\pi n_a^2\left(s_z + \frac{d}{2}\right)}{n_o\lambda}\rho^2,$$
$$U_{z_{02}}(x, y) = \int_0^1 J_0\left(\frac{2\pi n_a}{\lambda}\sqrt{x^2 + y^2}\rho\right)\cos\left(W_{z_{02}}(\rho)\right)\rho d\rho,$$
$$V_{z_{02}}(x, y) = \int_0^1 J_0\left(\frac{2\pi n_a}{\lambda}\sqrt{x^2 + y^2}\rho\right)\sin\left(W_{z_{02}}(\rho)\right)\rho d\rho, \quad W_{z_{02}}(\rho) = \frac{-\pi n_a^2\left(s_z - \frac{d}{2}\right)}{n_o\lambda}\rho^2.$$

Now suppose we need to estimate both the separation distance d and the midpoint coordinate s_z. Then, given that $\theta = (d, s_z)$, and with $u = \frac{x}{M}$ and $v = \frac{y}{M}$, the partial derivatives of f_θ with respect to the elements of θ are given by

$$\begin{aligned}\frac{\partial f_\theta(x,y)}{\partial d} = \frac{1}{M^2}\cdot\frac{4\pi n_a^2}{\lambda^2}\Bigg[&-U_{z_{01}}(u,v)\int_0^1 J_0\left(\frac{2\pi n_a\rho}{\lambda}\sqrt{u^2+v^2}\right)\sin\left(W_{z_{01}}(\rho)\right)\frac{\partial W_{z_{01}}(\rho)}{\partial d}\rho d\rho\\ &+V_{z_{01}}(u,v)\int_0^1 J_0\left(\frac{2\pi n_a\rho}{\lambda}\sqrt{u^2+v^2}\right)\cos\left(W_{z_{01}}(\rho)\right)\frac{\partial W_{z_{01}}(\rho)}{\partial d}\rho d\rho\\ &-U_{z_{02}}(u,v)\int_0^1 J_0\left(\frac{2\pi n_a\rho}{\lambda}\sqrt{u^2+v^2}\right)\sin\left(W_{z_{02}}(\rho)\right)\frac{\partial W_{z_{02}}(\rho)}{\partial d}\rho d\rho\\ &+V_{z_{02}}(u,v)\int_0^1 J_0\left(\frac{2\pi n_a\rho}{\lambda}\sqrt{u^2+v^2}\right)\cos\left(W_{z_{02}}(\rho)\right)\frac{\partial W_{z_{02}}(\rho)}{\partial d}\rho d\rho\Bigg],\end{aligned}$$

$$\begin{aligned}\frac{\partial f_\theta(x,y)}{\partial s_z} = \frac{1}{M^2}\cdot\frac{4\pi n_a^2}{\lambda^2}\Bigg[&-U_{z_{01}}(u,v)\int_0^1 J_0\left(\frac{2\pi n_a\rho}{\lambda}\sqrt{u^2+v^2}\right)\sin\left(W_{z_{01}}(\rho)\right)\frac{\partial W_{z_{01}}(\rho)}{\partial s_z}\rho d\rho\\ &+V_{z_{01}}(u,v)\int_0^1 J_0\left(\frac{2\pi n_a\rho}{\lambda}\sqrt{u^2+v^2}\right)\cos\left(W_{z_{01}}(\rho)\right)\frac{\partial W_{z_{01}}(\rho)}{\partial s_z}\rho d\rho\\ &-U_{z_{02}}(u,v)\int_0^1 J_0\left(\frac{2\pi n_a\rho}{\lambda}\sqrt{u^2+v^2}\right)\sin\left(W_{z_{02}}(\rho)\right)\frac{\partial W_{z_{02}}(\rho)}{\partial s_z}\rho d\rho\\ &+V_{z_{02}}(u,v)\int_0^1 J_0\left(\frac{2\pi n_a\rho}{\lambda}\sqrt{u^2+v^2}\right)\cos\left(W_{z_{02}}(\rho)\right)\frac{\partial W_{z_{02}}(\rho)}{\partial s_z}\rho d\rho\Bigg].\end{aligned}$$

Substituting f_θ, $\frac{\partial f_\theta(x,y)}{\partial d}$, and $\frac{\partial f_\theta(x,y)}{\partial s_z}$ into Eq. (20.3) then yields the 2×2 Fisher information matrix $\mathbf{I}(\theta)$ with entries given by

$$\begin{aligned}[\mathbf{I}(\theta)]_{1,1} &= N_{photons}\int_{\mathbb{R}^2}\frac{1}{f_\theta(x,y)}\left(\frac{\partial f_\theta(x,y)}{\partial d}\right)^2 dxdy\\ &= \frac{2N_{photons}}{\pi}\int_{\mathbb{R}^2}\frac{1}{U_{z_{01}}^2(x,y)+V_{z_{01}}^2(x,y)+U_{z_{02}}^2(x,y)+V_{z_{02}}^2(x,y)}\\ &\quad\Bigg[-U_{z_{01}}(x,y)\int_0^1 J_0\left(\rho\sqrt{x^2+y^2}\right)\sin\left(W_{z_{01}}(\rho)\right)\frac{\partial W_{z_{01}}(\rho)}{\partial d}\rho d\rho\\ &\quad+V_{z_{01}}(x,y)\int_0^1 J_0\left(\rho\sqrt{x^2+y^2}\right)\cos\left(W_{z_{01}}(\rho)\right)\frac{\partial W_{z_{01}}(\rho)}{\partial d}\rho d\rho\\ &\quad-U_{z_{02}}(x,y)\int_0^1 J_0\left(\rho\sqrt{x^2+y^2}\right)\sin\left(W_{z_{02}}(\rho)\right)\frac{\partial W_{z_{02}}(\rho)}{\partial d}\rho d\rho\\ &\quad+V_{z_{02}}(x,y)\int_0^1 J_0\left(\rho\sqrt{x^2+y^2}\right)\cos\left(W_{z_{02}}(\rho)\right)\frac{\partial W_{z_{02}}(\rho)}{\partial d}\rho d\rho\Bigg]^2 dxdy,\end{aligned}$$

$$\begin{aligned}[\mathbf{I}(\theta)]_{2,2} &= N_{photons}\int_{\mathbb{R}^2}\frac{1}{f_\theta(x,y)}\left(\frac{\partial f_\theta(x,y)}{\partial s_z}\right)^2 dxdy\\ &= \frac{2N_{photons}}{\pi}\int_{\mathbb{R}^2}\frac{1}{U_{z_{01}}^2(x,y)+V_{z_{01}}^2(x,y)+U_{z_{02}}^2(x,y)+V_{z_{02}}^2(x,y)}\\ &\quad\Bigg[-U_{z_{01}}(x,y)\int_0^1 J_0\left(\rho\sqrt{x^2+y^2}\right)\sin\left(W_{z_{01}}(\rho)\right)\frac{\partial W_{z_{01}}(\rho)}{\partial s_z}\rho d\rho\\ &\quad+V_{z_{01}}(x,y)\int_0^1 J_0\left(\rho\sqrt{x^2+y^2}\right)\cos\left(W_{z_{01}}(\rho)\right)\frac{\partial W_{z_{01}}(\rho)}{\partial s_z}\rho d\rho\end{aligned}$$

$$- U_{z_{02}}(x,y)\int_0^1 J_0\left(\rho\sqrt{x^2+y^2}\right)\sin\left(W_{z_{02}}(\rho)\right)\frac{\partial W_{z_{02}}(\rho)}{\partial s_z}\rho d\rho$$
$$\left. +V_{z_{02}}(x,y)\int_0^1 J_0\left(\rho\sqrt{x^2+y^2}\right)\cos\left(W_{z_{02}}(\rho)\right)\frac{\partial W_{z_{02}}(\rho)}{\partial s_z}\rho d\rho\right]^2 dxdy,$$

$$\begin{aligned}
[\mathbf{I}(\theta)]_{1,2} = [\mathbf{I}(\theta)]_{2,1} &= N_{photons}\int_{\mathbb{R}^2}\frac{1}{f_\theta(x,y)}\frac{\partial f_\theta(x,y)}{\partial d}\frac{\partial f_\theta(x,y)}{\partial s_z}dxdy \\
&= \frac{2N_{photons}}{\pi}\int_{\mathbb{R}^2}\frac{1}{U^2_{z_{01}}(x,y)+V^2_{z_{01}}(x,y)+U^2_{z_{02}}(x,y)+V^2_{z_{02}}(x,y)} \\
&\quad \left[-U_{z_{01}}(x,y)\int_0^1 J_0\left(\rho\sqrt{x^2+y^2}\right)\sin\left(W_{z_{01}}(\rho)\right)\frac{\partial W_{z_{01}}(\rho)}{\partial d}\rho d\rho\right. \\
&\quad + V_{z_{01}}(x,y)\int_0^1 J_0\left(\rho\sqrt{x^2+y^2}\right)\cos\left(W_{z_{01}}(\rho)\right)\frac{\partial W_{z_{01}}(\rho)}{\partial d}\rho d\rho \\
&\quad - U_{z_{02}}(x,y)\int_0^1 J_0\left(\rho\sqrt{x^2+y^2}\right)\sin\left(W_{z_{02}}(\rho)\right)\frac{\partial W_{z_{02}}(\rho)}{\partial d}\rho d\rho \\
&\quad \left. +V_{z_{02}}(x,y)\int_0^1 J_0\left(\rho\sqrt{x^2+y^2}\right)\cos\left(W_{z_{02}}(\rho)\right)\frac{\partial W_{z_{02}}(\rho)}{\partial d}\rho d\rho\right] \\
&\quad \left[-U_{z_{01}}(x,y)\int_0^1 J_0\left(\rho\sqrt{x^2+y^2}\right)\sin\left(W_{z_{01}}(\rho)\right)\frac{\partial W_{z_{01}}(\rho)}{\partial s_z}\rho d\rho\right. \\
&\quad + V_{z_{01}}(x,y)\int_0^1 J_0\left(\rho\sqrt{x^2+y^2}\right)\cos\left(W_{z_{01}}(\rho)\right)\frac{\partial W_{z_{01}}(\rho)}{\partial s_z}\rho d\rho \\
&\quad - U_{z_{02}}(x,y)\int_0^1 J_0\left(\rho\sqrt{x^2+y^2}\right)\sin\left(W_{z_{02}}(\rho)\right)\frac{\partial W_{z_{02}}(\rho)}{\partial s_z}\rho d\rho \\
&\quad \left. +V_{z_{02}}(x,y)\int_0^1 J_0\left(\rho\sqrt{x^2+y^2}\right)\cos\left(W_{z_{02}}(\rho)\right)\frac{\partial W_{z_{02}}(\rho)}{\partial s_z}\rho d\rho\right] dxdy,
\end{aligned}$$

where in each case the expression is obtained by making the change of variables $u' = \frac{2\pi n_a}{\lambda}u = \frac{2\pi n_a}{\lambda M}x$ and $v' = \frac{2\pi n_a}{\lambda}v = \frac{2\pi n_a}{\lambda M}y$ in the integration over $\mathbb{R}^2$ and renaming the variables u' and v' back to x and y.

For ease of presentation, we introduce, using $r = \sqrt{x^2+y^2}$, the abbreviations

$$\begin{aligned}
\tilde{U}_{z_{01}} &= \int_0^1 J_0(\rho r)\cos\left(\frac{\pi n_a^2 z_{01}}{n_o\lambda}\rho^2\right)\rho d\rho, & \tilde{U}'_{z_{01}} &= \int_0^1 J_0(\rho r)\sin\left(\frac{\pi n_a^2 z_{01}}{n_o\lambda}\rho^2\right)\rho^3 d\rho, \\
\tilde{V}_{z_{01}} &= \int_0^1 J_0(\rho r)\sin\left(\frac{\pi n_a^2 z_{01}}{n_o\lambda}\rho^2\right)\rho d\rho, & \tilde{V}'_{z_{01}} &= \int_0^1 J_0(\rho r)\cos\left(\frac{\pi n_a^2 z_{01}}{n_o\lambda}\rho^2\right)\rho^3 d\rho, \\
\tilde{U}_{z_{02}} &= \int_0^1 J_0(\rho r)\cos\left(\frac{\pi n_a^2 z_{02}}{n_o\lambda}\rho^2\right)\rho d\rho, & \tilde{U}'_{z_{02}} &= \int_0^1 J_0(\rho r)\sin\left(\frac{\pi n_a^2 z_{02}}{n_o\lambda}\rho^2\right)\rho^3 d\rho, \\
\tilde{V}_{z_{02}} &= \int_0^1 J_0(\rho r)\sin\left(\frac{\pi n_a^2 z_{02}}{n_o\lambda}\rho^2\right)\rho d\rho, & \tilde{V}'_{z_{02}} &= \int_0^1 J_0(\rho r)\cos\left(\frac{\pi n_a^2 z_{02}}{n_o\lambda}\rho^2\right)\rho^3 d\rho,
\end{aligned}$$

where, again, $z_{01} = s_z + d/2$ and $z_{02} = s_z - d/2$. Then, putting in the explicit expressions for $W_{z_{01}}(\rho)$ and $W_{z_{02}}(\rho)$, and noting that $\frac{\partial W_{z_{01}}(\rho)}{\partial d} = -\frac{\pi n_a^2}{2n_o\lambda}\rho^2$, $\frac{\partial W_{z_{01}}(\rho)}{\partial s_z} = -\frac{\pi n_a^2}{n_o\lambda}\rho^2$, $\frac{\partial W_{z_{02}}(\rho)}{\partial d} =$

$\frac{\pi n_a^2}{2n_o\lambda}\rho^2$, and $\frac{\partial W_{z02}(\rho)}{\partial s_z} = -\frac{\pi n_a^2}{n_o\lambda}\rho^2$, we obtain

$$[\mathbf{I}(\theta)]_{1,1} = \frac{N_{photons}\pi n_a^4}{2n_o^2\lambda^2}\int_{\mathbb{R}^2}\frac{\left[\left(\tilde{U}_{z01}\tilde{U}'_{z01} - \tilde{V}_{z01}\tilde{V}'_{z01}\right) - \left(\tilde{U}_{z02}\tilde{U}'_{z02} - \tilde{V}_{z02}\tilde{V}'_{z02}\right)\right]^2}{\tilde{U}^2_{z01} + \tilde{V}^2_{z01} + \tilde{U}^2_{z02} + \tilde{V}^2_{z02}}dxdy,$$

$$[\mathbf{I}(\theta)]_{2,2} = \frac{2N_{photons}\pi n_a^4}{n_o^2\lambda^2}\int_{\mathbb{R}^2}\frac{\left[\left(\tilde{U}_{z01}\tilde{U}'_{z01} - \tilde{V}_{z01}\tilde{V}'_{z01}\right) + \left(\tilde{U}_{z02}\tilde{U}'_{z02} - \tilde{V}_{z02}\tilde{V}'_{z02}\right)\right]^2}{\tilde{U}^2_{z01} + \tilde{V}^2_{z01} + \tilde{U}^2_{z02} + \tilde{V}^2_{z02}}dxdy,$$

$$[\mathbf{I}(\theta)]_{1,2} = [\mathbf{I}(\theta)]_{2,1}$$

$$= \frac{N_{photons}\pi n_a^4}{n_o^2\lambda^2}\int_{\mathbb{R}^2}\frac{\left(\tilde{U}_{z01}\tilde{U}'_{z01} - \tilde{V}_{z01}\tilde{V}'_{z01}\right)^2 - \left(\tilde{U}_{z02}\tilde{U}'_{z02} - \tilde{V}_{z02}\tilde{V}'_{z02}\right)^2}{\tilde{U}^2_{z01} + \tilde{V}^2_{z01} + \tilde{U}^2_{z02} + \tilde{V}^2_{z02}}dxdy.$$

To arrive at the lower bound δ_d on the standard deviation for estimating the distance d, we first obtain the inverse of the Fisher information matrix as

$$\mathbf{I}^{-1}(\theta) = \frac{1}{[\mathbf{I}(\theta)]_{1,1}[\mathbf{I}(\theta)]_{2,2} - \left([\mathbf{I}(\theta)]_{1,2}\right)^2}\cdot\begin{bmatrix}[\mathbf{I}(\theta)]_{2,2} & -[\mathbf{I}(\theta)]_{1,2}\\ -[\mathbf{I}(\theta)]_{1,2} & [\mathbf{I}(\theta)]_{1,1}\end{bmatrix}. \tag{20.10}$$

Since d is the first parameter in θ, δ_d is the square root of the first main diagonal entry of $\mathbf{I}^{-1}(\theta)$, given by

$$\delta_d = \sqrt{[\mathbf{I}^{-1}(\theta)]_{1,1}} = \sqrt{\frac{[\mathbf{I}(\theta)]_{2,2}}{[\mathbf{I}(\theta)]_{1,1}[\mathbf{I}(\theta)]_{2,2} - \left([\mathbf{I}(\theta)]_{1,2}\right)^2}}$$

$$= \sqrt{\frac{2\Gamma_0(d)}{\pi N_{photons}}\cdot\frac{\lambda n_o}{n_a^2}}, \tag{20.11}$$

where

$$\Gamma_0(d) = \frac{B(d)}{A(d)\cdot B(d) - C^2(d)},$$

$$A(d) = \int_{\mathbb{R}^2}\frac{\left[\left(\tilde{U}_{z01}\tilde{U}'_{z01} - \tilde{V}_{z01}\tilde{V}'_{z01}\right) - \left(\tilde{U}_{z02}\tilde{U}'_{z02} - \tilde{V}_{z02}\tilde{V}'_{z02}\right)\right]^2}{\tilde{U}^2_{z01} + \tilde{V}^2_{z01} + \tilde{U}^2_{z02} + \tilde{V}^2_{z02}}dxdy,$$

$$B(d) = \int_{\mathbb{R}^2}\frac{\left[\left(\tilde{U}_{z01}\tilde{U}'_{z01} - \tilde{V}_{z01}\tilde{V}'_{z01}\right) + \left(\tilde{U}_{z02}\tilde{U}'_{z02} - \tilde{V}_{z02}\tilde{V}'_{z02}\right)\right]^2}{\tilde{U}^2_{z01} + \tilde{V}^2_{z01} + \tilde{U}^2_{z02} + \tilde{V}^2_{z02}}dxdy,$$

$$C(d) = \int_{\mathbb{R}^2}\frac{\left(\tilde{U}_{z01}\tilde{U}'_{z01} - \tilde{V}_{z01}\tilde{V}'_{z01}\right)^2 - \left(\tilde{U}_{z02}\tilde{U}'_{z02} - \tilde{V}_{z02}\tilde{V}'_{z02}\right)^2}{\tilde{U}^2_{z01} + \tilde{V}^2_{z01} + \tilde{U}^2_{z02} + \tilde{V}^2_{z02}}dxdy.$$

Equation (20.11) demonstrates, by the inverse square root dependence of the lower bound on the mean photon count $N_{photons}$ detected from the pair of point sources, that arbitrarily small distances in 3D space can in fact be determined. In other words, as is the case with the distance between two in-focus point sources, the distance between two point sources in 3D space, no matter how small, can in principle be estimated with the desired standard deviation as long as a sufficient number of photons are detected from the point sources.

Equation (20.11) is specific to the scenario in which the two point sources are aligned parallel to the microscope's optical axis, and can be compared with the classical 3D resolution limit. As discussed in Section 14.5.5.3, the classical limit specifies the minimum distance by which two like point sources, aligned parallel to the optical axis, can be separated and yet still be distinguishable. This minimum distance is given by

$$d_{min} = \frac{2\lambda n_o}{n_a^2}, \tag{20.12}$$

and is the distance along the optical axis from the center to the first zero of the 3D point spread function considered in this example. The classical 3D resolution limit is therefore defined in an analogously heuristic way as Rayleigh's criterion is defined for the in-focus case.

To give an example comparison, the classical limit specifies that two point sources aligned parallel to the optical axis, emitting photons of wavelength $\lambda = 655$ nm and imaged with an objective of numerical aperture $n_a = 1.45$ and an immersion oil of refractive index $n_o = 1.515$, are indistinguishable if they are separated by a distance of less than $d_{min} \approx 944$ nm. Therefore, by the classical limit, separation distances of less than ~944 nm are unresolvable. According to Eq. (20.11), however, a much smaller distance of $d = 200$ nm between the two point sources can in fact be potentially determined with a standard deviation of $\delta_d = 19.65$ nm, provided that 20000 photons are on average detected from each point source (i.e., provided that $N_{photons} = 40000$). Moreover, Eq. (20.11) predicts that an even smaller distance of $d = 100$ nm can potentially be resolved with a standard deviation of $\delta_d = 20.78$ nm, if a higher photon count of 62500 photons can on average be detected from each point source (i.e., if $N_{photons} = 125000$). These values for δ_d correspond to the scenario in which the pair of point sources is centered on the optical axis at $s_z = 350$ nm above the focal plane.

Note that δ_{s_z}, the lower bound on the standard deviation for estimating the axial coordinate s_z of the midpoint between the two point sources, can be obtained in similar fashion as δ_d. We will leave its derivation as an exercise for the reader (Exercise 20.1).

20.5 Cramér-Rao lower bound for distance estimation — practical data

Cramér-Rao lower bounds for distance estimation under the practical data models can be calculated by starting with the general Fisher information matrix expression of Eq. (17.21). Equation (17.16) provides the general expression for the mean photon count $\nu_{\theta,k}$ in the kth pixel, and we adapt it to the distance estimation problem as formulated in Section 20.2 by defining the photon detection rate as $\Lambda_\theta(\tau) = \Lambda_1(\tau) + \Lambda_2(\tau)$, $\tau \geq t_0$, and by defining the photon distribution profile $f_{\theta,\tau}$ to be that of Eq. (20.1). Using $\nu_{\theta,k}$ as described, we can calculate Cramér-Rao lower bounds under specific scenarios by providing more precise definitions for Λ_θ and $f_{\theta,\tau}$ as we did in the examples of the above sections.

We first illustrate Cramér-Rao lower bounds pertaining to distance estimation from practical data by considering a pair of identical in-focus point-like objects which, similar to the object pair in the example of Section 20.3, lies parallel to the x-axis of the object space coordinate system. Here, however, we assume that in addition to not knowing the separation distance d, we know neither the coordinates (s_x, s_y) of the midpoint between the two objects, nor the object pair's orientation as given by the angle ϕ between the line segment connecting the two objects and the positive x-axis (Fig. 20.4(a)). We therefore

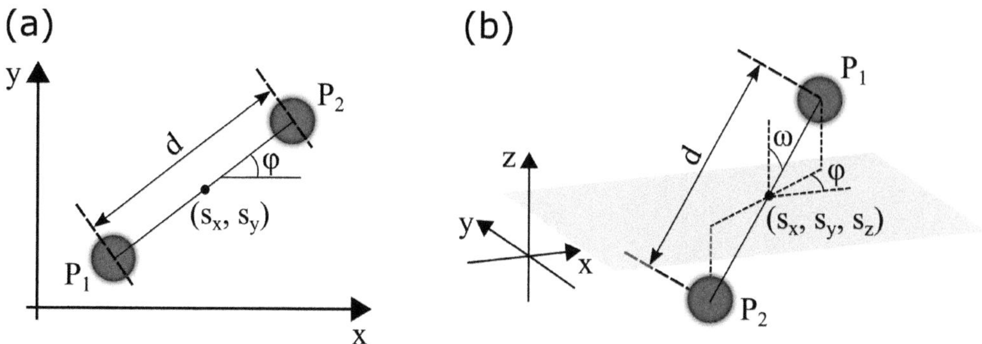

FIGURE 20.4
General geometries for the distance estimation problem. (a) In the general problem formulation for estimating the distance d between two in-focus objects P_1 and P_2, the other parameters relating to the geometry, potentially also unknown, are the midpoint (s_x, s_y) of the line segment connecting P_1 and P_2 and the angle ϕ that line segment forms with the positive x-axis. (b) For the problem in the 3D context, the parameters that describe the geometry, besides the distance d separating the objects P_1 and P_2, are the midpoint (s_x, s_y, s_z) of the line segment connecting P_1 and P_2, the angle ϕ which the xy-plane projection of that line segment forms with the positive x-axis, and the angle ω which that line segment forms with the positive z-axis.

set the objects' positional coordinates to $(x_{01}, y_{01}) = (s_x - d\cos(\phi)/2, s_y - d\sin(\phi)/2)$ and $(x_{02}, y_{02}) = (s_x + d\cos(\phi)/2, s_y + d\sin(\phi)/2)$, and the vector of unknown parameters to $\theta = (s_x, s_y, d, \phi)$. Again using the Airy profile to describe the image of each object, and assuming the use of a CCD detector in the presence of uniformly distributed background noise, lower bounds were computed for the case where the midpoint between the objects is located at the center of the pixel array from which θ is to be estimated. In Fig. 20.5(a), results are shown for the lower bound δ_d on the standard deviation for estimating the distance d. As a function of the distance d itself, we see that δ_d follows the same pattern as the corresponding δ_d under the fundamental data model, which has been reproduced from Fig. 20.2(a) for ease of comparison. As expected, for a given distance d, δ_d corresponding to a CCD image is always larger than δ_d corresponding to a fundamental data image. For relatively large distances, however, the two bounds are comparable in value. This shows that the estimation of larger distances of separation is not as severely affected by pixelation and extraneous noise, and demonstrates in a different way that larger distances of separation are easier to estimate.

As an illustration of the bound for a parameter besides the distance d, the lower bound δ_ϕ on the standard deviation for estimating the orientation angle $\phi = 0°$ is shown in Fig. 20.5(b), also as a function of d. Here we see that δ_ϕ is poor at the smallest distances, improves significantly with increasing distance initially, and continues to improve at a much lower rate with further increases in the distance. This is the same behavior that is exhibited by δ_d, and can be attributed to the fact when the distance of separation is so small such that there is substantial overlap between the images of the two objects, it is extremely difficult to determine the angle that the pair forms with the positive x-axis. Note that, as a function of d, the lower bounds δ_{s_x} and δ_{s_y} exhibit different patterns from each other and from δ_d and δ_ϕ. We leave it as an exercise for the reader to provide an interpretation of their patterns (Exercise 20.2).

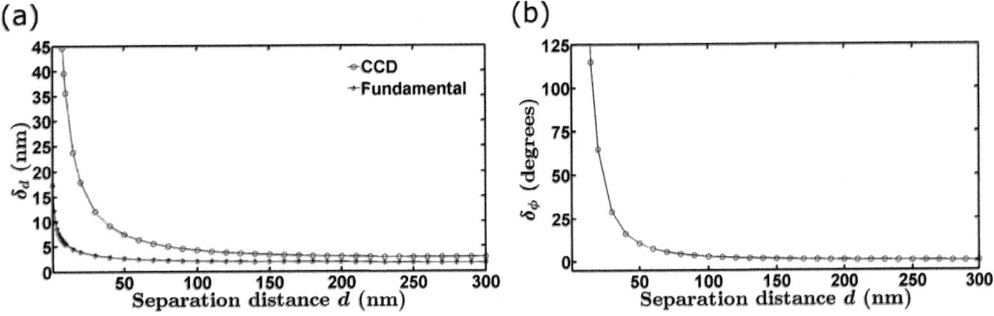

FIGURE 20.5
Dependence of the lower bounds for separation distance and orientation angle estimation under the CCD/CMOS data model. Lower bounds δ_d and δ_ϕ on the standard deviations for estimating (a) the distance d between two identical stationary and in-focus point-like objects and (b) the orientation angle ϕ of the object pair with respect to the positive x-axis are calculated under the CCD/CMOS data model. Both δ_d and δ_ϕ are shown as a function of d, which ranges from 1 nm to 300 nm. The image is assumed to be captured at a magnification of $M = 100$ by a CCD detector with a pixel size of 13 μm $\times$ 13 μm, and consists of a 17$\times$17 pixel array. The object pair is positioned such that the midpoint between the objects coincides with the center of the image, i.e., $(s_x, s_y) = (1105 \text{ nm}, 1105 \text{ nm})$ with the origin of the coordinate system taken to be the upper left corner of the image. The objects lie parallel to the x-axis, so that the orientation angle $\phi = 0°$. A background component is assumed that is uniformly distributed over the image with a mean of $\beta_k = \beta_0 = 100$ photons for each pixel k. The mean and standard deviation of the readout noise in each pixel k are $\eta_k = 0$ electrons and $\sigma_k = 6$ electrons. All remaining details are as specified in Fig. 20.2(a). In (a), the corresponding lower bound δ_d under the fundamental data model (*) is reproduced from Fig. 20.2(a) for comparison.

For the case of an object pair in 3D space, we consider here a more general problem in which we need to estimate five other parameters in addition to the distance of separation d. These parameters are the coordinates (s_x, s_y, s_z) of the midpoint between the objects, the orientation angle ϕ which the xy-plane projection of the line segment connecting the objects forms with the positive x-axis, and the orientation angle ω which the same line segment forms with the positive z-axis (Fig. 20.4(b)). Therefore, we have $\theta = (d, \phi, \omega, s_x, s_y, s_z)$, and unlike the more simplistic expressions in the example of Section 20.4, we have for the objects' positional coordinates the expressions $(x_{01}, y_{01}, z_{01}) = (s_x + d\sin(\omega)\cos(\phi)/2, s_y + d\sin(\omega)\sin(\phi)/2, s_z + d\cos(\omega)/2)$ and $(x_{02}, y_{02}, z_{02}) = (s_x - d\sin(\omega)\cos(\phi)/2, s_y - d\sin(\omega)\sin(\phi)/2, s_z - d\cos(\omega)/2)$. For our illustration we again look at a pair of identical point-like objects and use the same 3D point spread function model as in Section 20.4 to describe their images. Moreover, we again assume the use of a CCD detector and the presence of a uniform background component. If we were to compute the lower bound δ_d for a range of values for d, we would find that it exhibits similar behavior as a function of the distance d as we have seen before. Instead, we will examine how δ_d is affected by the object pair's orientation angle ω and axial position s_z by calculating it for different values of these parameters.

In Fig. 20.6(a), the lower bound δ_d is shown as a function of the angle ω, which ranges from 0° to 90°. When $\omega = 0°$, the object pair is aligned parallel to the z-axis so that one object resides directly in front of the other when viewed through the microscope. On the other extreme, when $\omega = 90°$, the object pair is aligned perpendicular to the z-axis so that the two objects are located side-by-side in the same xy-plane (at axial position s_z) when observed through the microscope. We see from the plot that δ_d improves significantly as ω is increased from 0° to 90°, with a value of around 125 nm when ω is close to 0° and a much smaller value of around 4 nm when ω is close to 90°. This behavior of δ_d reflects the fact that, for a given distance of separation, a larger ω value corresponds to less of an overlap between the images of the two objects when viewed along the z-axis.

In Fig. 20.6(b), δ_d is plotted as a function of the axial coordinate s_z of the midpoint between the two objects, which ranges from 2 μm below to 2 μm above the focal plane $s_z = 0$ nm. We see that just as we would expect it to be more difficult to determine the distance separating two objects when they are farther away, δ_d worsens as the object pair is moved away from the focal plane in either direction along the z-axis. This is at least the pattern seen when looking at axial positions about half a micron from the focal plane and beyond. From within approximately half a micron of the focal plane, we see two sharp deteriorations of δ_d, one below and one above the focal plane, when the object pair is moved

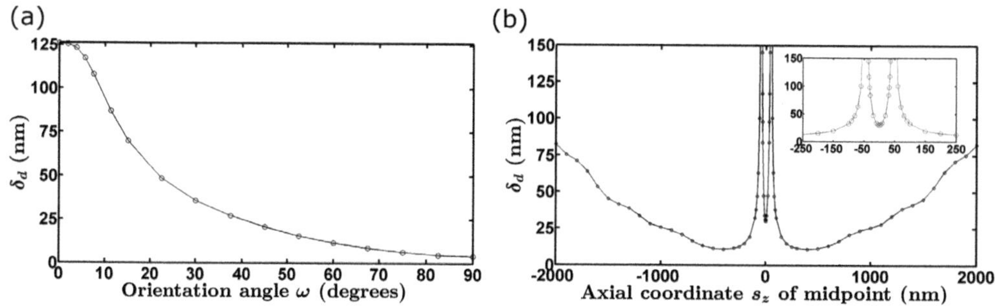

FIGURE 20.6
Lower bound for estimating the distance between two out-of-focus point sources under the CCD/CMOS data model. The lower bound δ_d on the standard deviation for estimating the distance $d = 180$ nm between two identical stationary and out-of-focus point-like objects is calculated under the CCD/CMOS data model and shown as a function of (a) the orientation angle ω which the line segment connecting the two objects forms with the positive z-axis and (b) the axial coordinate s_z of the midpoint between the objects. The image of each object is described by the 3D point spread function of Eq. (19.4), with phase function W_{z_0} as given in Eq. (19.10). Photons of wavelength $\lambda = 573$ nm are assumed to be collected by an objective with a numerical aperture of $n_a = 1.4$ and a magnification of $M = 100$, used in conjunction with immersion oil of refractive index $n_o = 1.515$. The photons are assumed to be captured using a CCD detector with a pixel size of 6.5 μm × 6.5 μm. The CCD image consists of a 21×21 pixel array with the object pair positioned such that the midpoint between the objects coincides laterally with the center of the array, i.e., $(s_x, s_y) = (682.5 \text{ nm}, 682.5 \text{ nm})$ with the origin of the coordinate system taken to be the upper left corner of the pixel array. The xy-plane projection of the line segment connecting the objects forms an angle of $\phi = 45°$ with the positive x-axis, and the mean number of photons $N_{photons}$ detected from the object pair over the detector plane is 9000 (i.e., 4500 per object). A background component is assumed that is uniformly distributed over the image with a mean of $\beta_k = \beta_0 = 40$ photons for each pixel k. The mean and standard deviation of the detector readout noise in each pixel k are $\eta_k = 0$ electrons and $\sigma_k = 3$ electrons. In (a), the axial coordinate of the midpoint between the objects is $s_z = 300$ nm. In (b), the orientation angle of the object pair with respect to the positive z-axis is $\omega = 60°$, and the inset corresponds to the portion of the plot delimited by $s_z = -250$ nm and $s_z = 250$ nm.

toward the focal plane (see figure inset for a clearer view of the two sharp deteriorations). This may seem unexpected at first, but is really just another manifestation of the poor depth discrimination capability of the optical microscope, which we discussed in Chapter 19. Estimating the distance of separation, as one would expect, is equivalent to determining the locations of both objects. Therefore, if one object is situated so close to the focal plane such that its z position cannot be estimated with an acceptable standard deviation, then we will also not be able to determine the distance between the objects with a reasonable standard deviation. Each near-focus sharp deterioration in Fig. 20.6(b) thus corresponds to one of the objects being close to the focal plane, while the sharp improvement centered at the focal plane accounts for the cases where the objects are equidistant, or approximately equidistant, from the focal plane, and neither object is close enough to the focal plane to render it very difficult to localize in the z dimension.

21

Deconvolution

As seen in the discussion of the IgG and transferrin trafficking experiments in Chapters 9 and 11, the imaging of 3D samples plays an important role in the investigation of cellular transport phenomena. In Section 11.2.2 where we introduced out-of-focus haze, and in Chapter 14 where we looked in detail at the 3D image formation process, we saw, however, that the optical properties of a microscopy system can lead to significant blurring of the images acquired of a 3D sample due to out-of-focus emission. The purpose of this chapter is to introduce deconvolution, a computational approach for removing this deterioration of the acquired images.

21.1 The deconvolution problem

We know that the optical properties of a microscope can lead to a significantly blurred image of the object. As we have seen in Sections 14.5.2 and 14.6, the image of the object is given as the convolution of the object with the point spread function. This means that each point of the image is in fact an average of the neighborhood of that point in the object, whereby the point spread function provides a weighting function. The 3D point spread function of a microscope is much broader along the optical axis than in the lateral direction. Therefore, while the point spread function's width in the xy-plane is not insignificant, the blurring effects are much more pronounced along the optical axis than in the focal plane. This means that, if we are focused on a certain focal plane, light emanating from objects above and below that focal plane can have a significant negative effect on the image. The reason for introducing TIRFM (Section 11.4) was in fact to deal with this situation when we are imaging in proximity to the cover glass. When imaged using a conventional fluorescence microscope, events on the plasma membrane adjacent to the cover glass are often obscured by strong fluorescent signals emanating from highly labeled compartments deeper in the cell. TIRFM significantly reduces this problem as the fluorescence excitation levels exponentially decay with increasing distance to the cover glass. For imaging away from the cover glass, however, TIRFM cannot be employed and a more general solution is needed.

Here we will introduce a computational approach, called *deconvolution*, that seeks to remove the blurring of the acquired data. Mathematically speaking, deconvolution attempts to undo or invert the convolution operation on the object to obtain a reconstruction of the object that led to the image.

To stay with the mathematical and engineering conventions that use the difference between the independent variables, we use a formulation by which we reconstruct the mirror image of the object rather than the object itself. The actual object can be easily obtained from the results of the deconvolution process. Formally we achieve this in the following way. In Section 14.6, we showed that the intensity image recorded on the (infinite and unpixelated) detector is given, for $(x', y', z') \in \mathbb{R}^3$, by

$$I'(x',y',z') = \int_{-\infty}^{\infty}\int_{-\infty}^{\infty}\int_{-\infty}^{\infty} psf\left(\frac{x'}{M} + x'_o, \frac{y'}{M} + y'_o, \frac{z'}{M^2 n_i} - \frac{z'_o}{n_o}\right) O'(x'_o, y'_o, z'_o)dx'_o dy'_o dz'_o,$$

where psf is the intensity point spread function, O' is the intensity object function, M is the magnification for $z' = 0$ and $z'_o = 0$, and n_i and n_o are the image space and object space refractive indices, respectively.

Performing the substitutions

$$(x_o, y_o, z_o) := \left(-x'_o, -y'_o, \frac{z'_o}{n_o}\right), \quad (x, y, z) := \left(\frac{x'}{M}, \frac{y'}{M}, \frac{z'}{M^2 n_i}\right),$$

$$O(x_o, y_o, z_o) := O'(-x_o, -y_o, n_o z_o), \quad \text{and} \quad I(x, y, z) = \frac{1}{n_o} I'(Mx, My, M^2 n_i z),$$

we obtain

$$\begin{aligned} I(x,y,z) &= \int_{-\infty}^{\infty}\int_{-\infty}^{\infty}\int_{-\infty}^{\infty} psf(x - x_o, y - y_o, z - z_o) O(x_o, y_o, z_o) dx_o dy_o dz_o \\ &= \int_{-\infty}^{\infty}\int_{-\infty}^{\infty}\int_{-\infty}^{\infty} psf(x_o, y_o, z_o) O(x - x_o, y - y_o, z - z_o) dx_o dy_o dz_o \\ &= (psf \star O)(x, y, z), \quad (x, y, z) \in \mathbb{R}^3, \end{aligned} \tag{21.1}$$

where we use $\star$ to denote the convolution of the intensity point spread function psf and the intensity object O.

In the deconvolution problem, we have the image I, assume that we know the point spread function psf, and seek to reconstruct the object O from this knowledge. In theory this problem is straightforward, as we can see if we apply the 3D Fourier transform to the image I. Using the fundamental property that the Fourier transform of the convolution of two functions is equal to the point-wise multiplication of the Fourier transforms of the two functions, we obtain

$$\mathcal{F}(I) = \mathcal{F}(psf \star O) = \mathcal{F}(psf)\mathcal{F}(O).$$

Therefore, simply solving for $\mathcal{F}(O)$ and applying the inverse Fourier transform gives us the desired result

$$O(x, y, z) = \mathcal{F}^{-1}\left(\frac{\mathcal{F}(I)}{\mathcal{F}(psf)}\right)(x, y, z), \quad (x, y, z) \in \mathbb{R}^3.$$

While this provides us a nice theoretical solution to the problem posed above, applying this approach to practically relevant situations poses significant problems. An important issue is that here we have assumed that the image is available as a precisely known function. An acquired image, however, is not a deterministic image but a stochastic realization as discussed in detail in Chapter 15. In practice the available image is therefore characterized by the probability distributions that are inherent due to the stochastic nature of the image acquisition process. A second problem is that we cannot assume that $\mathcal{F}(psf)$ is nonzero at all frequency points and this puts the inversion process in question. A further issue is that the continuous formulation of the image is obviously not suitable for the analysis of a practical image. We will address this point in the next section by deriving an analogous deconvolution formulation for the case of pixelated image data.

21.2 Discretization

The image that we acquire with a standard image detector or with a confocal scanning system is pixelated. In order to investigate deconvolution approaches we therefore need to consider convolution and deconvolution from the point of view of a pixelated image.

As a first step we assume the object to be constant over *voxels*, or 3D cubes. The image of such a voxel $V_o := [x_{o,1}, x_{o,2}] \times [y_{o,1}, y_{o,2}] \times [z_{o,1}, z_{o,2}]$ is then given by

$$I(x_i, y_i, z_i) = \int_{z_{o,1}}^{z_{o,2}} \int_{y_{o,1}}^{y_{o,2}} \int_{x_{o,1}}^{x_{o,2}} psf(x_i - x_o, y_i - y_o, z_i - z_o) O(x_o, y_o, z_o) dx_o dy_o dz_o$$
$$= O_{V_o} \int_{z_{o,1}}^{z_{o,2}} \int_{y_{o,1}}^{y_{o,2}} \int_{x_{o,1}}^{x_{o,2}} psf(x_i - x_o, y_i - y_o, z_i - z_o) dx_o dy_o dz_o,$$

$(x_i, y_i, z_i) \in \mathbb{R}^3$, where O_{V_o} is the value of the intensity object over the voxel V_o in object space. The value I_{V_i,V_o} of the image voxel $V_i := [x_{i,1}, x_{i,2}] \times [y_{i,1}, y_{i,2}] \times [z_{i,1}, z_{i,2}]$ in image space due to the object voxel V_o is obtained, by integrating I over the voxel V_i in the image space, as

$$I_{V_i,V_o} := \int_{z_{i,1}}^{z_{i,2}} \int_{y_{i,1}}^{y_{i,2}} \int_{x_{i,1}}^{x_{i,2}} I(x_i, y_i, z_i) dx_i dy_i dz_i$$
$$= O_{V_o} \int_{z_{i,1}}^{z_{i,2}} \int_{y_{i,1}}^{y_{i,2}} \int_{x_{i,1}}^{x_{i,2}} \int_{z_{o,1}}^{z_{o,2}} \int_{y_{o,1}}^{y_{o,2}} \int_{x_{o,1}}^{x_{o,2}} psf(x_i - x_o, y_i - y_o, z_i - z_o) dx_o dy_o dz_o dx_i dy_i dz_i$$
$$= psf^d_{V_i,V_o} O_{V_o},$$

where we have set

$$psf^d_{V_i,V_o} := \int_{z_{i,1}}^{z_{i,2}} \int_{y_{i,1}}^{y_{i,2}} \int_{x_{i,1}}^{x_{i,2}} \int_{z_{o,1}}^{z_{o,2}} \int_{y_{o,1}}^{y_{o,2}} \int_{x_{o,1}}^{x_{o,2}} psf(x_i - x_o, y_i - y_o, z_i - z_o) dx_o dy_o dz_o dx_i dy_i dz_i$$

as the point spread function-based convolution kernel integrated with respect to object voxel V_o and image voxel V_i. The value I_{V_i} of the image voxel V_i, given the object, is then given by summing I_{V_i,V_o} over all the object voxels, i.e.,

$$I_{V_i} = \sum_{V_o} psf^d_{V_i,V_o} O_{V_o}.$$

Introducing a standard coordinate system and voxels of uniform size, we denote the voxel intensities in object space by $O_d(o_x, o_y, o_z)$, $1 \leq o_x \leq N_x^o$, $1 \leq o_y \leq N_y^o$, $1 \leq o_z \leq N_z^o$, where N_x^o, N_y^o, and N_z^o specify boundaries for the object. Likewise, we denote the voxel intensities in image space by $I_d(i_x, i_y, i_z)$, $1 \leq i_x \leq N_x^i$, $1 \leq i_y \leq N_y^i$, $1 \leq i_z \leq N_z^i$, where N_x^i, N_y^i, and N_z^i specify the boundaries up to which we consider the image. We then have, for $1 \leq i_x \leq N_x^i, 1 \leq i_y \leq N_y^i$, and $1 \leq i_z \leq N_z^i$,

$$I_d(i_x, i_y, i_z) = \sum_{o_x=1}^{N_x^o} \sum_{o_y=1}^{N_y^o} \sum_{o_z=1}^{N_z^o} psf_d(i_x - o_x, i_y - o_y, i_z - o_z) O_d(o_x, o_y, o_z), \tag{21.2}$$

where $psf_d(p_x, p_y, p_z)$ is the discretized point spread function with values defined at least for $1 - N_x^o \leq p_x \leq N_x^i - 1$, $1 - N_y^o \leq p_y \leq N_y^i - 1$, and $1 - N_z^o \leq p_z \leq N_z^i - 1$. This shows that in the pixelated formulation we have a discrete convolution relation between the object and the image.

The deconvolution problem in this formulation is to find the pixelated object O_d given an experimentally acquired image and knowledge of the pixelated point spread function psf_d. Different approaches exist to the solution of this problem. They depend to a large extent on the assumptions on the data model for the acquired image.

21.2.1 Linear algebra formulation

We can also write the discrete convolution expression of Eq. (21.2) in matrix form as

$$\mathbb{I} = \mathbb{P}\mathbb{O}, \tag{21.3}$$

where $\mathbb{I}$, $\mathbb{P}$, and $\mathbb{O}$ are the matrices associated with the intensity image I_d, the intensity point spread function psf_d, and the intensity object O_d. We will not state these matrices explicitly. In a typical 3D problem these matrices turn out to be so large that they are not useful for practical computations. We simply discuss them here to illustrate that deconvolution can be formulated as a problem in linear algebra, even though we will use different, computationally more suitable, tools to perform the calculations.

A second reason for introducing the matrix formulation is that we will illustrate some of the numerical problems and approaches to overcome them with an analogous, but much more straightforward, problem from linear algebra. However, in order to do this, it is helpful to introduce a further reformulation so that it becomes clear that the deconvolution is nothing but a problem in solving a linear equation. We first write the matrix $\mathbb{I}$ as a vector y by using the vec operation. For a matrix $S = [s_1, s_2, \dots]$ where s_1, s_2, $\dots$ are column vectors, we define the vectorized form $\mathrm{vec}(S)$, which results from the stacking of the column vectors, as

$$\mathrm{vec}(S) = \begin{pmatrix} s_1 \\ s_2 \\ \vdots \end{pmatrix}.$$

Then the above equation can be written as

$$\mathrm{vec}\,(\mathbb{I}) = \mathrm{diag}\,(\mathbb{P}, \mathbb{P}, \dots)\,\mathrm{vec}\,(\mathbb{O}),$$

where for d the number of columns in $\mathbb{P}$, $\mathrm{diag}\,(\mathbb{P}, \mathbb{P}, \dots)$ is the matrix composed of $d \times d$ submatrices each of the same dimensions as $\mathbb{P}$, of which submatrices $(1,1), (2,2), \dots, (d,d)$ are copies of $\mathbb{P}$ and all other submatrices are zero matrices.

If we set $y = \mathrm{vec}\,(\mathbb{I})$, $A = \mathrm{diag}\,(\mathbb{P}, \mathbb{P}, \dots)$, and $x = \mathrm{vec}\,(\mathbb{O})$, then we can write Eq. (21.3) in vector form as

$$y = Ax.$$

Depending on the modeling assumptions that we place on the acquired data (Section 15.2), we can interpret the deconvolution problem as a specific parameter estimation problem, if we consider the vector x to be the vector θ of parameters we are interested in. This will be discussed in Sections 21.3 and 21.4.

21.3 Linear least squares algorithm

The deconvolution problem is often posed for the deterministic data model in which the Gaussian noise component of the data does not depend on the deterministic signal, as in this case the solution is particularly tractable. As we have seen in Section 16.4, the maximum likelihood estimator for this form of the deterministic data model is given by the least squares estimator due to the nature of the noise model that is employed. In fact, as our estimation problem is linear, it is the linear least squares problem that we analyze here.

We have shown above that in the deterministic formulation we can write the deconvolution problem as $y = Ax$. In an experimental situation, the acquired image is stochastic and corrupted by noise. We therefore write

$$y_\varepsilon = Ax + E, \tag{21.4}$$

where y_ε is the vectorized version of the acquired image and E is a vector of measurement errors.

The deconvolution task is to obtain an estimate of x from the measured data $y_\varepsilon = y + E$, assuming that A is known. If y were known, x would be given directly by $x = A^{-1}y$ (assuming A is invertible). Using this approach in our problem, if we replace y by y_ε, we would encounter severe inaccuracies due to the fact that A is typically badly conditioned. This means that using the estimate

$$\hat{x} = A^{-1}\, y_\varepsilon \tag{21.5}$$

typically produces highly unreliable results. Therefore we formulate the problem as the linear regression problem

$$\hat{x}_{LR} := \arg\min_x \left\| Ax - y_\varepsilon \right\|^2, \tag{21.6}$$

where $\| \;\; \|$ is the Euclidean norm, i.e.,

$$\left\| \begin{pmatrix} v_1 \\ \vdots \\ v_n \end{pmatrix} \right\| := \sqrt{\sum_{i=1}^{n} v_i^2}.$$

The solution for this problem is given by (Online Appendix Section D.1)

$$\hat{x}_{LR} = \left(A^T A\right)^{-1} A^T y_\varepsilon, \tag{21.7}$$

if $A^T A$ is invertible. In practice, the sizes of the vector y_ε and the matrix A are far too large to perform the computation as stated. Therefore the expression is typically computed using Fourier transform techniques, as we will see in Section 21.3.3. However, the formulation here nevertheless captures the essence of the computation we need to undertake.

Invertibility of $A^T A$ is not sufficient to guarantee that the solution, i.e., Eq. (21.7), is numerically well behaved. We use the following example to illustrate some of the numerical problems that may occur.

21.3.1 Condition number of a matrix

We first introduce the notion of the condition number of a matrix, which indicates how easily a matrix can be inverted. The condition number cond(Q) of a matrix Q is given by

$$cond(Q) = \frac{\sqrt{\lambda_{max}(Q^T Q)}}{\sqrt{\lambda_{min}(Q^T Q)}},$$

where $\lambda_{max}(Q^T Q)$ is the largest, and $\lambda_{min}(Q^T Q)$ is the smallest, eigenvalue of the matrix $Q^T Q$. The matrix Q is called *well-conditioned* if this number is relatively small, and *ill-conditioned* if it is large. A large condition number is normally associated with numerical problems in the inversion of the matrix. Now consider $A = \begin{pmatrix} 1 & 0 \\ 0 & \alpha \end{pmatrix}$ with α a nonzero number in the optimization problem of Eq. (21.6), such that $A^T A = \begin{pmatrix} 1 & 0 \\ 0 & \alpha^2 \end{pmatrix}$. Then A has the condition number $cond(A) = \sqrt{\lambda_{max}(A^T A)}/\sqrt{\lambda_{min}(A^T A)} = \sqrt{1/\alpha^2} = \alpha^{-1}$. If α is small, then A will be an ill-conditioned matrix, as the condition number α^{-1} will be large. It will have entries of very different magnitudes, and will be close to being non-invertible, potentially leading to a solution that is far from correct.

21.3.1.1 Example of an ill-conditioned least squares problem

For example, let $y = \begin{pmatrix} 1 \\ 0 \end{pmatrix}$ and noise vector $E = \beta \begin{pmatrix} 1 \\ 1 \end{pmatrix}$ with β real, such that $y_\varepsilon = \begin{pmatrix} 1+\beta \\ \beta \end{pmatrix}$. Then the solution to the least squares problem, i.e., Eq. (21.7), is given by

$$\hat{x}_{LR} = (A^T A)^{-1} A^T y_\varepsilon = \begin{pmatrix} 1 & 0 \\ 0 & \alpha^2 \end{pmatrix}^{-1} \begin{pmatrix} 1 & 0 \\ 0 & \alpha \end{pmatrix} \begin{pmatrix} 1+\beta \\ \beta \end{pmatrix} = \begin{pmatrix} 1+\beta \\ \alpha^{-1}\beta \end{pmatrix}.$$

Obviously the solution to the noise-free least squares problem ($\beta = 0$) is given by $\hat{x}_{LR} = \begin{pmatrix} 1 \\ 0 \end{pmatrix}$. To consider the impact of a small amount of noise in the data vector, assume that β is very small, e.g., $\beta = 10^{-2}$. Also assume that α is very small, e.g., $\alpha = 10^{-10}$, so that $cond(A) = 10^{10}$ and A is very close to being non-invertible. Then the solution to the optimization problem is

$$\hat{x}_{LR} = \begin{pmatrix} 1.01 \\ 10^8 \end{pmatrix} \neq \begin{pmatrix} 1 \\ 0 \end{pmatrix}.$$

Clearly this solution is far from the correct solution for the noise-free case. It is the second entry that is highly problematic. This illustrates that in a linear regression problem with an ill-conditioned A matrix, even very small numerical perturbations can lead to extreme errors in the solution.

21.3.2 Regularization of the least squares problem

As we have seen in our example, if the matrix A is not well-conditioned, then noise will get amplified and numerical problems will be magnified. We therefore use an approach called *regularization*. By regularization we mean that the optimization problem of Eq. (21.6) is modified to penalize large values in the solution. We change the problem formulation to

$$\hat{x}_{RLR} := \arg\min_x \left(\left\| Ax - y_\varepsilon \right\|^2 + \left\| Wx \right\|^2 \right), \tag{21.8}$$

where $W = W^T > 0$ is a weighting matrix that is positive definite. In other words, W is a symmetric matrix whose eigenvalues are all positive or, equivalently, is such that $z^T W z > 0$ for all $z \neq 0$. The weighting matrix is often used as a further parameter to modify the properties of the solution. For the current discussion we do not, however, need to be concerned about the specific choice of this matrix.

Equation (21.8) can be rewritten as

$$\hat{x}_{RLR} := \arg\min_x \left\| \begin{pmatrix} Ax - y_\varepsilon \\ Wx \end{pmatrix} \right\|^2 = \arg\min_x \left\| \begin{pmatrix} A \\ W \end{pmatrix} x - \begin{pmatrix} y_\varepsilon \\ 0 \end{pmatrix} \right\|^2.$$

This shows that the weighted linear regression problem can be solved in the same way as the standard linear regression problem in Eq. (21.6) (see also Online Appendix Section D.1). The solution is thus given by

$$\hat{x}_{RLR} = \left[\begin{pmatrix} A \\ W \end{pmatrix}^T \begin{pmatrix} A \\ W \end{pmatrix} \right]^{-1} \begin{pmatrix} A \\ W \end{pmatrix}^T \begin{pmatrix} y_\varepsilon \\ 0 \end{pmatrix}$$

$$= \left[\begin{pmatrix} A^T & W^T \end{pmatrix} \begin{pmatrix} A \\ W \end{pmatrix} \right]^{-1} \begin{pmatrix} A^T & W^T \end{pmatrix} \begin{pmatrix} y_\varepsilon \\ 0 \end{pmatrix}$$

$$= \left(A^T A + W^T W\right)^{-1} \left(A^T y_\varepsilon + W^T 0\right) = \left(A^T A + W^T W\right)^{-1} A^T y_\varepsilon. \tag{21.9}$$

The difference between the least squares solution $\hat{x}_{LR}$ and its regularized weighted form $\hat{x}_{RLR}$ is the addition of the term $W^T W$ to $A^T A$. It is this addition that "regularizes" the inversion of $A^T A$. In particular, if $A^T A$ is close to non-invertible, the regularization $W^T W$ can be used to produce a solution with much better numerical stability.

21.3.2.1 Example continued: regularization of the ill-conditioned least squares problem

The continuation of our example illustrates the role of regularization. For the regularized version of the optimization problem, we choose a regularization matrix $W = \begin{pmatrix} w_1 & 0 \\ 0 & w_2 \end{pmatrix}$. The solution to the regularized problem is then given by

$$\hat{x}_{RLR} = (A^T A + W^T W)^{-1} A^T y_\varepsilon = \left[\begin{pmatrix} 1 & 0 \\ 0 & \alpha^2 \end{pmatrix} + \begin{pmatrix} w_1^2 & 0 \\ 0 & w_2^2 \end{pmatrix} \right]^{-1} \begin{pmatrix} 1 & 0 \\ 0 & \alpha \end{pmatrix} \begin{pmatrix} 1+\beta \\ \beta \end{pmatrix}$$

$$= \begin{pmatrix} (1+w_1^2)^{-1}(1+\beta) \\ (\alpha^2 + w_2^2)^{-1}\alpha\beta \end{pmatrix}. \tag{21.10}$$

Choosing the same parameters as above with $w_1 = w_2 = 1$, we obtain

$$\hat{x}_{RLR} = \begin{pmatrix} 0.5(1.01) \\ \left(10^{-20}+1\right)^{-1} 10^{-12} \end{pmatrix} \approx \begin{pmatrix} 0.5 \\ 0 \end{pmatrix}.$$

The regularization thus resolved the problem with the second entry as it is now approximately correct. However, the first entry is now half the correct value of 1. This shows that regularization needs to be used carefully as it can lead to significant distortion of the results if the regularization matrices are not appropriately chosen.

Using a weighting matrix that is adapted to the properties of $A^T A$ produces significantly better results. In $A^T A$ there are no inversion problems associated with the $(1,1)$ entry. Therefore, we should choose w_1 to be small, e.g., $w_1 = 10^{-2}$. In contrast, the $(2,2)$ entry of $A^T A$ is highly problematic as it is very small, therefore we should pick a larger value for w_2, e.g., $w_2 = 1$ as chosen above. With these values, we obtain for the solution to the regularized problem

$$\hat{x}_{RLR} = \begin{pmatrix} \left(1+10^{-4}\right)^{-1}(1.01) \\ \left(10^{-20}+1\right)^{-1} 10^{-12} \end{pmatrix} \approx \begin{pmatrix} 1 \\ 0 \end{pmatrix},$$

which shows that in this case the regularized solution is a good approximation to the correct solution $x = \begin{pmatrix} 1 \\ 0 \end{pmatrix}$. Note that the condition number of $\begin{pmatrix} A \\ W \end{pmatrix}$ is

$$cond\left(\begin{pmatrix} A \\ W \end{pmatrix}\right) = \frac{\sqrt{\lambda_{max}(A^T A + W^T W)}}{\sqrt{\lambda_{min}(A^T A + W^T W)}} = \sqrt{\frac{1+w_1^2}{\alpha^2 + w_2^2}}.$$

We therefore have an important interpretation of the role of the regularization matrix. It perturbs the matrix A^TA to improve the condition number by increasing $\lambda_{min}(A^TA)$. With the current choice of $w_1 = 10^{-2}$ and $w_2 = 1$, we have

$$cond\left(\left(\begin{array}{c} A \\ W \end{array}\right)\right) = \sqrt{\frac{1+10^{-4}}{10^{-20}+1}} \approx 1,$$

demonstrating that the regularization remedies the severe ill conditioning of the matrix A, which by itself has the very large condition number $\sqrt{1/\alpha^2} = \alpha^{-1} = 10^{10}$.

21.3.3 A Fourier transform approach

In Eq. (21.9), we stated a solution to the vector formulation of the deconvolution problem. Although this formulation provides important insights, it is, as mentioned earlier, computationally not easily tractable. We therefore return to the original formulation. We know from Eq. (21.2) that $I_d = psf_d \star O_d$. Our task is to estimate the entries of the array O_d given the measured values of I_d. The data array I_d is assumed to have been acquired, for example, by a z-stack experiment (Section 11.2.1). We assume a deterministic data model where the Gaussian noise has zero mean and identical variance for all the data points. We use here a least squares criterion to formulate the deconvolution problem. Again, based on the arguments in Section 16.4, this least squares problem formulation provides the maximum likelihood estimate as the data model assumed matches the form of the deterministic data model considered there. Written concisely, our deconvolution problem is the least squares optimization problem

$$\hat{O}_d = \arg\min_{O_d} \|I_d - psf_d \star O_d\|^2 .$$

Here the norm $\| \ \|$ takes the square root of the sum of the squares of the entries of the 3D array that results from the subtraction. To address the numerical issues discussed above, we consider a regularized version of the problem. We therefore assume that we have an array W that is used as a weighting function in the regularization formulation. The regularized linear least squares problem that we are trying to solve is therefore given by

$$\hat{O}_d = \arg\min_{O_d} \left(\|I_d - psf_d \star O_d\|^2 + \|W \star O_d\|^2\right).$$

As we have shown in Online Appendix Section D.2, the solution to this problem is given by

$$\hat{O}_d = IDFT\left[\left(|DFT(psf_d)|^2 + |DFT(W)|^2\right)^{-1}\overline{DFT(psf_d)}DFT(I_d)\right],$$

where DFT denotes the multidimensional discrete Fourier transform and $IDFT$ its inverse.

We have stated the solution using discrete Fourier transforms that are computationally tractable even when large data sets are involved. We should point out that the non-regularized solution can be very easily obtained by setting $|DFT(W)| = 0$. The formulation of the solution in terms of Fourier transforms also gives an interesting observation into the role of regularization. As $|DFT(psf_d)|^2 + |DFT(W)|^2$ is the denominator of the solution, numerical problems can be expected for those parts of the spatial frequency space where $|DFT(psf_d)|$ is close to 0 or even 0. If the weighting term is designed such that $|DFT(W)|$ is nonzero when $|DFT(psf_d)|$ is close to zero, numerical problems can be reduced. A straightforward and often successful approach to the regularized linear least squares problem is simply to let $|DFT(W)|^2 := \gamma^2$, where $\gamma > 0$ is a positive constant.

In what we have discussed so far, we assume that we know the point spread function. In practice, the point spread function that is used is based on a theoretical point spread function model or an experimentally obtained point spread function. Point spread functions can be obtained experimentally by imaging a point source such as a small fluorescent bead (Note 21.2). Compared to theoretical point spread functions, experimentally acquired point spread functions have the advantage that they reflect the true optical properties of the microscope rather than a theoretical ideal. However, experimentally obtained point spread functions are noisy images. These noisy representations of the point spread function can lead to significant deteriorations of the deconvolved images, and therefore careful denoising is important (Note 21.3). Regardless of whether a theoretical or an experimental point spread function is used, errors can occur that need to be considered. To address the issue of the determination of the point spread function, algorithms have been introduced that estimate the point spread function together with the object. These algorithms are typically called *blind deconvolution* algorithms.

21.4 Maximum likelihood formulation

We can also formulate the deconvolution problem as a maximum likelihood estimation problem for the Poisson data model. If we assume that we want to estimate a pixelated version of the object of interest, we can consider the parameter θ that we seek to estimate as the parameter made up of the pixel values of the object. The data vector z is of course given by the measured pixel values of the image of the object. Depending on the stochastic model we assume on the data, we can then define a log-likelihood function $\mathcal{L}(\theta \mid z)$ and set up a standard maximum likelihood estimation problem in which we obtain the object estimate by maximizing $\mathcal{L}$ with respect to θ. Due to the typically large size of θ in a deconvolution problem, however, a direct numerical maximization of $\mathcal{L}$ is often impractical and a different approach to finding the maximum likelihood estimate of θ is needed.

21.4.1 Expectation maximization algorithm

As we mentioned in Section 16.3, it is often the case that finding the maximum likelihood estimate for a particular estimation problem is not analytically tractable. In such cases numerical approaches are typically required. However, even a direct numerical optimization of the log-likelihood function is often computationally too complex to lead to a practically useful numerical solution. Nevertheless, the computation of the maximum likelihood estimate using the *expectation maximization* (EM) formulation has proved very powerful. The EM algorithm assumes that the problem has a particular structure that is then exploited to obtain an elegant approach to the computation of the maximum likelihood estimate.

The EM algorithm assumes that, in addition to the unknown parameter vector θ and the known observed data Z in the space $\mathcal{Z}$, which is often referred to as the *incomplete data space*, the data model can be interpreted to include unknown data Y in the *complete data space* $\mathcal{Y}$ that helps to formulate the problem.

We assume that there is a function $h : \mathcal{Y} \to \mathcal{Z}$ that describes which data in the complete data space $\mathcal{Y}$ corresponds to which data in the incomplete data space $\mathcal{Z}$. With this function we can also relate the probability distribution $p_{incomp,\theta}$ on the incomplete data space $\mathcal{Z}$ to the probability distribution $p_{comp,\theta}$ on the complete data space $\mathcal{Y}$ by

$$p_{incomp,\theta}(z) = \sum_{y \in h^{-1}(z)} p_{comp,\theta}(y). \tag{21.11}$$

This relationship importantly shows that multiple data points y in the complete data space $\mathcal{Y}$ can give rise to the same data point z in the incomplete data space $\mathcal{Z}$, and that the sum of the probabilities of all such data points y must equal the probability that the data point z is observed.

The EM algorithm consists of two parts, the *expectation step* and the *maximization step.* These two steps make use of two different log-likelihood functions that can be associated with our problem, one for each of the two data classes. These functions are the log-likelihood $\mathcal{L}_{incomp}(\theta \mid Z)$ for the data $Z \in \mathcal{Z}$ from the incomplete data space and the log-likelihood $\mathcal{L}_{comp}(\theta \mid Y)$ for the data $Y \in \mathcal{Y}$ from the complete data space.

Although the formalism of the EM algorithm can appear complex at first, the basic idea behind the approach is quite intuitive. There are a number of important examples for which the maximum likelihood estimator would be fairly straightforward to compute if the more detailed data Y was available. Obviously, we do not have the complete data Y available and therefore cannot compute the log-likelihood for Y. Instead we do the next best thing which is to calculate the expected value of the log-likelihood for the complete data Y conditional on the observed incomplete data Z. This is the expression that is maximized. The EM algorithm is particularly suited for situations in which this maximization step is obtained in a straightforward way.

The EM algorithm is iterative. Starting with an initial parameter estimate $\hat{\theta}^{(0)}$, a sequence of parameter estimates $\hat{\theta}^{(n)}$, $n = 1, 2, \ldots$, is obtained by performing at each iteration both the expectation step and the maximization step of the EM algorithm. The basic purpose of the EM algorithm is to find estimates $\hat{\theta}^{(n)}$, $n = 1, 2, \ldots$, such that the log-likelihood functions for the acquired data, i.e., the incomplete log-likelihood functions $\mathcal{L}_{incomp}\left(\hat{\theta}^{(n)} \mid Z\right)$, $n = 0, 1, \ldots$, are monotonically increasing.

Expectation step: Given a parameter estimate $\hat{\theta}^{(n)}$, the purpose of the expectation step is to compute the conditional expectation of the complete log-likelihood function $\mathcal{L}_{comp}(\theta \mid Y)$ given the incomplete data Z and assuming that $\hat{\theta}^{(n)}$ is the "true" parameter. Expressed succinctly, we want to compute

$$Q\left(\theta \mid \hat{\theta}^{(n)}\right) := \mathrm{E}_{\hat{\theta}^{(n)}}\left[\mathcal{L}_{comp}(\theta \mid Y) \mid Z\right].$$

Here the conditional expectation $\mathrm{E}[A \mid B]$ for discrete random variables A and B is calculated as

$$\mathrm{E}[A \mid B] = \sum_{x \in range(A)} x P(A = x \mid B = y) = \sum_{x \in range(A)} x \frac{P(A = x, B = y)}{P(B = y)}.$$

Maximization step: The maximization step produces the next parameter estimate by finding the parameter that maximizes the conditional expectation Q of the complete log-likelihood function computed in the expectation step, i.e., it finds the next parameter estimate

$$\hat{\theta}^{(n+1)} := \arg\max_{\theta} \left(Q\left(\theta \mid \hat{\theta}^{(n)}\right)\right).$$

Analyzing the convergence of the EM algorithm is difficult. There are the issues that plague the analysis of maximum likelihood estimators, such as that convergence may be to a local maximum, that the global maximum might not be unique, or that the estimated parameter converges to a value outside the set Θ of allowable parameter values. Nevertheless, it can be shown that the algorithm does what is expected, meaning that for the sequence of estimates $\hat{\theta}^{(n)}$, $n = 0, 1, \ldots$, the incomplete log-likelihood functions $\mathcal{L}_{incomp}\left(\hat{\theta}^{(n)} \mid Z\right)$, $n = 0, 1, \ldots$, are nondecreasing.

21.5 Positron emission tomography

As an application of the EM algorithm, we are going to discuss a classical problem that originates in positron emission tomography (PET), a clinical imaging modality. Radiolabeled tracers are targeted, for example, to tumors where they accumulate. The emitted positrons are recorded by the detectors of the clinical imaging instrument. The task at hand is to deduce, from the detected signals, where in the patient the signals originate from, thereby localizing the tumor in the patient. For this purpose, we assume that the patient's body is made up of L different volume elements.

Let there be K pixels in the detector in which we collect photons that are emitted by the L sources. We assume that we have observed incomplete data $z_1, \ldots, z_K$, described by the random variables $Z_1, \ldots, Z_K$. The observed data represent the photon counts registered in the K pixels during the acquisition time interval. The number of photons emitted from source j and detected in one of the pixels is assumed to be Poisson-distributed with rate λ_j, $j = 1, \ldots, L$. Our task is to estimate the parameter $\lambda := (\lambda_1, \ldots, \lambda_L)$. The unobserved complete data is assumed to be given by $z_{k,j}$, $k = 1, \ldots, K$, $j = 1, \ldots, L$, which represents the number of photons that are emitted by source j and detected in the kth pixel during the acquisition time interval. These photon counts are described by the random variables $Z_{k,j}$, $k = 1, \ldots, K$, $j = 1, \ldots, L$. For $k = 1, \ldots, K$, we then have that $Z_k = \sum_{j=1}^{L} Z_{k,j}$, where for $j = 1, \ldots, L$, $Z_{k,j}$ is Poisson-distributed with mean

$$\mathrm{E}\left[Z_{k,j}\right] = l(k,j)\lambda_j,$$

with $l(k,j)$ denoting the fraction of photons emitted by source j that are registered in the kth pixel. Due to the assumption that we are only considering photons that are detected in a pixel, we have that all photons due to source j that are detected during the acquisition time interval are accounted for by the K pixels of the detector, i.e., we have that $\sum_{k=1}^{K} l(k,j) = 1$. Note that here the mapping h from the complete data $z_{k,j}$, $k = 1, \ldots, K$, $j = 1, \ldots, L$, to the incomplete data z_k, $k = 1, \ldots, K$, can be written as the vector-valued function

$$\begin{aligned} h\left(z_{1,1}, \ldots, z_{1,L}, z_{2,1}, \ldots, z_{2,L}, \ldots, z_{K,1}, \ldots, z_{K,L}\right) &= \left(\sum_{j=1}^{L} z_{1,j}, \sum_{j=1}^{L} z_{2,j}, \ldots, \sum_{j=1}^{L} z_{K,j}\right) \\ &= \left(z_1, z_2, \ldots, z_K\right). \end{aligned}$$

Using the above, we have that

$$\mathrm{E}\left[Z_k\right] = \mathrm{E}\left[\sum_{j=1}^{L} Z_{k,j}\right] = \sum_{j=1}^{L} \mathrm{E}\left[Z_{k,j}\right] = \sum_{j=1}^{L} l(k,j)\lambda_j, \quad k = 1, \ldots, K. \tag{21.12}$$

By the mutual independence of $Z_{k,j}$, $k = 1, \ldots, K$, $j = 1, \ldots, L$, the probability distribution for the complete data is given by the product description

$$p_{comp,\lambda}\left(z_{1,1}, \ldots, z_{1,L}, \ldots, z_{K,1}, \ldots, z_{K,L}\right) = \prod_{k=1}^{K} \prod_{j=1}^{L} \frac{1}{z_{k,j}!} e^{-l(k,j)\lambda_j} \left(l(k,j)\lambda_j\right)^{z_{k,j}}. \tag{21.13}$$

As $Z_{k,j}$, $k = 1, \ldots, K$, $j = 1, \ldots, L$, are mutually independent and Poisson-distributed, the random variables $Z_1, \ldots, Z_K$ are also mutually independent and Poisson-distributed. The probability distribution of the observed incomplete data is therefore given by the product description

$$\begin{aligned} p_{incomp,\lambda}(z_1,\ldots,z_K) &= \prod_{k=1}^{K} \frac{1}{z_k!} e^{-\mathrm{E}[Z_k]} \left(\mathrm{E}[Z_k]\right)^{z_k} \\ &= \prod_{k=1}^{K} \frac{1}{z_k!} e^{-\sum_{j=1}^{L} l(k,j)\lambda_j} \left(\sum_{j=1}^{L} l(k,j)\lambda_j\right)^{z_k}, \end{aligned} \tag{21.14}$$

where we have used Eq. (21.12) for the expectation of the incomplete data in the kth pixel. One can verify with concrete examples that the relationship specified by Eq. (21.11) is satisfied by these two probability distributions. Specifically, with a given realization of the observed data, one can demonstrate that Eq. (21.14) is obtained by summing the results from the evaluation of Eq. (21.13) for all possible realizations of the complete data that give rise to the given observed data (Exercise 21.2). We next consider the expectation step of the EM algorithm.

Expectation step: Denote by $\hat{\lambda}^{(n)} := \left(\hat{\lambda}_1^{(n)}, \hat{\lambda}_2^{(n)}, \ldots, \hat{\lambda}_L^{(n)}\right)$ the parameter estimate determined by the nth iteration of the EM algorithm, where we have set an initial condition $\hat{\lambda}^{(0)} := \left(\hat{\lambda}_1^{(0)}, \hat{\lambda}_2^{(0)}, \ldots, \hat{\lambda}_L^{(0)}\right)$. We need to compute

$$Q\left(\lambda \mid \hat{\lambda}^{(n)}\right) = \mathrm{E}_{\hat{\lambda}^{(n)}}\left[\mathcal{L}_{comp}\left(\lambda \mid z_{1,1},\ldots,z_{1,L},\ldots,z_{K,1},\ldots,z_{K,L}\right) \mid z_1,\ldots,z_K\right],$$

which is the expectation of the log-likelihood function for the complete data conditional on the observed incomplete data. From Eq. (21.13), we have that the log-likelihood of the complete data is given by

$$\begin{aligned} &\mathcal{L}_{comp}\left(\lambda \mid z_{1,1},\ldots,z_{1,L},\ldots,z_{K,1},\ldots,z_{K,L}\right) \\ &\quad = \ln p_{comp,\lambda}\left(z_{1,1},\ldots,z_{1,L},\ldots,z_{K,1},\ldots,z_{K,L}\right) \\ &\quad = \ln \prod_{k=1}^{K}\prod_{j=1}^{L} \frac{1}{z_{k,j}!} e^{-l(k,j)\lambda_j} \left(l(k,j)\lambda_j\right)^{z_{k,j}} \\ &\quad = \sum_{k=1}^{K}\sum_{j=1}^{L} \left[-\ln\left(z_{k,j}!\right) - l(k,j)\lambda_j + z_{k,j}\ln\left(l(k,j)\lambda_j\right)\right] \\ &\quad = -\sum_{k=1}^{K}\sum_{j=1}^{L} \ln(z_{k,j}!) - \sum_{j=1}^{L}\left(\lambda_j \sum_{k=1}^{K} l(k,j) - \sum_{k=1}^{K} z_{k,j}\ln\left(l(k,j)\right) - \ln\left(\lambda_j\right)\sum_{k=1}^{K} z_{k,j}\right) \\ &\quad = \left(\sum_{k=1}^{K}\sum_{j=1}^{L} z_{k,j}\ln\left(l(k,j)\right) - \sum_{k=1}^{K}\sum_{j=1}^{L}\ln(z_{k,j}!)\right) - \sum_{j=1}^{L}\left(\lambda_j \sum_{k=1}^{K} l(k,j) - \ln\left(\lambda_j\right)\sum_{k=1}^{K} z_{k,j}\right). \end{aligned}$$

Note that the expression in the left bracket does not depend on the parameter λ, and will therefore not play a role in the maximization of Q with respect to λ. In what follows we will denote this expression by C. Also recall that, by assumption, we have that $\sum_{k=1}^{K} l(k,j) = 1$ for $j = 1,\ldots,L$. We can therefore write the log-likelihood of the complete data as

$$\mathcal{L}_{comp}\left(\lambda \mid z_{1,1},\ldots,z_{1,L},\ldots,z_{K,1},\ldots,z_{K,L}\right) = C - \sum_{j=1}^{L}\left(\lambda_j - \ln\left(\lambda_j\right)\sum_{k=1}^{K} z_{k,j}\right),$$

and obtain

$$
\begin{aligned}
Q\left(\lambda \mid \hat{\lambda}^{(n)}\right) &= \mathrm{E}_{\hat{\lambda}^{(n)}}\left[\mathcal{L}_{comp}\left(\lambda \mid z_{1,1},\ldots,z_{1,L},\ldots,z_{K,1},\ldots,z_{K,L}\right) \mid z_1,\ldots,z_K\right] \\
&= \mathrm{E}_{\hat{\lambda}^{(n)}}\left[C-\sum_{j=1}^{L}\left(\lambda_j-\ln\left(\lambda_j\right)\sum_{k=1}^{K} z_{k,j}\right) \mid z_1,\ldots,z_K\right] \\
&= \mathrm{E}_{\hat{\lambda}^{(n)}}\left[C \mid z_1,\ldots,z_K\right]-\mathrm{E}_{\hat{\lambda}^{(n)}}\left[\sum_{j=1}^{L}\left(\lambda_j-\ln\left(\lambda_j\right)\sum_{k=1}^{K} z_{k,j}\right) \mid z_1,\ldots,z_K\right] \\
&= \mathrm{E}_{\hat{\lambda}^{(n)}}\left[C \mid z_1,\ldots,z_K\right]-\sum_{j=1}^{L}\left(\lambda_j-\ln(\lambda_j)\sum_{k=1}^{K}\mathrm{E}_{\hat{\lambda}^{(n)}}\left[z_{k,j} \mid z_1,\ldots,z_K\right]\right) \\
&= \mathrm{E}_{\hat{\lambda}^{(n)}}\left[C \mid z_1,\ldots,z_K\right]-\sum_{j=1}^{L}\left(\lambda_j-\ln(\lambda_j)\sum_{k=1}^{K}\mathrm{E}_{\hat{\lambda}^{(n)}}\left[z_{k,j} \mid z_k\right]\right),
\end{aligned}
$$

where we have used in the last step that $Z_{k,j}$ is independent of $Z_{k'}$ for all $k' \neq k$. We postpone the computation of the conditional expectation term $\mathrm{E}_{\hat{\lambda}^{(n)}}(k,j) := \mathrm{E}_{\hat{\lambda}^{(n)}}\left[z_{k,j} \mid z_k\right]$ until we have considered the maximization step.

Maximization step: To carry out the maximization step, we need to find the parameter λ that maximizes $Q\left(\lambda \mid \hat{\lambda}^{(n)}\right)$. To do this, we differentiate $Q\left(\lambda \mid \hat{\lambda}^{(n)}\right)$ with respect to λ, set the derivative to 0, and solve for λ. We begin with

$$
\begin{aligned}
0 = \frac{\partial}{\partial\lambda} Q\left(\lambda \mid \hat{\lambda}^{(n)}\right) &= \frac{\partial}{\partial\lambda}\left(\mathrm{E}_{\hat{\lambda}^{(n)}}\left[C \mid z_1,\ldots,z_K\right]-\sum_{j=1}^{L}\left(\lambda_j-\ln\left(\lambda_j\right)\sum_{k=1}^{K}\mathrm{E}_{\hat{\lambda}^{(n)}}(k,j)\right)\right) \\
&= -\frac{\partial}{\partial\lambda}\sum_{j=1}^{L}\left(\lambda_j-\ln\left(\lambda_j\right)\sum_{k=1}^{K}\mathrm{E}_{\hat{\lambda}^{(n)}}(k,j)\right).
\end{aligned}
$$

Considering the j_0th component, $j_0 = 1,\ldots,L$, of λ, we have

$$
0 = -\frac{\partial}{\partial\lambda_{j_0}}\sum_{j=1}^{L}\left(\lambda_j-\ln\left(\lambda_j\right)\sum_{k=1}^{K}\mathrm{E}_{\hat{\lambda}^{(n)}}(k,j)\right) = -1+\frac{1}{\lambda_{j_0}}\sum_{k=1}^{K}\mathrm{E}_{\hat{\lambda}^{(n)}}(k,j_0).
$$

Therefore, for $j_0 = 1,\ldots,L$, the j_0th component $\hat{\lambda}_{j_0}^{(n+1)}$ of the parameter estimate $\hat{\lambda}^{(n+1)}$ for the next iteration is given by the solution of this equation, i.e.,

$$
\hat{\lambda}_{j_0}^{(n+1)} := \sum_{k=1}^{K}\mathrm{E}_{\hat{\lambda}^{(n)}}(k,j_0).
$$

To obtain an explicit expression, we now need to evaluate $\mathrm{E}_{\hat{\lambda}^{(n)}}(k,j_0) = \mathrm{E}_{\hat{\lambda}^{(n)}}\left[Z_{k,j_0} \mid Z_k = z_k\right]$, the step that had been postponed at the end of the expectation step. Since $Z_{k,1},\ldots,Z_{k,L}$ are mutually independent and Poisson-distributed with means $l(k,1)\hat{\lambda}_1^{(n)},\ldots,l(k,L)\hat{\lambda}_L^{(n)}$, and since $Z_k = \sum_{j=1}^{L} Z_{k,j}$, we have, by the result derived in Section A.2 of the Online Appendix, that

$$
\mathrm{E}_{\hat{\lambda}^{(n)}}(k,j_0) = \mathrm{E}_{\hat{\lambda}^{(n)}}\left[Z_{k,j_0} \mid Z_k = z_k\right] = \frac{l(k,j_0)\hat{\lambda}_{j_0}^{(n)} z_k}{\sum_{j=1}^{L} l(k,j)\hat{\lambda}_j^{(n)}}.
$$

We therefore obtain, for the EM algorithm, the iterative expression

$$\hat{\lambda}_{j_0}^{(n+1)} = \hat{\lambda}_{j_0}^{(n)} \sum_{k=1}^{K} \frac{l(k, j_0) z_k}{\sum_{j=1}^{L} l(k,j) \hat{\lambda}_j^{(n)}}.$$

This algorithm is known as the Shepp-Vardi algorithm, named after the researchers who described it (Note 21.4). We can now interpret this algorithm in the context of the deconvolution problem.

21.5.1 Deconvolution for the Poisson data model

The least squares approach to deconvolution was based on a data model in which the data is modeled as a deterministic image that is corrupted by additive Gaussian noise. The Poisson data model is more realistic, but can require more complex analysis steps.

We will show here that for the Poisson data model, the deconvolution problem can be posed as a special case of the model that we used to analyze the PET example. We will therefore immediately have a solution to the problem in the EM algorithm approach that was introduced above.

We assume there are L voxels that make up the object that is imaged by K pixels in possibly different images. As in the case of the PET example, we assume that $Z_1, \ldots, Z_K$ are Poisson-distributed random variables that represent the numbers of photons $z_1, \ldots, z_K$ that are collected during the exposure time interval by the K pixels.

The probability that a photon emitted by the jth object voxel is captured in the kth pixel is, in fact, given by the normalized discrete intensity point spread function $l(k,j) := \frac{1}{C_j} psf(k,j)$ where $C_j = \sum_{k=1}^{K} psf(k,j)$ is a normalization constant to guarantee that $\sum_{k=1}^{K} l(k,j) = 1$ for $j = 1, \ldots, L$. Here, $psf(k,j)$ refers to the mean value of the kth image pixel due to photons detected from a point source located at the jth object voxel. Our task is then to estimate $\lambda_1, \ldots, \lambda_L$, the means of the Poisson-distributed numbers of photons that are emitted by the L object voxels and detected in one of the K pixels. With this formulation, we immediately see that the deconvolution problem is identical to the PET estimation problem we solved with the Shepp-Vardi algorithm. We therefore obtain the iterative solution for the j_0th object voxel as

$$\hat{\lambda}_{j_0}^{(n+1)} = \hat{\lambda}_{j_0}^{(n)} \sum_{k=1}^{K} \frac{l(k, j_0) z_k}{\sum_{j=1}^{L} l(k,j) \hat{\lambda}_j^{(n)}} = \hat{\lambda}_{j_0}^{(n)} \sum_{k=1}^{K} \frac{\frac{psf(k,j_0) z_k}{\sum_{i=1}^{K} psf(i,j_0)}}{\sum_{j=1}^{L} \frac{psf(k,j) \hat{\lambda}_j^{(n)}}{\sum_{i=1}^{K} psf(i,j)}}. \tag{21.15}$$

Although we will not discuss it here, as in the case of the least squares approach to deconvolution, regularization may also be of importance when the deconvolution problem is formulated as a maximum likelihood estimation problem for the Poisson data model.

A problem that is often encountered manifests itself in a very noisy appearance of the estimated object. The reason for this phenomenon can be understood to be due to the fact that often the number of parameters, i.e., the number of voxels in the object, is not significantly lower than the number of data points, i.e., the number of pixels in the images. As the noisy appearance typically becomes more prominent when the algorithm is executed past a certain number of iterations, it is sometimes advisable to reduce the number of iterations used (Note 21.5).

In terms of implementation, Eq. (21.15) can be computed efficiently in the Fourier domain. To see this, we note that the image to be deconvolved, the point spread function, and the object estimate produced by the nth iteration of the algorithm, represented respectively by z_k, $psf(k,j)$, and $\hat{\lambda}_j^{(n)}$ for $k = 1, \ldots, K$ and $j = 1, \ldots, L$ in Eq. (21.15), are in fact 3D arrays as presented in Section 21.2. We can therefore write Eq. (21.15) explicitly

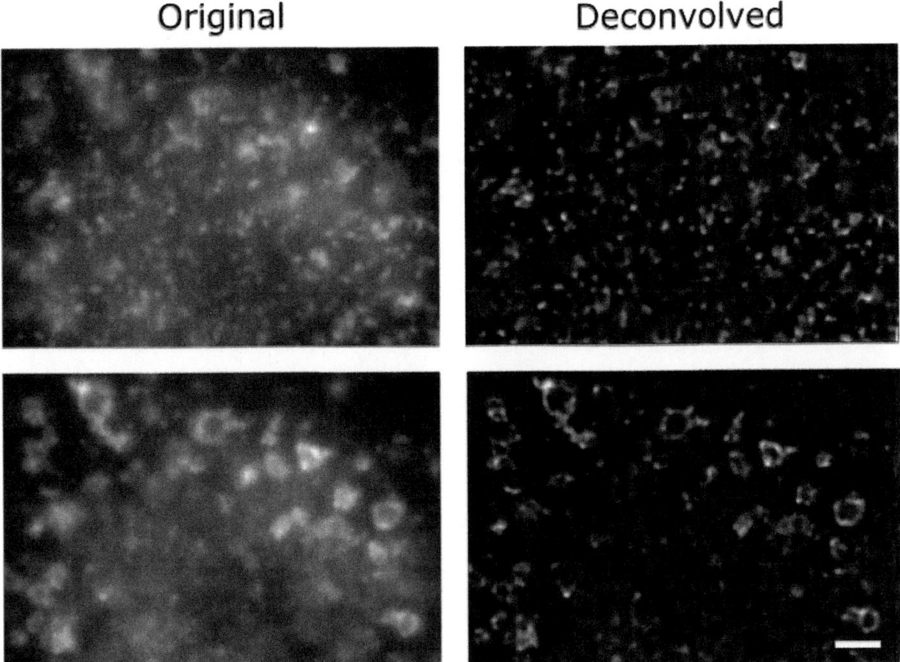

FIGURE 21.1
Images of a fixed HMEC-1 cell deconvolved using the Shepp-Vardi algorithm. The images shown correspond to two z-slices that are 800 nm apart. The HMEC-1 cell is transfected with FcRn-GFP and fixed prior to the acquisition of the z-stack. The point spread function used for the deconvolution is simulated using Eq. (15.17) and the image function of Eq. (15.11), with parameters matching those of the experimental setup used for the acquisition of the HMEC-1 cell z-stack. Scale bar: 2 μm.

in terms of the element-wise multiplication and division, as well as the convolution and cross-correlation, of 3D arrays, and utilize the fact that both the convolution and the cross-correlation can be computed as the inverse transform of products in the Fourier domain. This is the implementation that we will use for the example in Section 21.5.2.

21.5.2 An illustrative example

Here we give an example in which we use the Shepp-Vardi approach to deconvolve a z-stack of images. For our data, we have a z-stack of an endothelial cell that is transfected with FcRn-GFP. The cell was fixed prior to imaging, and the z-stack was acquired in steps of 100 nm. The z-stack spans a 10.8-μm axial range that more than covers the thickness of the cell. For our point spread function, we have a z-stack of a point source simulated according to Eq. (15.17) and the image function of Eq. (15.11), with values for the various parameters based on the experimental setup used to acquire the data. The point spread function z-stack spans an axial range of -2 μm to 2 μm, and has an axial step size of 100 nm to match that of the acquired data. We apply the Shepp-Vardi algorithm to a lateral sub-region of the data z-stack and obtain a deconvolved z-stack. Two z-slices from the deconvolved z-stack are shown in Fig. 21.1 alongside the corresponding z-slices from the data z-stack. In both cases, we can see a significant reduction of blur after deconvolution, allowing FcRn-positive subcellular structures to be visualized with improved clarity.

22

Spatial Statistics

In Section 11.6, we discussed single molecule microscopy and its two prominent applications of single molecule tracking and localization-based super-resolution microscopy. A core aspect of these techniques is the determination of the locations of individual molecules from the images of the acquired data set. When studying the spatial distributions of these molecule locations, the question arises as to whether an apparent spatial pattern is completely random or represents an actual pattern such as a cluster. The study of such questions is the topic of the field of spatial statistics, which we will briefly introduce in the current chapter.

Considering the point patterns shown in Fig. 22.1, the pattern on the left appears to show a tendency for the points to cluster, whereas the pattern in the middle appears to demonstrate a tendency for the points to separate from one another. The pattern on the right exhibits a behavior that appears to be in between. The methods presented in this chapter are designed to analyze such random point patterns and to give us quantitative tools to assess whether patterns are "purely random" or whether there is "clustering" or "inhibition".

22.1 Formal definitions

We will now introduce a technical framework for the discussion of spatial statistics. The content of this section is not necessary for a significant part of what follows. A rigorous treatment does, however, necessitate a careful introduction of the mathematical framework which requires advanced concepts from measure-theoretic probability theory. A reader who is not interested in the rigorous mathematical foundations may want to omit this section.

We consider a space X to be a subset of $\mathbb{R}^d$ and let $\mathcal{X}$ be the Borel σ-algebra of X. A mathematically useful way of characterizing events in the set X is through a counting measure ϕ on X. For each set $B \in \mathcal{X}$ the value $\phi(B)$ equals the number of events in B. We will assume that ϕ is locally finite, meaning that $\phi(B) < \infty$ for all bounded sets $B \in \mathcal{X}$. The measure ϕ characterizes all possible spatial locations of events $s_1, s_2, \cdots \in X$, as for all $B \in \mathcal{X}$, we have that $\phi(B) = \sum_{i=1}^{\infty} I_B(s_i)$, where I_B is the indicator function for the set B. Also, if s is an event then $\phi(\{s\}) > 0$.

We can now introduce a random spatial pattern by introducing random counting measures. For this purpose let $(\Omega, \mathcal{A}, P)$ be a probability space and let Φ be a collection of locally finite counting measures on X. On this collection of counting measures we introduce a σ-algebra $\mathcal{N}$ as the smallest σ-algebra containing the sets of counting measures $\{\phi \in \Phi \mid \phi(B) = n\}$ for all $B \in \mathcal{X}$ and all $n = 0, 1, \ldots$. A *spatial point process* N on X is then defined as a measurable function from $(\Omega, \mathcal{A})$ into $(\Phi, \mathcal{N})$. A spatial point process N defines a probability measure Π_N defined on $(\Phi, \mathcal{N})$ by

$$\Pi_N(Y) := P(N \in Y) := P(\omega \mid N(\omega) \in Y), \quad Y \in \mathcal{N}.$$

We can now define two point processes N_1 and N_2 on $(\Phi, \mathcal{N})$ to be *identically distributed*

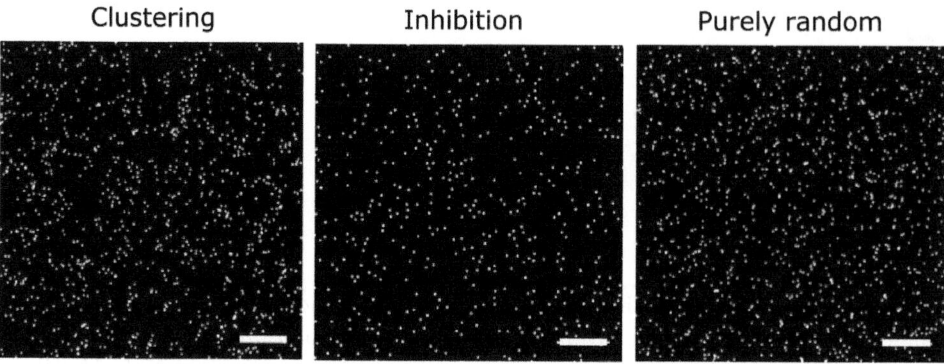

FIGURE 22.1
Three different types of point patterns. The image on the left shows a pattern in which the points exhibit a tendency to cluster. The middle image shows a pattern in which the points show a tendency to be repelled from one another. The image on the right shows a pattern in which the points exhibit a "purely random" spatial distribution. Scale bars: 5 μm.

if $\Pi_{N_1}(Y) = \Pi_{N_2}(Y)$, for all $Y \in \mathcal{N}$. We call a point process N a *simple point process* if $\phi(\{s\}) \in \{0,1\}$ for all $s \in X$ and almost all $\phi \in \Phi$.

22.1.1 Spatial Poisson processes

Poisson point processes are central to our discussion of spatial point processes in a similar manner as Poisson processes are central to the discussion of the temporal aspect of the photon detection process (Section 15.1.1).

Definition 22.1.1 *A spatial process N is an* (inhomogeneous) Poisson process with mean measure μ_N, *where μ_N is a measure that is finite for any compact subset $K \in \mathcal{X}$, if*

1. $P(N(B) \in \{0,1,2,\dots\}) = 1$, *for any $B \in \mathcal{X}$,*
2. *for any collection of disjoint sets $B_1, B_2, \dots, B_k \in \mathcal{X}$, the random variables $N(B_1), N(B_2), \dots, N(B_k)$ are independent, and*
3. *for all $s \in X$, and for any infinitesimal region ds at s,*

$$P(N(ds) = 0) = 1 - \mu_N(ds) + o(\mu_N(ds)),$$
$$P(N(ds) = 1) = \mu_N(ds) + o(\mu_N(ds)),$$
$$P(N(ds) > 1) = o(\mu_N(ds)).$$

It follows from this definition that for $B \in \mathcal{X}$, a Poisson process $N(B)$ with mean measure μ_N has a Poisson distribution with mean $\mu_N(B)$, i.e.,

$$P(N(B) = n) = \frac{(\mu_N(B))^n e^{-\mu_N(B)}}{n!}, \quad n = 0, 1, \dots.$$

If the measure μ_N is such that it is a scalar multiple of the Lebesgue measure ν, i.e., there exists a $\lambda > 0$ such that $\mu_N(B) = \lambda\nu(B)$ for all $B \in \mathcal{X}$, then the Poisson process is called a *homogeneous* Poisson process. A homogeneous spatial Poisson process is also referred to as *completely spatially random* (CSR). As we will see, it describes spatial patterns without further structure. In fact, the "purely random" image of Fig. 22.1 shows data that is generated to be completely spatially random.

Moments of random variables are important quantities to characterize the stochastic behavior of a random variable. Moment measures play a similar role for spatial point processes. The *kth moment measure of the spatial process* N is defined as

$$\mu_N^{(k)}(B_1 \times \cdots \times B_k) := \mathrm{E}\left[N(B_1)N(B_2)\cdots N(B_k)\right] = \int_{\Phi} \phi(B_1)\phi(B_2)\cdots\phi(B_k)\Pi_N(d\phi),$$

where $B_1, B_2, \ldots, B_k \in \mathcal{X}$. For simplicity of notation, we set $\mu_N := \mu_N^{(1)}$.

22.2 Intensity functions of spatial processes

In the prior section, we introduced the notion of a spatial random process, which necessitates a relatively high level of mathematical abstraction to obtain a precise mathematical setting. However, to be able to follow the ideas behind the derivations in this section, we only need two concepts. One is the notion of a spatial process N which "counts" the number of "events" in a set A, i.e., $N(A)$ is a random variable that counts the number of random events in the set A. The other notion is the expected value $\mathrm{E}\left[N(A)\right]$ for this random variable, which is computed in the same way that the expectation of a random variable with discrete values is computed, i.e.,

$$\mathrm{E}\left[N(A)\right] = 0 \cdot P\left(N(A) = 0\right) + 1 \cdot P\left(N(A) = 1\right) + 2 \cdot P\left(N(A) = 2\right) + \cdots .$$

We also define the first moment as the expected value, i.e., $\mu_N^{(1)}(A) := \mathrm{E}\left[N(A)\right]$. Similarly, we can define the second moment as $\mu_N^{(2)}(A \times B) = \mathrm{E}\left[N(A)N(B)\right]$ for two "measurable" sets A and B. For our purposes, we do not need to be concerned about the definition of a measurable set, as all the sets that we will encounter satisfy this technical condition of measurability.

Using the above, we can define quantities that will play an important role in our later developments. The first quantity allows us to compare the "density" of the spatial distribution with a uniform one given by the Lebesgue measure. The *first-order intensity* is given by

$$\lambda(s) := \lim_{\nu(ds)\to 0} \frac{\mu_N(ds)}{\nu(ds)}, \quad s \in X, \tag{22.1}$$

where $\nu(A)$ is the Lebesgue measure of the set A. We should note here that the Lebesque measure of a subset of $\mathbb{R}^2$ is, for sets such as rectangles and disks, nothing other than the area of the set. Here, by the limit $\nu(ds) \to 0$, we mean a sequence of compact sets, such as balls around s, whose extent (e.g., radius) and therefore measure (area or volume) tends to 0. The intensity measures the density of the points, i.e., the expected number of points of the point process N in an infinitesimally small neighborhood of s normalized by the area of the neighborhood.

The *second-order intensity* and *pair correlation function* for the locations s and u give a measure of the stochastic dependence of the numbers of points in infinitesimally small neighborhoods of s and u. These notions are defined as follows. The second-order intensity is given by

$$\lambda_2(s,u) := \lim_{\substack{\nu(ds)\to 0 \\ \nu(du)\to 0}} \frac{\mu_N^{(2)}(ds \times du)}{\nu(ds)\nu(du)}, \quad s,u \in X. \tag{22.2}$$

The pair correlation function g is then defined by

$$g(s,u) := \frac{\lambda_2(s,u)}{\lambda(s)\lambda(u)}, \quad s,u \in X. \tag{22.3}$$

22.2.1 Computing the intensity functions

There are important examples of spatial processes. One such example is a Poisson process, which has a Poisson probability distribution. For a set A, a Poisson process $N(A)$ has the distribution function

$$P(N(A) = n) = \frac{(\mu_N(A))^n e^{-\mu_N(A)}}{n!}, \quad n = 0, 1, \ldots. \tag{22.4}$$

The reader may recall an analogous definition for a temporal Poisson process in Section 15.1.1. In fact, from a more abstract point of view, these notions are essentially identical. The high degree of similarity between the definitions is therefore not coincidental.

We would like to calculate the spatial first- and second-order intensities for a Poisson process. By definition, the first-order intensity is given by

$$\lambda(s) = \lim_{\nu(ds)\to 0} \frac{\mu_N(ds)}{\nu(ds)} = \lim_{\nu(ds)\to 0} \frac{\mathrm{E}\left[N(ds)\right]}{\nu(ds)}, \quad s \in X.$$

We now assume orderliness of the process, which means that for any ds that is a small enough neighborhood of $s \in X$, we have that at most one point is in that neighborhood, i.e., $N(ds) \leq 1$. For such ds we therefore have that

$$\mathrm{E}\left[N(ds)\right] = 1 \cdot P\left(N(ds) = 1\right) + 0 \cdot P\left(N(ds) = 0\right) = P\left(N(ds) = 1\right).$$

This implies that

$$\lambda(s) = \lim_{\nu(ds)\to 0} \frac{\mathrm{E}\left[N(ds)\right]}{\nu(ds)} = \lim_{\nu(ds)\to 0} \frac{P\left(N(ds) = 1\right)}{\nu(ds)} = \lim_{\nu(ds)\to 0} \frac{\mu_N(ds)e^{-\mu_N(ds)}}{\nu(ds)}, \quad s \in X.$$

An important special case of a Poisson process is a homogeneous Poisson process. Here we have that for all "measurable" sets A, there exists a constant $\lambda > 0$ such that $\mu_N(A) = \lambda\nu(A)$. An important aspect of this identity is that the probabilistic properties of the random events depend only on the "size" of the set A and not where A is located, as, for example, shifting or translating A does not change its Lebesque measure. This means that all throughout the space we are studying, the random patterns have the same characteristics and therefore are homogeneous. In fact we call such a random process completely spatially random to suggest that there is no pattern to the spatial events. Such processes play a role similar to Gaussian white noise in the theory of temporal random processes.

For a homogeneous Poisson process, we therefore have, for $s \in X$,

$$\lambda(s) = \lim_{\nu(ds)\to 0} \frac{\mu_N(ds)e^{-\mu_N(ds)}}{\nu(ds)} = \lim_{\nu(ds)\to 0} \frac{\lambda\nu(ds)e^{-\lambda\nu(ds)}}{\nu(ds)} = \lim_{\nu(ds)\to 0} \lambda e^{-\lambda\nu(ds)} = \lambda.$$

This shows that for a homogeneous Poisson process, the intensity function is simply the parameter λ that characterizes the underlying Poisson distribution.

For the second-order intensity, we proceed in analogous fashion and begin with its definition

$$\lambda_2(s,u) = \lim_{\substack{\nu(ds)\to 0\\ \nu(du)\to 0}} \frac{\mu_N^{(2)}(ds \times du)}{\nu(ds)\nu(du)} = \lim_{\substack{\nu(ds)\to 0\\ \nu(du)\to 0}} \frac{\mathrm{E}\left[N(ds)N(du)\right]}{\nu(ds)\nu(du)}, \quad s,u \in X.$$

We have that

$$\begin{aligned}
&\mathrm{E}\left[N(ds)N(du)\right] \\
&\quad = 0 \cdot P\left(\{N(ds)=0\} \cup \{N(du)=0\}\right) + 1 \cdot P\left(\{N(ds)=1\} \cap \{N(du)=1\}\right) + \cdots \\
&\quad = P\left(\{N(ds)=1\} \cap \{N(du)=1\}\right) + \cdots .
\end{aligned}$$

For $s \neq u$, the neighborhoods can be assumed to be disjoint, i.e., $ds \cap du = \{\}$. Moreover, by orderliness of the process, we can assume that for small enough neighborhoods there is at most one event in the neighborhood, i.e., $N(ds) \leq 1$ and $N(du) \leq 1$. For such neighborhoods, we have that $P\left(\{N(ds) > 1\} \cup \{N(du) > 1\}\right) = 0$ and hence

$$\begin{aligned}
\mathrm{E}\left[N(ds)N(du)\right] &= P\left(\{N(ds)=1\} \cap \{N(du)=1\}\right) + \cdots \\
&= P\left(\{N(ds)=1\} \cap \{N(du)=1\}\right),
\end{aligned}$$

which implies, for $s \neq u$ and assuming an orderly process, that

$$\lambda_2(s,u) = \lim_{\substack{\nu(ds)\to 0\\ \nu(du)\to 0}} \frac{\mu_N^{(2)}(ds \times du)}{\nu(ds)\nu(du)} = \lim_{\substack{\nu(ds)\to 0\\ \nu(du)\to 0}} \frac{P\left(\{N(ds)=1\} \cap \{N(du)=1\}\right)}{\nu(ds)\nu(du)}.$$

For a homogeneous Poisson process, we have that $N(ds)$ and $N(du)$ are independent when ds and du are disjoint. Therefore, for $s \neq u$, $s, u \in X$,

$$\begin{aligned}
\lambda_2(s,u) &= \lim_{\substack{\nu(ds)\to 0\\ \nu(du)\to 0}} \frac{P\left(\{N(ds)=1\} \cap \{N(du)=1\}\right)}{\nu(ds)\nu(du)} \\
&= \lim_{\substack{\nu(ds)\to 0\\ \nu(du)\to 0}} \frac{P\left(N(ds)=1\right) P\left(N(du)=1\right)}{\nu(ds)\nu(du)} \\
&= \lim_{\nu(ds)\to 0} \frac{P\left(N(ds)=1\right)}{\nu(ds)} \lim_{\nu(du)\to 0} \frac{P\left(N(du)=1\right)}{\nu(du)} \\
&= \lambda^2.
\end{aligned}$$

The pair correlation function for a homogeneous Poisson process is thus given by

$$g(s,u) = \frac{\lambda_2(s,u)}{\lambda(s)\lambda(u)} = \frac{\lambda^2}{\lambda \cdot \lambda} = 1, \quad s, u \in X, \quad s \neq u.$$

22.2.2 Stationary point processes

Stationarity captures the idea that the stochastic behavior of a spatial process is "identical" over the full space and does not depend on a specific location. There are different notions of stationarity available. A spatial process is called *stationary* if translated versions of the process are identically distributed. We call a spatial point process with second-order intensity λ_2 *second-order stationary* if the first-order intensity λ is constant, i.e., $\lambda(s) = \lambda$, $s \in X$, and the second-order intensity only depends on the difference of the variables, i.e., if there exists a function λ_2^* such that

$$\lambda_2(s,u) = \lambda_2^*(s-u), \quad s, u \in X.$$

It is called *isotropic* if the second-order intensity only depends on the distance between the variables, i.e., if there exists a function λ_2^o such that

$$\lambda_2(s,u) = \lambda_2^o\left(\|s-u\|\right), \quad s, u \in X.$$

Clearly, a homogeneous Poisson process is second-order stationary and is isotropic as $\lambda(s) = \lambda$, $s \in X$, and $\lambda_2(s,u) = \lambda^2$, $s, u \in X$. Homogeneous Poisson processes are the prototypes of "spatially uniform" processes, and as mentioned above are said to be completely spatially random. The existence of patterns in a spatial process is often defined as a deviation of the process from complete spatial randomness.

22.3 K function and L function

One of the basic questions that can be approached with the theory of spatial statistics is whether a point pattern is completely spatially randomly distributed or has some characteristics such as clustering. An important tool for the analysis of such questions is *Ripley's K function*. This function analyzes the distances between a random point and neighboring points. Various characteristics of the underlying spatial process, such as whether or not the process exhibits clustering, can be determined by comparing Ripley's K function for the specific process that is being analyzed with Ripley's K function for a completely spatially random process.

The (probabilistic) K function of a stationary spatial point process with first-order intensity λ is given by

$$K(r) := \frac{1}{\lambda}\mathrm{E}\left[\text{number of additional events within distance } r \text{ of a randomly chosen event}\right],$$

$r \geq 0$.

There is an important relationship between the K function and the second-order intensity λ_2^o for an isotropic process with intensity λ. It can be shown, if $X = \mathbb{R}^2$ and if the process is isotropic, that

$$\lambda K(r) = \frac{2\pi}{\lambda}\int_0^r u\lambda_2^o(u)du, \quad r \geq 0. \tag{22.5}$$

In derivative form, we obtain for the second-order intensity λ_2^o, in terms of the derivative of the K function, the expression

$$\lambda_2^o(r) = \frac{\lambda^2}{2\pi r}K'(r), \quad r > 0.$$

If the spatial process is completely spatially random, i.e., a homogeneous Poisson process with intensity λ, we know that $\lambda_2^o = \lambda^2$. Therefore we can immediately compute the integral expression for K and obtain

$$K(r) = \pi r^2, \quad r \geq 0.$$

A simple translation of the K function is obtained by the *L function*

$$L(r) := \sqrt{\frac{K(r)}{\pi}}, \quad r \geq 0. \tag{22.6}$$

The L function is often used in preference to the K function as its graph can lead to a simple "visual" test for complete spatial randomness. If the process is completely spatially random with intensity λ, then

$$L(r) - r = \sqrt{\frac{K(r)}{\pi}} - r = \sqrt{\frac{\pi r^2}{\pi}} - r = 0.$$

Therefore plotting $L(r)-r$ for a spatial process gives an immediate test for complete spatial randomness as other processes would yield a nonzero plot.

If we have clustering in a spatial process, we will have a higher proportion of events within certain distances of one another than would be expected under complete spatial randomness. Therefore, from the definition of the K function, we expect that in this case the plot of $L(r)-r$ will be above 0. Conversely, if we have inhibition (Section 22.3.1) in a spatial process, whereby a lower proportion of events exist within certain distances of one another, we expect that the plot of $L(r)-r$ will be characterized by negative values. The L function can thus be used to infer properties of a spatial process.

22.3.1 An example of an inhibition process

In many practical situations, the points of a spatial process are constrained by how close they can be to one another due to physical constraints. This leads to *inhibition processes* whose specific characteristics depend on the particular ways in which the inhibition is defined. We describe here a classical example, the Matérn I process.

The model for the Matérn I process starts with a homogeneous Poisson process N on $X=\mathbb{R}^d$ with intensity λ. We then remove all pairs of points that are separated by a distance strictly less than $\delta>0$. The remaining events define the Matérn I process N_I.

We now want to determine the intensity and second-order intensity of this inhibition process N_I. We have shown in Section 22.2.1 that the intensity λ_I can be determined as

$$\lambda_I(s)=\lim_{\nu(ds)\to 0}\frac{P\left(N_I(ds)=1\right)}{\nu(ds)},\qquad s\in X.$$

By the definition of this process, we know that there is only one event in all small neighborhoods of s if there is an event at s and no other event within a distance of δ from s, since otherwise the two events would be removed. We can therefore write

$$\begin{aligned}P\left(N_I(ds)=1\right)&=P\left(\{N(ds)=1\}\cap\{N\left(B_\delta(s)\right)=1\}\right)\\&=P\left(N\left(B_\delta(s)\right)=1\mid N(ds)=1\right)P\left(N(ds)=1\right),\end{aligned}$$

where $B_\delta(s)=\{s':\|s'-s\|<\delta\}$. Hence, for $s\in X$,

$$\begin{aligned}\lambda_I(s)&=\lim_{\nu(ds)\to 0}\frac{P\left(N_I(ds)=1\right)}{\nu(ds)}\\&=\lim_{\nu(ds)\to 0}\frac{P\left(N\left(B_\delta(s)\right)=1\mid N(ds)=1\right)P\left(N(ds)=1\right)}{\nu(ds)}\\&=P\left(N\left(B_\delta(s)\right)=1\mid N\left(\{s\}\right)=1\right)\lambda\\&=P\left(N\left(B_\delta(s)\setminus\{s\}\right)=0\right)\lambda\\&=P\left(N\left(B_\delta(s)\right)=0\right)\lambda\\&=\lambda e^{-\lambda\nu(B_\delta(s))}.\end{aligned}$$

In $\mathbb{R}^2$, the density of the Matérn I process is therefore $\lambda_I=\lambda e^{-\lambda\pi\delta^2}$, as the area of a disk with radius δ is $\pi\delta^2$.

The second-order intensity $\lambda_{2,I}$ of the Matérn I process can be calculated as follows. We know from Section 22.2.1 that it is given as

$$\lambda_{2,I}(s,u)=\lim_{\substack{\nu(ds)\to 0\\ \nu(du)\to 0}}\frac{P\left(\{N_I(ds)=1\}\cap\{N_I(du)=1\}\right)}{\nu(ds)\nu(du)}.$$

We distinguish three cases depending on the distance between the points u and s. If this distance is less than δ, i.e., $\|u - s\| < \delta$, we have that the corresponding events in the base process would be removed by the definition of the inhibition property. Therefore, the probability of such an event occurring in the inhibition process is 0, i.e., for $\|u - s\| < \delta$, we have that

$$\begin{aligned}\lambda_{2,I}(s,u) &= \lim_{\substack{\nu(ds)\to 0\\ \nu(du)\to 0}} \frac{P\left(\{N_I(ds)=1\}\cap\{N_I(du)=1\}\right)}{\nu(ds)\nu(du)}\\ &= \lim_{\substack{\nu(ds)\to 0\\ \nu(du)\to 0}} \frac{P(\{\})}{\nu(ds)\nu(du)} = \lim_{\substack{\nu(ds)\to 0\\ \nu(du)\to 0}} \frac{0}{\nu(ds)\nu(du)}\\ &= 0.\end{aligned}$$

This shows the expected result that the correlation between two spatial locations is 0 if their distance is less than the repulsion distance δ.

If the distance between u and s is at least 2δ, i.e., $\|u - s\| \geq 2\delta$, we have that the balls with centers u and s and radius δ have empty intersection, i.e., $B_\delta(u) \cap B_\delta(s) = \{\}$. Therefore, the random variables $N_I(B_\delta(u))$ and $N_I(B_\delta(s))$ are independent. This implies that the expression for the probability of the joint event factors, and hence

$$\begin{aligned}\lambda_{2,I}(s,u) &= \lim_{\substack{\nu(ds)\to 0\\ \nu(du)\to 0}} \frac{P\left(\{N_I(ds)=1\}\cap\{N_I(du)=1\}\right)}{\nu(ds)\nu(du)}\\ &= \lim_{\substack{\nu(ds)\to 0\\ \nu(du)\to 0}} \frac{P\left(N_I(ds)=1\right)P\left(N_I(du)=1\right)}{\nu(ds)\nu(du)}\\ &= \lim_{\nu(ds)\to 0} \frac{P\left(N_I(ds)=1\right)}{\nu(ds)} \lim_{\nu(du)\to 0} \frac{P\left(N_I(du)=1\right)}{\nu(du)}\\ &= \lambda_I(s)\lambda_I(u)\\ &= \lambda^2 e^{-\lambda v(B_\delta(s)\cup B_\delta(u))}.\end{aligned}$$

It now remains to consider the case where $\delta \leq \|u - s\| < 2\delta$. In this case we have

$$\begin{aligned}\lambda_{2,I}(s,u) &= \lim_{\substack{\nu(ds)\to 0\\ \nu(du)\to 0}} \frac{P\left(\{N_I(ds)=1\}\cap\{N_I(du)=1\}\right)}{\nu(ds)\nu(du)}\\ &= \lim_{\substack{\nu(ds)\to 0\\ \nu(du)\to 0}} \frac{P\left(N_I(ds)=1 \mid N_I(du)=1\right)P\left(N_I(du)=1\right)}{\nu(ds)\nu(du)}\\ &= \lim_{\nu(ds)\to 0} \frac{1}{\nu(ds)} P\left(N_I(ds)=1 \mid N_I\left(\{u\}\right)=1\right)\lambda_I(u).\end{aligned}$$

Now consider $P\left(N_I(ds)=1 \mid N_I(\{u\})=1\right)$. Since $N_I(\{u\}) = 1$, we know that u is an event and that there can be no further events in the ball around u with radius δ. If $N_I(ds) = 1$, for all small neighborhoods of s this means that s is an event and that it is the only event in a ball $B_\delta(s)$ around s of radius δ. Since $\|u - s\| < 2\delta$, the intersection between the two balls is non-empty. Since we know that there is no event other than u in the ball $B_\delta(u)$, there can be no event in the subset of the ball $B_\delta(s)$ that intersects with $B_\delta(u)$. We can therefore eliminate the intersection $B_\delta(s) \cap B_\delta(u)$ from consideration, and the probability that s is an event for N_I, provided that u is an event for N_I, is just the probability that the base process N has an event at s and has no event in $(B_\delta(s) \setminus \{s\}) \setminus B_\delta(u)$, the subset of the ball $B_\delta(s)$ that does not intersect with the ball $B_\delta(u)$. Using that the probability of N having no event in $(B_\delta(s) \setminus \{s\}) \setminus B_\delta(u)$ is given by

$$P\left(N\left((B_\delta(s)\setminus\{s\})\setminus B_\delta(u)\right)=0\right) = e^{-\lambda\nu((B_\delta(s)\setminus\{s\})\setminus B_\delta(u))} = e^{-\lambda\nu(B_\delta(s)\setminus B_\delta(u))},$$

we have that

$$\begin{aligned}
\lambda_{2,I}(s,u) &= \lim_{\nu(ds)\to 0} \frac{1}{\nu(ds)} P\left(N_I(ds)=1 \mid N_I(\{u\})=1\right)\lambda_I(u) \\
&= \lim_{\nu(ds)\to 0} \frac{1}{\nu(ds)} P\left(\{N(ds)=1\} \cap \{N\left((B_\delta(s)\setminus\{s\})\setminus B_\delta(u)\right)=0\}\right)\lambda_I(u) \\
&= \lim_{\nu(ds)\to 0} \frac{1}{\nu(ds)} P\left(N(ds)=1\right) P\left(N\left((B_\delta(s)\setminus\{s\})\setminus B_\delta(u)\right)=0\right)\lambda_I(u) \\
&= \lambda e^{-\lambda\nu(B_\delta(s)\setminus B_\delta(u))}\lambda_I(u) \\
&= \lambda e^{-\lambda\nu(B_\delta(s)\setminus B_\delta(u))}\lambda e^{-\lambda\nu(B_\delta(u))} \\
&= \lambda^2 e^{-\lambda\nu(B_\delta(s)\cup B_\delta(u))}.
\end{aligned}$$

This is the same expression that we obtained for the case where $\|u-s\| \geq 2\delta$. Therefore, summarizing, we have that

$$\lambda_{2,I}(s,u) = \begin{cases} 0, & 0 \leq \|u-s\| < \delta, \\ \lambda^2 e^{-\lambda\nu(B_\delta(s)\cup B_\delta(u))}, & \delta \leq \|u-s\|, \end{cases} \qquad s,u \in X.$$

22.3.2 Estimating the K function

In a concrete experimental situation, we obviously do not know Ripley's K function and need to estimate it based on the measured data, i.e., the measured point locations. Given estimated point locations $s_1, s_2, \ldots, s_N$ in a set A, we define an elementary estimator $\hat{K}_e$ of the K function as

$$\hat{K}_e(r) := \frac{1}{\hat{\lambda}} \sum_{i=1}^{N} \sum_{j=1, i\neq j}^{N} \frac{1}{N} I_{\{x\in\mathbb{R} \mid 0\leq x\leq r\}}\left(\|s_i - s_j\|\right), \quad r \geq 0. \tag{22.7}$$

Here $\hat{\lambda} := \frac{N}{|A|}$, where $|A|$ is the area of A.

This estimator is based on "counting" the number of points that are within the specified distance r of each other. As such, it is the sample version of the probabilistic K function. However, there is a serious problem with this definition that is due to the set A and the number of estimated locations being, by necessity, finite. These issues, referred to as *edge effects*, have to do with the fact that the number of points within the given distance r at the "edges" of the set A are underestimated, as there are no estimated points beyond the set A. To correct for edge effects, various corrections have been proposed in the literature. One such correction offered by Ripley is given as

$$\hat{K}_R(r) := \frac{1}{\hat{\lambda}} \sum_{i=1}^{N} \sum_{j=1; i\neq j}^{N} \frac{1}{w\left(s_i, s_j\right)} \frac{1}{N} I_{\{x\in\mathbb{R} \mid 0\leq x\leq r\}}\left(\|s_i - s_j\|\right), \quad r \geq 0, \tag{22.8}$$

where $w\left(s_i, s_j\right)$ is the proportion of the circumference of a circle centered at s_i and passing through s_j that is inside the set A.

Using the estimator $\hat{K}_R$, we define the L function estimator $\hat{L}_R(r) := \sqrt{\hat{K}_R(r)/\pi}$, $r \geq 0$, and use it to compute the $L(r)-r$ function for the three point patterns of Fig. 22.1. Plotted in Fig. 22.2, we see that, consistent with expectation, the $\hat{L}_R(r)-r$ function for the "purely random", i.e., CSR, point pattern has a value close to zero for any given distance r. For the point pattern that exhibits clustering, the $\hat{L}_R(r)-r$ function stays close to zero for distances r up to 500 nm, shoots up significantly at $r = 500$ nm, and decreases towards zero when r

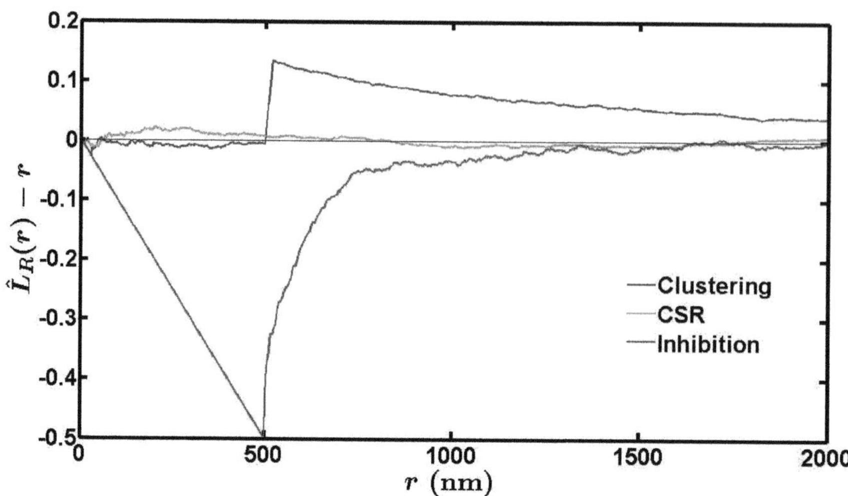

FIGURE 22.2
The $L(r) - r$ function for the three point patterns of Fig. 22.1. For each point pattern, $\hat{L}_R$ is computed using the K function estimator $\hat{K}_R$ of Eq. (22.8).

increases further. The positive value of $\hat{L}_R(r) - r$ for r greater than 500 nm is characteristic of a pattern with clustering. The initial closeness to zero and the peak at $r = 500$ nm are attributable to this particular point pattern consisting of a base set of completely spatially

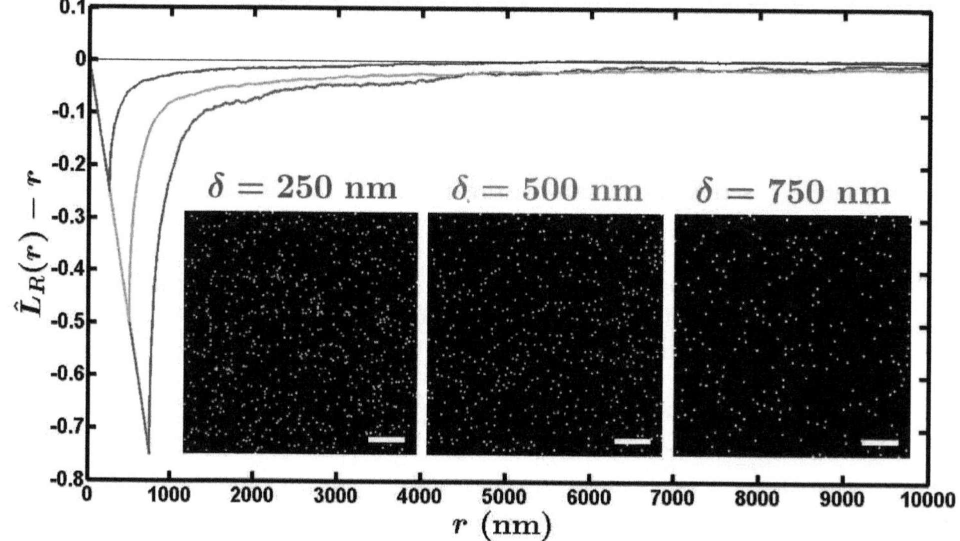

FIGURE 22.3
The $L(r) - r$ function for realizations of Matérn I processes with repulsion distances $\delta = 250$ nm (blue), $\delta = 500$ nm (green), and $\delta = 750$ nm (red). In each case, $\hat{L}_R$ is computed using the K function estimator $\hat{K}_R$ of Eq. (22.8). Small portions of the point patterns used to compute the three curves are also shown. Scale bars: 5 μm.

randomly distributed points with additional points introduced that are a distance of 500 nm to 520 nm away from some of the base points. Lastly, the $\hat{L}_R(r) - r$ function for the pattern that exhibits inhibition can be seen to take on characteristically negative values. The curve is a negatively sloped line between $r = 0$ nm and $r = 500$ nm, increases sharply at $r = 500$ nm, and continues to increase more gradually towards zero with increasing r. The specific behavior of this curve is due to this particular pattern having only points that are spaced more than 500 nm apart from one another. This point pattern is in fact a realization of a Matérn I process with $\delta = 500$ nm.

In Fig. 22.3, a comparison of the $L(r) - r$ function for realizations of three Matérn I processes with different repulsion distances δ is shown. We can see that the three curves are qualitatively similar, each comprising an initial negatively sloped line followed by a sharp and then more gradual increase towards zero. The value of r at which the negatively sloped line ends in each curve, however, tells us the repulsion distance δ of the underlying Matérn I process. We can see that the δ values are 250 nm, 500 nm, and 750 nm for the three Matérn I processes. From the images shown, we can visually verify that the corresponding point patterns are accordingly characterized by different levels of separation between the points.

Notes

Chapter 15

1. The material in Chapter 15 is based on the approach to modeling microscopy experiments introduced in [75, 93, 85], where the fundamental and practical data models are defined. The fundamental data model is based on the theory of spatio-temporal random processes (e.g., [107]).
2. Poisson processes are used in a wide range of disciplines to model the occurrences of random events. For a comprehensive reference on the topic of Poisson processes, see, e.g., [107, 97].
3. In our modeling of an image, we impose the condition that at least one photon is detected during the acquisition time interval, since the locations and time points of photon detection can only be recorded if at least one photon arrives at the detector during the time interval. The mean number of photons comprising an image is therefore determined under this condition. A situation in a completely different application area where a similar problem occurs is illustrated by the following. Assume we are asked to estimate the average speed of cars passing on a sparsely traveled stretch of road late in the night to analyze adherence to a speed limit. We are tasked to measure the speeds of cars driving on the stretch of road between 10 pm and 11 pm each Friday for ten weeks, and to report the average speed during the one-hour time interval. We assume that the arrival times of the cars are distributed according to a Poisson process. We might also assume that the speeds are Gaussian-distributed with unknown mean and variance. It is our task to estimate the mean and the variance.

   A problem then arises if no car drives through the stretch of road during the one-hour interval. This is certainly part of our model as there is a non-zero probability, given a Poisson arrival process (especially one that models a low count), that a zero count will be observed. If no car drives through the stretch of road during the time period, then we cannot measure the speed of the car.

   It is therefore appropriate to only consider the time intervals during which a car was in fact driving on the stretch of road. This amounts to conditioning in the same fashion as we introduced in our image analysis problem.
4. We have here modeled the random variable U_τ and its density f_τ as potentially dependent on the time parameter τ. This importantly allows one to describe the spatial distribution of photons for a moving object, which varies over time with the location of the object. This is explicitly captured in Eq. (15.8), which shows f_τ to be the image function shifted by the potentially time-dependent x_0 and y_0 coordinates of the object. In general, U_τ and f_τ might also depend on other parameters, irrespective of time. Through the image functions presented in Section 15.1.2.1, one can see that they can potentially depend on parameters such as the numerical aperture of the objective, the wavelength of the detected photons, and the axial position z_0 of an out-of-focus point source.

5. A proof for Eqs. (15.12) and (15.13), which give the photon detection rate and distribution profile of the superposition of two independent photon detection processes, is given in [85].
6. The mathematical foundation that underlies all practical data models was introduced in [75, 93, 85]. The modeling of the photon count in a pixel as a Poisson random variable and the result of its corruption by detector readout noise as a Poisson-Gaussian mixture is found in [106, 75, 93, 85]. For the modeling of the data in an EMCCD pixel, the geometric model of multiplication and its high gain and Gaussian approximations were introduced in [30]. The high gain approximation is also found in [11] in a different, but equivalent representation.
7. See [85] for a proof of Eq. (15.15), the mean of the Poisson-distributed number of photons detected in a pixel.
8. Even though in a practical situation the detector $C = \bigcup_{k=1}^{N_p} C_k$ is a finite-sized detector, for each pixel k of the mean image of Fig. 15.10(a), μ_k of Eq. (15.17) is calculated using the Airy image function by way of the time-independent version of the photon distribution profile f_τ of Eq. (15.8), which is defined over $\mathbb{R}^2$ instead of C. To understand why this can be done, see Exercise 15.9.
9. The importance of accurately integrating the photon distribution profile to determine the mean pixel photon counts in a region of interest is investigated in [27] in the context of single molecule localization. In that study, it is shown that the numerical integration needs to be carried out with a sufficient level of accuracy to ensure that the maximum likelihood location estimator closely recovers the true location of the single molecule with a standard deviation that is comparable to the square root of the Cramér-Rao lower bound.
10. The use of the error function to express the mean pixel photon count in the case of a 2D Gaussian image function was introduced in [92]. The same work also shows that the derivative of the mean pixel photon count with respect to some parameter of interest θ can likewise be expressed in terms of the error function when the image of the object of interest is described by the 2D Gaussian profile. Computations of such derivatives can therefore also be sped up by using the error function representation. This is of practical importance for the maximum likelihood estimation of θ, in which the gradient of the log-likelihood function, which depends on such derivatives, needs to be computed iteratively for different values of θ.
11. To exclude the scenario of no photons being detected from the object of interest during the acquisition time interval $[t_0, t]$ from the practical data models, an analysis equivalent to that in the case of the fundamental data model could be carried out by considering, for $z_1, \ldots, z_{N_p}$ representing the numbers of photons detected in the N_p pixels comprising the detector C, the joint probability distribution $p\left(z_1, \ldots, z_{N_p} \mid \mathcal{N}_C(t) > 0\right)$, where $\mathcal{N}_C(t)$ is the Poisson-distributed number of photons emitted by the object of interest and detected by the detector C during the interval $[t_0, t]$. Here $p\left(z_1, \ldots, z_{N_p} \mid \mathcal{N}_C(t) > 0\right)$ is the multivariate Poisson distribution conditioned on there being at least one photon detected by the detector C during the interval $[t_0, t]$.
12. For standard references on the subject of branching processes, see [45, 7].

Chapter 16

1. There are many excellent textbooks on parameter estimation that include significant discussion of maximum likelihood estimation. See, for example, [58, 62, 14].

2. The stochastic independence of the measurements in different pixels is a consequence of the derivation of the practical data model from the fundamental data model as described in Section 15.2.1. In particular, this follows if the photon detection rate Λ and photon distribution profile f_τ are deterministic functions. If, however, we generalized these functions to be stochastic, then independence of the pixel measurements would in general no longer follow. This might, for example, be the case if we consider models of stochastic processes involving the imaged object.
3. See [107] for a treatment of the probability density function for the arrival times of a Poisson process.
4. Maximum likelihood estimation for both the fundamental and the practical data scenario was introduced in [75]. In [1], an extensive study of maximum likelihood estimation in the practical data scenario is presented. The investigation in [1] includes an assessment of estimation in the practically relevant context in which the image of the object of interest is not correctly or accurately described by the image function used by the estimator.
5. The data model presented in Section 16.6.3, in which a fixed (i.e., nonrandom) number of photons is detected in an image, is described in [75]. In the same work, the maximum likelihood location estimator under this data model, in the case of a 2D Gaussian image profile, is also given along with its mean and covariance matrix. The material presented in Sections 16.5.3 and 16.6.2 on the center of mass location estimator in the context of the fundamental data model is new.

Chapter 17

1. A rigorous statement of the Cramér-Rao inequality which includes the regularity assumptions can be found in, e.g., [130, 81]. A standard textbook treatment of the Fisher information matrix and the Cramér-Rao lower bound is given in, e.g., [58, 81, 62, 118]. See [107] for material related to the Fisher information matrix for spatio-temporal processes.
2. The noise coefficient for the comparison of the Fisher information content afforded by different detectors as a function of the signal level in a pixel was first introduced in [30]. For a data model described as a function of a Poisson random variable, a formal proof is given in [30] that shows the value of the noise coefficient to range from 0 to 1.
3. For more information on the excess noise factor associated with stochastic electron multiplication, such as that found in an EMCCD detector, see [66, 52, 54]. For a noise coefficient-based analysis that relates the halving of the maximum Fisher information content to the excess noise factor when an EMCCD detector operated at a high EM gain setting is used to detect relatively large signals in its pixels, see [30].
4. For more details on the Ultrahigh Accuracy Imaging Modality (UAIM), the EMCCD detector-based imaging modality that yields images with high Fisher information content by the use of a high magnification and/or a small exposure time to obtain very low pixel photon counts, see [29]. In the same work, noise coefficient-based analyses are also presented that demonstrate the use of a higher EM gain setting to increase the information content of the data in an EMCCD pixel when a very small signal is expected.
5. Examples of Fisher information analysis involving multi-image data can be found

in [127] for parameter estimation pertaining to the trajectory of a moving object, in [91, 28] for the estimation of the distance separating two objects, and in [86, 88] for 3D localization using a MUM setup.

Chapter 18

1. For additional results on the performance of maximum likelihood localization when applied to simulated single molecule image data, see, e.g., [93, 30, 29].
2. The Cramér-Rao lower bound for the localization of an object in two dimensions, under both the fundamental data model and the practical data models, was first described, derived, and studied in [75] and [93]. These works include general results that can be customized with any image profile, as well as specific expressions involving the Airy and 2D Gaussian image profiles.
3. The Cramér-Rao lower bound or its square root is often referred to in the literature as a bound on, or a limit of, the "accuracy" or "precision" with which the parameter in question can be estimated. The usage of "accuracy" by some and "precision" by others is due to the fact that these terms can mean different things to different people, especially those who come from different backgrounds. For example, Ronald Fisher, after whom the Fisher information is named, uses the term "intrinsic accuracy" to refer to the Fisher information [38]. Others, however, use "accuracy" to refer to the bias of an estimator (e.g., [31, 121]), while some statisticians assess the "accuracy" of estimations in terms of variance-based expressions (e.g., [36]). In terms of "precision", some scientists use it synonymously with standard deviation (e.g., [31, 121]), whereas some statisticians use it to refer to the inverse of the variance (e.g., [51]).

   In works such as [75, 93, 91], the Cramér-Rao lower bound or its square root is described as a limit of the "accuracy" with which the parameter in question can be estimated. This usage of the term "accuracy", which equates it with variance or standard deviation, is based on the definition specified by the International Organization for Standardization (ISO). Given repeated measurements of a quantity, ISO defines "accuracy" as a combination of the closeness of the mean of the measurements to the true value and the closeness of agreement between the measurements. A common way to quantify this definition of "accuracy" is through the mean squared error (e.g., [123, 76]), which is equal to the sum of the square of the bias of the estimator and the variance of the estimator. An estimator is therefore highly "accurate" only if it has a small bias and a small variance. Moreover, for an unbiased estimator, the mean squared error is just the variance of the estimator. In other words, the variance of an unbiased estimator determines its "accuracy". It therefore follows that as the Cramér-Rao lower bound is a bound on the variance of any unbiased estimator of the parameter in question, by the mean squared error definition of "accuracy" it is in fact a limit of the "accuracy" for the estimation of the parameter [76].

   For clarity, we avoid the use of either "accuracy" or "precision" in this book when referring to the Cramér-Rao lower bound or its square root. We simply adhere to the mathematical definitions and refer to the bound and its square root as a lower bound on the variance and standard deviation, respectively, with which the parameter in question can be estimated. Similarly, we use the statistical terms bias and variance to describe the basic characteristics of statistical estimators.
4. Examples of the exploitation of nonuniform illumination, in possible conjunction

with time-varying excitation to increase the Fisher information for the estimation of the location of a single molecule, are found in [10, 33].

5. For expressions of Fisher information matrices pertaining to the estimation of parameters associated with non-stationary objects in the context of the fundamental data model, see [128, 113].

6. In Section 18.10.3, our analysis involving the sum image implicitly made use of the fact that the summing of independent EMCCD images produces an EMCCD image. More precisely, we used the fact that the EMCCD data model and its approximations are applicable to the sum image, and that in the modeling of the sum image and an individual image actually acquired with an EMCCD camera, the only difference lies in the increased values for the molecule and background photon counts and the readout noise standard deviation in the case of the sum image. Referring to Eq. (15.25), this relies on two points — one, the fact that the sum of two independent amplified signals $S_{amp,k,1} + B_{amp,k,1}$ and $S_{amp,k,2} + B_{amp,k,2}$, each distributed according to the probability mass function of Eq. (15.26) with μ_k replaced by $\nu_{k,1}$ and $\nu_{k,2}$, respectively, is an amplified signal $S_{amp,k,sum} + B_{amp,k,sum}$ distributed according to the same probability mass function with μ_k replaced by $\nu_{k,sum} = \nu_{k,1} + \nu_{k,2}$, and two, the fact that the sum of two independent Gaussian random variables $W_{k,1}$ and $W_{k,2}$ with means $\eta_{k,1}$ and $\eta_{k,2}$ and variances $\sigma^2_{k,1}$ and $\sigma^2_{k,2}$ is a Gaussian random variable $W_{k,sum}$ with mean $\eta_{k,sum} = \eta_{k,1} + \eta_{k,2}$ and variance $\sigma^2_{k,sum} = \sigma^2_{k,1} + \sigma^2_{k,2}$. The second result is a well-known property of independent Gaussian random variables, while the first result essentially states that the sum of two independent amplified signals, each originating from an initial Poisson signal, is an amplified signal of the same stochastic description, originating from an initial Poisson signal that is the sum of the initial Poisson signals for the two independent amplified signals. Putting the two results together, we have that the summing of independent EMCCD images yields an image that retains the same underlying stochastic description, and therefore is also an EMCCD image. We will not provide a proof for the first result, and will simply give the intuitive explanation that since our modeling of electron multiplication (Online Appendix Section G.1) assumes that an EMCCD detector multiplies each electron independently of all other electrons according to the same stochastic process, the sum of two separately multiplied batches of electrons should be no different from the electron count that would have been obtained if the two batches of electrons were first combined and then multiplied. (For the simpler case of the CCD/CMOS data model, we leave it as an exercise for the reader (Exercise 18.6) to show that the sum of two CCD/CMOS images is also a CCD/CMOS image.)

Chapter 19

1. The Cramér-Rao lower bound for the 3D localization of an object was first described, derived, and investigated in [90] under both the fundamental and the practical data scenarios. The terminology *depth discrimination*, used in the sense of the optical microscope's ability to discern the axial locations of objects, was introduced in [86].

2. The extension of the Cramér-Rao lower bound for 3D localization to the case of image data acquired of multiple focal planes using a MUM setup was first presented and studied in [86] and [88]. In [88], the result was used as a benchmark for assessing the performance of an implementation of MUMLA used to localize point sources from simulated and experimental MUM data. It was demonstrated

in these works that MUM can be employed to overcome the depth discrimination problem. In [111], an approach that utilizes the Cramér-Rao lower bound for 3D localization in the context of MUM was presented for the determination of the number of focal planes and the inter-plane spacings to use in a MUM setup.

3. MUMLA was introduced in [88]. For examples of its use in analyzing simulated and experimental (i.e., live cell single molecule) MUM data, see [88, 87].

Chapter 20

1. The approach to consider Rayleigh's two-point resolution problem as a parameter estimation problem analyzed using the Fisher information was proposed in [91]. The presentation of this chapter follows [91] for the 2D case and [26] for the 3D case. In [28], different imaging modalities such as multifocal plane microscopy and localization-based super-resolution microscopy are compared from the point of view of the 3D resolution problem. The two-object resolution problem as presented here is closely related to the problem of the localization of two objects that are closely spaced. The relationship between these problems is studied in detail in [94].
2. See [91] for the estimation of small distances of separation from experimental images of Cy5 molecules.

Chapter 21

1. The presentation of the EM algorithm and the Shepp-Vardi algorithm follows that in [107] and [110].
2. A fluorescent bead is a good approximation of a point source only if it is small enough. In [25], this is demonstrated by the finding that the maximum likelihood localization of a bead via the fitting of an Airy profile is suitable only for beads up to 100 nm in diameter. The assessment of suitability is based on the estimator's ability to recover a bead's true location with a standard deviation that is comparable to the square root of the Cramér-Rao lower bound.
3. In [60] the influence of noise corruption of point spread functions on deconvolved images is investigated. In this work, a method of noise reduction is proposed for the denoising of experimentally obtained point spread functions. The method is based on a state space realization of the given 3D image data set, and uses singular value decompositions to reduce the noise level without significantly suppressing the peak shape of the point spread function.
4. The Shepp-Vardi algorithm [104] is a re-derivation of the Richardson-Lucy deconvolution algorithm in the context of positron emission tomography. The Richardson-Lucy algorithm derives from the application of Bayes' theorem, and was described separately by William Richardson [96] and Leon Lucy [63] in the 1970s.
5. For methods that can be used to reduce the noise artifacts that are often present in the object estimated by the Shepp-Vardi algorithm, see [107].

Chapter 22

1. The presentation of this chapter follows page 619 and onwards in the book by N. A. C. Cressie [34].

Exercises

Chapter 15

1. Let X be a Poisson random variable with mean λ. Show that for $k = 1, 2, \ldots,$

$$P(X = k \mid X > 0) = \frac{e^{-\lambda}\lambda^k}{k!\,(1 - e^{-\lambda})}.$$

   Use this conditional probability mass function to show that

$$\mathrm{E}[X \mid X > 0] = \frac{\lambda}{1 - e^{-\lambda}}.$$

2. Use the expression for the Airy image function and the first zero of J_1 to derive Rayleigh's criterion.
3. Use a computational software program to simulate an Airy image function for typical values of the numerical aperture n_a and the wavelength λ.
4. For several α values, simulate an Airy image function. Let $\sigma_g = 1.323/\alpha$. Simulate a 2D Gaussian image function with parameter σ_g and overlay it with the Airy image function simulated with α. How do the Gaussian and Airy image functions compare?
5. For both the Airy image function and the 2D Gaussian image function, verify that the second condition in the definition of an image function is satisfied, i.e., verify that $\int_{\mathbb{R}^2} q(x, y)dxdy = 1$.
6. Show that for the 2D Gaussian image function $q(x, y) = \frac{1}{2\pi\sigma_g^2} e^{-\frac{1}{2\sigma_g^2}(x^2+y^2)}$, $(x, y) \in \mathbb{R}^2$,

$$\int_{\mathbb{R}}\int_{\mathbb{R}} x\; q(x, y)\; dxdy = 0 \quad \text{and} \quad \int_{\mathbb{R}}\int_{\mathbb{R}} y\; q(x, y)\; dxdy = 0.$$

   Use this result to show that if a photon has an arrival point on the detector modeled by the random vector $U = [X\ Y]$ with probability density $f(x, y) = \frac{1}{M^2} q(\frac{x}{M} - x_0, \frac{y}{M} - y_0)$, $(x, y) \in \mathbb{R}^2$, $(x_0, y_0) \in \mathbb{R}^2$, $M > 0$, then

$$\mathrm{E}[U] = (Mx_0, My_0), \tag{22.9}$$

   i.e., the expected position of the impact point of the photon is given by the location parameter (x_0, y_0) multiplied by the lateral magnification M. Also, show that

$$\mathrm{cov}(U) = M^2\sigma_g^2 \begin{bmatrix} 1 & 0 \\ 0 & 1 \end{bmatrix}. \tag{22.10}$$

7. We generalize the preceding problem by considering an image function q that is symmetric, i.e., $q(x, y) = q(-x, y) = q(x, -y)$ for $(x, y) \in \mathbb{R}^2$. Assume that the

first and second moments exist for the probability density functions of X and Y (i.e., $\mathrm{E}[X]$, $\mathrm{E}\left[X^2\right]$, $\mathrm{E}[Y]$, and $\mathrm{E}\left[Y^2\right]$ all exist), and that X and Y are uncorrelated (i.e., $\mathrm{E}[XY] = \mathrm{E}[X]\cdot\mathrm{E}[Y]$). Show that results analogous to Eqs. (22.9) and (22.10) are obtained, i.e., show that

$$\mathrm{E}[U] = (Mx_0, My_0)$$

and

$$\mathrm{cov}(U) = \begin{bmatrix} \mathrm{E}\left[X^2\right] - M^2x_0^2 & 0 \\ 0 & \mathrm{E}\left[Y^2\right] - M^2y_0^2 \end{bmatrix}.$$

8. Verify that Eq. (15.13) is indeed a photon distribution profile.
9. Let $\Lambda(\tau)$, $\tau \geq t_0$, and $f_\tau(x,y)$, $(x,y) \in \mathbb{R}^2$, $\tau \geq t_0$, be the photon detection rate and photon distribution profile that describe photon detection over an infinitely large detector $C = \mathbb{R}^2$. Let $C^{rd} \subseteq \mathbb{R}^2$ be an open subset of the infinitely large detector that may represent a "reduced" detector with a possibly finite size. Then it can be shown that photon detection over C^{rd} is described by photon detection rate Λ^{rd} and photon distribution profile f_τ^{rd}, given by

$$\Lambda^{rd}(\tau) = \rho_\tau \Lambda(\tau), \quad \tau \geq t_0, \quad \text{and} \quad f_\tau^{rd}(x,y) = \frac{1}{\rho_\tau} f_\tau(x,y), \quad (x,y) \in C^{rd}, \quad \tau \geq t_0,$$

where $\rho_\tau := \int_{C^{rd}} f_\tau(x,y)dxdy$. Suppose C^{rd} is made up of a collection of N_p non-overlapping pixels $\{C_1, \ldots, C_{N_p}\}$. By starting with Λ^{rd} and f_τ^{rd}, show that for $k = 1, \ldots, N_p$, the expression for μ_k, the mean number of photons detected in the kth pixel and over the time interval $[t_0, t]$, simplifies to the expression of Eq. (15.15).

Chapter 16

1. Show that the maximum likelihood estimator $\hat{\theta}_{MLE}$ of Eq. (16.6) indeed maximizes the log-likelihood function of Eq. (16.5).
2. Verify that the maximum likelihood estimator $\hat{\theta}_{MLE}$ of Eq. (16.9) indeed maximizes the log-likelihood function of Eq. (16.8).
3. Suppose an image of a stationary object located at $\theta = (x_0, y_0) \in \mathbb{R}^2$ is acquired under the fixed photon count data model described in Section 16.6.3. It therefore consists of the coordinates $(x_k, y_k) \in \mathbb{R}^2$, $k = 1, \ldots, N_0$, of N_0 photons detected from the object. Let $[X\ Y]$ be the random vector that describes the location of a detected photon, and let $\hat{\theta}_{CM}$ be an estimator of the object location θ, given by

$$\hat{\theta}_{CM} = \begin{bmatrix}\hat{x}_0 & \hat{y}_0\end{bmatrix} = \begin{bmatrix}\frac{1}{MN_0}\sum_{k=1}^{N_0} x_k & \frac{1}{MN_0}\sum_{k=1}^{N_0} y_k\end{bmatrix},$$

where M is the magnification of the imaging system. Show that if $\mathrm{E}\left[[X\ Y]\right] = [Mx_0\ My_0]$, then $\hat{\theta}_{CM}$ is an unbiased estimator of θ.
4. Calculate the variance of the sample mean $\hat{m}$ of Eq. (16.15).

Chapter 17

1. The derivative, with respect to parameter θ, of the log-likelihood function of θ, given the measurement z, is called the *score*. Show that the expectation of the score, i.e., $\mathrm{E}\left[\frac{\partial \mathcal{L}(\theta|z)}{\partial \theta}\right]$, is zero.

2. For a scalar parameter θ, demonstrate the additivity property of the Fisher information matrix.

3. Due to the arrival of photons according to a Poisson process, repeat acquisitions using the same exposure time will produce images consisting of different numbers of detected photons. Suppose, however, that a hypothetical detector is used that guarantees the detection of exactly N_0 photons per image, N_0 being some positive integer. Further suppose that the imaged object is stationary. In this case, if the detector records the locations at which the photons are detected, then the log-likelihood function corresponding to an image is given by

$$\mathcal{L}(\theta|r_1, \ldots, r_{N_0}) = \sum_{k=1}^{N_0} \ln f_\theta(r_k),$$

where $r_k = (x_k, y_k) \in C$ denotes the location on the detector C at which the kth photon is detected, and $f_\theta(x,y)$, $(x,y) \in C$, is the photon distribution profile. Using this log-likelihood function, show that the Cramér-Rao lower bound is given by

$$\mathbf{I}^{-1}(\theta) = \frac{1}{N_0}\left(\int_C \frac{1}{f_\theta(x,y)}\left(\frac{\partial f_\theta(x,y)}{\partial\theta}\right)^T\left(\frac{\partial f_\theta(x,y)}{\partial\theta}\right) dxdy\right)^{-1},$$

which is just the expression of Eq. (17.13) with $N_{photons}$ replaced by N_0.

4. Assume that a translation-invariant imaging system with an infinitely large detector $C = \mathbb{R}^2$ is used to image a stationary object located at $\theta = (x_0, y_0) \in \mathbb{R}^2$, whose image is described by the 2D Gaussian image function. The photon distribution profile for this scenario is given by

$$f_\theta(x,y) = \frac{1}{M^2}\frac{1}{2\pi\sigma_g^2} e^{-\frac{1}{2\sigma_g^2}\left(\left(\frac{x}{M}-x_0\right)^2+\left(\frac{y}{M}-y_0\right)^2\right)}, \quad (x,y) \in \mathbb{R}^2.$$

Use the result of Exercise 17.3 to show that under the data model that guarantees the detection of exactly $N_0 > 0$ photons in each repeat image acquisition, the Cramér-Rao lower bound pertaining to the estimation of this object's positional coordinates is given by

$$\mathbf{I}^{-1}(\theta) = \frac{1}{N_0}\cdot\begin{bmatrix}\sigma_g^2 & 0\\ 0 & \sigma_g^2\end{bmatrix}.$$

5. Using Eq. (17.18), show that for a given $\nu_{\theta,k}$, an increase in the value of the expectation term results in a decrease in the lower bound on the variance for estimating θ (i.e., a smaller Cramér-Rao lower bound).

Chapter 18

1. For different choices of numerical aperture and wavelength, calculate δ_{x_0}, the square root of the Cramér-Rao lower bound on the variance with which the x_0 coordinate of an in-focus point source can be estimated from an image acquired under the fundamental data scenario in which the detector has an infinitely large photon detection area. Assume that on average 100 photons are detected in an image, and that the image of the point source is given by the Airy image function.

2. Assume that an objective with numerical aperture $n_a = 1.4$ is used to image an in-focus GFP molecule which has an emission wavelength of $\lambda = 520$ nm. At a

minimum, how many photons on average need to be detected in an image in order to potentially determine the location of the molecule with a standard deviation as small as 50 nm? Assume that images of the molecule are acquired under the fundamental data scenario in which the detector has an infinitely large photon detection area. Also assume that the image of the molecule is described by the Airy image function. Perform the same calculation for standard deviations of 25 nm, 5 nm, and 1 nm.

3. Assume that an objective with numerical aperture $n_a = 1.4$ is used to image an in-focus fluorescent molecule. By what percentage does δ_{x_0}, the square root of the Cramér-Rao lower bound on the variance with which the molecule's x_0 coordinate can be estimated, worsen if the objective is replaced by one with numerical aperture $n_a = 1.2$? Assume that the molecule is imaged under the fundamental data scenario in which the detector has an infinitely large photon detection area, and that its image is given by the Airy image function. Perform the same calculation for replacement objectives with numerical apertures $n_a = 1$, $n_a = 0.9$, and $n_a = 0.7$.

4. Suppose an average of 300 photons are detected per image from a QD with an emission wavelength of 655 nm. Further suppose that in a second experiment another QD is imaged which has an emission wavelength of 525 nm. Assuming that the same objective is used in the two experiments, on average how many photons need to be detected per image in the second experiment in order to obtain the same lower bound δ_{x_0} on the standard deviation for estimating the QD's x_0 coordinate? Assume that both experiments are carried out under the fundamental data scenario in which the detector has an infinitely large photon detection area. Assume also that the point spread function is given by the Airy image function.

5. Let J be the temporal matrix of Eq. (17.12), i.e.,

$$J = \int_{t_0}^{t} \frac{1}{\Lambda_\theta(\tau)} \left(\frac{\partial \Lambda_\theta(\tau)}{\partial \theta} \right)^T \frac{\partial \Lambda_\theta(\tau)}{\partial \theta} d\tau.$$

Verify that J is a positive semi-definite matrix by showing that for every row vector w of real numbers the same length as the parameter vector θ, the scalar wJw^T is nonnegative.

6. Consider two independent N_p-pixel images each described by the CCD/CMOS data model. Based on Eq. (15.20), let the first image be described by $\mathcal{I}_{k,1} = S_{k,1} + B_{k,1} + W_{k,1}$, $k = 1, \ldots, N_p$, where $S_{k,1} + B_{k,1}$ is a Poisson random variable with mean $\nu_{k,1}$, and $W_{k,1}$ is a Gaussian random variable with mean $\eta_{k,1}$ and variance $\sigma_{k,1}^2$. Similarly, let the second image be described by $\mathcal{I}_{k,2} = S_{k,2} + B_{k,2} + W_{k,2}$, $k = 1, \ldots, N_p$, where $S_{k,2} + B_{k,2}$ is a Poisson random variable with mean $\nu_{k,2}$, and $W_{k,2}$ is a Gaussian random variable with mean $\eta_{k,2}$ and variance $\sigma_{k,2}^2$. Show that the N_p-pixel sum image formed by adding the two images is also described by the CCD/CMOS data model.

7. Suppose that N_{im} repeat images of a stationary object are acquired under unchanging conditions, and suppose that the acquired images are described by the Poisson data model. Show that the Fisher information matrix corresponding to the set of N_{im} images is identical to the Fisher information matrix corresponding to the sum image formed by adding the N_{im} repeat images.

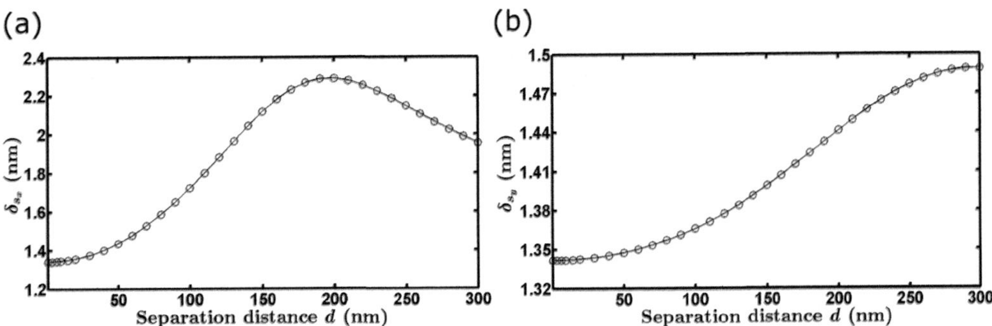

FIGURE B
Dependence of the lower bounds for midpoint estimation under the CCD/CMOS data model. Lower bounds δ_{s_x} and δ_{s_y} on the standard deviations for estimating (a) the coordinate s_x and (b) the coordinate s_y of the midpoint between two identical stationary and in-focus point-like objects that lie parallel to the x-axis are calculated under the CCD/CMOS data model. Both δ_{s_x} and δ_{s_y} are shown as a function of the distance d that separates the two objects, ranging from 1 nm to 300 nm. See Fig. 20.5 for more details on the settings under which δ_{s_x} and δ_{s_y} are computed.

Chapter 19

1. Show that for an in-focus point source, i.e., a point source axially located at $z_0 = 0$, the image function q_{z_0} of Eq. (19.4), with phase function W_{z_0} given by Eq. (19.10), reduces to the Airy image function of Eq. (15.9). You may find useful the Bessel function identities $\int_0^c tJ_0(t)dt = cJ_1(c)$, $c > 0$, and $\int_0^\infty \frac{J_1^2(t)}{t}dt = \frac{1}{2}$.

2. Consider the image function q_{z_0} of Eq. (19.4), with phase function W_{z_0} given by Eq. (19.10). Show that for an in-focus point source, i.e., a point source axially located at $z_0 = 0$, the square root of the Cramér-Rao lower bound on the variance for estimating its lateral coordinates, i.e., δ_{x_0} and δ_{y_0} of Eq. (19.7), reduces to δ_{x_0} and δ_{y_0} of Eq. (18.17). You may find useful the result of Exercise 19.1 and the Bessel function identities $\int_0^c t^2 J_1(t)dt = c^2 J_2(c)$, $c > 0$, and $\int_0^\infty \frac{J_2^2(t)}{t}dt = \frac{1}{4}$.

Chapter 20

1. Derive, within the context of the example of Section 20.4, an expression for δ_{s_z}, the lower bound on the standard deviation with which the axial location s_z of the midpoint between two point sources on the optical axis can be estimated.

2. For the example involving two in-focus point sources in Section 20.5, the lower bounds δ_{s_x} and δ_{s_y} on the standard deviations for estimating the s_x and s_y coordinates of the midpoint between the point sources are shown in Fig. B as functions of the distance d that separates the point sources. Provide an explanation for the patterns exhibited by δ_{s_x} and δ_{s_y}.

Chapter 21

1. Given the convolution expression in Eq. (21.1) and assuming that the object is an infinitesimally small point object, what is the true image of this object?

2. Suppose there are $K = 2$ pixels comprising the detector with which we collect photons emitted by $L = 2$ sources. Given the observed data $z_1 = 1$ photon

and $z_2 = 2$ photons, show that the probability distribution of Eq. (21.14), i.e., $p_{incomp,\lambda}(z_1, z_2)$, is the sum of the results obtained by evaluating the probability distribution of Eq. (21.13), i.e., $p_{comp,\lambda}(z_{1,1}, z_{1,2}, z_{2,1}, z_{2,2})$, for all realizations of the complete data $(z_{1,1}, z_{1,2}, z_{2,1}, z_{2,2})$ that give rise to the incomplete data $(z_1, z_2) = (1, 2)$.

Figure Credits

Chapter 1

Figure 1.1(a) Image courtesy of Rita Greer under Free Art License 1.3 (https://artlibre.org/licence/lal/en), obtained from Wikimedia Commons.

Figure 1.1(b) Photograph courtesy of the Historical Division of the National Museum of Health and Medicine ("Microscope belonging to Dr. Robert Hooke, 1665" (190301-D-MP902-0033). John Mayall Jr. Collection, [Object ID: M-030.00276]). Drawing courtesy of Wellcome Collection under the Creative Commons Attribution 4.0 International Public License (CC BY 4.0), obtained from https://wellcomecollection.org/works/e2bynyv4 and used here in unmodified form. The full text of the CC BY 4.0 license is available at https://creativecommons.org/licenses/by/4.0/legalcode.

Figure 1.1(c) Image courtesy of Wellcome Collection under the Creative Commons Attribution 4.0 International Public License (CC BY 4.0), obtained from https://wellcomecollection.org/works/jcm8kb66 and used here in unmodified form. The full text of the CC BY 4.0 license is available at https://creativecommons.org/licenses/by/4.0/legalcode.

Figure 1.2(a) Image courtesy of Wellcome Collection under the Creative Commons Attribution 4.0 International Public License (CC BY 4.0), obtained from https://wellcomecollection.org/works/ft6mf62b and used here in unmodified form. The full text of the CC BY 4.0 license is available at https://creativecommons.org/licenses/by/4.0/legalcode.

Figure 1.2(b) Image courtesy of Wellcome Collection under the Creative Commons Attribution 4.0 International Public License (CC BY 4.0), obtained from https://wellcomecollection.org/works/z953keyc and used here in modified form. The image shown is a cropped version of the original that excludes labels and other content. The full text of the CC BY 4.0 license is available at https://creativecommons.org/licenses/by/4.0/legalcode.

Figure 1.2(c) Image courtesy of Wellcome Collection under the Creative Commons Attribution 4.0 International Public License (CC BY 4.0), obtained from https://wellcomecollection.org/works/tfb74a8u and used here in unmodified form. The full text of the CC BY 4.0 license is available at https://creativecommons.org/licenses/by/4.0/legalcode.

Chapter 2

Figure 2.2 The data shown are part of a study that investigates the sites of FcRn function in mice (H. P. Montoyo, C. Vaccaro, M. Hafner, R. J. Ober, W. Mueller, and E. S. Ward. Conditional deletion of the MHC Class I-related receptor FcRn reveals the sites of IgG homeostasis in mice. *Proceedings of the National Academy of Sciences USA*, 106:2788-2793, 2009).

Figure 2.3 Adapted from figure originally published in *The Journal of Immunology* (R. J. Ober, C. Martinez, C. Vaccaro, J. Zhou, and E. S. Ward. 2004. Visualizing the site and dynamics of IgG salvage by the MHC class I-related receptor, FcRn. *J. Immunol.* 172:2021-2029. Copyright © 2004 The American Association of Immunologists, Inc.).

Figure 2.6 Adapted from figures originally published in *The Journal of Immunology* (R. J. Ober, C. Martinez, C. Vaccaro, J. Zhou, and E. S. Ward. 2004. Visualizing the site and dynamics of IgG salvage by the MHC class I-related receptor, FcRn. *J. Immunol.* 172:2021-2029. Copyright © 2004 The American Association of Immunologists, Inc.).

Figure 2.7 Image courtesy of Ramraj Velmurugan.

Figure 2.9 This image is a modified version of a drawing created by Ramraj Velmurugan.

Figure 2.10 Adapted from figure originally published in *The Journal of Immunology* (R. J. Ober, C. Martinez, C. Vaccaro, J. Zhou, and E. S. Ward. 2004. Visualizing the site and dynamics of IgG salvage by the MHC class I-related receptor, FcRn. *J. Immunol.* 172:2021-2029. Copyright © 2004 The American Association of Immunologists, Inc.).

Chapter 3

Figure 3.9(a) Image courtesy of Carl Zeiss Microscopy GmbH, Jena, Germany.

Figure 3.10(a) Data courtesy of Semrock, Inc.

Figure 3.10(b) Data courtesy of Hamamatsu Photonics K.K.

Chapter 5

Figure 5.2 This illustration is a relabeled version of a public domain image from the U.S. National Library of Medicine's Genetics Home Reference.

Figure 5.3 This illustration is a slightly modified version of an image created and released into the public domain by Mad Price Ball.

Figure 5.9 This illustration is a relabeled version of an image created and released into the public domain by Mariana Ruiz Villarreal.

Figure 5.12 The image shown, in each case, is that of PDB ID 1AO6 (S. Sugio, A. Kashima, S. Mochizuki, M. Noda, and K. Kobayashi. Crystal structure of human serum albumin at 2.5 Å resolution. *Protein Engineering*, 12:439-446, 1999) from the Research Collaboratory for Structural Bioinformatics Protein Data Bank (RCSB PDB), created with the JSmol viewer.

Figure 5.13 This illustration is a relabeled version of an image created and released into the public domain by Mariana Ruiz Villarreal.

Figure 5.14 The image shown is a slightly modified version of an image created by Zsolt Bikádi.

Figure 5.17(b) Image courtesy of Jane Shelby Richardson.

Figure 5.18 Figure adapted from C. Löw, U. Weininger, P. Neumann, M. Klepsch, H. Lilie, M. T. Stubbs, and J. Balbach, Structural insights into an equilibrium folding intermediate of an archaeal ankyrin repeat protein, *Proceedings of the National Academy of Sciences USA*, 105:3779-3784, 2008, Copyright 2008 National Academy of Sciences, U.S.A.

Figure 5.19 The image shown is that of PDB ID 1BL8 (D. A. Doyle, J. Morais Cabral, R. A. Pfuetzner, A. Kuo, J. M. Gulbis, S. L. Cohen, B. T. Chait, and R. MacKinnon. The structure of the potassium channel: molecular basis of K^+ conduction and selectivity. *Science*, 280:69-77, 1998) from the Research Collaboratory for Structural Bioinformatics Protein Data Bank (RCSB PDB), created with the JSmol viewer.

Figure 5.20(b) Image courtesy of Jeff Dahl.

Figure 5.21 Image courtesy of Jena Bioscience GmbH.

Figure 5.22 The image shown is that of PDB ID 1CX8 (C. M. Lawrence, S. Ray, M. Babyonyshev, R. Galluser, D. W. Borhani, and S. C. Harrison. Crystal structure of the ectodomain of human transferrin receptor. *Science*, 286:779-782, 1999) from the Research Collaboratory for Structural Bioinformatics Protein Data Bank (RCSB PDB), created with the JSmol viewer.

Figure 5.23 The image shown is that of PDB ID 1EXU (A. P. West Jr. and P. J. Bjorkman. Crystal structure and immunoglobulin G binding properties of the human major histocompatibility complex-related Fc receptor. *Biochemistry*, 39:9698-9708, 2000) from the Research Collaboratory for Structural Bioinformatics Protein Data Bank (RCSB PDB), created with the JSmol viewer.

Figure 5.24(a) Image reprinted from *Journal of Magnetic Resonance*, volume 177, R. Fu, W. W. Brey, K. Shetty, P. Gor'kov, S. Saha, J. R. Long, S. C. Grant, E. Y. Chekmenev, J. Hu, Z. Gan, M. Sharma, F. Zhang, T. M. Logan, R. Brüschweller, A. Edison, A. Blue, I. R. Dixon, W. D. Markiewicz, and T. A. Cross, Ultra-wide bore 900 MHz high-resolution NMR at the National High Magnetic Field Laboratory, pages 1-8, Copyright 2005, with permission from Elsevier.

Chapter 6

Figure 6.2(c) The unlabeled ribbon structure is an image created and released into the public domain by Tim Vickers.

Figure 6.5 The image shown is that of PDB ID 1IGT (L. J. Harris, S. B. Larson, K. W. Hasel, and A. McPherson. Refined structure of an intact IgG2a monoclonal antibody. *Biochemistry*, 36:1581-1597, 1997) from the Research Collaboratory for Structural Bioinformatics Protein Data Bank (RCSB PDB), created with the JSmol viewer.

Figure 6.6(b) The image shown is that of PDB ID 1MLC (B. C. Braden, H. Souchon, J.-L. Eiselé, G. A. Bentley, T. N. Bhat, J. Navaza, and R. J. Poljak. Three-dimensional structures of the free and the antigen-complexed Fab from monoclonal anti-lysozyme antibody D44.1. *Journal of Molecular Biology*, 243:767-781, 1994) from the Research Collaboratory for Structural Bioinformatics Protein Data Bank (RCSB PDB), created with the JSmol viewer.

Chapter 8

Figure 8.5 Data courtesy of Life Technologies Corporation, a part of Thermo Fisher Scientific Inc. (www.thermofisher.com); © 2020 Thermo Fisher Scientific Inc., used under permission.

Figure 8.6 Data courtesy of Life Technologies Corporation, a part of Thermo Fisher Scientific Inc. (www.thermofisher.com); © 2020 Thermo Fisher Scientific Inc., used under permission.

Figure 8.11(b) Image courtesy of the National Aeronautics and Space Administration.

Figure 8.12 Data courtesy of Life Technologies Corporation, a part of Thermo Fisher Scientific Inc. (www.thermofisher.com); © 2020 Thermo Fisher Scientific Inc., used under permission.

Figure 8.14 The image shown is that of PDB ID 1EMA (M. Ormö, A. B. Cubitt, K. Kallio, L. A. Gross, R. Y. Tsien, and S. J. Remington. Crystal structure of the *Aequorea victoria* green fluorescent protein. *Science*, 273:1392-1395, 1996) from the Research Collaboratory for Structural Bioinformatics Protein Data Bank (RCSB PDB), created with the JSmol viewer.

Figure 8.15 Data courtesy of Chroma Technology Corp.

Chapter 9

Figure 9.6 Adapted from figure originally published in *The Journal of Immunology* (R. J. Ober, C. Martinez, C. Vaccaro, J. Zhou, and E. S. Ward. 2004. Visualizing the site and dynamics of IgG salvage by the MHC class I-related receptor, FcRn. *J. Immunol.* 172:2021-2029. Copyright © 2004 The American Association of Immunologists, Inc.).

Figure 9.8 Adapted from figures originally published in *The Journal of Immunology* (R. J. Ober, C. Martinez, C. Vaccaro, J. Zhou, and E. S. Ward. 2004. Visualizing the site and dynamics of IgG salvage by the MHC class I-related receptor, FcRn. *J. Immunol.* 172:2021-2029. Copyright © 2004 The American Association of Immunologists, Inc.).

Figure 9.9 Figure reprinted from *Biophysical Journal*, volume 95, S. Ram, P. Prabhat, J. Chao, E. S. Ward, and R. J. Ober, High accuracy 3D quantum dot tracking with multifocal plane microscopy for the study of fast intracellular dynamics in live cells, pages 6025-6043, Copyright 2008 Biophysical Society, with permission from Elsevier.

Figure 9.10 Adapted from figure originally published in *The Journal of Immunology* (R. J. Ober, C. Martinez, C. Vaccaro, J. Zhou, and E. S. Ward. 2004. Visualizing the site and dynamics of IgG salvage by the MHC class I-related receptor, FcRn. *J. Immunol.* 172:2021-2029. Copyright © 2004 The American Association of Immunologists, Inc.).

Chapter 10

Figure 10.9 The plot shown is a version of an image provided by Carl Zeiss Microscopy GmbH, Jena, Germany, modified only in terms of font properties and line thickness and color.

Figure 10.11 Data courtesy of Semrock, Inc.

Figure 10.12(a) The image shown is a cropped version of a photograph taken by Sungyong You.

Figure 10.12(b) Image courtesy of Semrock, Inc.

Figure 10.13 Data courtesy of Chroma Technology Corp.

Figure 10.14 Data courtesy of Semrock, Inc.

Figure 10.15 Data courtesy of Life Technologies Corporation, a part of Thermo Fisher Scientific Inc. (www.thermofisher.com); © 2020 Thermo Fisher Scientific Inc., used under permission.

Figure 10.16 Data courtesy of Semrock, Inc.

Figure 10.18 Data courtesy of Semrock, Inc.

Figure 10.20 The images shown are cropped and rotated versions of photographs taken by Sungyong You.

Figure 10.22 Data courtesy of Semrock, Inc.

Chapter 11

Figure 11.1 GFP data courtesy of Chroma Technology Corp. Alexa Fluor 568 data courtesy of Life Technologies Corporation, a part of Thermo Fisher Scientific Inc. (www.thermofisher.com); © 2020 Thermo Fisher Scientific Inc., used under permission.

Figure 11.2 Data courtesy of Semrock, Inc.

Figure 11.3 Data courtesy of Chroma Technology Corp.

Figure 11.4 Adapted from figure originally published in *The Journal of Immunology* (R. J. Ober, C. Martinez, C. Vaccaro, J. Zhou, and E. S. Ward. 2004. Visualizing the site and dynamics of IgG salvage by the MHC class I-related receptor, FcRn. *J. Immunol.* 172:2021-2029. Copyright © 2004 The American Association of Immunologists, Inc.).

Figure 11.10 Figure adapted from R. J. Ober, C. Martinez, X. Lai, J. Zhou, and E. S. Ward, Exocytosis of IgG as mediated by the receptor, FcRn: an analysis at the single-molecule level, *Proceedings of the National Academy of Sciences USA*, 101:11076-11081, 2004, Copyright 2004 National Academy of Sciences, U.S.A.

Figure 11.11 The image shown is adapted from an image distributed under the Creative Commons Attribution 4.0 International Public License (CC BY 4.0; https://creativecommons.org/licenses/by/4.0/legalcode). The modifications consist of the cropping of the original image and the removal of other curves and the graph legend. The original image is courtesy of Zhou Han et al., available as a figure from the article Z. Han, L. Jin, F. Chen, J. J. Loturco, L. B. Cohen, A. Bondar, J. Lazar, and V. A. Pieribone, Mechanistic studies of the genetically encoded fluorescent protein voltage probe ArcLight, *PLoS ONE*, 9:e113873, 2014, https://doi.org/10.1371/journal.pone.0113873.

Figure 11.12 Figure reprinted from *Methods in Cell Biology*, volume 126, J. Canton and S. Grinstein, Measuring lysosomal pH by fluorescence microscopy, pages 85-99, Copyright 2015, with permission from Elsevier.

Figure 11.15 Figure adapted from R. J. Ober, C. Martinez, X. Lai, J. Zhou, and E. S. Ward, Exocytosis of IgG as mediated by the receptor, FcRn: an analysis at the single-molecule level, *Proceedings of the National Academy of Sciences USA*, 101:11076-11081, 2004, Copyright 2004 National Academy of Sciences, U.S.A.

Figure 11.20 Figure reprinted from *Biophysical Journal*, volume 103, S. Ram, D. Kim, R. J. Ober, and E. S. Ward, 3D single molecule tracking with multifocal plane microscopy reveals rapid intercellular transferrin transport at epithelial cell barriers, pages 1594-1603, Copyright 2012 Biophysical Society, with permission from Elsevier.

Figure 11.22 Figure adapted from P. Prabhat, Z. Gan, J. Chao, S. Ram, C. Vaccaro, S. Gibbons, R. J. Ober, and E. S. Ward, Elucidation of intracellular recycling pathways leading to exocytosis of the Fc receptor, FcRn, by using multifocal plane microscopy, *Proceedings of the National Academy of Sciences USA*, 104:5889-5894, 2007, Copyright 2007 National Academy of Sciences, U.S.A.

Chapter 12

Figure 12.8 Data for the Andor detectors courtesy of Andor Technology Ltd. Data for the Hamamatsu detectors courtesy of Hamamatsu Photonics K.K.

Bibliography

[1] A. V. Abraham, S. Ram, J. Chao, E. S. Ward, and R. J. Ober. Quantitative study of single molecule location estimation techniques. *Optics Express*, 17:23352–23373, 2009.

[2] S. Abrahamsson, J. Chen, B. Hajj, S. Stallinga, A. Y. Katsov, J. Wisniewski, G. Mizuguchi, P. Soule, F. Mueller, C. D. Darzacq, X. Darzacq, C. Wu, C. I. Bargmann, D. A. Agard, M. Dahan, and M. G. L. Gustafsson. Fast multicolor 3D imaging using aberration-corrected multifocus microscopy. *Nature Methods*, 10:60–63, 2013.

[3] M. Abramowitz and I. A. Stegun, editors. *Handbook of mathematical functions*. Dover, 1965.

[4] K. Anand, J. Ziebuhr, P. Wadhwani, J. R. Mesters, and R. Hilgenfeld. Coronavirus main proteinase ($3CL^{pro}$) structure: basis for design of anti-SARS drugs. *Science*, 300:1763–1767, 2003.

[5] R. Ando, H. Mizuno, and A. Miyawaki. Regulated fast nucleocytoplasmic shuttling observed by reversible protein highlighting. *Science*, 306:1370–1373, 2004.

[6] E. C. Arnspang, P. Kulatunga, and B. C. Lagerholm. A single molecule investigation of the photostability of quantum dots. *PLoS ONE*, 7:e44355, 2012.

[7] K. B. Athreya and P. E. Ney. *Branching Processes*. Dover, 2004.

[8] D. Axelrod. Total internal reflection fluorescence microscopy in cell biology. *Traffic*, 2:764–774, 2001.

[9] D. Axelrod, T. P. Burghardt, and N. L. Thompson. Total internal reflection fluorescence. *Annual Review of Biophysics and Bioengineering*, 13:247–268, 1984.

[10] F. Balzarotti, Y. Eilers, K. C. Gwosch, A. H. Gynnå, V. Westphal, F. D. Stefani, J. Elf, and S. W. Hell. Nanometer resolution imaging and tracking of fluorescent molecules with minimal photon fluxes. *Science*, 355:606–612, 2017.

[11] A. G. Basden, C. A. Haniff, and C. D. Mackay. Photon counting strategies with low-light-level CCDs. *Monthly Notices of the Royal Astronomical Society*, 345:985–991, 2003.

[12] A. Bashan and A. Yonath. Ribosome crystallography: catalysis and evolution of peptide-bond formation, nascent chain elongation and its co-translational folding. *Biochemical Society Transactions*, 33:488–492, 2005.

[13] E. Betzig, G. H. Patterson, R. Sougrat, O. W. Lindwasser, S. Olenych, J. S. Bonifacino, M. W. Davidson, J. Lippincott-Schwartz, and H. F. Hess. Imaging intracelular fluorescent proteins at nanometer resolution. *Science*, 313:1642–1645, 2006.

[14] P. J. Bickel and K. A. Doksum. *Mathematical Statistics: Basic Ideas and Selected Topics*, volume I. CRC Press, 2nd edition, 2015.

[15] P. M. Blanchard and A. H. Greenaway. Simultaneous multiplane imaging with a distorted diffraction grating. *Applied Optics*, 38:6692–6699, 1999.

[16] M. Born and E. Wolf. *Principles of Optics*. Cambridge University Press, 1999.

[17] J. Braman, editor. *In Vitro Mutagenesis Protocols*, volume 182 of *Methods in Molecular Biology*. Humana Press, 2nd edition, 2002.

[18] P. Bruhns, B. Iannascoli, P. England, D. A. Mancardi, N. Fernandez, S. Jorieux, and M. Daëron. Specificity and affinity of human Fcγ receptors and their polymorphic variants for human IgG subclasses. *Blood*, 113:3716–3725, 2009.

[19] T. P. Burghardt and N. L. Thompson. Evanescent intensity of a focused Gaussian light beam undergoing total internal reflection in a prism. *Optical Engineering*, 23:230162, 1984.

[20] R. E. Campbell, O. Tour, A. E. Palmer, P. A. Steinbach, G. S. Baird, D. A. Zacharias, and R. Y. Tsien. A monomeric red fluorescent protein. *Proceedings of the National Academy of Sciences USA*, 99:7877–7882, 2002.

[21] J. Canton and S. Grinstein. Measuring lysosomal pH by fluorescence microscopy. In *Methods in Cell Biology*, volume 126. Elsevier, 2015.

[22] M. Chalfie, Y. Tu, G. Euskirchen, W. W. Ward, and D. C. Prasher. Green fluorescent protein as a marker for gene expression. *Science*, 263:802–805, 1994.

[23] D. K. Challa, R. Velmurugan, R. J. Ober, and E. S. Ward. FcRn: from molecular interactions to regulation of IgG pharmacokinetics and functions. In M. Daëron and F. Nimmerjahn, editors, *Current Topics in Microbiology and Immunology*, volume 382. Springer, 2014.

[24] D. C. Champeney. *A Handbook of Fourier Theorems*. Cambridge University Press, 1987.

[25] J. Chao, T. Lee, E. S. Ward, and R. J. Ober. Fluorescent microspheres as point sources: a localization study. *PLoS ONE*, 10:e0134112, 2015.

[26] J. Chao, S. Ram, A. V. Abraham, E. S. Ward, and R. J. Ober. A resolution measure for three-dimensional microscopy. *Optics Communications*, 282:1751–1761, 2009.

[27] J. Chao, S. Ram, T. Lee, E. S. Ward, and R. J. Ober. Investigation of the numerics of point spread function integration in single molecule localization. *Optics Express*, 23:16866–16883, 2015.

[28] J. Chao, S. Ram, E. S. Ward, and R. J. Ober. A comparative study of high resolution microscopy imaging modalities using a three-dimensional resolution measure. *Optics Express*, 17:24377–24402, 2009.

[29] J. Chao, S. Ram, E. S. Ward, and R. J. Ober. Ultrahigh accuracy imaging modality for super-localization microscopy. *Nature Methods*, 10:335–338, 2013.

[30] J. Chao, E. S. Ward, and R. J. Ober. Fisher information matrix for branching processes with application to electron-multiplying charge-coupled devices. *Multidimensional Systems and Signal Processing*, 23:349–379, 2012.

[31] M. K. Cheezum, W. F. Walker, and W. H. Guilford. Quantitative comparison of algorithms for tracking single fluorescent particles. *Biophysical Journal*, 81:2378–2388, 2001.

[32] H. Chen, H. Gai, and E. S. Yeung. Inhibition of photobleaching and blue shift in quantum dots. *Chemical Communications*, pages 1676–1678, 2009.

[33] J. Cnossen, T. Hinsdale, R. Ø. Thorsen, M. Siemons, F. Schueder, R. Jungmann, C. S. Smith, B. Rieger, and S. Stallinga. Localization microscopy at doubled precision with patterned illumination. *Nature Methods*, 17:59–63, 2020.

[34] N. A. C. Cressie. *Statistics for Spatial Data.* John Wiley & Sons, Inc., 1993.

[35] J. M. Di Noia and M. S. Neuberger. Molecular mechanisms of antibody somatic hypermutation. *Annual Review of Biochemistry*, 76:1–22, 2007.

[36] B. Efron. *Large-Scale Inference: Emprical Bayes Methods for Estimation, Testing, and Prediction.* Cambridge University Press, 2010.

[37] A. Einstein. Über einen die Erzeugung und Verwandlung des Lichtes betreffenden heuristischen Gesichtspunkt. *Annalen der Physik*, 322:132–148, 1905.

[38] R. A. Fisher. Theory of statistical estimation. *Mathematical Proceedings of the Cambridge Philosophical Society*, 22:700–725, 1925.

[39] Z. Gan, S. Ram, C. Vaccaro, R. J. Ober, and E. S. Ward. Analyses of the recycling receptor, FcRn, in live cells reveal novel pathways for lysosomal delivery. *Traffic*, 10:600–614, 2009.

[40] S. F. Gibson and F. Lanni. Experimental test of an analytical model of aberration in an oil-immersion objective lens used in three-dimensional light microscopy. *Journal of the Optical Society of America A*, 8:1601–1613, 1991.

[41] J. W. Goodman. *Introduction to Fourier Optics.* Roberts & Company Publishers, 3rd edition, 2005.

[42] G. R. Grimmett and D. R. Stirzaker. *Probability and Random Processes.* Oxford University Press, 3rd edition, 2001.

[43] H. Gross, F. Blechinger, and B. Achtner. *Handbook of Optical Systems*, volume 4. Wiley-VCH, 2008.

[44] M. Gu. *Advanced Optical Imaging Theory.* Springer-Verlag, 2000.

[45] T. E. Harris. *The Theory of Branching Processes.* Prentice-Hall, 1963.

[46] E. Hecht. *Optics.* Addison Wesley, 4th edition, 2002.

[47] M. Heilemann, S. van de Linde, M. Schüttpelz, R. Kasper, B. Seefeldt, A. Mukherjee, P. Tinnefeld, and M. Sauer. Subdiffraction-resolution fluorescence imaging with conventional fluorescent probes. *Angewandte Chemie International Edition*, 47:6172–6176, 2008.

[48] R. Heim, A. B. Cubitt, and R. Y. Tsien. Improved green fluorescence. *Nature*, 373:663–664, 1995.

[49] S. W. Hell and J. Wichmann. Breaking the diffraction resolution limit by stimulated emission: stimulated-emission-depletion fluorescence microscopy. *Optics Letters*, 19:780–782, 1994.

[50] S. T. Hess, T. P. K. Girirajan, and M. D. Mason. Ultra-high resolution imaging by fluorescence photoactivation localization microscopy. *Biophysical Journal*, 91:4258–4272, 2006.

[51] P. D. Hoff. *A First Course in Bayesian Statistical Methods.* Springer, 2009.

[52] J. N. Hollenhorst. A theory of multiplication noise. *IEEE Transactions on Electron Devices*, 37:781–788, 1990.

[53] J. Hynecek. Impactron - a new solid state image intensifier. *IEEE Transactions on Electron Devices*, 48:2238–2241, 2001.

[54] J. Hynecek and T. Nishiwaki. Excess noise and other important characteristics of low light level imaging using charge multiplying CCDs. *IEEE Transactions on Electron Devices*, 50:239–245, 2003.

[55] S. Inoué. Foundations of confocal scanned imaging in light microscopy. In J. B. Pawley, editor, *Handbook of Biological Confocal Microscopy.* Springer Science+Business Media, 3rd edition, 2006.

[56] J. R. Janesick. *Scientific Charge-Coupled Devices.* SPIE Press, 2001.

[57] M. F. Juette, T. J. Gould, M. D. Lessard, M. J. Mlodzianoski, B. S. Nagpure, B. T. Bennett, S. T. Hess, and J. Bewersdorf. Three-dimensional sub-100 nm resolution fluorescence microscopy of thick samples. *Nature Methods*, 5:527–529, 2008.

[58] S. M. Kay. *Fundamentals of Statistical Signal Processing: Estimation Theory.* Prentice Hall PTR, 1993.

[59] R. E. Kontermann and S. Dübel, editors. *Antibody Engineering*, volume 1. Springer-Verlag, 2nd edition, 2010.

[60] X. Lai, Z. Lin, E. S. Ward, and R. J. Ober. Noise suppression of point spread functions and its influence on deconvolution of three-dimensional fluorescence microscopy image sets. *Journal of Microscopy*, 217:93–108, 2005.

[61] J. R. Lakowicz. *Principles of Fluorescence Spectroscopy.* Springer, 2nd edition, 1999.

[62] E. L. Lehmann and G. Casella. *Theory of Point Estimation.* Springer, 2nd edition, 1998.

[63] L. B. Lucy. An iterative technique for the rectification of observed distributions. *The Astronomical Journal*, 79:745–754, 1974.

[64] K. A. Lukyanov, D. M. Chudakov, S. Lukyanov, and V. V. Verkhusha. Photoactivatable fluorescent proteins. *Nature Reviews Molecular Cell Biology*, 6:885–890, 2005.

[65] L. V. Marks. *The Lock and Key of Medicine: Monoclonal Antibodies and the Transformation of Healthcare.* Yale University Press, 2015.

[66] K. Matsuo, M. C. Teich, and B. E. A. Saleh. Noise properties and time response of the staircase avalanche photodiode. *IEEE Transactions on Electron Devices*, 32:2615–2623, 1985.

[67] M. V. Matz, A. F. Fradkov, Y. A. Labas, A. P. Savitsky, A. G. Zaraisky, M. L. Markelov, and S. A. Lukyanov. Fluorescent proteins from nonbioluminescent Anthozoa species. *Nature Biotechnology*, 17:969–973, 1999.

[68] G. Miesenböck, D. A. De Angelis, and J. E. Rothman. Visualizing secretion and synaptic transmission with pH-sensitive green fluorescent proteins. *Nature*, 394:192–195, 1998.

[69] W. E. Moerner and L. Kador. Optical detection and spectroscopy of single molecules in a solid. *Physical Review Letters*, 62:2535–2538, 1989.

[70] D. B. Murphy. *Fundamentals of light microscopy and electronic imaging.* Wiley-Liss, 2001.

[71] K. Murphy and C. Weaver. *Janeway's Immunobiology.* Garland Science, Taylor & Francis Group, 9th edition, 2017.

[72] M. Newberry. Pixel response effects on CCD camera gain calibration. Technical report, Mirametrics, Inc., 1998.

[73] R. J. Ober, C. Martinez, X. Lai, J. Zhou, and E. S. Ward. Exocytosis of IgG as mediated by the receptor, FcRn: an analysis at the single-molecule level. *Proceedings of the National Academy of Sciences USA*, 101:11076–11081, 2004.

[74] R. J. Ober, C. Martinez, C. Vaccaro, J. Zhou, and E. S. Ward. Visualizing the site and dynamics of IgG salvage by the MHC class I-related receptor, FcRn. *Journal of Immunology*, 172:2021–2029, 2004.

[75] R. J. Ober, S. Ram, and E. S. Ward. Localization accuracy in single-molecule microscopy. *Biophysical Journal*, 86:1185–1200, 2004.

[76] V. M. Panaretos. *Statistics for Mathematicians: A Rigorous First Course.* Birkhäuser, 2016.

[77] G. Patterson, M. Davidson, S. Manley, and J. Lippincott-Schwartz. Superresolution imaging using single-molecule localization. *Annual Review of Physical Chemistry*, 61:345–367, 2010.

[78] G. H. Patterson and J. Lippincott-Schwartz. A photoactivatable GFP for selective photolabeling of proteins and cells. *Science*, 297:1873–1877, 2002.

[79] M. Pluta. *Advanced Light Microscopy, vol. 1: Principles and Basic Properties.* Elsevier, 1988.

[80] M. Pluta. *Advanced Light Microscopy, vol. 2: Specialized Methods.* Elsevier, 1989.

[81] H. V. Poor. *An Introduction to Signal Detection and Estimation.* Springer, 2nd edition, 1994.

[82] P. Prabhat, Z. Gan, J. Chao, S. Ram, C. Vaccaro, S. Gibbons, R. J. Ober, and E. S. Ward. Elucidation of intracellular recycling pathways leading to exocytosis of the Fc receptor, FcRn, by using multifocal plane microscopy. *Proceedings of the National Academy of Sciences USA*, 104:5889–5894, 2007.

[83] P. Prabhat, S. Ram, E. S. Ward, and R. J. Ober. Simultaneous imaging of different focal planes in fluorescence microscopy for the study of cellular dynamics in three dimensions. *IEEE Transactions on Nanobioscience*, 3:237–242, 2004.

[84] M. Pyzik, T. Rath, W. I. Lencer, K. Baker, and R. S. Blumberg. FcRn: the architect behind the immune and nonimmune functions of IgG and albumin. *Journal of Immunology*, 194:4595–4603, 2015.

[85] S. Ram. *Resolution and localization in single molecule microscopy*. PhD thesis, University of Texas at Arlington, 2007.

[86] S. Ram, J. Chao, P. Prabhat, E. S. Ward, and R. J. Ober. A novel approach to determining the three-dimensional location of microscopic objects with applications to 3D particle tracking. *Proceedings of the SPIE*, 6443:64430D, 2007.

[87] S. Ram, D. Kim, R. J. Ober, and E. S. Ward. 3D single molecule tracking with multifocal plane microscopy reveals rapid intercellular transferrin transport at epithelial cell barriers. *Biophysical Journal*, 103:1594–1603, 2012.

[88] S. Ram, P. Prabhat, J. Chao, E. S. Ward, and R. J. Ober. High accuracy 3D quantum dot tracking with multifocal plane microscopy for the study of fast intracellular dynamics in live cells. *Biophysical Journal*, 95:6025–6043, 2008.

[89] S. Ram, P. Prabhat, E. S. Ward, and R. J. Ober. Improved single particle localization accuracy with dual objective multifocal plane microscopy. *Optics Express*, 17:6881–6898, 2009.

[90] S. Ram, E. S. Ward, and R. J. Ober. How accurately can a single molecule be localized in three dimensions using a fluorescence microscope? *Proceedings of the SPIE*, 5699:426–435, 2005.

[91] S. Ram, E. S. Ward, and R. J. Ober. Beyond Rayleigh's criterion: a resolution measure with application to single-molecule microscopy. *Proceedings of the National Academy of Sciences USA*, 103:4457–4462, 2006.

[92] S. Ram, E. S. Ward, and R. J. Ober. A novel resolution measure for optical microscopes: stochastic analysis of the performance limits. *Proceedings of the IEEE International Symposium on Biomedical Imaging: Nano to Macro*, pages 770–773, 2006.

[93] S. Ram, E. S. Ward, and R. J. Ober. A stochastic analysis of performance limits for optical microscopes. *Multidimensional Systems and Signal Processing*, 17:27–57, 2006.

[94] S. Ram, E. S. Ward, and R. J. Ober. A stochastic analysis of distance estimation approaches in single molecule microscopy: quantifying the resolution limits of photon-limited imaging systems. *Multidimensional Systems and Signal Processing*, 24:503–542, 2013.

[95] V. Ramakrishnan. The ribosome emerges from a black box. *Cell*, 159:979–984, 2014.

[96] W. H. Richardson. Bayesian-based iterative method of image restoration. *Journal of the Optical Society of America*, 62:55–59, 1972.

[97] S. M. Ross. *Applied Probability Models with Optimization Applications*. Dover, 1992.

[98] R. Roy, S. Hohng, and T. Ha. A practical guide to single-molecule FRET. *Nature Methods*, 5:507–516, 2008.

[99] M. J. Rust, M. Bates, and X. Zhuang. Sub-diffraction-limit imaging by stochastic optical reconstruction microscopy STORM. *Nature Methods*, 3:793–795, 2006.

[100] A. Sayre. *Rosalind Franklin and DNA*. W. W. Norton & Company, 1975.

[101] B. Schuler. Single-molecule FRET of protein structure and dynamics - a primer. *Journal of Nanobiotechnology*, 11:S2, 2013.

[102] N. C. Shaner, R. E. Campbell, P. A. Steinbach, B. N. G. Giepmans, A. E. Palmer, and R. Y. Tsien. Improved monomeric red, orange and yellow fluorescent proteins derived from *discosoma* sp. red fluorescent protein. *Nature Biotechnology*, 22:1567–1572, 2004.

[103] D. M. Shcherbakova, P. Sengupta, J. Lippincott-Schwartz, and V. V. Verkhusha. Photocontrollable fluorescent proteins for superresolution imaging. *Annual Review of Biophysics*, 43:303–329, 2014.

[104] L. A. Shepp and Y. Vardi. Maximum likelihood reconstruction for emission tomography. *IEEE Transactions on Medical Imaging*, 1:113–122, 1982.

[105] E. M. Slayter. *Optical methods in biology*. John Wiley & Sons, Inc., 1970.

[106] D. L. Snyder, C. W. Helstrom, A. D. Lanterman, M. Faisal, and R. L. White. Compensation for readout noise in CCD images. *Journal of the Optical Society of America A*, 12:272–283, 1995.

[107] D. L. Snyder and M. I. Miller. *Random Point Processes in Time and Space*. Springer Verlag, 2nd edition, 1991.

[108] T. A. Steitz. A structural understanding of the dynamic ribosome machine. *Nature Reviews Molecular Cell Biology*, 9:242–253, 2008.

[109] A. C. Stiel, M. Andresen, H. Bock, M. Hilbert, J. Schilde, A. Schönle, C. Eggeling, A. Egner, S. W. Hell, and S. Jakobs. Generation of monomeric reversibly switchable red fluorescent proteins for far-field fluorescence nanoscopy. *Biophysical Journal*, 95:2989–2997, 2008.

[110] R. L. Streit. *Poisson Point Processes: Imaging, Tracking, and Sensing*. Springer, 2010.

[111] A. Tahmasbi, S. Ram, J. Chao, A. V. Abraham, F. W. Tang, E. S. Ward, and R. J. Ober. Designing the focal plane spacing for multifocal plane microscopy. *Optics Express*, 22:16706–16721, 2014.

[112] E. Toprak, H. Balci, B. H. Blehm, and P. R. Selvin. Three-dimensional particle tracking via bifocal imaging. *Nano Letters*, 7:2043–2045, 2007.

[113] M. R. Vahid, B. Hanzon, and R. J. Ober. Fisher information matrix for single molecules with stochastic trajectories. *SIAM Journal on Imaging Sciences*, 13:234–264, 2020.

[114] S. van de Linde, M. Heilemann, and M. Sauer. Live-cell super-resolution imaging with synthetic fluorophores. *Annual Review of Physical Chemistry*, 63:519–540, 2012.

[115] S. van de Linde, I. Krstić, T. Prisner, S. Doose, M. Heilemann, and M. Sauer. Photoinduced formation of reversible dye radicals and their impact on super-resolution imaging. *Photochemical & Photobiological Sciences*, 10:499–506, 2011.

[116] S. van de Linde, A. Löschberger, T. Klein, M. Heidbreder, S. Wolter, M. Heilemann, and M. Sauer. Direct stochastic optical reconstruction microscopy with standard fluorescent probes. *Nature Protocols*, 6:991–1009, 2011.

[117] W. G. J. H. M. van Sark, P. L. T. M. Frederix, D. J. Van den Heuvel, H. C. Gerritsen, A. A. Bol, J. N. J. van Lingen, C. de Mello Donegá, and A. Meijerink. Photooxidation and photobleaching of single CdSe/ZnS quantum dots probed by room-temperature time-resolved spectroscopy. *Journal of Physical Chemistry B*, 105:8281–8284, 2001.

[118] H. L. Van Trees, K. L. Bell, and Z. Tian. *Detection, Estimation, and Modulation Theory - Part 1.* Wiley, 2nd edition, 2013.

[119] V. V. Verkhusha and A. Sorkin. Conversion of the monomeric red fluorescent protein into a photoactivatable probe. *Chemistry & Biology*, 12:279–285, 2005.

[120] G. D. Victora and M. C. Nussenzweig. Germinal centers. *Annual Review of Immunology*, 30:429–457, 2012.

[121] A. von Diezmann, Y. Shechtman, and W. E. Moerner. Three-dimensional localization of single molecules for super-resolution imaging and single-particle tracking. *Chemical Reviews*, 117:7244–7275, 2017.

[122] M. A. Walling, J. A. Novak, and J. R. E. Shepard. Quantum dots for live cell and in vivo imaging. *International Journal of Molecular Sciences*, 10:441–491, 2009.

[123] B. A. Walther and J. L. Moore. The concepts of bias, precision and accuracy, and their use in testing the performance of species richness estimators, with a literature review of estimator performance. *Ecography*, 28:815–829, 2005.

[124] J. D. Watson. *The Double Helix: A Personal Account of the Discovery of the Structure of DNA.* W. W. Norton & Company, 1980.

[125] J. D. Watson, T. A. Baker, S. P. Bell, A. Gann, M. Levine, and R. Losick. *Molecular Biology of the Gene.* Pearson, 7th edition, 2013.

[126] K. I. Willig, R. R. Kellner, R. Medda, B. Hein, S. Jakobs, and S. W. Hell. Nanoscale resolution in GFP-based microscopy. *Nature Methods*, 3:721–723, 2006.

[127] Y. Wong, J. Chao, Z. Lin, and R. J. Ober. Effect of time discretization of the imaging process on the accuracy of trajectory estimation in fluorescence microscopy. *Optics Express*, 22:20396–20420, 2014.

[128] Y. Wong, Z. Lin, and R. J. Ober. Limit of the accuracy of parameter estimation for moving single molecules imaged by fluorescence microscopy. *IEEE Transactions on Signal Processing*, 59:895–911, 2011.

[129] L. S. Yeap, J. K. Hwang, Z. Du, R. M. Meyers, F. L. Meng, A. Jakubauskaitė, M. Liu, V. Mani, D. Neuberg, T. B. Kepler, J. H. Wang, and F. W. Alt. Sequence-intrinsic mechanisms that target AID mutational outcomes on antibody genes. *Cell*, 163:1124–1137, 2015.

[130] S. Zacks. *The Theory of Statistical Inference.* John Wiley & Sons, Inc., 1971.

List of Symbols

σ_k	Standard deviation of detector readout noise in the kth image pixel	285
θ	Vector of unknown parameters	293
θ_c	Critical angle	164
θ_g	Angle of incidence at the cover glass-sample interface	165
θ_s	Angle of refraction at the cover glass-sample interface	208
θ_{im}	Angle of incidence at the immersion medium-cover glass interface	208
$\mathrm{Var}(X)$	Variance of random variable X	A-2
φ	Phase shift of wave	223
$\|\cdot\|$	Euclidean norm	223
$\lvert\cdot\rvert$	Modulus	229
$\{\}$	Empty set	431
A	Amplitude of wave	223
b	Brightness	96
C	Molar concentration	96
c	Speed of light in vacuum	91
c_k	Electron-count-to-DU conversion factor for the kth image pixel	199
$D_i(x, y)$	Distance in image space between the lens and the plane behind the lens at coordinate point (x, y)	232
$D_l(x, y)$	Thickness of lens at coordinate point (x, y)	232
$D_o(x, y)$	Distance in object space between the lens and the plane before the lens at coordinate point (x, y)	232
$D_{l,i}$	Thickness of lens at the optical axis on the part attributed to the image space	232
$D_{l,o}$	Thickness of lens at the optical axis on the part attributed to the object space	232
DFT	Discrete Fourier transform	A-13
E	Energy of a photon	91
f_τ	Photon distribution profile	273
F_i	Image focal point	211
f_i	Image focal length	211
F_o	Object focal point	211
f_o	Object focal length	211
$f_{ms,i}$	Image focal length of microscopy system	220

Index of Names

Index